P9-DDI-042

INDUSTRIAL SUPERVISION
In the Age of High Technology

DAVID L. GOETSCH

Merrill, an imprint of
MACMILLAN PUBLISHING COMPANY
New York

MAXWELL MACMILLAN CANADA
Toronto

MAXWELL MACMILLAN INTERNATIONAL
New York Oxford Singapore Sydney

Cover art/photo: Douglas Fraser
Editor: Stephen Helba
Production Editor: Christine M. Harrington
Art Coordinator: Ruth A. Kimpel
Text Designer: Anne Flanagan
Cover Designer: Cathleen Norz
Production Buyer: Patricia A. Tonneman

This book was set in Clearface by Bi-Comp, Inc. and was printed and bound by R. R. Donnelley & Sons Company. The cover was printed by Lehigh Press, Inc.

Printed in the United States of America

Macmillan Publishing Company
866 Third Avenue
New York, NY 10022

Macmillan Publishing Company is part of the
Maxwell Communication Group of Companies.

Maxwell Macmillan Canada, Inc.
1200 Eglinton Avenue East, Suite 200
Don Mills, Ontario M3C 3N1

Library of Congress Cataloging-in-Publication Data
Goetsch, David L.
Industrial supervision: in the age of high technology / David L. Goetsch.
p. cm.
Includes index.
ISBN 0-675-22137-4
1. Supervision of employees. 2. Employees—Effect of technological innovations on. I. Title.
HF5549.12.G64 1992
658.3'02—dc20

91–5098
CIP

Printing: 1 2 3 4 5 6 7 8 9 Year: 2 3 4 5

Dedicated to the memory of Larry Lance Goetsch
and to his children, Stephanie and Lance.

MERRILL'S INTERNATIONAL SERIES IN ENGINEERING TECHNOLOGY

ADAMSON	*Applied Pascal for Technology,* 0-675-20771-1 *The Electronics Dictionary for Technicians,* 0-02-300820-2 *Microcomputer Repair,* 0-02-300825-3 *Structured BASIC Applied to Technology,* 0-675-20772-X *Structured C for Technology,* 0-675-20993-5 *Structured C for Technology* (w/disks), 0-675-21289-8
ANTONAKOS	*The 68000 Microprocessor: Hardware and Software Principles and Applications,* 0-675-21043-7
ASSER/STIGLIANO/ BAHRENBURG	*Microcomputer Servicing: Practical Systems and Troubleshooting,* 0-675-20907-2 *Microcomputer Theory and Servicing,* 0-675-20659-6 *Lab Manual to Accompany Microcomputer Theory and Servicing,* 0-675-21109-3
ASTON	*Principles of Biomedical Instrumentation and Measurement,* 0-675-20943-9
BATESON	*Introduction to Control System Technology, Third Edition,* 0-675-21010-0
BEACH/JUSTICE	*DC/AC Circuit Essentials,* 0-675-20193-4
BERLIN	*Experiments in Electronic Devices to Accompany Floyd's Electronic Devices and Electronic Devices: Electron Flow Version, Third Edition,* 0-02-308422-7 *The Illustrated Electronics Dictionary,* 0-675-20451-8
BERLIN/GETZ	*Experiments in Instrumentation and Measurement,* 0-675-20450-X *Fundamentals of Operational Amplifiers and Linear Integrated Circuits,* 0-675-21002-X *Principles of Electronic Instrumentation and Measurement,* 0-675-20449-6
BERUBE	*Electronic Devices and Circuits Using MICRO-CAP II,* 0-02-309160-6
BOGART	*Electronic Devices and Circuits, Second Edition,* 0-675-21150-6
BOGART/BROWN	*Experiments in Electronic Devices and Circuits, Second Edition,* 0-675-21151-4
BOYLESTAD	*DC/AC: The Basics,* 0-675-20918-8 *Introductory Circuit Analysis, Sixth Edition,* 0-675-21181-6
BOYLESTAD/ KOUSOUROU	*Experiments in Circuit Analysis, Sixth Edition,* 0-675-21182-4 *Experiments in DC/AC Basics,* 0-675-21131-X
BREY	*Microprocessors and Peripherals: Hardware, Software, Interfacing, and Applications, Second Edition,* 0-675-20884-X *The Intel Microprocessors—8086/8088, 80186, 80286, 80386, and 80846—Architecture, Programming and Interfacing, Second Edition,* 0-675-21309-6
BROBERG	*Lab Manual to Accompany Electronic Communication Techniques, Second Edition,* 0-675-21257-X
BUCHLA	*Digital Experiments: Emphasizing Systems and Design, Second Edition,* 0-675-21180-8 *Experiments in Electric Circuits Fundamentals, Second Edition,* 0-675-21409-2 *Experiments in Electronics Fundamentals: Circuits, Devices and Applications, Second Edition,* 0-675-21407-6
BUCHLA/McLACHLAN	*Applied Electronic Instrumentation and Measurement,* 0-675-21162-X
CICCARELLI	*Circuit Modeling: Exercises and Software, Second Edition,* 0-675-21152-2
COOPER	*Introduction to VersaCAD,* 0-675-21164-6
COX	*Digital Experiments: Emphasizing Troubleshooting, Second Edition,* 0-675-21196-4

CROFT	*Getting a Job: Resume Writing, Job Application Letters, and Interview Strategies,* 0-675-20917-X
DAVIS	*Technical Mathematics,* 0-675-20338-4 *Technical Mathematics with Calculus,* 0-675-20965-X *Study Guide to Accompany Technical Mathematics,* 0-675-20966-8 *Study Guide to Accompany Technical Mathematics with Calculus,* 0-675-20964-1
DELKER	*Experiments in 8085 Microprocessor Programming and Interfacing,* 0-675-20663-4
FLOYD	*Digital Fundamentals, Fourth Edition,* 0-675-21217-0 *Electric Circuits Fundamentals, Second Edition,* 0-675-21408-4 *Electronic Devices, Third Edition,* 0-675-22170-6 *Electronic Devices: Electron Flow Version,* 0-02-338540-5 *Electronics Fundamentals: Circuits, Devices, and Applications Second Edition,* 0-675-21310-X *Fundamentals of Linear Circuits,* 0-02-338481-6 *Principles of Electric Circuits, Electron Flow Version, Second Edition,* 0-675-21292-8 *Principles of Electric Circuits, Third Edition,* 0-675-21062-3
FULLER	*Robotics: Introduction, Programming, and Projects,* 0-675-21078-X
GAONKAR	*Microprocessor Architecture, Programming, and Applications with the 8085/8080A, Second Edition,* 0-675-20675-8 *The Z80 Microprocessor: Architecture, Interfacing, Programming, and Design,* 0-675-20540-9
GILLIES	*Instrumentation and Measurements for Electronic Technicians,* 0-675-20432-1
GOETSCH	*Industrial Supervision: In the Age of High Technology,* 0-675-22137-4
GOETSCH/RICKMAN	*Computer-Aided Drafting with AutoCAD,* 0-675-20915-3
GOODY	*Programming and Interfacing the 8086/8088 Microprocessor,* 0-675-21312-6
HUBERT	*Electric Machines: Theory, Operation, Applications, Adjustment, and Control,* 0-675-21136-0
HUMPHRIES	*Motors and Controls,* 0-675-20235-3
HUTCHINS	*Introduction to Quality: Management, Assurance and Control,* 0-675-20896-3
KEOWN	*PSpice and Circuit Analysis,* 0-675-22135-8
KEYSER	*Materials Science in Engineering, Fourth Edition,* 0-675-20401-1
KIRKPATRICK	*The AutoCAD Book: Drawing, Modeling and Applications, Second Edition,* 0-675-22288-5 *Industrial Blueprint Reading and Sketching,* 0-675-20617-0
KRAUT	*Fluid Mechanics for Technicians,* 0-675-21330-4
KULATHINAL	*Transform Analysis and Electronic Networks with Applications,* 0-675-20765-7
LAMIT/LLOYD	*Drafting for Electronics,* 0-675-20200-0
LAMIT/WAHLER/HIGGINS	*Workbook in Drafting for Electronics,* 0-675-20417-8
LAMIT/PAIGE	*Computer-Aided Design and Drafting,* 0-675-20475-5
LAVIANA	*Basic Computer Numerical Control Programming, Second Edition,* 0-675-21298-7
MACKENZIE	*The 8051 Microcontroller, 0-02-373650-X*
MARUGGI	*Technical Graphics: Electronics Worktext, Second Edition,* 0-675-21378-9 *The Technology of Drafting,* 0-675-20762-2 *Workbook for the Technology of Drafting,* 0-675-21234-0

McCALLA	*Digital Logic and Computer Design,* 0-675-21170-0
McINTYRE	*Study Guide to Accompany Electronic Devices, Third Edition,* 0-02-379296-5
	Study Guide to Accompany Electronics Fundamentals, Second Edition, 0-675-21406-8
MILLER	*The 68000 Microprocessor Family: Architecture, Programming, and Applications, Second Edition,* 0-02-381560-4
MONACO	*Essential Mathematics for Electronics Technicians,* 0-675-21172-7
	Introduction to Microwave Technology, 0-675-21030-5
	Laboratory Activities in Microwave Technology, 0-675-21031-3
	Preparing for the FCC General Radiotelephone Operator's License Examination, 0-675-21313-4
	Student Resource Manual to Accompany Essential Mathematics for Electronics Technicians, 0-675-21173-5
MONSEEN	*PSPICE with Circuit Analysis,* 0-675-21376-2
MOTT	*Applied Fluid Mechanics, Third Edition,* 0-675-21026-7
	Machine Elements in Mechanical Design, Second Edition, 0-675-22289-3
NASHELSKY/BOYLESTAD	*BASIC Applied to Circuit Analysis,* 0-675-20161-6
PANARES	*A Handbook of English for Technical Students,* 0-675-20650-2
PFEIFFER	*Proposal Writing: The Art of Friendly Persuasion,* 0-675-20988-9
	Technical Writing: A Practical Approach, 0-675-21221-9
POND	*Introduction to Engineering Technology,* 0-675-21003-8
QUINN	*The 6800 Microprocessor,* 0-675-20515-8
REIS	*Digital Electronics Through Project Analysis,* 0-675-21141-7
	Electronic Project Design and Fabrication, Second Edition, 0-02-399230-1
	Laboratory Manual for Digital Electronics Through Project Analysis, 0-675-21254-5
ROLLE	*Thermodynamics and Heat Power, Third Edition,* 0-675-21016-X
ROSENBLATT/FRIEDMAN	*Direct and Alternating Current Machinery, Second Edition,* 0-675-20160-8
ROZE	*Technical Communication: The Practical Craft,* 0-675-20641-3
SCHOENBECK	*Electronic Communications: Modulation and Transmission, Second Edition,* 0-675-21311-8
SCHWARTZ	*Survey of Electronics, Third Edition,* 0-675-20162-4
SELL	*Basic Technical Drawing,* 0-675-21001-1
SMITH	*Statistical Process Control and Quality Improvement,* 0-675-21160-3
SORAK	*Linear Integrated Circuits: Laboratory Experiments,* 0-675-20661-8
SPIEGEL/LIMBRUNNER	*Applied Statics and Strength of Materials,* 0-675-21123-9
STANLEY, B.H.	*Experiments in Electric Circuits, Third Edition,* 0-675-21088-7
STANLEY, W.D.	*Operational Amplifiers with Linear Integrated Circuits, Second Edition,* 0-675-20660-X
SUBBARAO	*16/32-Bit Microprocessors: 68000/68010/68020 Software, Hardware, and Design Applications,* 0-675-21119-0
TOCCI	*Electronic Devices: Conventional Flow Version, Third Edition,* 0-675-20063-6
	Fundamentals of Pulse and Digital Circuits, Third Edition, 0-675-20033-4
	Introduction to Electric Circuit Analysis, Second Edition, 0-675-20002-4
TOCCI/OLIVER	*Fundamentals of Electronic Devices, Fourth Edition,* 0-675-21259-6

WEBB	*Programmable Logic Controllers: Principles and Applications, Second Edition,* 0-02-424970-X
WEBB/GRESHOCK	*Industrial Control Electronics,* 0-675-20897-1
WEISMAN	*Basic Technical Writing, Sixth Edition,* 0-675-21256-1
WOLANSKY/AKERS	*Modern Hydraulics: The Basics at Work,* 0-675-20987-0
WOLF	*Statics and Strength of Materials: A Parallel Approach,* 0-675-20622-7

PREFACE

BACKGROUND

The field of industrial supervision has changed significantly over the past two decades. There are many reasons for this. Some of the more prominent include the following: tremendous technological advances; increased pressures from an intensely competitive marketplace; emergence of the global economy; a steadily growing body of legislation and common law relating to employee rights, health, and safety; a growing potential for human conflict in the workplace; the rising tide of substance abuse; the changing nature of labor/management relations; a new emphasis on quality; and a growing interest in ethics.

All of these factors, when taken together, have made the job of the modern industrial supervisor more challenging and more important than it has ever been. They have also created a need for a book that focuses strictly on industrial supervision, as opposed to the generic approach used in other supervision books.

WHY WAS THIS BOOK WRITTEN AND FOR WHOM?

This book was written in response to the need for a practical teaching resource that focuses on the specific needs of industrial supervisors as they have evolved in the modern world of high technology. It is intended for use in universities, colleges, community colleges, and corporate training settings that offer programs, courses, workshops, and/or seminars in industrial supervision. Educators in such instructional disciplines as industrial technology, manufacturing technology, industrial psychology, and industrial management and supervision may find this book both valuable and easy to use.

The direct, straight-forward presentation of material focuses on making the theories and principles of supervision practical and useful in a real-world setting. Up-to-date research has been integrated throughout in a down-to-earth manner. Each chapter includes both a real-world case study and simulation activities to help prospective supervisors learn to apply what they are studying to the solution of the kind of problems faced in a live work setting.

ORGANIZATION OF THE BOOK

The text is divided into five parts, each focusing on a major component of the overall responsibility of modern industrial supervision. Part One presents the theories, principles, and practices of supervision relating to leadership. Part Two covers the management roles of the supervisor including planning, organization, staffing control, and decision making. Part Three describes the supervisor's counseling responsibilities including performance appraisal, dealing with problem employees, managing and resolving conflict, handling employee complaints, and disciplining employees. Part Four focuses on the supervisor as an innovator. Topics covered include training, productivity/competitiveness, and quality/competitiveness. Part Five deals with the legal concerns of the modern industrial supervisor relating to equal employment opportunity, employee health and safety, and labor/management relations.

A standard format is used throughout the book. Each chapter begins with a list of the major objectives and ends with a comprehensive summary. Following the summary, each chapter contains end material consisting of review questions, key terms and phrases, a case study, and simulation activities. These end materials are provided to encourage review, stimulate additional thought, and provide practical application opportunities.

HOW THIS BOOK DIFFERS FROM OTHERS

There are many excellent books available on supervision. However, most of them attempt to reach the widest possible segment of the market by taking the "generic approach." The philosophy of this approach is that "supervision is supervision." This book was written because in the age of high technology, supervision in the industrial sector can no longer be adequately taught using the generic approach. Many issues, concerns, and factors relating *specifically* to the modern industrial environment must be given more attention, greater depth of coverage, and more illumination by specific industrial examples than can be provided in a generic test. Some of the areas deserving more attention and requiring specific industrial examples are the following:

- Handling technology-related problems
- Facilitating human and technological change
- Encouraging ethical behavior in a competitive environment
- Conflict management and conflict resolution
- Building a world-class workforce
- Productivity, quality, and competitiveness
- Diversity in the workplace
- Substance abuse in the workplace
- Changing nature of labor relations
- AIDS and other health and safety concerns
- Creativity in decision making
- Workforce training and the supervisor's role in providing it.

ANCILLARY MATERIAL

Also available from your Merrill/Macmillan representative is an Instructor's Resource Manual that contains a test bank of questions, suggested answers to text review questions, and a set of 65 transparency masters. Many thanks to Ronald Tucker for his assistance in preparing the test bank.

ABOUT THE AUTHOR

David L. Goetsch is provost of the Okaloosa-Walton Community College/University of West Florida Campus in Fort Walton Beach, where he also is director of the Center for Manufacturing Competitiveness (CMC). The CMC is Florida's designated representative to the Southern Technology Council's Consortium for Manufacturing Competitiveness, a consortium of colleges and universities with statewide missions in the area of technology transfer and industrial modernization.

Dr. Goetsch is also president of the Institute for Corporate Development (ICD), a private institute dedicated to the improvement of productivity, quality, and competitiveness in American business and industry. The Institute conducts research; provides consultation services, and conducts customized, on-site training to help business and industrial firms improve their productivity, quality, and competitiveness by building a world-class workforce at all levels.

ACKNOWLEDGMENTS

The author wishes to gratefully acknowledge the contribution of numerous individuals whose assistance was invaluable in the development of this book. My appreciation goes to James R. Richburg, president of Okaloosa-Walton Community College, and J. Lamar Roberts, business executive and community leader, for their contributions, particularly in the area of leadership. Special thanks to Faye Crawford for her excellent work in the formatting and word processing of the manuscript. Thanks also to the following professionals for the improvements they suggested while reviewing the manuscript at various stages of its development: Gerald Arffa, Purdue University; Glenn Horkheimer, Southwest Wisconsin Tech; Alva Jared, University of Wisconsin, Platteville; J. Phillip McGrath, Purdue University; James P. Orr, Southern Illinois University; James Robertson, South Florida Community College; Wendell Stoye, Palomar College; Fred Sutton, Cuyahoga Community College; Alfred Travers, Indiana Vocational Technical College; Ronald Tucker, Morehead State University; and Robert Yancy, Southern College of Technology.

CONTENTS

PART ONE

The Supervisor as a Leader

INTRODUCTION
What is Supervision?

CHAPTER OBJECTIVES

After studying this chapter, you will be able to define or explain the following topics:

- Supervision
- The Transition to Supervisor Process
- Characteristics of Good Supervisors
- Responsibility as it Relates to Supervision
- What Is Expected of a Supervisor
- Typical Duties of Supervisors
- The Changing Role of the Modern Supervisor

CHAPTER OUTLINE

- Supervision Defined
- Making the Transition to Supervisor
- Characteristics of Good Supervisors
- Responsibility and the Supervisor
- What Is Expected of Supervisors?
- Duties of Supervisors
- Changing Role of the Modern Supervisor
- Endnotes

Supervisors bridge the gap between employees and management (Figure I–1). They are the vital link between that component of a company that plans and organizes work and the component that actually does the work. Often people become supervisors based on their outstanding performance in a technical or skilled position, but they have little or no supervisory training.

This can cause problems because the skills necessary to succeed as a supervisor are vastly different from those needed to succeed as a technician or skilled worker. This book was designed to help prospective and practicing supervisors develop the skills needed to succeed in the increasingly complex and challenging age of high technology.

SUPERVISION DEFINED

Supervision is the first line of management (Figure I–2). Supervisors carry a variety of different titles, such as *foreman, team leader, group leader, unit head,* and so on. Regardless of what they are called, supervisors provide the first line of leadership to employees who perform the work necessary to produce a company's product(s). They also serve as the liaison with management for these workers (Figure I–3).

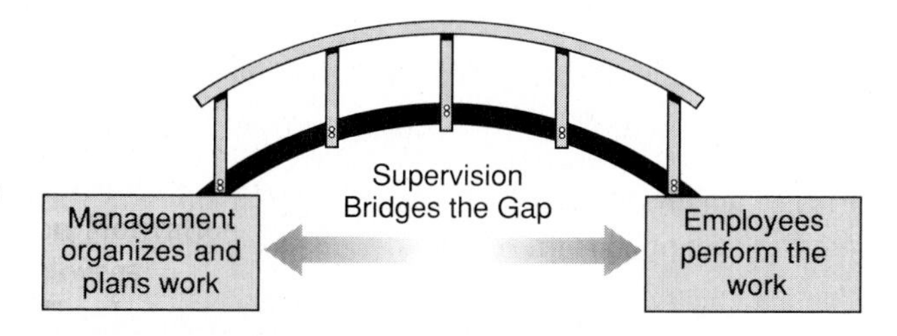

Figure I–1
Supervision bridges the gap between managers who plan and organize work and employees who do the work.

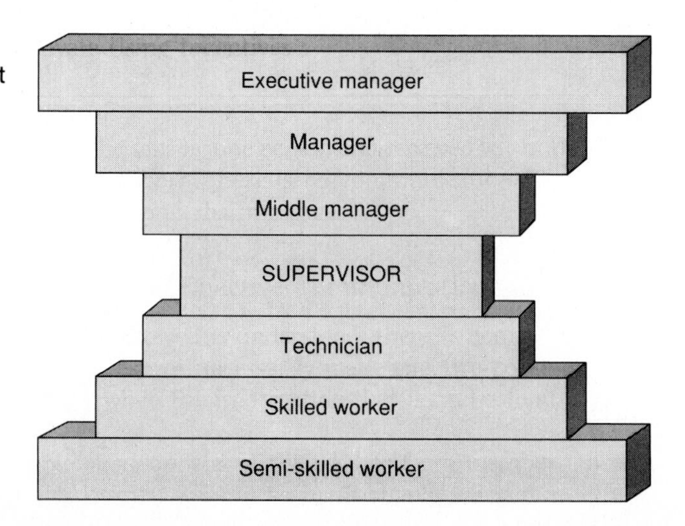

Figure I–2
Supervisors see the first level of management.

Figure I–3
Supervisors work with management and labor and serve as a communication channel between them.

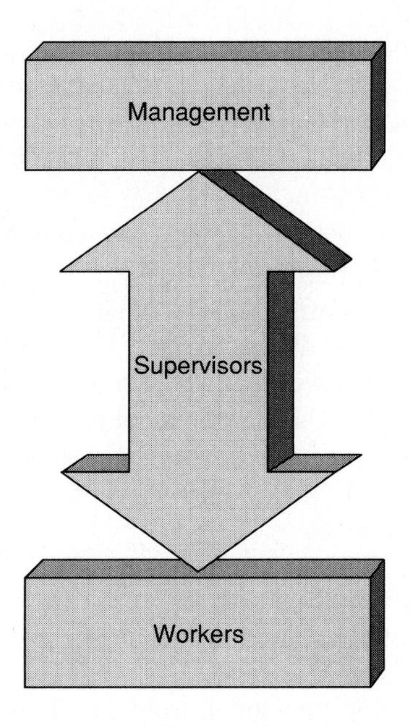

The role of the supervisor was clearly defined by the Taft-Hartley Act of 1947 which amended the National Labor Relations Act of 1935. The Taft-Hartley Act defines a supervisor as follows:

> . . . any individual having authority, in the interest of the employer, to hire, transfer, suspend, lay off, recall, promote, discharge, assign, reward, or discipline other employees or responsibility to direct them, or to adjust their grievances, or effectively to recommend such action if in connection with the foregoing the exercise of such authority is not of a merely routine or clerical nature, but requires the use of independent judgement.[1]

The job of supervisor is one of the most challenging in the workplace. Managers above the supervisor level have their expectations and workers below supervisors have theirs. It is not uncommon for these expectations to be different. "Caught in the middle" is how supervisors sometimes describe their position. Whether the priorities, values, and goals of management and labor coincide or not, the supervisor's challenge is the same—to work with both to produce positive results. In other words, regardless of the circumstances, supervisors are expected to get the job done. According to Robert Albanese, adjectives that describe the supervisor's job are *important, tough, changing, challenging, essential, controversial, undervalued,* and *stressful.*[2]

MAKING THE TRANSITION TO SUPERVISOR

How do supervisors become supervisors? There is no one single answer to this question. However, a typical scenario follows:

> Jorge Camacho began working at Southwestern Plastic Recycling Company as a laborer right out of high school. He worked hard and was highly respected by his co-workers. As a result, Camacho was promoted rapidly, first to machine feeder, then to machine operator, then to process technician, and finally to line supervisor.

This is a typical scenario showing how industrial workers become supervisors. Unfortunately, nothing in such scenarios prepares new supervisors to be good leaders, motivators, communicators, planners, organizers, staffers, controllers, decision makers, trainers, disciplinarians, or any of the many other roles they will have to play. Perhaps the most difficult part of the transition comes when new supervisors realize that they are held accountable not just for their performance, but also for that of others.

New supervisors are sometimes surprised to learn of the rights they lose as a result of their promotion to the first level of management. Earnest Archer summarizes these rights as follows:[3]

- Right to lose their temper
- Right to hobnob and be *one of the crowd*
- Right to shut the desk
- Right to bring personal problems to work
- Right to speak freely
- Right to be against change
- Right to pass the buck
- Right to *get even*
- Right to choose favorites
- Right to think of themselves first
- Right to ask an employee to do something they would not do
- Right to expect immediate reward for their work

To Archer's list the following additional lost rights might be added:

- Right to keep recognition to themselves
- Right to procrastinate in decision making
- Right to neglect setting a good example
- Right to apply personal biases in dealing with workers

Clearly, a successful transition from worker to supervisor will not happen automatically. It will require education, training, and experience coupled with patience and a continual, sincere effort to improve.

The role of the supervisor has evolved over the years. In the past, supervisors had absolute authority and control. They could use the "Do what I say or you're fired!" approach in dealing with people. This is no longer the case. Fair labor practice laws, the advent of participatory management, and the realization that people produce better when they are led rather than pushed have changed the approaches available to supervisors.

To this add the complications brought on by rapid and continual changes in technology, and you can see that the job of modern supervisors is more complex and demanding than at any time in the past. The list of characteristics needed by people who want to be supervisors has grown as the complexity of the job has increased.

CHARACTERISTICS OF GOOD SUPERVISORS

Good supervisors look just like the rest of us. In fact, in a crowd it would be impossible to pick out the supervisors from other people. This is because the characteristics of good supervisors manifest themselves internally rather than externally. What follows is a list of characteristics needed by supervisors for success in today's highly technical, highly complex world of industry. People who want to be supervisors should have the potential to develop all of the following skills, abilities, and characteristics:

- Technical job skills
- Leadership skills
- Communication skills
- Ability to adapt to change
- Ability to facilitate change
- Motivational skills
- High ethical standards
- Problem-solving skills
- Planning, organizing, and monitoring skills
- Staffing skills
- Budgeting skills
- Training skills
- Ability to appraise job performance
- Counseling skills
- Ability to discipline appropriately
- Ability to handle complaints in a positive manner
- Ability to innovate
- Ability to keep up with continually changing rules and regulations, both internal and external
- Dependability and flexibility
- Ability to accept responsibility

This list represents a pretty tall order, and it would be the rare individual who began his or her first supervisory job having already developed all of these characteristics. However, most of them can be learned. The more of these characteristics supervisors learn, the more likely they are to be effective and successful in a modern industrial setting. The remainder of this book contains information and activities that can help prospective and practicing supervisors develop the characteristics necessary to become good supervisors.

According to Roger Fritz, president of Organizational Development Consultants, supervisors can enhance their effectiveness by getting to know themselves better. Fritz lists the following rules of thumb for enhancing the effectiveness of supervisors:[4]

1. People are motivated by different factors. The supervisor's task is to determine what motivates each employee he or she supervises.
2. Employees have their own personal goals. The supervisor's task is to learn the personal goals of each subordinate and determine how closely they match those of the organization.
3. Focusing too intently on a personal strength can cause problems. For example, say a supervisor is always punctual and takes great pride in this, but she has a subordinate who is less punctual. If the supervisor lets this one factor cause her to overlook the valuable qualities of the occasionally tardy employee, her personal strength (punctuality) may turn into a supervisory weakness.
4. Only horses respond to a carrot on a stick. Supervisors who wish to motivate employees should give more attention to creating a work environment that promotes self-motivation.
5. By knowing staffers well, supervisors can do a better job of using communication as a tool to enhance effectiveness. For example, such knowledge will give supervisors a feel for how far they can take a new idea without having it rejected overnight.

RESPONSIBILITY AND THE SUPERVISOR

Perhaps the word that best characterizes the job of the modern supervisor is *responsibility.* Supervisors are responsible to management, to the employees they supervise, to related support personnel, to fellow supervisors, to users of their company's products, and to their profession (Figure I–4).

Supervisors are responsible to management for interacting with the people, technology, and resources available to them so that they produce a quality product in a competitive manner on or ahead of schedule. Supervisors are responsible to the employees they supervise for ensuring that they have (1) the resources needed to get the job done; (2) a safe, clean, harmonious working environment; (3) the guidance needed to do a good job; (4) the training and education needed to stay productive; and (5) fair and equitable treatment.

Supervisors are responsible to related support personnel to be cognizant of and sensitive to their needs. Supervisors are responsible to their fellow supervisors to be mutually supportive. Supervisors are responsible to customers to do their part to ensure a quality product at a competitive price. Finally, supervisors are responsible to their profession to stay up to date and to perform in a way that speaks well of the profession.

WHAT IS EXPECTED OF SUPERVISORS?

What is expected of modern supervisors can be summarized briefly and succinctly. Supervisors are expected to produce a quality product/service in a competitive manner

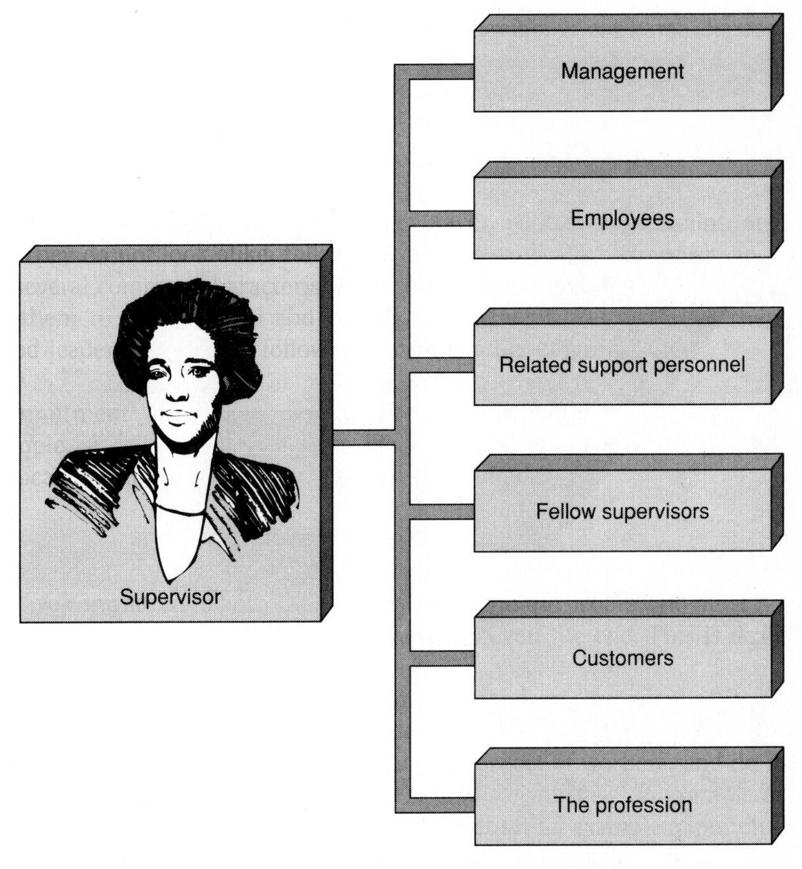

Figure I–4
Supervisors are responsible to a variety of people.

on or ahead of schedule. This means that supervisors are expected to make the most efficient and effective use possible of the people, technology, and resources available to them. This is how modern industrial firms succeed in an increasingly competitive marketplace.

Making efficient and effective use of people requires that supervisors develop and maintain positive attitudes and motivate employees to make the organization's goals their own. Making efficient and effective use of technology requires that supervisors continually improve the use of tools, machines, equipment, systems, processes, and facilities available to them. Making efficient and effective use of resources requires that supervisors properly marshal the materials, supplies, information, funds, time, and energy available to them.

DUTIES OF SUPERVISORS

Regardless of the type of industry, the actual duties of supervisors are fairly similar. Depending on the industry, supervisors will devote varying amounts of time to these duties. The most common duties of modern supervisors can be summarized as follows:

- Planning work
- Organizing work
- Assigning work
- Monitoring work
- One-on-one verbal communication
- Group verbal communication
- Written communication
- Electronic communication
- Training/teaching/updating
- Attending/conducting meetings
- Record keeping and reporting
- Personal/professional development
- Appraising, evaluating, and giving feedback
- Counseling/advising
- Community relations activities
- Liaison activities
- Enhancing quality
- Improving productivity
- Maintaining a safe and healthy work environment
- Selecting and orienting employees
- Maintaining positive union-management relations

CHANGING ROLE OF THE MODERN SUPERVISOR

The traditional "I'm the boss, so do as I say" style of supervision is rapidly becoming obsolete. Modern industrial supervisors spend more time empowering, enabling, and coordinating than giving orders. According to Tom Peters, the modern supervisor needs to be competent in the areas of participative management, group facilitation and problem solving.[5]

The trend toward the use of *quality circles* and other types of self-managed work teams in which employees suggest ways to continually improve quality, enhance productivity, and lower costs; the need to develop and maintain a world-class workforce in order to compete in a global marketplace; the rapid diversification of the workforce with regard to minorities and women; and the changing nature of labor-management relations make competence in such areas as employee training, human relations, conflict resolution, and team-building essential. Finally, the modern supervisor must be able to grow and change continually while simultaneously serving as a positive change agent for those he or she supervises. Is the job of the modern supervisor more difficult than ever? Yes, but it is also more challenging, more important, and more filled with opportunity than ever. This book was developed to help prospective and practicing supervisors meet the challenges of supervising effectively in the rapidly changing age of high technology.

ENDNOTES

1. Taft-Hartley Act. *United States Statutes at Large,* Volume 61, p. 138, PL 80-120, Section 2(11), 1947.
2. Albanese, R. *Management* (Cincinnati: South-Western Publishing Company, 1988), p. 725.
3. Archer, E. R. "Things You Lose the Right to Do When You Become a Manager," *Supervisory Management,* July 1990, p. 8.
4. Fritz, R. "What Effective Supervisors Do—and Don't Do," *Supervisory Management,* January 1991, p. 8.
5. Peters, T. *Thriving on Chaos* (New York: Harper & Row, 1987), p. 360.

CHAPTER ONE

Technology and the Supervisor

CHAPTER OBJECTIVES

After studying this chapter, you will be able to define or explain the following topics:

- Technology
- High Technology
- Widely Used Modern Industrial Technologies
- The Impact of Technology in the Workplace
- How to Help Workers Handle Change
- The Relationship of Industry, Technology, and the Future

CHAPTER OUTLINE

- What Is Technology?
- What Is High Technology?
- Modern Industrial Technologies
- Impact of Technology in the Workplace
- Helping Workers Handle Change
- A Supervisor's Model for Change
- Summary
- Key Terms and Phrases
- Case Study Application Problem
- Review Questions
- Simulation Activities
- Endnotes

In almost every aspect of our lives, at work and at home, we depend on technology. At the same time, the technologies on which we depend are becoming more and more advanced. Consider the following modern technological advances:

- In 1990, Tokyo-based Sony Corporation announced it had developed the world's most powerful microprocessor chip. The chip can process 150 million instructions per second. Sony plans to release a new personal computer based on this chip that can outperform contemporary mainframe computers.[1]
- In 1990, scientists at IBM's Almaden Research Center in San Jose, California, announced a major technological breakthrough. After many years of trying, they had finally produced a working blue semiconductor laser. Compared with the traditional red laser, the blue laser can make much smaller pinpricks in optical recording materials, thereby quadrupling storage capacity.[2]
- In 1990, Graphics Communication Technologies Ltd., a consortium of Japanese companies, announced the development of a high-resolution video telephone. GCT produced a set of semiconductor chips for color videophones that are used in conjunction with high-capacity fiber-optic networks in the United States, Europe, and Asia. The result is a videophone that produces a clear, well-defined screen image and high-quality voice transmission.[3]
- In 1990, A.C.T., Inc. in Santa Ana, California, announced the development of its A.C.T. ATM. Resembling a bank's automated teller machine, the A.C.T. ATM allows travelers to pick up airline tickets in minutes, twenty-four hours a day. Ticket information is sent by telephone to a travel agent. The A.C.T. ATM prints the ticket when travelers identify themselves with a credit card. The machines will be located in hotel lobbies, office buildings, and grocery stores.[4]
- In 1990, *Popular Science* announced the development of a "Studious TV." While you watch this TV, it watches you. Using artificial intelligence, the TV monitors your viewing habits, determines your favorite shows, and lists them on a menu for reference when you turn the TV on.[5]

These are but a few examples of how technology continues to evolve. As it does, it becomes more and more a part of our lives. Consequently, modern supervisors must be technologically literate. That is, they must be able to function successfully in a technology-dependent environment. In addition they must be able to help those they supervise function successfully in such an environment. Of particular interest to the modern industrial supervisor is the workplace environment and the impact technology is having there.

Modern industry is that enterprise through which materials are converted into usable products. The work necessary to accomplish this conversion is done by people and, increasingly, in conjunction with technology. Supervisors are catalysts in the process.

Industry in the United States is navigating through troubled waters. Producers of industrial goods must compete in a global marketplace where the competition is fierce. Long the world's leader in the industrial arena, the United States is being challenged on all fronts and, in many cases, the challengers are winning.

In almost every type of industry, U.S. companies have lost market share to foreign competitors whose products are better designed, more reliable, and less expensive. Competition is not limited to one nation. Taiwan, Korea, and several European countries have emerged as tough competitors. But the best example of a competitor whose emergence has cut into markets traditionally controlled by the United States is Japan.

U.S. manufacturers led the way in the development and original manufacturing of automobiles, radios, television sets, and then computers. Japan now dominates these markets worldwide. Why did this happen? There are a number of reasons, but three stand out as the most important: (1) the Japanese government worked with the Japanese private sector to create an economic environment that is supportive of business; (2) Japanese industry is making better use of technology; and (3) Japanese industry is making better use of people.

Efforts are currently underway to improve the business environment in which U.S. industrial firms operate in this country. American industry is making progress toward more effective use of technology. However, less attention is being given to the better use of people, and in the long run this may be the most important of all the reemergence strategies for U.S. industry. For this reason, making more effective and efficient use of people is the central focus of this text.

Even in the age of high technology, the highly skilled, highly motivated, productive worker is still industry's greatest asset. The person best able to make the most efficient and effective use of this asset is the well-trained, knowledgeable supervisor. This book is devoted to helping people develop the knowledge and skills needed to be successful industrial supervisors in the increasingly complex, increasingly competitive age of high technology.

Historically, supervision has been viewed as a process concerned with accomplishing work through people, and this concept is still valid. However, modern industries have become so technology dependent that the impact of technology on productivity and on people in the workplace cannot be ignored. Supervisors are still responsible for ensuring that the work is accomplished through people, but more and more people use technology to do their work. And the technology is becoming increasingly sophisticated and increasingly complex. The modern industrial worker depends on technology and technology depends on people. The relationship is symbiotic.

This means that supervisors must be able to bring out the best from both people and technology and learn to make optimum use of the people/technology partnership. To do so, supervisors must: (1) understand technology as a concept; (2) be familiar with the latest technological developments in their industry; (3) understand the impact of technology in the workplace; (4) be familiar with technology/people problems and understand how to deal with them; and (5) understand how to deal with the rapid and continual changes associated with modern technology.

WHAT IS TECHNOLOGY?

Technology is what results when people apply scientific principles to the solution of everyday problems. In other words, technology is **applied science**. People apply scientific principles to the solution of everyday problems using tools, processes, and resources. These are the key elements of technology (Figure 1–1).

Notice the **people connection** with each element of technology. People design, make, and use tools. People plan, develop, and apply processes. People identify, secure, and make use of resources. People themselves are a resource. In fact, people's knowledge and skills represent the most important resource associated with technology.

WHAT IS HIGH TECHNOLOGY?

The term **high technology** has become an entrenched buzzword. Its first widespread use corresponded with the development of programmable integrated circuits, the key tech-

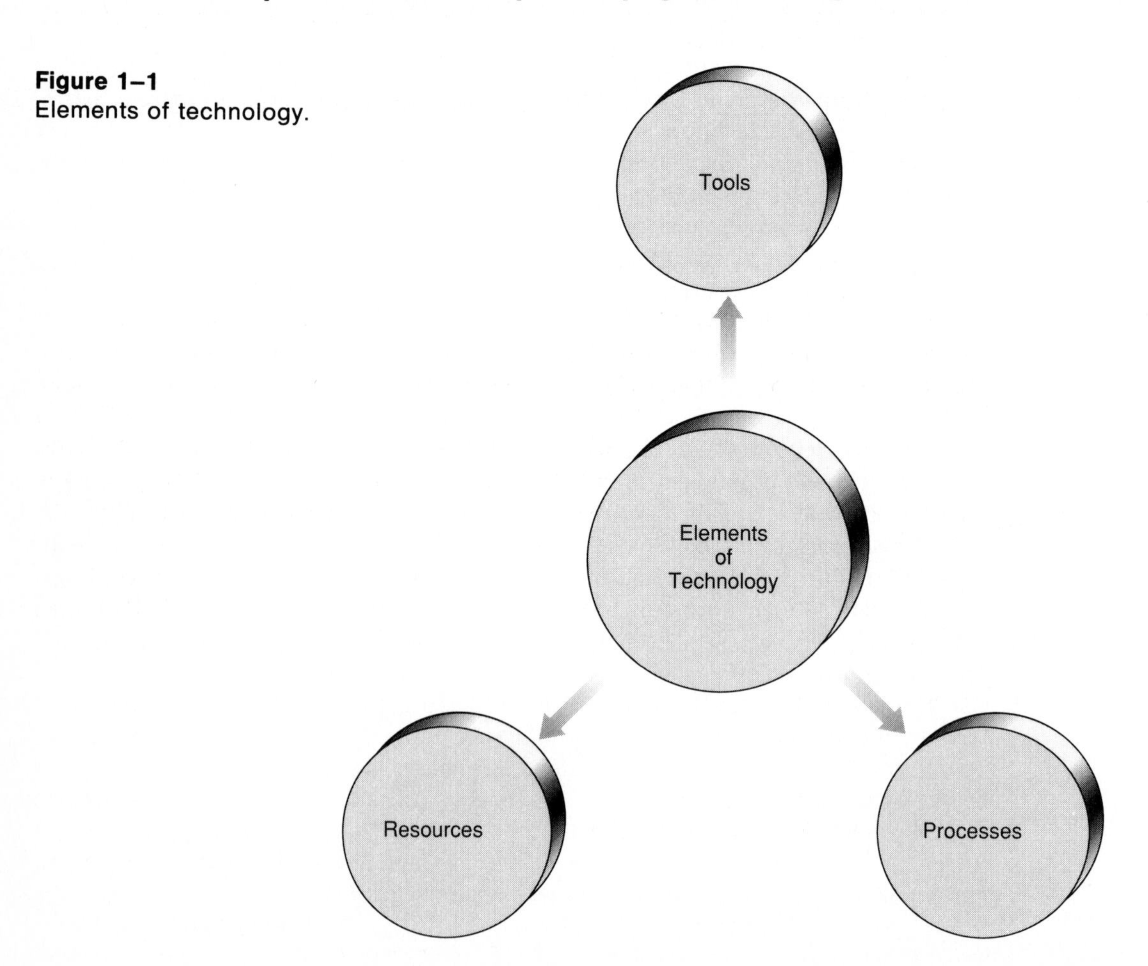

Figure 1–1
Elements of technology.

nology enabling the downsizing of computers. However, it is used to describe an infinite list of technologies.

It is a catchy term but not very descriptive. After all, if today's most modern developments are high technology, what will we call tomorrow's? A less catchy but more descriptive term is *advanced technology.*

We know that advanced technologies are the most advanced, most modern developments in a given field at a given time. This is also what is meant by the buzzword *high technology.* When used in this book, the term high technology should be taken to mean the most advanced technology of the time in a given field.

MODERN INDUSTRIAL TECHNOLOGIES

Supervisors cannot possibly be experts on all the different technologies used in a modern industrial setting. However, you can and should be at least familiar with the more common of these technologies. This section provides a brief overview of technologies with which modern industrial supervisors should be familiar. They are divided into two broad categories.

1. Material technologies
2. Process and production technologies

Modern Industrial Materials

The materials used in industry have changed over the years and they continue to change. Ferrous and nonferrous metals and their numerous alloys are still widely used engineering materials. However, plastics are now the most often used material. Ceramics and elastomers are also used in the production of manufactured products.

The most rapidly growing material technology is that dedicated to the development of composites. A **composite** is a heterogeneous solid that consists of two or more different materials. The materials used to form a composite are bonded together either mechanically or metallurgically to form a solid.

Some composites are actually stronger than steel and lighter than aluminum. Certain composites available for industrial uses offer several advantageous characteristics including, but not limited to, the following:

- Low thermal conductivity
- Good heat resistance
- Long fatigue life
- High resistance to corrosion
- Good wear resistance

Because of these characteristics, composites are being used in the manufacture of a variety of industrial products ranging from aircraft to sporting equipment.

Modern Production, Planning, and Process Technologies

Perhaps the best example of how technology has changed in the workplace is the typical office telephone. Not so many years ago an office telephone's capabilities were limited to ringing, receiving calls, and sending calls. Compare this with today's typical office telephone that has repertory dialing, station speed call, queuing, conference calling, system speed call, consultation call, camp-on, save and repeat, park, call forwarding, pickup, connect, transfer, hold, message recording, and many other options on its list of capabilities.

Another example of a technological development that is having a universal impact across all types of industries is the facsimile machine. Faxed letters now rival mailed letters in daily volume. These are examples of technological advances in the field of telecommunication.

The laser, artificial intelligence, robotics, machine vision, and of course the computer are having a marked effect on all types of industries. The modern technologies discussed here relate specifically to such industrial processes as design, planning, control, and production. Other processes could have been included, but these will give supervisors a sufficient overview of technologies that are changing the way work is done in an industrial setting.

The various processes used in a modern industrial setting are summarized graphically in Figure 1–2. Most of these processes fall under three broad headings:

- Design processes
- Planning and control processes
- Production processes

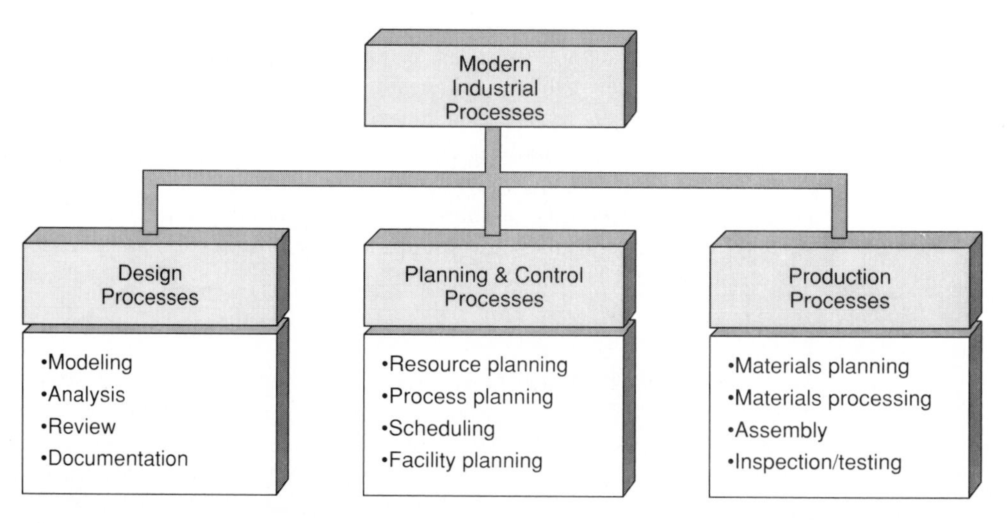

Figure 1–2
Modern industrial processes.

Design Processes

The computer has had a significant impact on how people design products. All four of the principal design processes—modeling, analysis, review, and documentation—have been changed by the advent of **CAD/CAM** (computer-aided design/computer-aided manufacturing) technology.

Design modeling used to involve actually building a scale model or a full-sized prototype. With CAD/CAM the model is built mathematically in the CAD/CAM database and stored there. It can be displayed on the screen of a computer terminal and easily edited and manipulated while displayed.

Modern CAD/CAM systems are capable of building and displaying both two-dimensional and three-dimensional geometric models of a part (Figures 1–3 and 1–4). Color display capabilities enhance the designer's ability to communicate clearly even more. Analyzing a design to determine how it will perform under the conditions to which it will be eventually subjected has also been simplified by the computer (Figure 1–5). Typical analysis tasks such as stress-strain calculations were traditionally very time consuming. However, computers can perform such calculations in seconds.

Finite element analysis software has simplified the analysis process even further. Such software allows the computer to subdivide the geometric model of a part into numerous rectangular or triangular segments or elements. It can then determine how each element performs when subjected to simulated conditions. Finite element analysis allows designers to pinpoint clearly the locations of failures or potential problems (Figure 1–6).

Design review involves checking the accuracy of all aspects of a design. Computers and CAD/CAM software have simplified the process by making available such capabilities as semiautomatic dimensioning, layering, and interference checking. With some CAD/CAM systems certain dimensions are calculated automatically. By displaying different parts of a design in different colored layers, designers can quickly and easily identify alignment problems. Interference checking allows designers to connect mating parts on the screen of a CAD/CAM terminal to see if interference problems exist.

Design documentation has also been simplified by CAD/CAM. Data needed to produce drawings, specifications, building materials, and parts lists are created and stored in the CAD/CAM database during the modeling phase of the design process and then edited and modified as a result of the analysis and review phases. Various methods exist for printing and plotting hard copies of design documentation (Figure 1–7).

Planning and Control Processes

Two new technologies that are having a significant impact on the planning and control processes are **computer-aided process planning** or **CAPP** and **manufacturing resources planning** or **MRP**. CAPP is a technology developed to bridge the gap between design and production. Once a product has been designed, the processes needed to manufacture it must be planned. There are so many different sequences of processes that can be used to produce a product that determining the optimum sequence used to be difficult if not impossible.

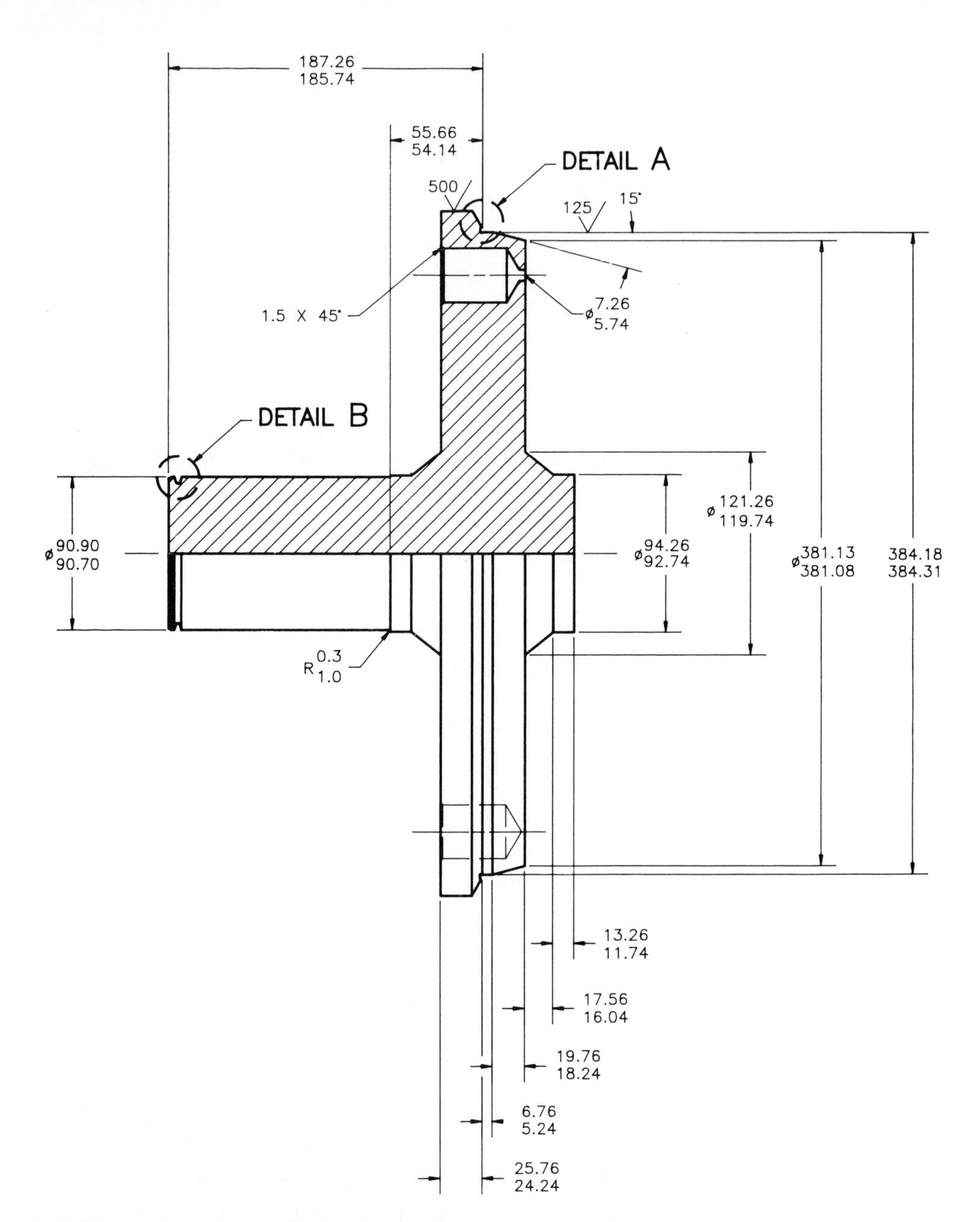

Figure 1–3
Two-dimensional CAD drawing.

Figure 1–4
Solid model.
Courtesy of Computervision (A Prime Company).

Figure 1–5
CAD workstation.
Courtesy of CADKEY.

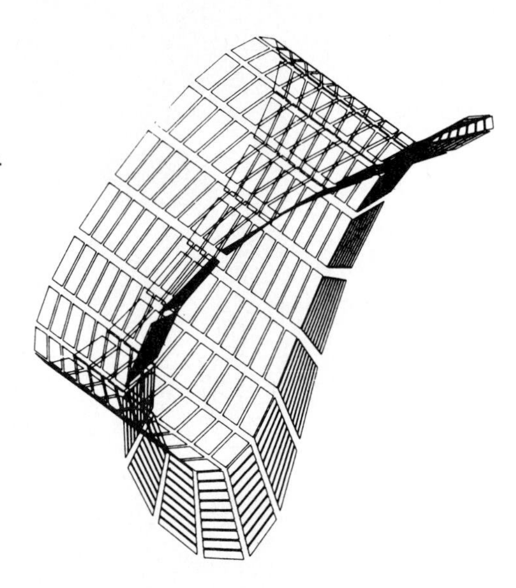

Figure 1–6
Finite element analysis drawing.
Courtesy of McDonnell Douglas Automation Company.

However, the advent of computers, particularly microcomputers, gave manufacturing personnel an opportunity to solve this problem. CAPP systems are expert computer systems used to collect and store engineering principles, manufacturing processes information, and information about the capabilities of a specific production facility.

Once a part has been designed, a CAPP system compares the design criteria against the information stored in it to determine the optimum plan for producing the part. Such a plan contains a list of the machines to be used, the tooling required, optimum feed and speed rates for tools, and the optimum sequence of operation.

CAPP has changed the way process plans are developed. MRP has changed the way resource plans are developed. MRP involves the use of specialized software to determine the amount of raw materials needed to produce a batch of a given product. It also encompasses financial tracking and accounting. With these capabilities, MRP can be a powerful tool for inventory planning and control.

Production Processes

Producing manufactured products has always involved material handling, material processing, assembly, and inspection/testing. It still does. However, the ways in which these processes are accomplished are continually evolving. Some of the new and emerging production technologies with which modern industrial supervisors should be familiar are the following:

- Automated material handling
- Computer numerical control (CNC) of machines
- Automated assembly
- Statistical process control (SPC)

- Just-in-time delivery (JIT)
- Flexible manufacturing
- Computer-integrated manufacturing (CIM)
- Artificial intelligence
- Laser applications

Material handling is the movement of raw materials and/or workpieces from station to station in a sequence of production processes and presenting workpieces to machines for processing. In modern industrial settings the process might be accomplished

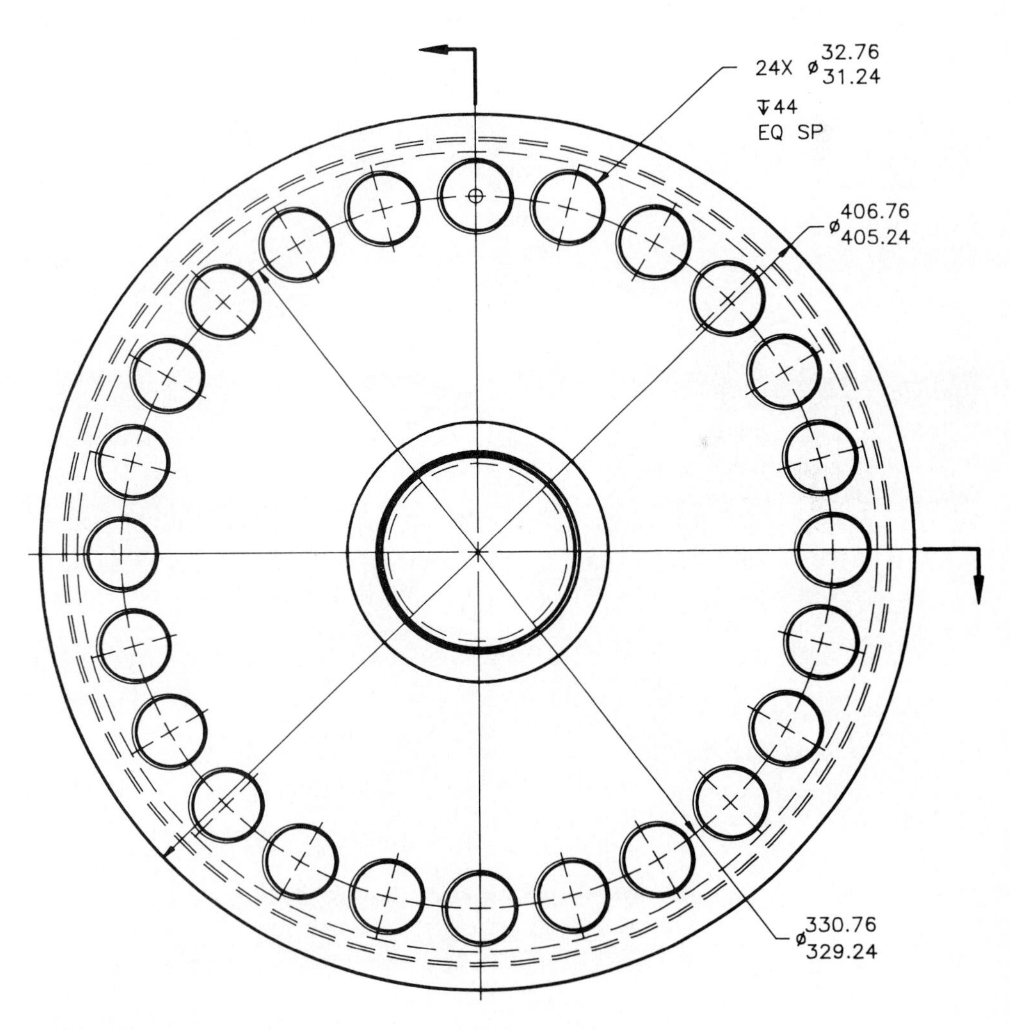

Figure 1–7
Two-dimensional CAD drawing.

through the use of computer-controlled **automated material handling** systems. Such systems typically use industrial robots for loading and unloading material and in presenting work to machines (Figure 1–8). They use conveyor systems and/or automated guided vehicles (AGVs) for moving material from point to point.

Control of modern industrial processes is often accomplished using computers. **Computer numerical control** (CNC) and a derivative technology, distributive numerical control (DNC), are becoming the norm for modern industrial settings. Both take advantage of and depend on personal computer technology.

With CNC, a computer program for controlling a specific machine during specific operations is written and loaded into a microcomputer, which executes the program. The microcomputer is typically *factory hardened* (especially designed to withstand a

Figure 1–8
IRb2000 FMS cell with Orbit 1600A positioner.
Courtesy of ESAB Automation, Inc.

factory environment) and located at the machine site. Part programs, or programs that guide the machine in the processing of a specific part, are stored in *random access memory* (RAM) so they can be easily accessed and edited. Static programs such as operating programs are stored in *read only memory* (ROM).

With DNC there are microcomputers at the machine site and a central host computer that may be located off the shop floor. Part programs are stored in the host computer until needed and then downloaded via a data transmission link to the local microcomputers for execution (Figure 1–9).

Assembly has always involved such operations as fastening, inspecting, labeling, packaging, and final preparation. It still does. But in many modern industrial settings, robots are being used for **automated assembly**, packaging, fastening, joining, and painting. Inspection operations are being aided by sensor and machine vision technology.

Inspection and testing have always been technology-dependent operations. This is even more the case in modern industrial settings. The concepts of **statistical process control** or **SPC** and **just-in-time delivery** or **JIT** are changing both how and when inspections take place.

SPC combines the statistical methods and the unique capabilities of a computer to ensure acceptable compliance with specifications. SPC was developed to provide a more

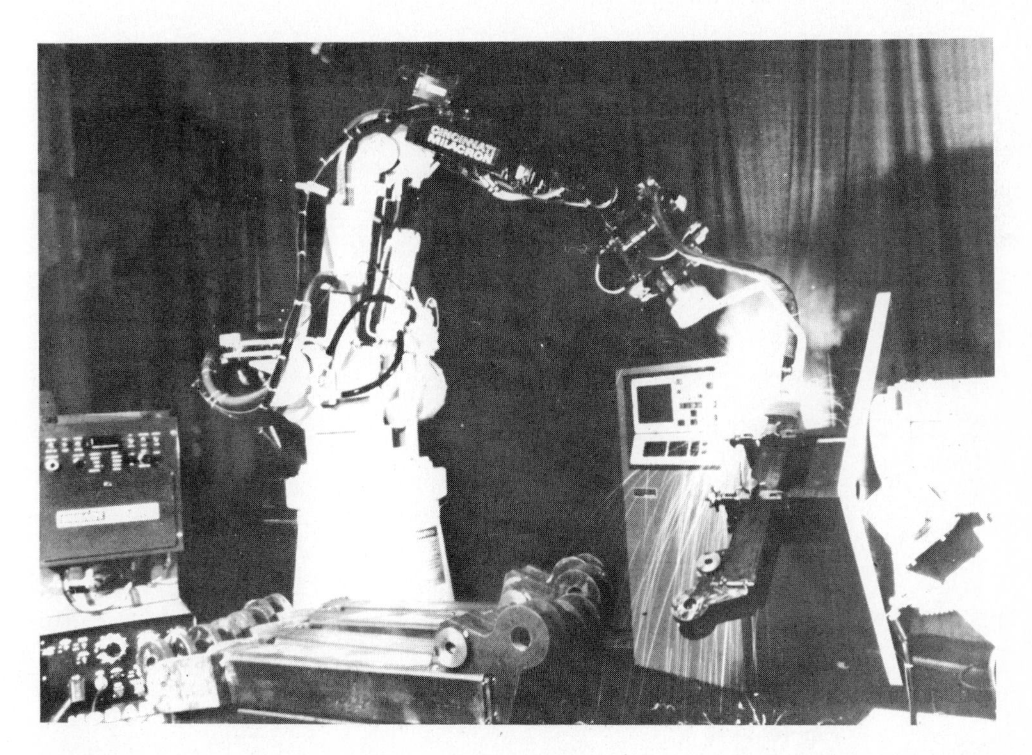

Figure 1–9
Distributed numerical control.

reliable quality assurance approach that can be achieved using inspection and testing of random samples. It allows production personnel to continually monitor the performance of a process, identify faults, and correct them immediately.

JIT involves receiving materials needed for production of a product where and when they are needed rather than relying on warehouses full of stacked inventory. The goal of JIT is to identify and eliminate any aspect of production that does not add value to the product.

Flexible manufacturing is not a new technology. Rather, it is a modern approach to producing a product that takes advantage of new technologies. Most manufactured products are produced in relatively small batches of less than 3,000 parts. This has always presented production personnel with a catch-22 dilemma.

Such batches are too large to produce competitively on stand-alone manually operated or CNC-controlled machines. On the other hand, they are too small to justify the time and expense of setting up a transfer line. What has been needed for years are production systems that are both fast and flexible. By pulling together such technologies as CNC, automated material handling, and machine vision, industrial personnel are now able to accomplish flexible manufacturing. Small, dedicated groups of workstations are called *flexible manufacturing cells.* A collection of integrated cells is called a *flexible manufacturing system* (Figure 1–10).

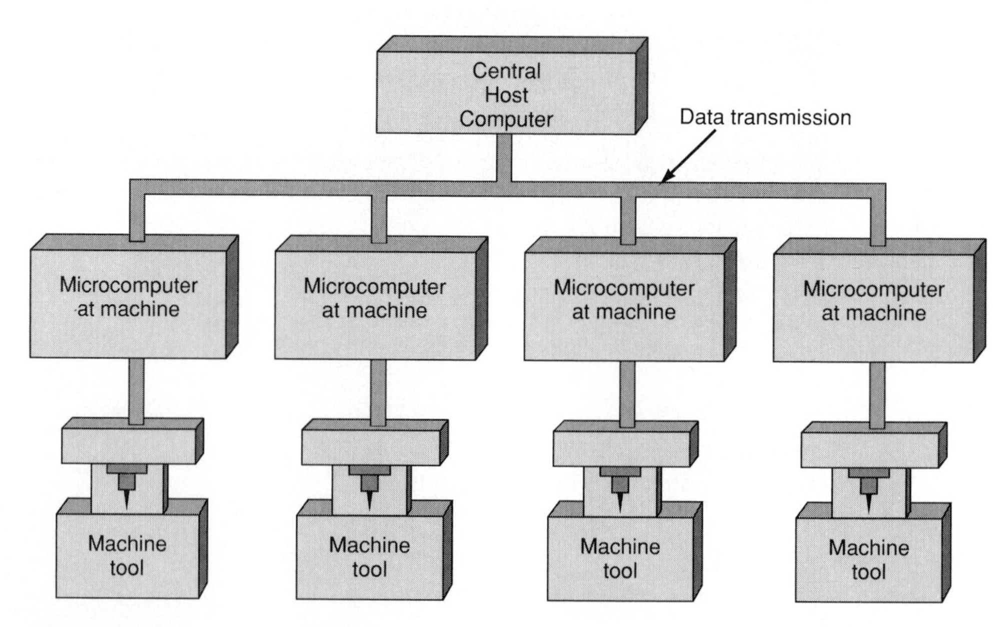

Figure 1–10
Flexible manufacturing system.
Courtesy of Cincinnati Milacron.

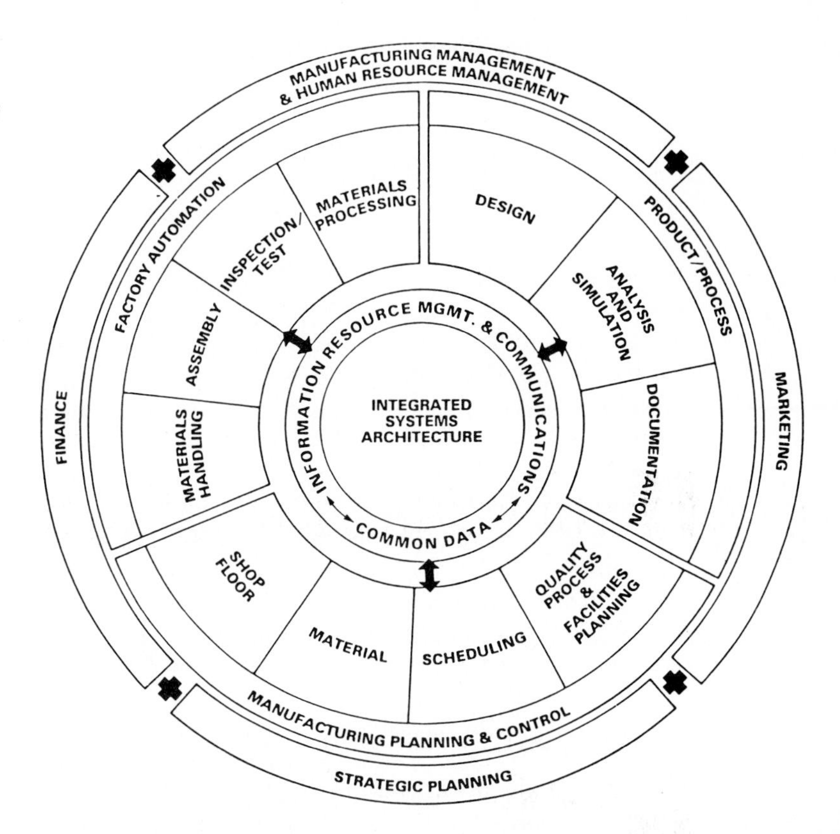

Figure 1–11
CIM wheel.
Copyright 1985, Society of Manufacturing Engineers, Dearborn, MI.

Computer integrated manufacturing (CIM), like flexible manufacturing, is not a new technology. Rather, it is a new way of doing business in industry that takes advantage of new technologies. A flexible manufacturing system accomplishes automation and integration of the production component of a manufacturing firm. CIM accomplishes automation and integration of *all* components of a company (Figure 1–11).

Artificial intelligence is a cornerstone technology that is still emerging but that is already having a pronounced impact in the workplace. Many of the future developments in industry, particularly in the production component, will depend on artificial intelligence. **Artificial intelligence** is the technology through which computers mimic human capabilities and thereby make intelligent decisions. Computer systems that apply artificial intelligence are called **expert systems**.

Burrows defines expert systems as being "repositories of human expertise bundled with software that makes decisions based on a given set of rules."[6] One of the fastest growing applications of artificial intelligence is in manufacturing. Diagnosis programs are being used more and more to make routine decisions. This frees workers' time, which they can then devote to making nonroutine decisions.

According to Automation Research Corporation, the manufacturing market segments for expert systems break down as follows:[7]

22.5 percent automotive
14.0 percent electronics
12.5 percent chemical
11.5 percent aerospace
9.0 percent food
7.0 percent machinery
16.5 percent other

Laser technology has also made inroads in the modern industrial sector. According to Neil Gross in *Business Week,* industrial applications of lasers are increasing rapidly.[8] Gross lists superconductor production and chip-making as two contemporary uses of lasers. Using lasers, high-quality superconductor materials can be produced for use in sensing devices and electronic circuits. With lasers, damaged microprocessor chips can be repaired rather than discarded.

Automated material handling, CNC, DNC, automated assembly, SPC, JIT, flexible manufacturing, CIM, artificial intelligence, and many other high technology developments are changing how work is done in industry. As a consequence, high technology is presenting industrial supervisors with a new set of challenges. This will continue as existing technologies are modernized and new technologies continue to emerge.

IMPACT OF TECHNOLOGY IN THE WORKPLACE

Modern technologies are having the same effects in the workplace today that they have always had, but the effects are now more pronounced. These effects fall into three broad categories:

- Cognitive versus psychomotor work
- Levels of basic intellectual skills
- Rate of change

One impact of technology is the continual shift from **psychomotor work** or hands-on work to **cognitive work** or brain work, a process that began in the industrial revolution and that continues today. It has intensified through such technological developments as automated material handling. Not only are machines processing parts, but also they are being loaded and unloaded by robots, and parts are being transported by AGVs and conveyor systems.

As the emphasis continues to shift from the psychomotor domain to the cognitive, the necessary level of basic skills in such areas as reading, writing, and mathematics increases. This means that the industrial worker of today must be better educated than the worker of yesterday, and the worker of tomorrow must be better educated than the worker of today. It also means that workers who selected a career field because of its emphasis on hands-on skills will have to make personal adjustments in order to stay happy and productive as their jobs become more cognitively oriented.

A final impact of technology in the workplace is a continual increase in the **rate of change**. No matter how positive, change is difficult for people. Reluctance to change is a human characteristic. With workers being reluctant to change on the one hand and the rate of technological change increasing on the other, modern industrial supervisors face new and difficult challenges.

HELPING WORKERS HANDLE CHANGE

One of the key roles modern industrial supervisors must play is that of the **change agent**. Technological advances have increased the rate of change and will continue to do so, but the natural human reluctance to change has remained constant. Consequently, modern industrial supervisors must possess the ability to help people change. What makes this responsibility even more challenging is that in order for change to bring long-term success, people must not only accept the change, but they must also come to claim ownership in it.

There are several approaches available to supervisors for helping people to accept change and take responsibility for its success. Figure 1–12 shows some of the more widely used approaches arranged along a continuum from the most negative approach to the most positive. The location of an approach on the continuum is based on the likelihood of its long-term positive results.

Coercion involves using power in a negative way to force people to change. The most common example of how coercion is applied is when workers are told, "Change or you are fired!" Although this approach may produce short-term results, in the long run it is likely to fail as the anger and resentment of employees begin to be manifested in nonproductive ways.

Manipulation is better than coercion, but not much. If employees feel they are being manipulated, they are not likely to accept the responsibility for making a change succeed.

Participation is a positive approach. By allowing employees to participate in the development of change implementation plans, supervisors can increase the likelihood that employees will accept the change. The downside of participation is that achieving consensus among employees can be time-consuming and difficult. In fact, it might occasionally prove to be impossible. However, employees who have participated in the development of a change implementation plan, even if they did not agree with all aspects of the plan, are more likely to work for the change than are employees who are excluded from the process.

Communication is a critical ingredient in helping people accept change. Misinformation and rumors will begin to circulate at even the first intimation of change. They

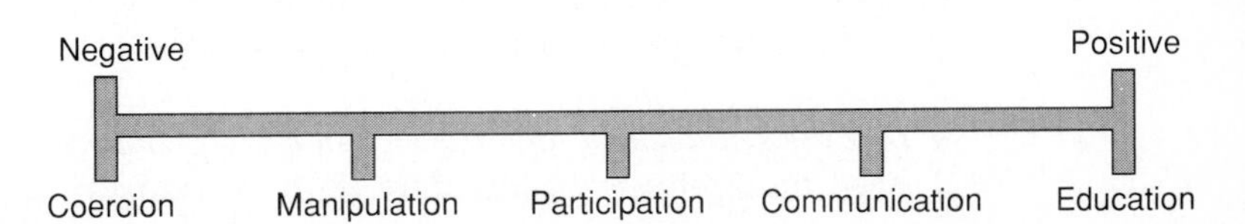

Figure 1–12
Continuum of approaches to facilitating change.

both grow out of and feed on the natural human fear of the unknown and, of course, change represents the unknown. Change can cause employees to feel that their security is threatened. If this happens, it can cause them to focus on their security concerns to such an extent that productivity falls off. Consequently, it is important to communicate continually and effectively.

Education is the most important ingredient in successfully implementing technology-based change. One of the reasons people are reluctant to accept a new technology is fear of the unknown. For example, people who have spent years operating machines manually and are experts at it naturally will be reluctant to accept new computer-controlled machines. Because of this new technology they suddenly find themselves beginners again. The new technology threatens their status as an expert, a status of which they are understandably proud. It also threatens their job security. "What if I can't learn?" is a question they will surely ask themselves.

For this reason, education should be a major component in a company's change implementation plan. Through education, experts in an old technology can become experts in a new technology and in the process can put to rest their feelings of insecurity.

An interesting phenomenon that often occurs in such situations might be called the **technology transition syndrome.** It works as follows:

Step One: Employees involved in a technological upgrade are enrolled in a training program to learn how to use the new technology.

Step Two: They begin the training program reluctantly, go along half-heartedly, and spend breaks belittling the new technology, swearing it will never work.

Step Three: Training continues and employees begin to make progress. It is limited, but nevertheless it is progress. During a break one brave soul takes a chance and says, "You know, this new system isn't all that bad."

Step Four: The training continues, and suddenly the light goes on for several employees. At break, one of them speaks for himself and others when he says, "I kind of like this new system."

Step Five: By now the trainees have gotten comfortable with the new system and good at operating it. At break, the employee who was originally the most reluctant is heard saying, "I really liked it from the start. We should have gone to this system years ago!"

Making New Technologies Work

Supervisors can play a key role in making sure that new technologies adopted by their company work to their maximum potential. Antoinette K. O'Connell recommends the following strategies supervisors can use to make new technologies work:[9]

- *Don't get blindsided.* Learn about new technologies that might be used in your unit. Read the professional literature, talk to technical personnel, and attend technical updating workshops and/or conferences. The more you know about new technologies, the better equipped you will be to help employees accept, adopt, and use them.

- *Maintain a positive attitude.* Problems are inevitable when adopting new technologies. This can complicate matters. Employees who are reluctant to change in the first place may seize upon problems as a way to justify their reluctance. Supervisors can help overcome such situations by maintaining a positive attitude and a sense of humor.
- *Don't try to be all things to all people.* Supervisors will eventually confront a situation in which they no longer know every operational detail of every technological system used in their unit. This is normal. It is better to maintain a broad general knowledge than to try to keep up with the intricate operational details of every piece of equipment under your supervision.
- *Keep emergency telephone numbers available.* When the inevitable technical problems do occur, it is a good idea to be able to get help fast. Supervisors should keep an updated directory of vendors, colleagues, and friends who can provide technical assistance by telephone.
- *Provide training for employees.* The most important strategy in making new technologies work is to train the employees who will use it. Planning for training should begin before, not after, the purchase of new technologies. Also, sufficient time should be built in to complete the necessary training before employees are expected to use the new technology in a live setting.

A SUPERVISOR'S MODEL FOR CHANGE

Figure 1–13 illustrates a model that modern industrial supervisors can use to help employees accept change and take responsibility for making it succeed. It involves the overlapping application of four approaches: participation, communication, education, and facilitation.

First, supervisors involve employees in the development of a change implementation plan. While the plan is being developed and throughout the other steps in the

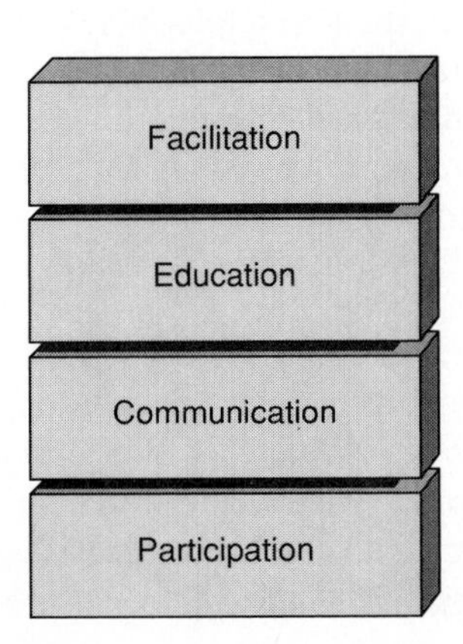

Figure 1–13
Building blocks of successful change.

model communication is continuous. The next step is education. A training program is set up so that employees can learn how to use the new technology. As training progresses, supervisors undertake an ongoing facilitation effort.

This involves providing one-on-one support to the employees, reassuring them when necessary, and arranging extra help for those who need it. It also involves identifying problems, preferably while they are small, and solving them before they grow large enough to upset the transition. Supervisors who invest the time and effort necessary to apply this model properly will be successful change agents.

Industry, Technology, and the Future

The following quote from *Management Digest* sheds some light on what the future holds for industry with regard to technology:

> In recent years, it seems that each day has witnessed exciting news from America's research laboratories. We've learned of synthetic materials that can produce magnets powerful enough to levitate a train, for example. And of computers that respond to light instead of electronic signals. The provost of one of the nation's leading scientific institutions, Dr. Angel G. Jordan, of Carnegie-Mellon University, argues that industry and academia must work together more closely. Only then will technological innovation effectively bolster the competitiveness of new products and services in the marketplace.[10]

The concept to which Dr. Jordan refers in this quote is **technology transfer**, an increasingly important concept in modern industry. Technology transfer is the process of transferring new technologies from the research laboratory into the workplace. It is a two-step process. The first step is known as *commercialization.* The commercialization of technology is a business development issue requiring cooperation between science and commerce. It involves investing in a new technological development and taking the risk that it will be a commercial success. Although this is a critical step in the process, it is not the step that concerns supervisors.

Modern industrial supervisors are concerned with the second step in the technology transfer process, *diffusion.* Diffusion has occurred when a new technology is accepted and widely used in the marketplace or, in the case of industrial technologies, in the workplace. Major stumbling blocks to diffusion are reluctance to change on the part of employees and inability to effectively and efficiently use the new technology. The change facilitation process described in this chapter will help modern supervisors accomplish the diffusion step.

As research laboratories, vendors, and users of technology continue to work together to refine the technology transfer process, the time span between when a new technology is invented and when it first reaches the workplace will continue to decrease. Correspondingly, new technologies will reach a state of **functional obsolescence** at an increasingly rapid rate. What this means to modern supervisors is that they will have to continually improve their ability to facilitate change.

Another significant trend for the future is the one toward more **integration** of technologies and processes. The automation movement of the 1980s led to the creation of **islands of automation** or stand-alone processes that are automated but are not electronically tied to related processes. Although some productivity gains can be realized through the creation of islands of automation, more often than not what results is the creation of bottlenecks at other stations when segregated work processes are not or cannot be properly coordinated. The integration of technologies will allow machines to "talk" to each other electronically so that work flows evenly and smoothly. In addition, workers at all levels will have instant access to a broad array of information about all processes, not just their own.

What the trend toward integration means to supervisors is that they will have to become increasingly broad-based and versatile with regard to their job knowledge. In order to take full advantage of the capabilities integration can bring to the workplace, supervisors will need to be familiar not only with the work processes of their units, but also with the processes of all units involved in producing the product or delivering the service. The ability to access information immediately at any time about any process or any component of the company won't help supervisors who do not understand the information and are not able to determine its significance.

For example, a supervisor of an automated printed circuit board assembly unit in a fully integrated manufacturing facility might regulate the daily production of new boards by electronically accessing information from the sales department concerning signed contracts and sales projections. Correspondingly, before promising a customer a specific delivery date, a sales representative might electronically access current production schedules and in-stock inventory figures.

A final technological trend for the future is the growing use of artificial intelligence in the workplace, discussed earlier in the chapter. Future supervisors will be less involved in routine day-to-day decisions. These will be made by expert systems that rely on artificial intelligence, giving supervisors more time to devote to the more complex people-oriented decisions that, owing to technological advances, will continue to become even more complex.

SUMMARY

1. U.S. industry is competing in a tough global marketplace. In almost every type of industry U.S. firms have lost market share to foreign competitors whose products are designed better, are more reliable, and are less expensive. Our competitors are winning by making better use of technology and people. Supervisors can play a key role in helping industrial firms make better use of people and technology.
2. Technology is applied science. It is what results when people apply scientific principles to the solution of everyday problems. People apply scientific principles using tools, processes, and resources.
3. High technology is the most advanced, most modern development in a given technological field at a given time. A more descriptive term is advanced technology.
4. The most modern material technology is composite technology. A composite is a heterogenous solid that consists of two or more materials. The materials used to form a composite are bonded together mechanically or metallurgically. Characteris-

tics of composites include but are not limited to low thermal conductivity, good heat resistance, long fatigue life, high resistance to corrosion, and good wear resistance.

5. Some of the most advanced industrial processes include but are not limited to CAD/CAM, finite element analysis, CAPP, MRP, automated assembly handling, CNC, DNC, automated assembly, SPC, JIT, flexible manufacturing, CIM, and artificial intelligence.
6. The impact of technology in modern industry can be summarized as: (a) more cognitive and less psychomotor work; (b) a need for a higher level of basic intellectual skills; and (c) an increase in the rate of change.
7. Modern supervisors must be effective change agents. The most effective approaches available to supervisors are participation, communication, and education.

KEY TERMS AND PHRASES

Applied science
Artificial intelligence
Automated assembly
Automated material handling
CAD/CAM
CAPP
Change agent
CIM
CNC
Coercion
Cognitive work
Composite
Communication
Education
Expert systems
Functional obsolescence
Finite element analysis
Flexible manufacturing
High technology
Integration
JIT
Manipulation
MRP
Participation
People connection
Psychomotor work
Rate of change
SPC
Technology
Technology transfer
Technology transition syndrome

CASE STUDY: Kodak Focuses on High Tech[11]

The following real-world case study contains a technology-related situation of the type that might be encountered by modern supervisors. Read the case study carefully and answer the accompanying questions.

After more than a century as the world's leader of the photography industry, Eastman Kodak Company entered the 1990s facing intense international competition. Kay Whitmore, a thirty-four-year-old Kodak veteran, assumed the role of CEO in June 1990 facing a tough challenge. What direction should Kodak take in a dynamic, unsure business environment?

Whitmore's plan? To cling tenaciously to the traditional camera and photographic film markets while simultaneously moving into high technology electronics markets. New products would include digital and hybrid copiers and desktop printers. Drugs and chemicals will also be added to Kodak's product line.

Can Kodak hold market share in its traditional markets and simultaneously enter new high technology markets? It remains to be seen. What is clear, however, is that this same scenario will be played out in many U.S. firms during the 1990s.

1. What problems might Kodak face in attempting to enter high technology markets?
2. What can Kodak management do to help make the transition to high technology products a positive experience?
3. What impact do you think the transition will have in the workplace at Kodak plants?

REVIEW QUESTIONS

1. Why are U.S. industrial firms losing market share to foreign competition?
2. Define the term *technology.*
3. Define the term *high technology.*
4. Explain the people connection with regard to high technology.
5. What is a composite?
6. List four characteristics of a composite.
7. Briefly explain the following modern industrial technologies: CAD/CAM; MRP; JIT; finite element analysis; CNC; CIM; CAPP; SPC; artificial intelligence.
8. Briefly explain the impact technology is having in the workplace.
9. Why is it important for supervisors to be effective change agents?
10. Briefly explain five approaches supervisors can use in trying to effect change.

SIMULATION ACTIVITIES

The following activities are provided to give you opportunities to develop your supervisory skills by applying the material presented in this chapter in solving simulated supervisory problems.

1. Construction Products Corporation manufactures a variety of building products including roof trusses, prefabricated kitchen cabinets, and prefabricated bathroom cabinets. The planning department has used a manual paper-and-pencil system for years. Management has decided it is time to purchase a computerized materials planning system. The head of the planning department is in favor of the change but is not sure that members of her department will go along. How would you handle this problem?
2. Morris Townsend has been the machine shop supervisor at Walton Manufacturing Company for five years. During this time he has taken several college courses and learned to write computer programs. Morris is convinced that it is time for the machine shop to retrofit its machines for CNC. He is not sure how to proceed. Develop a plan that will help Morris Townsend win the support of his team members.

ENDNOTES

1. Freundlich, N. "Sony's New Chip Is a Superfast Thinker," *Business Week,* May 28, 1990, p. 65.
2. Buderi, R. "Why Everybody's Aglow over Blue Lasers," *Business Week,* June 11, 1990, p. 75.
3. Freundlich, N. "Sony's New Chip Is a Superfast Thinker," *Business Week,* June 16, 1990, p. 167.
4. Yeaple, J. A. "What's New," *Popular Science,* August 1990, p. 15.
5. Hawkins, W. J. "What's New Electronics," *Popular Science,* August 1990, p. 16.
6. Burrows, P. "Artificial Intelligence Finds Work in Manufacturing," *Electronic Business,* February 5, 1990, p. 65.
7. Ibid.
8. Gross, N., Carey, J., McWilliams, G., and Brandt, R. "The New World's Lasers are Conquering," *Business Week,* July 16, 1990, pp. 160–161.
9. O'Connell, A. K. "Making the New Technology Work," *Supervisory Management,* March 1990, p. 8.
10. Kozlov, (Edited by). "Where Is the New Technology Taking Business?" *Management Digest,* January 1990, p. 6.
11. Appleman, H. "Kodak's Chief Putting Focus on High Tech," *Associated Press,* February 3, 1991.

CHAPTER TWO

Leadership, Change, and the Supervisor

CHAPTER OBJECTIVES

After studying this chapter, you will be able to define or explain the following topics:

- Leadership
- Born Leaders and Made Leaders
- Leadership, Motivation, and Inspiration
- Common Leadership Styles
- How to Select Appropriate Leadership Styles
- How to Solidify Follower Loyalty
- The Differences Between Leadership and Management
- Trust-Building as a Leadership Concept
- The Relationship Between Leadership and Change
- The Relationship of Leadership and Teamwork

CHAPTER OUTLINE

- What Is Leadership?
- Are Leaders Born or Made?
- Leadership, Motivation, and Inspiration
- Theory X, Theory Y, and Theory Z
- Leadership Styles
- Selecting the Appropriate Leadership Style
- Winning and Maintaining Followership
- Leadership vs. Management: Some Critical Differences
- Leadership and Change
- Leadership and Teamwork
- Summary
- Key Terms and Phrases
- Case Study Application Problem
- Review Questions
- Simulation Activities
- Endnotes

Leadership is an intangible concept that, when applied properly, can bring tangible results. Leadership is sometimes referred to as an art and other times as a science. In reality, it is both an art and a science.

The impact good leadership can have can be readily seen through even a cursory examination of an organization. Well-led organizations, whether they are large companies or small departments within a company, have several easily identifiable characteristics. These characteristics include high levels of productivity; a positive can-do attitude; a commitment to accomplishing organizational goals; effective, efficient use of resources; high levels of quality; and a mutually supportive, teamwork approach to getting work done.

These are characteristics all chief executive officers want and need. Supervisors can provide the leadership that will develop these characteristics in their organizational units. This chapter will help prospective and practicing supervisors learn how to be effective leaders.

WHAT IS LEADERSHIP?

There are enough different definitions of leadership to fill several chapters. Part of the reason that there is such a variety of definitions is that leadership is needed in so many different fields of endeavor. Leadership has been defined as it applies to the military, athletics, education, business, industry, and many other fields. For the purpose of this text, leadership is defined as it relates specifically to the modern supervisor. From this perspective, **leadership** can be defined as follows:

> ***Leadership is the ability to inspire people to make a total and willing commitment to accomplishing organizational goals.***

James R. Richburg, president of Okaloosa-Walton Community College, describes leadership as follows:

- Leadership is believing the mission and goals, not just knowing them.
- Leadership is consensus building, not just saying, "Follow me."
- Leadership is building ownership, not a unilateral action.
- Leadership is looking in all directions, not just looking straight ahead.
- Leadership provides individuals opportunities for fulfillment. Leadership does not exploit individuals.
- Leadership concerns image, but leadership is not believing your own press clippings.
- Leadership is uplifting, energizing, coaching, and cheerleading. Leadership is not evading, bullying, harassing, or forcing.
- Leadership builds the podium and shares recognition. Leadership is not playing king of the mountain.

- Leadership is caring, compassionate, and enabling. Leadership is not harsh, harassing, or debilitating.
- Leadership knows when to follow, but leadership is not all-knowing.

What Is a Good Leader?

Good leaders come in all shapes, sizes, genders, ages, races, political persuasions, and national origins. They do not look alike, talk alike, or even work alike. However, good leaders do share several common characteristics, including the ability to inspire people and to motivate them to make a total and willing commitment. Regardless of their backgrounds, good leaders exhibit the following characteristics (Figure 2–1):

- A balanced commitment
- A positive example
- Good communication skills
- Influence
- Persuasiveness

Good leaders are committed to both the job to be done and the people who must do the job. They are able to strike the appropriate balance between the two. This is dealt with in greater depth later in this chapter.

Good leaders project a positive example at all times. Supervisors who adopt a "Do as I say, not as I do" attitude will not be effective leaders. In order to inspire workers to follow them, supervisors must be willing to do what they expect of workers, and do it better. For example, if dependability is important, a supervisor must set an example of dependability. If punctuality is important, a supervisor must set an example of punctuality. To be a good leader, a supervisor must set an example of all the characteristics that are important on the job.

Good leaders are good communicators. They are willing, patient, skilled listeners. They are also able to communicate their ideas clearly, succinctly, and in a nonthreatening manner. They use their communication skills to establish and nurture rapport with workers.

Commenting on his company's attempts to institute a total quality management process, Bo McBee, director of quality for Armstrong Industries, stresses the need for a variety of approaches to communicating the message. According to McBee, "The nature of the stories and the communication change with time."[1] He explains that his company uses a variety of media for communicating the total quality message including newsletter stories, videos, and stories in other company publications. Supervisors should keep this point in mind as a way of promoting better communication: Vary the media to communicate the message.

Good leaders have influence with those they supervise. **Influence** is the art of using power to motivate people to do what you want them to do. Supervisors' **power** derives from the authority that goes with their jobs and the credibility that comes from having the advanced knowledge and skills necessary to become a supervisor. Power is an

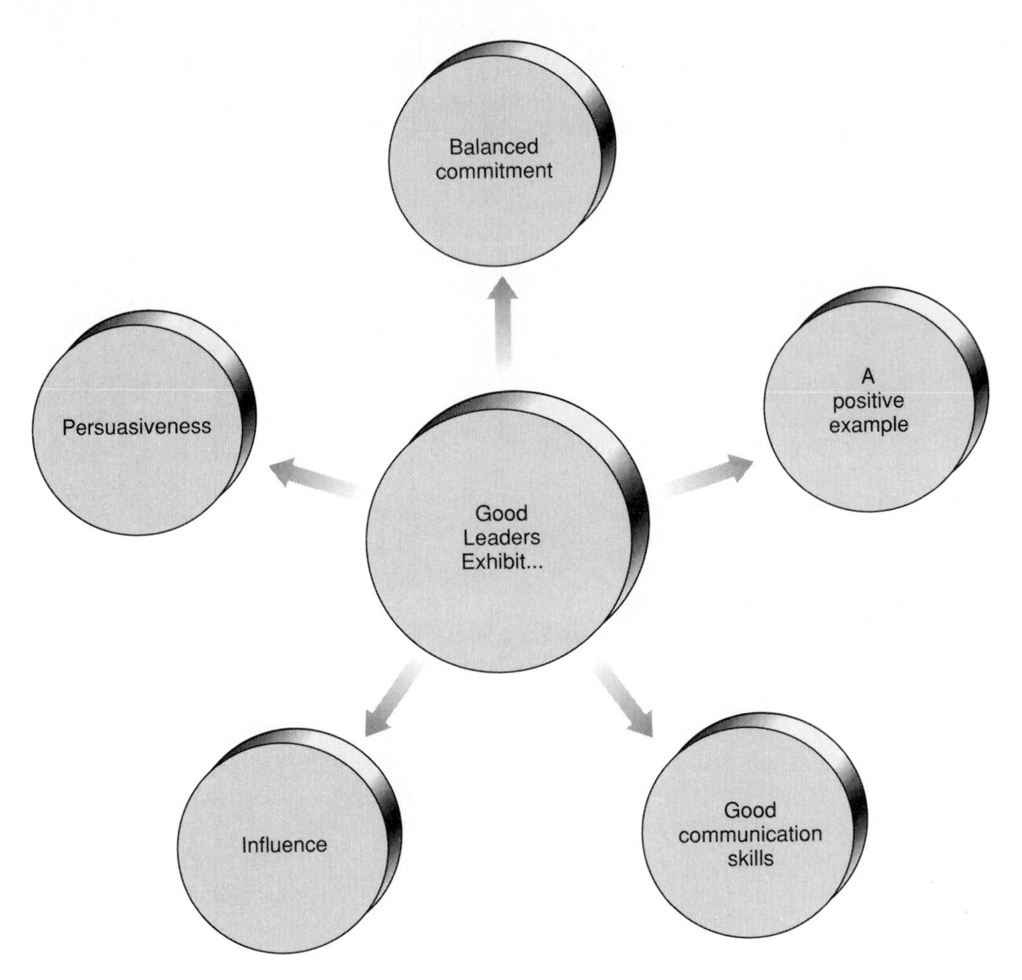

Figure 2–1
Characteristics of good leaders.

irrelevant concept unless a leader knows how to convert it to influence. Power that is properly, appropriately, and effectively applied becomes influence.

Finally, good leaders are **persuasive**. Supervisors who expect people to simply do what they are ordered to do will have limited success. Those who are able to use their communications skills to persuade workers to accept their point of view and to help workers develop ownership in that point of view can have unlimited success.

James A. Belasco gives the following advice to managers and supervisors who wish to persuade employees to accept their point of view: "You can achieve your vision by getting others to see it as a means of solving the biggest job-related issue confronting them. Demonstrate how using your vision deals with the realities of their daily activities."[2]

ARE LEADERS BORN OR MADE?

Perhaps the oldest debate relating to leadership revolves around the question, "Are leaders born or made?" This debate has never been settled and probably never will be. The point of view set forth in this book is that leaders are a lot like athletes. Some are born with a great deal of athletic potential while others develop their ability through determination and hard work. Inborn ability, or the lack of it, represents only the starting point. Success from that point forward depends on the individual's willingness and determination to develop and improve.

There are athletes born with tremendous natural ability who never live up to their potential. On the other hand, there are athletes with limited natural ability who, through hard work and determination, perform beyond their apparent potential. This is also true of supervisors who want to be good leaders. Some supervisors have more natural ability than others. However, regardless of their individual starting points, all supervisors can become good leaders through education, training, practice, determination, and effort (Figure 2–2).

Figure 2–2 illustrates an important concept. Supervisors with limited natural leadership ability can still become good leaders. Supervisors with more natural leadership ability can become even better supervisors. In both cases, the key lies in the willingness of the individual to do what is necessary to continually develop and improve.

LEADERSHIP, MOTIVATION, AND INSPIRATION

One of the characteristics good leaders share is the ability to inspire and motivate others to make a commitment. The key to **motivating** people lies in the ability to relate their personal needs to the organization's goals. The key to **inspiring** people lies in the ability to relate what they *believe* to the organizational goals. Implicit in both cases is the leader's need to know and understand workers.

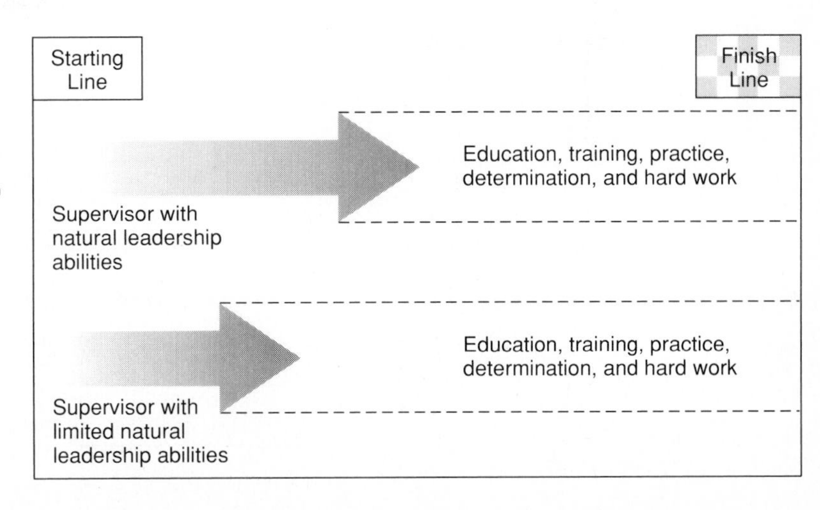

Figure 2–2
Through education, training, practice, determination, and hard work even people with limited natural leadership abilities can become successful leaders.

Understanding Individual Needs

Perhaps the best model for explaining individual human needs is the one developed by psychologist Abraham H. Maslow. Maslow's **hierarchy of needs** (Figure 2–3) displays the five successive levels of basic human needs.

The lowest level in the hierarchy encompasses our basic survival needs. We all need air to breathe, food to eat, water to drink, clothing to wear, and shelter in which to live. The next level encompasses our safety/security needs. We all need to feel safe from harm and secure in our personal world. To this end, we enact laws, pay taxes to employ police and military personnel, buy insurance, try to save and invest our money, and perhaps install security systems in our homes.

The next level encompasses our social needs. People are social animals by nature. We place great importance on our ties to families, friends, social organizations, civic groups, special clubs, and even work-based groups such as company sports teams.

The next level on the hierarchy encompasses esteem needs. Self-esteem is a key ingredient in the personal happiness of individuals. We all need to feel self-worth, dignity, and respect. We need to feel that we matter. This can be seen in the clothes we wear, the cars we drive, and the behavior we exhibit in public. It can also be seen in the job titles we use. When garbage collectors refer to themselves as sanitation engineers they are exhibiting the human need for respect and worth.

The highest level of the hierarchy encompasses our self-actualization needs. Complete self-fulfillment is a need that is rarely satisfied in people. The need for self-actualization manifests itself in a variety of ways. Some people seek to achieve it through their work, others through their hobbies and associations.

It is important for supervisors to understand how to apply Maslow's model if they hope to use it to motivate workers. Procedures for applying this model are as follows:

1. Needs must be satisfied in order from the bottom up.
2. Once a need is satisfied, it no longer works as a motivating factor. For example, people who have satisfied their need for financial security will not be motivated by a pay raise.

Figure 2–3
Hierarchy of human needs.

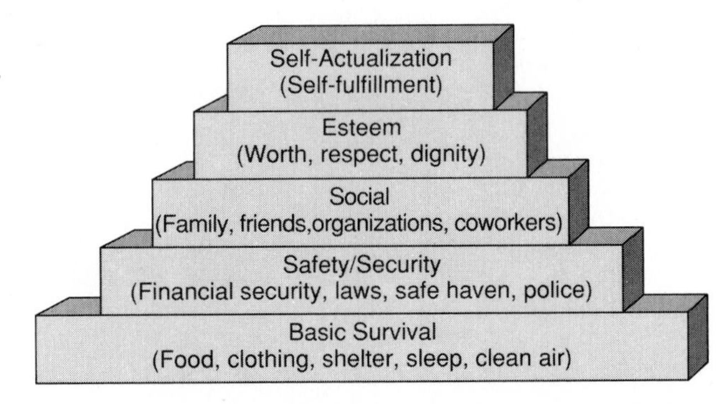

3. People will focus most intently on their lowest unmet need. For example, workers who have not met their security needs will not be motivated by factors relating to their social needs.

To understand how a supervisor can apply Maslow's model to motivate an employee consider the example of John, a manufacturing technician, and his supervisor, Diane. Their company recently won a new contract that is on a tight schedule. In order to meet deadlines, Diane's team must exceed its normal weekly production rate. Everyone is going along except John. John is doing a good job, but his production has not increased, and he has not responded to Diane's efforts to encourage him.

While examining Maslow's model, Diane comes up with an idea. She decides that John's survival and safety/security needs are being adequately met. However, she may be able to use a social need as a motivator. Diane knows that two of John's best friends are on the company bowling team and that John wants very badly to be on the team. Because slots on the team are limited, employees must have the support and recommendation of their supervisors before they are allowed to try out. Diane decides to use this factor to influence John to improve his performance. She calls John aside during a break and the following conversation takes place:

Diane: John, tryouts for the company bowling team are coming up and I would like to recommend someone from our unit to try out. Are you interested?

John: Are you kidding? I can't wait to try out! I was going to ask you to recommend me.

Diane: John, I want to recommend you, but I'm not comfortable doing so while your production rate is lower than that of other operators in our unit. If you can get your production rate up to our new level, I will recommend you.

John: I'll do it!

Understanding Individual Beliefs

All people hold within themselves a basic set of beliefs that give expression to their value system. If supervisors know their workers well enough to understand their basic beliefs, they can use this knowledge to inspire them on the job. Developing this level of understanding of workers comes from observing, listening, asking, and taking the time to establish trust (Figure 2–4). Supervisors who develop this level of understanding of workers can use it to inspire workers to higher levels of performance.

This is done by showing workers how the organization's goals relate to their beliefs. For example, if pride of workmanship is part of a worker's value system, that worker can be inspired to help achieve the organization's quality goals. Inspiration, as a level of leadership, is on an even higher plane than motivation. Supervisors who become good enough leaders to inspire their workers will achieve the best results.

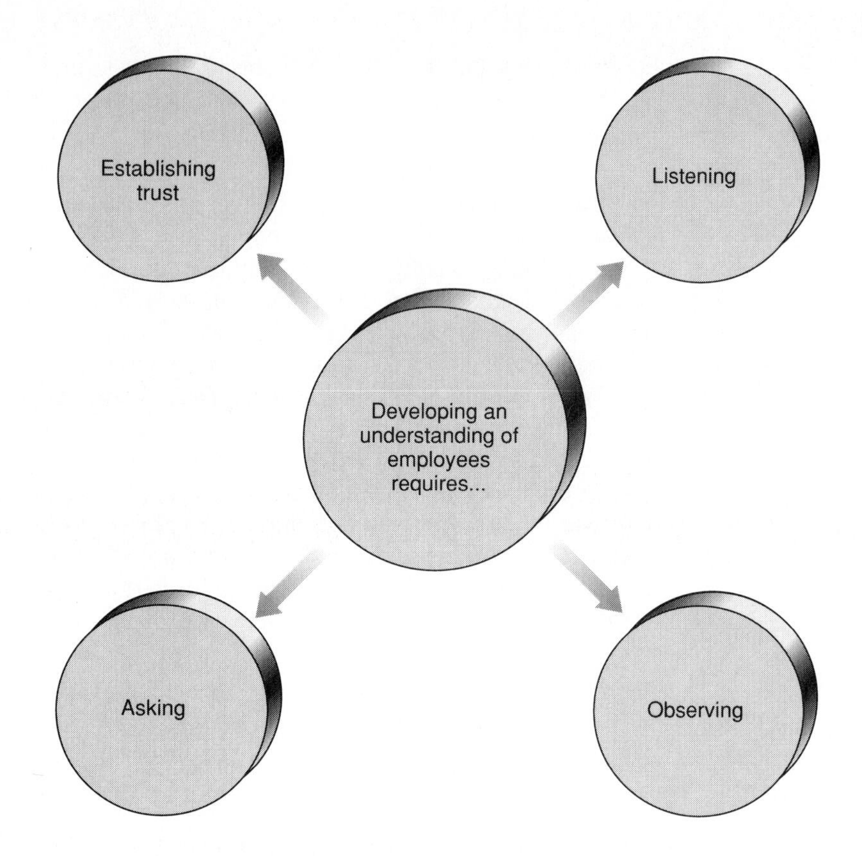

Figure 2–4
How to develop an understanding of employees.

THEORY X, THEORY Y, AND THEORY Z

Douglas McGregor, late professor of industrial management at Massachusetts Institute of Technology (MIT), became famous for classifying management and leadership styles as being examples of either Theory X or Theory Y. Theory X is a prescriptive, even dictatorial approach. Theory Y is based on a belief in the inherent positive potential of human beings. Modern supervisors should be familiar with Theory X, Theory Y, and Theory Z. Theory Z is a relatively new theory that has come to be associated with Japanese styles of management.

Theory X and Leadership

According to McGregor, traditional management/leadership styles are based on an invalid set of assumptions that leads to an overly directive approach; an approach that, in the long run, is counterproductive and has a negative impact on productivity. Such an approach was labeled **Theory X** by McGregor. It is based on the following assumptions (Figure 2–5):

1. Most people have an inherent dislike of work and will avoid it if possible.
2. Because most people dislike work, they must be coerced, threatened, and directed to put forth the effort necessary to achieve organizational goals.
3. Most people do not want responsibility, prefer to be directed, have little ambition, and want security more than anything else.

These assumptions have been at the heart of traditional labor/management relations in the United States since the industrial revolution. The assumptions themselves go back even further. Some theorists trace them all the way back to feudal days when management was a master and labor the servant.

Theory Y and Leadership

If Theory X represents one end of the spectrum of approaches to leadership, **Theory Y** represents the opposite end. It is based on an inherent belief in the almost unlimited potential of people and their capacity for work. The Theory Y approach to leadership encompasses the following assumptions (Figure 2–6):

1. The expenditure of energy on work is as natural as the expenditure of energy on play.
2. When committed to objectives, people will work to achieve them without external control or the threat of punishment.
3. People will commit to objectives if appropriate rewards are made, particularly rewards that relate to an individual's self-respect and personal improvement.
4. Under the right conditions, people will not just accept responsibility, they will seek it.
5. People have the capacity to exercise creativity and imagination in solving organizational problems.
6. In most industrial settings, workers' intellectual potential is only partially tapped.

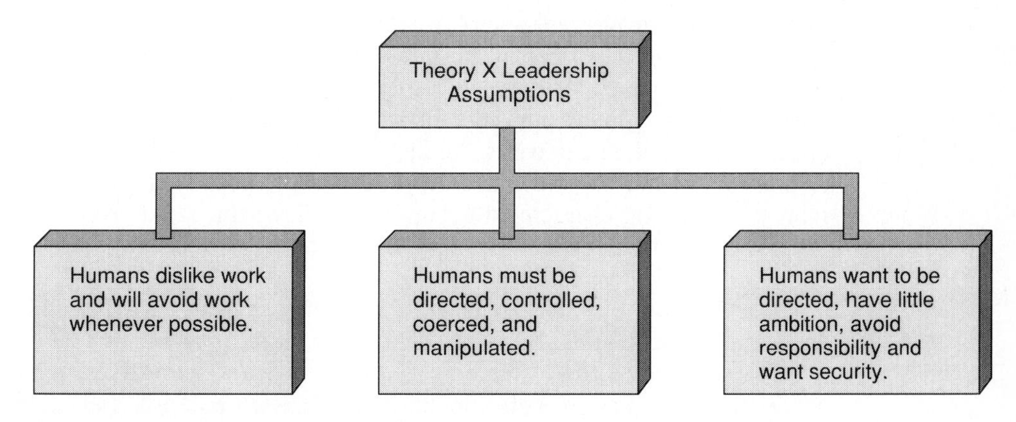

Figure 2–5
Theory X leadership.

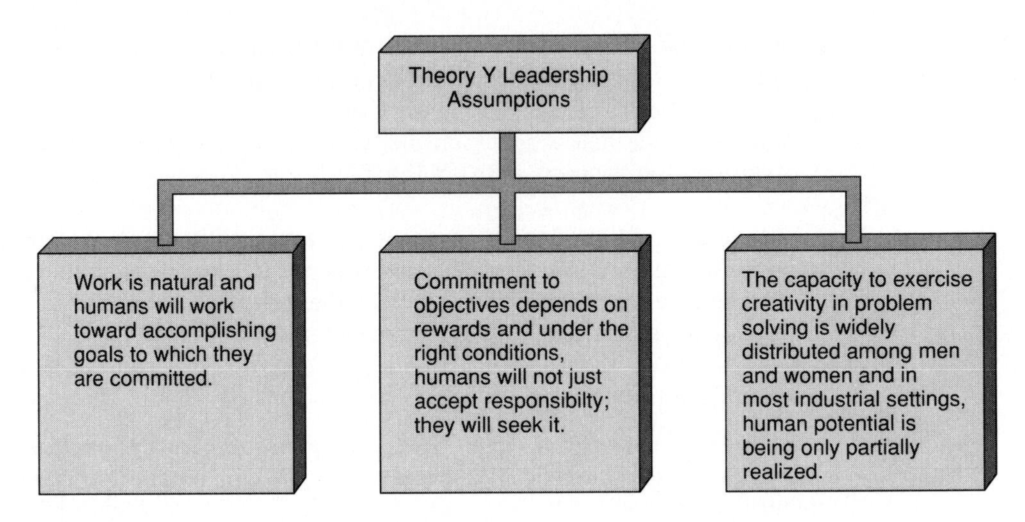

Figure 2–6
Theory Y leadership.

Theory Y represents a radical change from the traditional philosophy of Theory X. According to McGregor in his book *The Human Side of Enterprise,* a Theory Y manager is a teacher, coach, and helper rather than a boss or dictator. The days when the Theory X "Do what I say or you are fired" approach worked are past. Theory Y is more in keeping with the modern belief in workers' personal rights. However, in some cases it does not work as effectively as the modern supervisor might wish. Such cases led to the emergence of what in this book will be called Theory Z.

Theory Z and Leadership

Theory Z is a management approach set forth by William Ouchi of UCLA as a result of his research into the success of Japanese organizations.[3] Ouchi identified seven characteristic ways in which Japanese and American organizations differ. According to Ouchi, Japanese organizations exhibit the following characteristics: lifetime employment, slow promotion, broad career paths, implicit controls, consensus decision making, group responsibility, and holistic concern.[4] American organizations, on the other hand, exhibit the following characteristics: unsure employment, rapid promotion, highly specialized career paths, explicit controls, individual decision making, individual responsibility, and divided concern.[5]

Ouchi points out that there are American companies that exhibit Theory Z or Japanese characteristics. Such characteristics place organizations in the middle between the Theory X and Theory Y approaches. These companies, among the most successful in the United States, include IBM, Hewlett-Packard, and Eastman Kodak.

Although the Theory Z approach has come to be associated with Japanese firms, it is inaccurate to attribute this approach to all Japanese companies. The larger firms in

Japan do tend to apply Theory Z. However, those with fewer than 300 employees do not necessarily do so. Moreover, large companies in Japan often use women and temporary workers as buffers against business downturns.[6]

Regardless of the approach—Theory X, Y, or Z—supervisors should consider the job satisfaction model illustrated in Figure 2–7 and use it as a guide in their dealings with employees. This model shows that workers at all levels feel their jobs must meet three basic needs if they are to feel fulfilled: **financial security, personal satisfaction,** and **societal contributions.**

Different people prioritize these needs differently. But regardless of the order, all three needs must be met for people to be happy in their jobs. A worker's job satisfaction will be in proportion to the extent these needs are met in the individual's own order of priority. The opposite is also true.

To illustrate this point, consider the example of Pete Jones. Pete is a quality control inspector who is not happy with his job. His supervisor, Bart Ames, is a Theory Y–type leader who is widely respected by those he supervises, including Pete. Bart involves all members of the quality control team in setting goals and in planning strategies to accomplish them. Once the goals have been set, he expects team members to take responsibility for meeting their obligations on time or ahead of schedule. To encourage this, he gives team members the maximum possible authority in making decisions and backs them when they make mistakes.

This approach seems to work well with every member of the team except Pete. His participation in goal setting and strategy planning is half-hearted at best. Once the goals have been set, Pete is reluctant to take responsibility for accomplishing them. He would rather be told what to do and when to do it. He cannot be trusted to take the initiative and he is never a self-starter.

Figure 2–7
All workers at all levels have these needs in common.

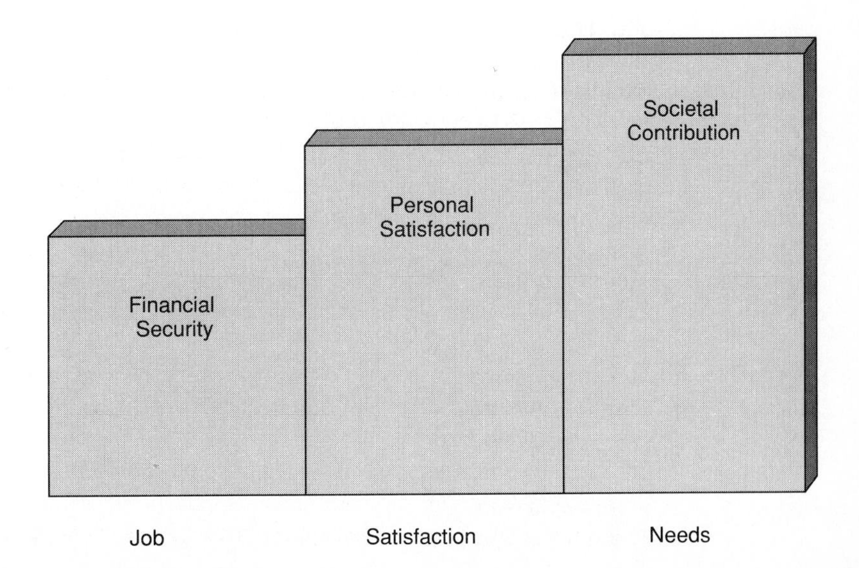

Bart's frustration with Pete has caused him to doubt the Theory Y approach. He decides to meet with Pete and talk. During the conversation he learns that Pete is unhappy with his job. He is satisfied with his pay, but does not enjoy his work, nor does he think it is important. Bart knows why Pete is not responding to his Theory Y approach, but what can he do about it? At this point, Bart has three options:

1. Work with Pete to help him enjoy and appreciate his work more.
2. Help Pete get training for another job he might enjoy and appreciate more.
3. Be more directive with Pete and monitor him more closely.

The first two options fall into the domain of Theory Y. The third option falls into the domain of Theory X. Theory Z is more middle-of-the-road. It occupies the space on the continuum between the two extremes represented by Theory X and Theory Y (Figure 2–8). Hence, while Theory X and Theory Y are absolute, Theory Z leaves more room for flexibility. The underlying assumptions are as follows:

1. Under the right circumstances, *most* people can be as happy at work as they are at play or even more so.
2. Under the right circumstances, *most* people will commit to organizational goals and exercise self-control in working to accomplish them.
3. Under the right circumstances, *most* people will accept and seek responsibility.
4. Under the right circumstances, *most* people will respond to rewards as long as the rewards are tied to their security, personal satisfaction, and societal contribution needs.

A key term in each of these assumptions is the word *most*. If, under the right circumstances, most people will respond in a certain way, it also follows that some will not. These assumptions also imply that there will be times when the right circumstances cannot be created. In either case, even the most devoted Theory Y leader may, on occasion, need to apply Theory X or Theory Z strategies.

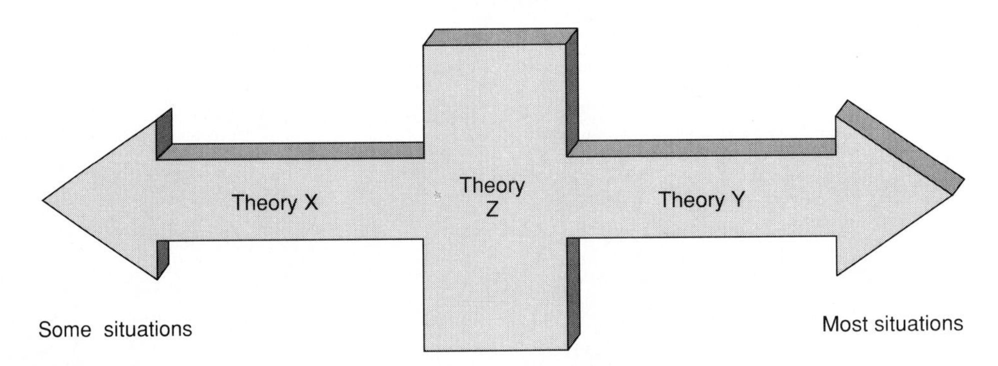

Figure 2–8
Theory Z proponents apply Theory Y or Theory X depending on the situation.

LEADERSHIP STYLES

Leadership styles grow out of the theories explained in the previous section. They have to do with how supervisors interact with the people they want to lead. Leadership styles have many different names. Most styles fall into the following five categories (Figure 2–9):

- Autocratic leadership
- Democratic leadership
- Participative leadership
- Goal-oriented leadership
- Situational leadership

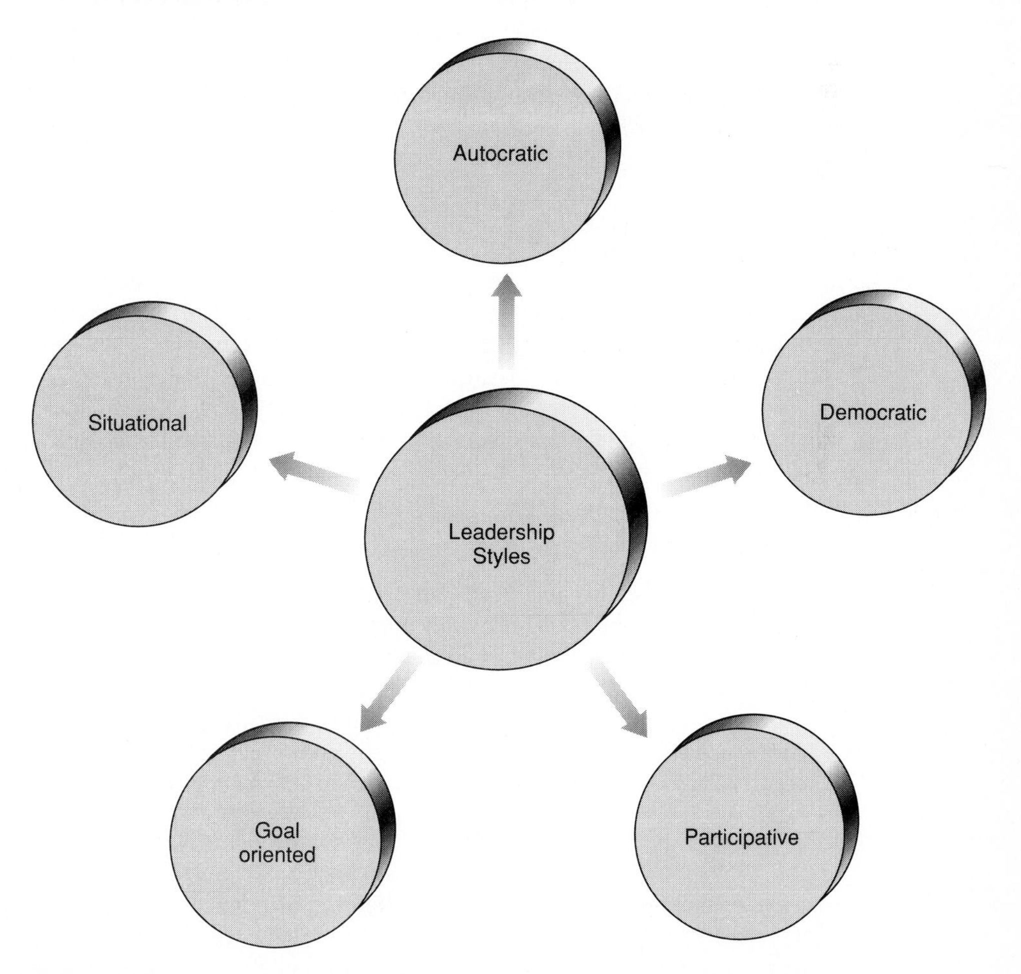

Figure 2–9
Leadership styles.

Autocratic Leadership

Autocratic leadership is also called **directive** or **dictatorial** leadership. Leaders who take this approach make decisions without consulting the workers who will have to implement them. They tell the workers what to do and expect them to comply obediently. Autocratic leaders subscribe to the assumptions encompassed in Theory X. Critics of this approach say that although it can work in the short run or in isolated instances, in the long run it is not effective.

Democratic Leadership

Leaders who take the **democratic leadership** approach, also called **consultive** or **consensus** leadership, involve the workers who will have to implement the decisions. The leader makes the final decision, but only after receiving the input and recommendations of all team members. However, democratic leadership can result in the selection of the most popular decision as opposed to the most appropriate decision, and the most popular decision is not necessarily the best one. It can also lead to ineffective compromises.

Participative Leadership

In **participative leadership,** also called **open, free-reign,** or **non-directive** leadership, leaders exert little control over the decision-making process. Rather, they provide information about the problem and allow team members to develop strategies and solutions. The underlying assumption of this style is that workers will more readily accept responsibility for the solutions, goals, and strategies they develop. But this approach breaks down fast if team members are not mature, responsible, and committed to the best interests of the organization.

Goal-Oriented Leadership

Leaders who take the **goal-oriented leadership** approach, also called **results-based** or **objective-based** leadership, ask team members to focus only on the goals at hand. Only strategies that make a definite and measurable contribution to accomplishing organizational goals are discussed. The influence of personalities and other factors unrelated to the specific goals of the organization are minimized. Critics of this approach say it can break down when team members focus so intently on specific goals that they overlook opportunities and/or potential problems that fall outside their narrow focus.

Situational Leadership

In **situational leadership,** also called **fluid** or **contingency** leadership, leaders select the style that is appropriate based on the circumstances as they exist at the time. In identifying these circumstances, leaders consider the following factors:

- Relationship of the supervisor and team members
- How precisely actions taken must comply with specific guidelines
- Amount of authority the leader actually has with team members

Depending on what is learned when these factors are considered, the supervisor decides whether to take the autocratic, democratic, participative, or goal-oriented approach. Under different circumstances, the same supervisor would apply a different leadership style.

Task-Oriented Styles

Task-oriented styles are the autocratic and goal-oriented styles. They focus intently on the task to the exclusion of the personal needs of team members. They work best in the following two situations that are actually the opposite of each other (Figure 2–10):

- Supervisors have clear authority, processes must adhere strictly to specified guidelines, and a positive relationship exists between the supervisor and team members.
- Supervisors do not have clear authority, processes are not strictly regulated, and a poor relationship exists between a supervisor and team members.

People-Oriented Styles

People-oriented styles are the **participative** and **democratic** styles. They focus more intently on the needs of the people doing the work than on the work itself. They work best in situations where the following factors apply (Figure 2–11):

- The supervisor's authority is not clearly understood.
- Work specifications and guidelines are flexible.
- A lukewarm relationship exists between the supervisor and team members.

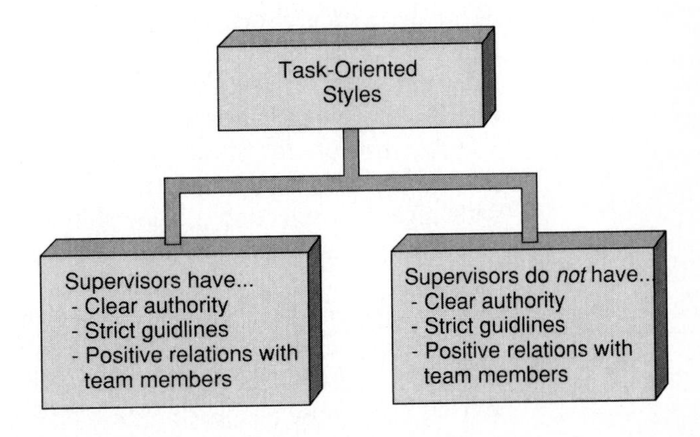

Figure 2–10
Task-oriented styles work in two situations that are on opposite ends of the spectrum.

Figure 2–11
Situational factors that favor the people-oriented styles.

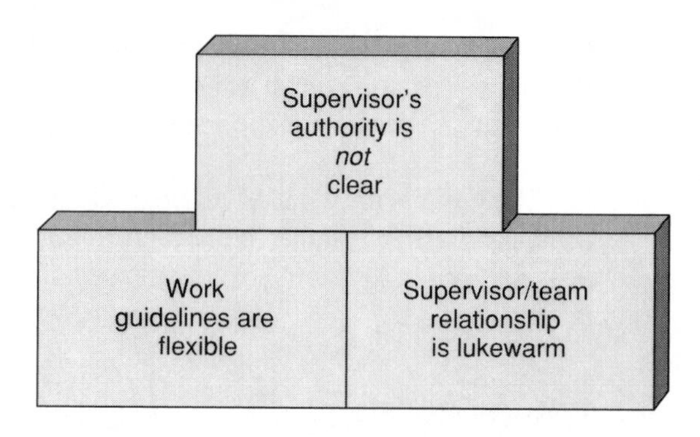

Building Blocks for People-Oriented Styles

SELECTING THE APPROPRIATE LEADERSHIP STYLE

Supervisors are responsible for continually enhancing the productivity of the people they supervise. One of keys to accomplishing this goal is selecting the appropriate leadership style. The most appropriate style is the one that allows supervisors to achieve the best results in both the short and the long run. Since the people doing the work and the conditions under which they can do it can and frequently do change, situational leadership may have the most potential for modern supervisors.

Leadership styles are like tools. The key is to select the right tool for the task at hand. For example, a hammer is an excellent tool if the problem is to drive in a nail. However, it does not work when the task is to change a tire. This rule of thumb also applies in selecting a leadership style.

Depending on the conditions that exist at a given time and the people involved, the same supervisor might apply one leadership style today and a different style tomorrow. Some situations allow for people-oriented styles; others require task-oriented styles. Still others require combinations of the elements from different styles. Situations faced by modern supervisors are increasingly fluid. Consequently, supervisors should be familiar with all the leadership styles set forth in this chapter, perceptive enough to determine which style a given situation calls for, and flexible enough to move from style to style as dictated by the situation.

The Institute for Social Research at the University of Michigan conducted a comprehensive study in the late 1960s that still has relevance for the modern supervisor with regard to selecting a leadership style.[7] The study, conducted by Rensis Likert, determined that the most productive supervisors had the following characteristics:

- They supervised but did not do the work of their employees.
- They gave employees room to work by not supervising too closely.
- They delegated appropriately and properly (both responsibility and authority).
- They were more people than task oriented.
- They maintained a harmonious working environment.

In another research project conducted in the late 1960s the University of Michigan studied the impact supervisory styles have on productivity. The results of this study were as follows:

- The most productive work units typically have a supervisor who is more people than task oriented.
- Task-oriented, autocratic supervision results in low morale and high turnover.
- Most supervisors adapt to the needs of the job rather than being exclusively task or people oriented.

WINNING AND MAINTAINING FOLLOWERSHIP

Supervisors can be good leaders only if the people they supervise follow them willingly and steadfastly. Followership must be won and, having been won, must be maintained. This section discusses how supervisors can win and maintain followership by the people they supervise.

Popularity and the Leader

Many new supervisors confuse popularity with followership. An important point to understand in supervising people is the difference between popularity and respect. Long-term followership grows out of respect, not popularity. Good leaders *may* be popular with those they supervise, but they *must* be respected. Not all good leaders are popular, but they all are respected.

All supervisors must occasionally make unpopular decisions. This is why leadership positions are sometimes described as lonely positions. Making an unpopular decision does not necessarily cause leaders to lose followership, provided the leader is seen as being fair, objective, and impartial (Figure 2–12). Correspondingly, leaders who make inappropriate decisions that are popular in the short run may actually lose followership in the long run. If the long-term consequences of a decision turn out to be detrimental to the team, team members will hold the supervisor responsible no matter how strongly they originally supported the decision that was made.

Leadership Characteristics that Win and Maintain Followership

Leaders win and maintain followership by engendering the respect of their followers. Some characteristics of leaders that build respect are as follows (Figure 2–13):

- A sense of purpose
- Self-discipline
- Honesty
- Credibility
- Common sense
- Stamina
- Commitment
- Steadfastness

Figure 2–12
It is more important for superiors to be respected than to be popular.

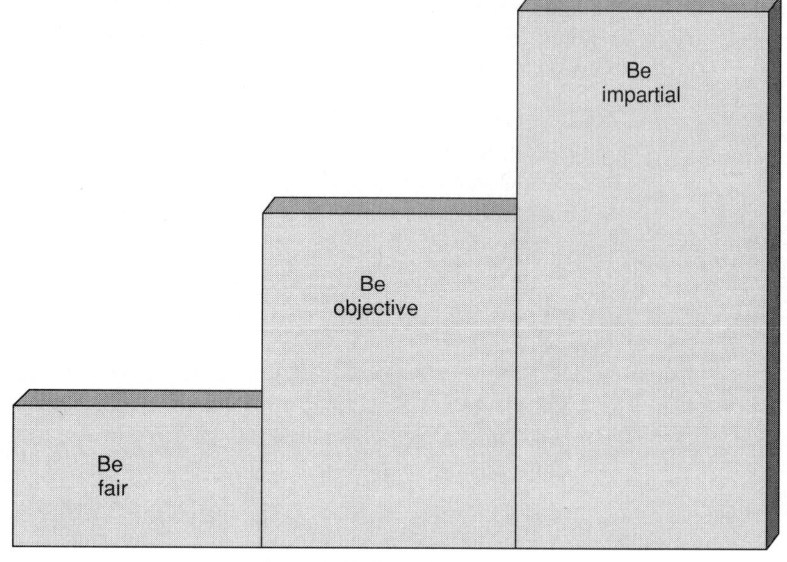

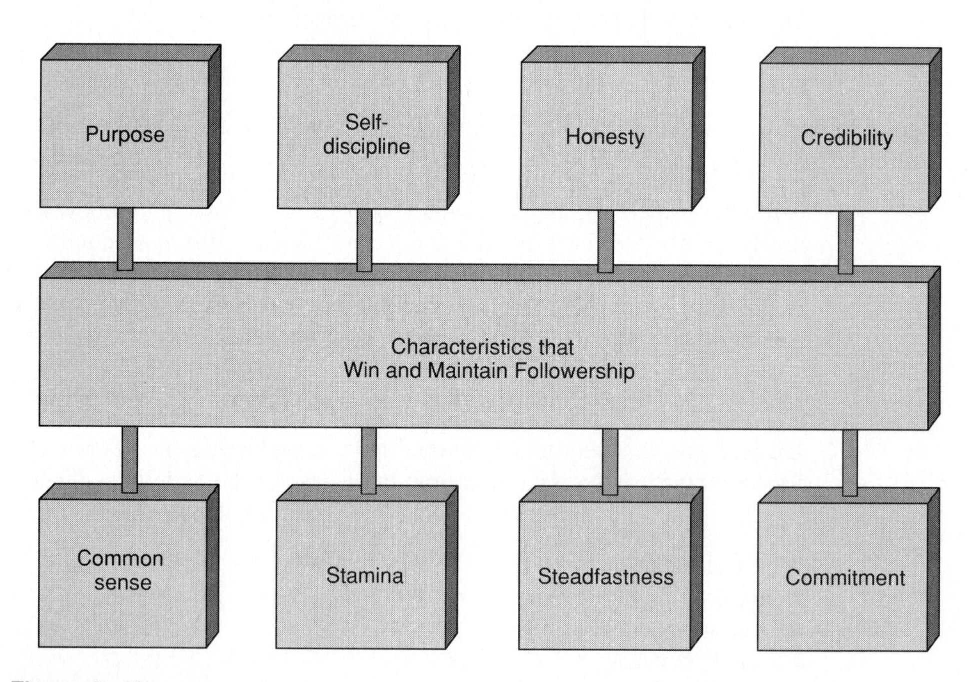

Figure 2–13
Supervisors who exhibit these characteristics will win the respect of their followers.

Sense of Purpose

Successful leaders have a strong sense of purpose. They know who they are, where their unit fits into the overall organization, and the contributions their area of responsibility makes to the success of the organization.

Self-Discipline

Successful leaders develop discipline and use it to set an example. Through **self-discipline** leaders avoid negative self-indulgences, inappropriate displays of emotion such as anger, and counterproductive responses to the everyday pressures of the job. They set an example of handling problems and pressures with equilibrium and a positive attitude.

Honesty

Successful leaders are trusted by their followers. This is because they are open, honest, and forthright with their team members and themselves. They can be depended on to make difficult decisions in unpleasant situations with steadfastness and sincerity.

Credibility

Good leaders have **credibility**. In an industrial setting this means having well-developed job skills. Supervisors with the most credibility are those who can show workers how to do a job better, rather than just telling them. One school of thought says a good supervisor can supervise people doing any job, even without having the knowledge and skills required to do the job. While this may be true up to a point, such supervisors will not have the credibility necessary to win and maintain the full respect of team members. This, in turn, will limit their long-term effectiveness.

Common Sense

Successful leaders have **common sense**. They know what is important in a given situation and what is not. They know that applying tact is important when dealing with people.

Stamina

Successful leaders must have **stamina**. Frequently they will be the first to arrive and the last to leave. The pressures they face will be more intent. Energy, stamina, and good health are important to those who lead.

Steadfastness

Successful leaders are **steadfast** and resolute. People will not follow a supervisor they perceive to be wishy-washy and noncommittal. Nor will they follow a supervisor whose resolve they question. Successful leaders must have the steadfastness to stay the course even when it becomes difficult.

Commitment

Successful leaders are **committed** to the goals of the organization, the people they supervise, and their own ongoing personal and professional development. They are willing to do everything within the limits of the law, professional ethics, and company policy to help their team succeed.

Pitfalls to Avoid

The previous section set forth the positive characteristics that will help supervisors win and maintain the respect and loyal followership of team members. In addition, supervisors should be aware of several common pitfalls to avoid. The pitfalls can quickly undermine the respect supervisors must work so hard to build.

- *Trying to be a buddy instead of the boss.* Positive relations and good rapport are important, but supervisors cannot be a buddy to those they hope to lead. The nature of the relationship will not allow it.
- *Having an intimate personal relationship with an employee.* This practice is both unwise and unethical. A positive supervisor-employee relationship cannot exist under such circumstances. Few people can succeed at being both the lover and the boss, and few things can damage the morale of a team so quickly and completely.
- *Trying to keep things the same when supervising former peers.* The supervisor-employee relationship, no matter how positive, is different from the peer-peer relationship. This can be a difficult fact to accept and an even more difficult adjustment to make. But it must be made if the supervisor is going to succeed in the long run.

LEADERSHIP VS. MANAGEMENT: SOME CRITICAL DIFFERENCES

Leadership and management, although both are badly needed in the modern workplace, are not the same thing. To be good leaders and good managers, supervisors must know the difference between the two concepts. According to John P. Kotter, leadership and management "are two distinctive and complementary systems of action."[8] Kotter lists the following differences between management and leadership:[9]

- *Management* is about coping with complexity; *leadership* is about coping with change.
- *Management* is about planning and budgeting for complexity; *leadership* is about setting the direction for change through the creation of a vision.
- *Management* develops the capacity to carry out plans through organizing and staffing; *leadership* aligns people to work toward the vision.
- *Management* ensures the accomplishment of plans through controlling and problem solving; *leadership* motivates and inspires people to want to accomplish the plan.

Warren Bennis quotes Field Marshall Sir William Slim, who led the British Army's brilliant reconquest of Burma during World War II, on drawing the distinction between management and leadership:

> Managers are necessary; leaders are essential. . . . Leadership is of the Spirit, compounded of personality and vision. . . . Management is of the mind, more a matter of accurate calculation, statistics, methods, timetables, and routine.[10]

Bennis compares leaders and managers:[11]

- Managers administer; leaders innovate.
- Managers are copies; leaders are originals.
- Managers maintain; leaders develop.
- Managers focus on systems and structure; leaders focus on people.
- Managers rely on control; leaders inspire.
- Managers take the short view; leaders take the long view.
- Managers ask how and when; leaders ask what and why.
- Managers accept the status quo; leaders challenge it.
- Managers do things right; leaders do the right thing.

Although Bennis takes a more critical view of management than does Kotter, the points he makes are worthy of consideration. In reality, the most successful supervisors will be those who can appropriately combine the characteristics of both managers and leaders.

Trust Building and Leadership

Trust is a necessary ingredient for success in the highly competitive modern workplace. It means, in the words of Zielinski and Busse, "employees who can make hard decisions, access key information, and take initiative without fear of recrimination from management, and managers who believe their people can make the right decisions."[12]

Building trust requires leadership on the part of supervisors. **Trust-building** strategies include the following:

- *Taking the blame, but sharing the credit.* Supervisors who point the finger of blame at their employees, even when the employees are at fault, will not build trust. Leaders must be willing to accept responsibility for the performance of people they hope to lead. Correspondingly, when credit is due, leaders must be prepared to spread it around appropriately. Such unselfishness on the part of supervisors will build trust among employees.
- *Pitching in and helping.* Supervisors can show leadership and build trust by rolling up their sleeves and helping out when a deadline is approaching. A willingness to get their hands dirty when circumstances warrant will help supervisors build trust among employees.
- *Being consistent.* People trust consistency. It lets them know what to expect. Even when employees disagree with supervisors they appreciate consistent behavior.
- *Being equitable.* Supervisors cannot play favorites and hope to build trust. Employees want to know that they are treated not just well, but as well as all other employees. Fair and equitable treatment of all employees will build trust.

LEADERSHIP AND CHANGE

In an unsure and rapidly changing marketplace, industrial companies are constantly involved in the development of strategies for keeping up, staying ahead, and/or setting new directions. What can supervisors do to play a positive role in the process? David Shanks recommends the following guidelines:[13]

- Have a clear vision and corresponding goals
- Exhibit a strong sense of responsibility
- Be an effective communicator
- Have a high energy level
- Have the will to change

Shanks developed these guidelines to help executives guide their companies through periods of corporate stress and change, but they also apply to supervisors. These characteristics of good leaders apply to any manager at any level who must help his or her organization deal with the uncertainty caused by change.

Facilitating Change as a Leadership Function

The following quote by management consultant Donna Deeprose carries a particularly relevant message for modern supervisors:

> In an age of rapidly accelerating technology, restructuring, repositioning, downsizing, and corporate takeovers, change may be the only constant. Is there anything you can do about it? Of course there is. You can make change happen, you can let it happen to you, or you can stand by while it goes on around you.[14]

Deeprose divides supervisors into three categories based on how they handle change: driver, rider, or spoiler.[15] Supervisors who are drivers lead their units in new directions as a response to change. Supervisors who just go along reacting to change as it happens rather than getting out in front of it are riders. Supervisors who actively resist change are spoilers. Deeprose gives examples of how a driver would behave in a variety of supervisory situations. These examples are summarized here:[16]

- In viewing the change taking place in an organization, drivers stay mentally prepared to take advantage of the change.
- When facing change about which they have misgivings, drivers step back and examine their own motivations.
- When a higher manager has an idea that has already been tried before and failed, drivers let the boss know what problems were experienced early and offer suggestions for avoiding the problems this time around.
- When a company announces major changes in direction, drivers find out all they can about the new plans, communicate what they learn with their employees, and solicit input to determine how to make a contribution to the achievement of the company's new goals.

- When permission to implement a change that will affect other departments is received, drivers go to these other departments and explain the change in their terms, solicit their input, and involve them in the implementation process.
- When demand for their unit's work declines, drivers solicit input from users and employees as to what changes and new products or services might be needed and include the input in a plan for updating and changing direction.
- When an employee suggests a good idea for change, drivers support the change by justifying it to higher management and using their influence to obtain resources for it while countering opposition to it.
- When their unit is assigned a new, unfamiliar responsibility, drivers delegate the new responsibilities to their employees and make sure that they get the support and training needed to succeed.

These examples show that a driver is a supervisor who exhibits the leadership characteristics necessary to play a positive, facilitating role in helping workers and organizations successfully adapt to change on a continual basis.

LEADERSHIP AND TEAMWORK

One of the most important leadership functions of supervisors is team building. More and more, work in modern industry is done in teams. This means that the input of one employee is the output of another, which creates a strong interdependency. Even in the rare case when employees work relatively independently, it is important that they be able to interact positively with other members of the larger team (i.e., the work unit, support employees, clients, etc.).

Modern industrial supervisors need to know how to build and maintain teams. Mel Schnake recommends eight strategies for encouraging teamwork, summarized as follows:[17]

- *Develop an expectation of teamwork.* Such expectations are created by example, by how rewards are given (i.e., to individuals or to teams), and through continual communication of the teamwork message. Teamwork goals should be set and measured.
- *Provide emotional support.* Encourage employees to make recommendations for improving individual and team performance and support them when they do. Supervisors should also encourage team members to support each other.
- *Emphasize common goals.* This will hold human conflict to a minimum. People working together toward a common goal are less likely to fall into conflict with each other. The key to having common goals is to involve all team members in the formulation of the goals. Once common goals are established, the goals rather than the supervisor will guide employee behavior.
- *Stand up for the team.* On occasion the supervisor will have to go to bat with higher management on behalf of the team. When this happens team members tend to unite behind the supervisor. This does not mean that supervisors should be in a perpetual

state of conflict with higher management. Rather, it means that they should know the needs of the team and be willing to work to meet them.

- *Compete against other work teams.* Friendly competition among teams will draw employees together in a positive way. When teams compete against each other, the only way to win is to work together.
- *Orient new members of the team.* One of the downsides of team building is that a team can become so close-knit that it can be difficult for new members to win acceptance. Orienting new employees and involving team members in the orientation can help break down barriers.
- *Encourage employee participation.* Involve employees in making decisions that will affect them and that they will have to carry out. This will build ownership in the decision, and ownership will translate into increased commitment to the team.
- *Encourage open communication.* People have a natural desire to know, to be informed. This is why newspapers and television news programs are so popular. Open, continual communication as a way to keep people informed will build teamwork.

Two of the most widely used examples of work teams in modern industry are quality circles and self-directed work teams. These examples are similar in many ways, but there are also differences. Modern industrial supervisors should be familiar with both of these concepts.

Quality Circles

Quality circles are groups of employees, typically from the same department but occasionally from several different departments, that meet regularly to discuss problems and develop solutions to them. Quality circles range in size from three or four to twelve or fifteen employees. Occasionally, quality circles select their own leaders. However, more often they are led by a manager or supervisor.[18]

Wayne, Griffin, and Bateman identified several characteristics of effective quality circles. Those characteristics are summarized as follows:[19]

- *Cohesion.* A cohesive quality circle consists of members who participate and want to continue to participate.
- *High performance standards.* Effective quality circles set their own performance standards and set them high.
- *Intrinsic satisfaction.* Effective quality circles are composed of members who are satisfied with external rewards (i.e., pay, working conditions).
- *Organizational commitment to the quality circle.* Effective quality circles are composed of members who believe the organization is committed to the quality circle concept.
- *Self-esteem.* Effective quality circles are composed of employees who have high levels of self-esteem. People with high self-esteem approach the solution of problems with greater confidence.

By examining these characteristics we see that in order for quality circles to work, they must be effectively led. Leadership in communicating organizational commitment, in developing mutually arrived at goals, in helping to continually enhance the self-esteem of participants, and in promoting group cohesion are all positive and appropriate activities for supervisors involved with quality circles.

Self-Directed Work-Teams

The **self-directed work team** is the work unit that has a great deal of autonomy in setting its own goals, solving its own problems, and developing its own work strategies. The team also interviews, hires, orients, trains, and evaluates its own employees; controls costs and quality; sets schedules; and handles all the other work tasks required for the group to function.

SUMMARY

1. Leadership is the ability to inspire people to make a total and willing commitment to accomplishing organizational goals. Good leaders exhibit such characteristics as balanced commitment, a positive example, good communication skills, influence, and persuasiveness.
2. Some supervisors have more natural ability than others. However, regardless of where they start, it is education, training, practice, determination, and effort that are the key to success as a supervisor.
3. The key to motivating people lies in the ability to relate their personal goals to the organizational goals. In order to do this, supervisors must understand the individual needs and personal beliefs of people they supervise.
4. Theory X leadership is based on the erroneous assumptions that: (1) most people have an inherent dislike of work and will avoid it if possible; (2) because most people dislike work they must be threatened and coerced; and (3) most people do not want responsibility.
5. Theory Y leadership is based on the following assumptions: (1) work is as natural as play; (2) when committed to objectives, humans will work to achieve them without external control; (3) people will commit to objectives if appropriate rewards are available; (4) under the right conditions people will seek responsibility; (5) men and women can be creative problem solvers; and (6) in most settings human potential is only partially tapped.
6. Theory Z was developed by William Ouchi of UCLA to describe companies that take a middle-of-the-road approach between Theory X and Theory Y. Theory Z has come to be associated with Japanese management styles.
7. Most leadership styles fall into one of five categories: autocratic, democratic, participative, goal-oriented, and situational.
8. Leadership characteristics that win and maintain followership include a sense of purpose, self-discipline, honesty, credibility, common sense, stamina, steadfastness, and commitment.

9. Research studies undertaken by the University of Michigan suggest that most productive supervisors use techniques that fall on a continuum between Theory X and Theory Y (Theory Z).
10. There are a number of key differences between management and leadership. Managers cope with complexity while leaders cope with change. Managers set budgets while leaders set direction and create vision. Managers develop capacity while leaders align people.
11. Leaders can build trust by taking the blame while sharing the credit, pitching in and helping, being consistent, and being equitable.
12. Supervisors can play a positive role in the change process by having a clear vision and corresponding goals, exhibiting a strong sense of responsibility, being effective communicators, having a high energy level, and having the will to change.
13. Strategies for encouraging teamwork include developing an expectation of teamwork, providing emotional support, emphasizing common goals, standing up for the team, competing against other teams, orienting new team members, encouraging employee participation, and encouraging open communication.

KEY TERMS AND PHRASES

Autocratic leadership
Commitment
Credibility
Democratic leadership
Financial security
Goal-oriented leadership
Hierarchy of needs
Influence
Inspiration
Leadership
Motivation
Participative leadership
People-oriented styles
Personal satisfaction
Persuasiveness
Power
Quality circles
Self-directed work teams
Self-discipline
Situational leadership
Societal contribution
Task-oriented styles
Theory X
Theory Y
Theory Z
Trust building

CASE STUDY: Building Trust Through Leadership

The following real-world case study contains a leadership challenge of the type modern industrial supervisors may face in a work setting. Read the case study and answer the accompanying questions.

Mutual trust between a company and its employees can go a long way toward improving productivity, quality, and competitiveness. However, trust doesn't happen automatically. It has to be built through leadership.

Brian McDermott, editor of the *Total Quality Newsletter,* tells a story of a consultant making a presentation to a group of senior corporate executives.[20] The consultant informs the executives that in order for a specific concept to work they will have to trust their employees. The senior executive interrupts and says, "No way. We can't trust the people on our payroll. We won't."

McDermott goes on to report the results of a recent Gallup poll in which only 25 percent of the workers surveyed said their executives do more than talk about trusting them to make good decisions. With so little trust between management and employees, how can a company empower employees to participate in the continual improvement of productivity and quality?

McDermott contrasts this attitude with that displayed at the Ford assembly plant in St. Paul, Minnesota, where management shut down the assembly line for ninety minutes to celebrate winning Ford's Quality One award, an accomplishment that would not have been possible without a high level of trust between labor and management. Building such trust is a leadership issue.

1. How would you evaluate the leadership skills of the executive in this case study who refused to trust his people?
2. What style of leadership do you think is best suited to building trust between supervisors and employees?
3. Explain how trust building will help solidify followership.

REVIEW QUESTIONS

1. Define the term *leadership.*
2. List five characteristics of good leaders.
3. Briefly state your opinion concerning the following statement: "Leaders are born, not made."
4. What are the key factors in becoming a good leader?
5. What is the key to motivating people?
6. What is the key to inspiring people?
7. What are the five levels on the hierarchy of needs?
8. Briefly explain three procedures for applying the hierarchy of needs model in motivating people.
9. Briefly explain the following theories of leadership: Theory X; Theory Y; Theory Z.
10. What are the three basic needs all people have with regard to their jobs?
11. List and briefly explain five leadership styles.
12. Explain the term *task-oriented styles.*
13. Explain the term *people-oriented styles.*
14. Briefly explain how to select the appropriate leadership style.
15. List eight leadership characteristics that win and maintain followership.

SIMULATION ACTIVITIES

The following activities are provided to give you opportunities to develop your supervisory skills by applying the material presented in this chapter and previous chapters in solving simulated supervisory problems.

1. Adams Parcel Delivery Service started with one truck and one driver five years ago. The company expanded rapidly and that original truck driver, Sue Jennings, is now the supervisor of twenty truck drivers. The company's growth has been so phenomenal that everyone is having to work at peak performance levels in order to keep up. Sue Jennings is under a lot of pressure. She has responded to the pressure by becoming increasingly abusive toward her team of drivers, threatening them and trying to coerce them into higher levels of performance. This approach is not working. Is there a better way for Sue to handle this situation? What advice would you give her?
2. John Crawford has been a respected and popular member of the assembly unit of Golf Products, Inc. for over five years. He and his team members assemble a variety of different grades of golf bags. This morning John was offered the vacant position of unit leader and he accepted. Tomorrow he will become the new supervisor of his old teammates. He wants to win and maintain their loyal followership, but is not sure how to go about it. How should he proceed?

ENDNOTES

1. "Diligent Communication Needed to Turn Lofty Ideas into Action," *Total Quality Newsletter,* November 1990, p. 2.
2. Belasco, J. A. "Practical Advice in Making Your Vision a Reality," *Supervisory Management,* December 1990, p. 6.
3. Ouchi, W. *Theory Z: How American Business Can Meet the Japanese Challenge* (New York: Avon Books, 1981).
4. Ibid.
5. Marsland, S. and Beer, M. "The Evolution of Japanese Management: Lessons for U.S. Managers," *Organizational Dynamics,* 1983, Volume 11, pp. 49–67.
6. Ibid.
7. Likert, R. *The Human Organization* (New York: McGraw-Hill, 1967).
8. Kotter, J. P. "What Leaders Really Do," *Harvard Business Review,* May-June 1990, pp. 103, 104.
9. Ibid., p. 104.
10. Bennis, W. "Leadership in the 21st Century," *Training,* May 1990, p. 44.
11. Ibid.
12. Zielinski, D. and Busse, C. "Quality Efforts Flourish When Trust Replaces Fear and Doubt," *Total Quality,* December 1990, p. 103.
13. Shanks, D. C. "The Role of Leadership in Strategy Development," *Journal of Business Strategy,* January/February 1989, p. 36.

14. Deeprose, D. "Change: Are You a Driver, Rider, or Spoiler?" *Supervisory Management,* February 1990, p. 3.
15. Ibid.
16. Ibid.
17. Schnake, M. E. *Human Relations* (Columbus, OH: Merrill, 1990), pp. 349–350.
18. Word, R., Hull, F., and Azumi, K. "Evaluating Quality Circles: The American Application," *California Management Review,* 1983, Volume 26, pp. 37–53.
19. Wayne, S., Griffin, R. W., and Bateman, T. S. "Improving the Effectiveness of Quality Circles," *Personal Administrator,* 1986, Volume 31, pp. 79–88.
20. McDermott, B. "Trust: Principles and and Practice Often Collide," *Total Quality Newsletter,* December 1990, p. 1.

CHAPTER THREE

Motivation and the Supervisor

CHAPTER OBJECTIVES

After studying this chapter, you will be able to explain the following topics:

- The Relationship of People, Personalities, and Motivation
- The Effect of the Work Ethic on Motivation
- The Effect of Job Satisfaction on Motivation
- The Effect of Expectancy on Motivation
- The Effect of Achievement on Motivation
- The Effect of Job Design on Motivation
- The Effect of Competition on Motivation
- The Effect of Communication on Motivation
- The Effect of Promotions on Motivation
- How to Motivate New Employees
- How to Motivate Problem Employees
- Empowerment as a Motivational Technique
- How to Motivate Part-Time Employees
- How to Motivate Using Incentives

CHAPTER OUTLINE

- People, Personalities, and Motivation
- The Work Ethic
- Job Satisfaction and Motivation
- Morale Maintenance and Motivation
- Expectancy and Motivation
- Achievement and Motivation
- Job Design and Motivation
- Competition and Motivation
- Communication and Motivation
- Promotions and Motivation
- New Employees and Motivation
- Problem Employees and Motivation
- Empowerment as a Motivational Strategy
- Motivating Part-Time Workers
- Incentive Programs and Motivation
- Summary
- Key Terms and Phrases
- Case Study Application Problem
- Review Questions
- Simulation Activities
- Endnotes

People can be extraordinarily creative, innovative, and effective at accomplishing goals they really want to obtain. America's history is replete with stories of individuals who overcame enormous adversity and countless obstacles to accomplish their personal goals. Invariably, the key to their success was their **motivation.** By coupling motivation with determination, people can accomplish almost anything.

Because of this, motivating people and helping them learn to be self-motivating are important responsibilities of the modern supervisor. This chapter will help you develop the knowledge necessary to be an effective motivator.

PEOPLE, PERSONALITIES, AND MOTIVATION

Every person is unique. We all have different backgrounds, different needs, different likes and dislikes, and different goals. We have different life experiences and we react differently to those experiences. We also have different emotional and psychological makeups. These are the factors that form our personalities and make us who we are. Personality is the outward manifestation of our unique inner responses to life.

Because we are all unique, does the supervisor with a twelve-person team have to develop twelve different strategies for motivating team members? Not really, although it is important always to remember that people are individuals. Instead, the modern supervisor should become familiar with the ways in which all individuals, in spite of their unique personalities, are alike and knowledgeable about what motivates people in general.

All working people have two separate but interrelated sets of needs. One set comprises general life needs, the other consists of job-specific needs. In order to understand how to motivate people on the job, supervisors must understand these two areas of needs.

General Human Needs Related to Work[1]

All individuals share two sets of common needs that relate to their work. The first set, which consists of five categories of need, is derived from Maslow's hierarchy of needs. As we saw in Chapter 2, these general human needs consist of the following (Figure 3–1):

- Basic survival needs
- Safety/security needs
- Social needs
- Esteem needs
- Self-actualization needs

These general areas of need relate either directly or indirectly to a person's job. Basic **survival needs** include air, water, food, clothing, and shelter. For most of us it is through work that we generate the income necessary to meet these needs.

Safety/security needs include safety from harm or pain and security from criminal or financial difficulties. Although safety/security needs are never pushed completely

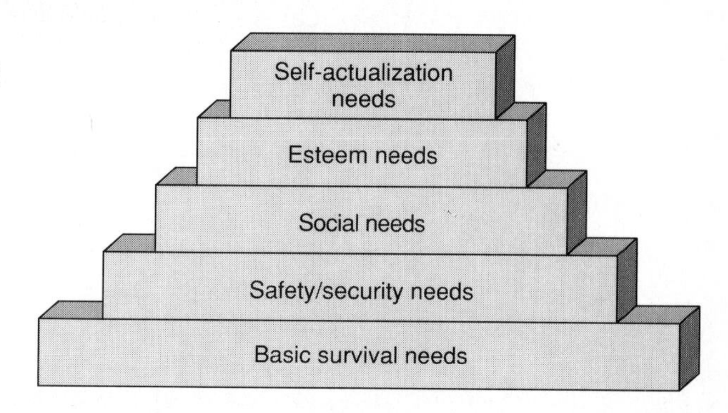

Figure 3–1
General human needs that relate to work.

from our minds, most working people feel relatively safe and secure. These needs relate directly to our jobs. By working we are able to provide a home in a safe neighborhood, install security systems, make financial investments, and purchase insurance.

Social needs include our needs for family, friends, and belongings. Our jobs have a great deal to do with meeting these needs. They give us the financial resources to marry and start a family. They determine where we are able to live and, in turn, who our neighbors will be. Many of our friendships are with people we meet either directly or indirectly through our jobs. In addition, our jobs can provide a sense of belonging.

Esteem needs relate to our need to have feelings of self-respect, worth, and dignity. One of the principal sources of those feelings is our work. One of the primary determinants of a person's status in the eyes of society is his or her job. Many people have the unfortunate tendency to let others define their worth and provide the basis for their feelings of self-respect. This results in people seeing themselves through the eyes of society. Counselors work hard to help people overcome this tendency and to call upon internal stimuli in developing feelings of self-worth. The counselors are right, but it is the rare individual who fully succeeds in this regard. Most of us need positive external stimuli and positive reinforcement to meet our esteem needs. Supervisors who understand this fact can use it when trying to motivate team members.

Supervisors can use praise as a way to help enhance the self-esteem of team members. However, in order for praise to serve this purpose, it must be meaningful. Robert Luke suggests six strategies supervisors can use to make their praise more meaningful. These strategies are summarized as follows:[2]

- *Be prepared.* Know exactly what actions you want to reinforce and be ready to cite specific examples of these actions.
- *Be specific.* Give specific examples when praising an employee. For example, rather than saying "Thank you for being so prompt," say "I've asked you to make five critical deliveries this week and you've made every one of them on time; thank you."
- *Be realistic.* Make sure the person you are praising actually performed the task for which he or she will be praised.
- *Be timely.* Give praise as soon as possible after the praiseworthy behavior has taken place. Do not save praise and give it later or only at the annual performance review.

- *Be discerning.* Let employees know when their performance is not acceptable. This will add significance to praise when you give it.
- *Be creative.* Look for ways to add emphasis to praise. Giving something tangible along with the praise can make it more meaningful, such as a day off, a free lunch, an employee-of-the-month nomination, or anything else that adds emphasis to praise.

According to Luke,

> Probably the worst action you can take is to withhold praise. Many simply see no news as no news. Some may even interpret it as bad news. Meaningful praise is one way of removing any uncertainty. Let people know what they are doing well. Reinforce the admirable qualities of their work.[3]

Self-actualization encompasses our need to be a whole person completely at ease with all aspects of our life. With work so large a part of our lives, we cannot achieve self-actualization unless we are happy in our jobs. Because of this, supervisors should be familiar with what people need in order to be happy in their jobs.

Specific Human Needs Related to Work

The needs explained in the previous section relate to work, but they also apply to life in general. Those explained in this section relate specifically to an individual's job. Specific job-related needs are as follows (Figure 3–2):

- Financial reward
- Personal satisfaction
- Societal contribution

For people to be completely happy on the job, their work must meet all three of these needs in their order of priority. If these needs are not met, individuals will be unhappy.

Identifying Employee Motivators

True motivation must come from within. However, supervisors can encourage the process by providing appropriate stimulators and a conducive work environment. Identifying what motivates individual employees is an important part of the process. Franklin Stein suggests using a matrix approach that divides employee motivators into three broad categories: (1) frequent motivators; (2) occasional motivators; and (3) annual motivators.[4] Under each of these categories are three subcategories: (1) money motivators; (2) personal motivators; and (3) policy motivators.

Under the category of **frequent motivators** small monetary rewards can be given to employees who exceed performance standards. Personal motivators such as birthday or anniversary cards and other types of personal recognition can be used. Policy motivators

Figure 3–2
Specific human needs that relate to work.

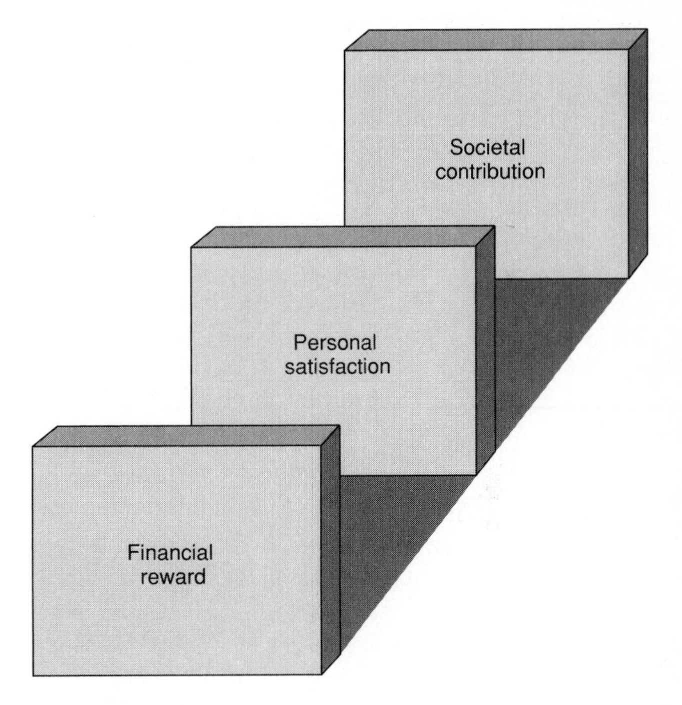

include such things as holding meetings on company rather than personal time or occasionally allowing longer lunch hours.

Occasional motivators might include giving raises and/or performance bonuses (money motivators) more frequently (i.e., quarterly or semi-annually). A personal motivator under this category might be giving a high-performing employee a long weekend or an opportunity to attend conferences, workshops, or seminars. Policy motivators under this category might be involving employees in developing a flextime schedule or finding other approaches to scheduling work.

Under the category of **annual motivators**, money motivators might include annual salary reviews, annual bonuses, annual position reviews for potentially upgrading positions, and taking over a higher level job for a year. Personal motivators might include an annual company family picnic, Christmas party, or other types of companywide celebrations. A policy motivator under this category could be developing a company child-care facility.

Stein stresses that these are only suggestions of motivators to promote innovative thinking on the part of supervisors. There are many other motivators under each of these categories that could be used. The key is for supervisors to build their own matrix of motivators and update it continually, paying special attention to what works best with individual employees.

THE WORK ETHIC

It has been said that one of the things that traditionally distinguished Americans from people in other countries was the work ethic. It is now often said that a large measure of Japan's economic success since World War II can be attributed to the Japanese work ethic. Many believe that American industrial productivity and competitiveness have declined because the work ethic in America has declined. But what is the work ethic? The **work ethic** is a philosophy that can be summarized as follows:

> *Work is intrinsically good and intrinsically rewarding.*

Is the Work Ethic Dead?

Daniel Yankelovich and John Immerwahr of the Public Agenda Foundation conducted a three-year study of the work ethic in America involving over 1,300 workers. Their findings are summarized below:[5]

- Fewer than 25 percent of the study subjects said they were working to their full potential.
- Fewer than 50 percent of the study subjects said they put forth effort beyond the minimum required.
- Almost 70 percent of the study subjects felt that American workmanship was better ten years ago.
- Almost 80 percent of the study subjects felt that workers took more pride in their work ten years ago.
- Almost 70 percent of the study subjects felt workers were more motivated toward their jobs ten years ago.

These findings suggest there is a problem with the work ethic in America. This presents modern supervisors with a challenge. What can supervisors do?

Improving the Work Ethic

When the work ethic seems to decline the natural reaction is to blame the worker. This is not always appropriate. The fault could be the company's. Supervisors cannot give workers a positive work ethic. However, they can help develop the work ethic in people and also recharge one that has run down. Some tips for enhancing the work ethic or helping to develop a better work ethic are contained in the following paragraphs.

Supervisors can help employees see a direct connection between the quality/quantity of their work and the amount of their pay. If salary/wage programs are not set up in a way that allows for a direct relationship, work with management to restructure the program so that financial incentives are available and are tied to job performance.

Do not confuse happy, satisfied, or comfortable employees with productive employees. Whereas productive employees may be happy in their jobs, it does not automatically follow that happy employees are productive employees. Employees might by happy with their jobs because the pay is good, working conditions are good, and nobody pushes them to work too hard.

Job satisfaction is not necessarily a motivator for better performance. But it cannot be overlooked because job dissatisfaction is definitely a *demotivator*. Consequently, modern supervisors must concern themselves with job satisfaction, not as an end unto itself but as one step toward the goal of motivating employees. The relationship between job satisfaction and motivation is presented in the next section.

JOB SATISFACTION AND MOTIVATION

In the previous section, it was mentioned that job satisfaction is the foundation upon which higher levels of productivity can be built. Consequently, it is an important element in the ongoing motivation of employees and supervisors have a key role to play. Factors relating to job satisfaction can be divided into two levels: those that keep employees happy and those that motivate them to higher performance. Factors that fall into the first category are shown in Figure 3–3. These factors should be thought of as morale maintenance factors.

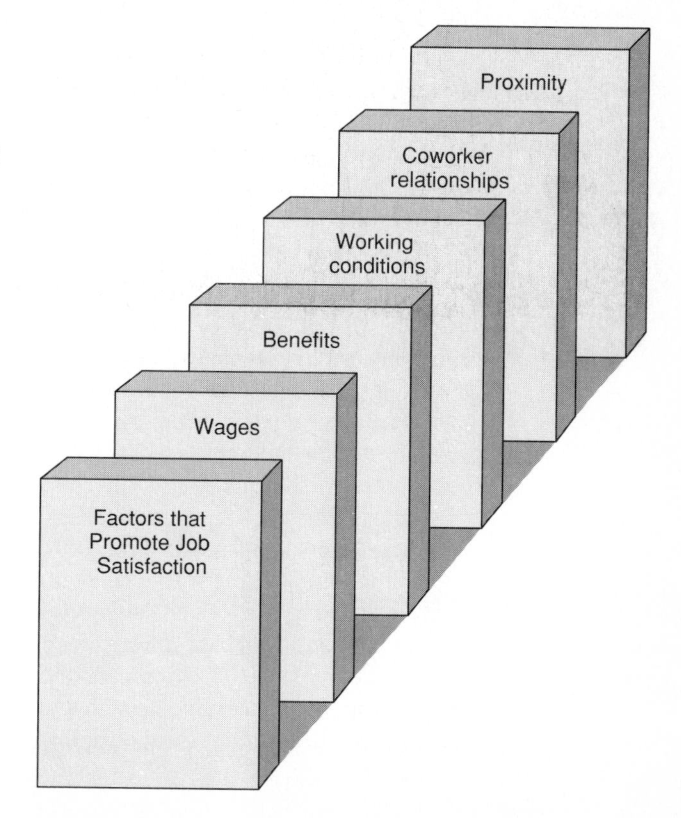

Figure 3–3
Factors that promote job satisfaction do not necessarily motivate workers to higher levels of performance.

MORALE MAINTENANCE AND MOTIVATION

Morale maintenance factors are those that will keep employees from quitting, but will not motivate them to higher levels of performance. This is why they can be considered only the foundation of an ongoing effort to motivate employees.

To understand this concept, consider the example of a fictitious organizational unit, PCB Assembly Unit Five, a printed circuit board assembly team in Elec-Tech, Inc. The unit is responsible for soldering discrete electronic components on high-density printed circuit boards. Employees in this unit love their jobs. The pay is good; they get a two-week paid vacation each year; their working environment is clean, orderly, and safe; and the plant is close to home for most team members. Predictably, morale in Unit Five is very high. Unfortunately, its production rate is very low. In fact, the weekly production rate for Unit Five is the lowest for all such units at Elec-Tech.

The supervisor, Martha Clinton, cannot understand it. She has worked hard to create an environment that would enhance productivity. Martha spent months convincing higher management to institute a more competitive wage and benefits package. She personally supervised the restructuring of her unit, doing away with the traditional assembly line approach that was repetitive and monotonous and replacing it with a teamwork approach that is more flexible and challenging. Since most of her employees are single mothers, Martha even convinced management to institute on-site child care on a six-month trial basis.

Morale in her unit is up and absenteeism is down. In fact, everything about the unit is different since Martha Clinton took over. Everything, that is, except productivity. It has improved a little, but not much.

What the supervisor in this example does not see is that all her efforts so far, although commendable, are only the first step. She is like a home builder who lays the foundation for a house, stops at that point, and then wonders why the house won't sell. The next step for Martha Clinton is to apply performance motivators.

Performance Motivators

Employees with reasonably good morale can be motivated to higher levels of productivity through the effective application of the **performance motivators** shown in Figure 3–4. Challenging work, advancement potential, on-going opportunities for personal/professional development, and competition are effective motivators. Let's examine how these motivators could be used by the supervisor at Elec-Tech.

Martha Clinton has already taken steps to make the work more challenging by replacing the boring assembly line with a teamwork approach. This alone may account for the small improvement she has seen in production rates. To build on this she might consider breaking the jobs she supervises into progressive levels to create opportunities for advancement and base advancement on performance. She might also institute a training program for her employees or tie them into an existing program to give them opportunities for continuing growth and development. Finally, Martha should consider establishing a program of competition with appropriate incentives and rewards built in.

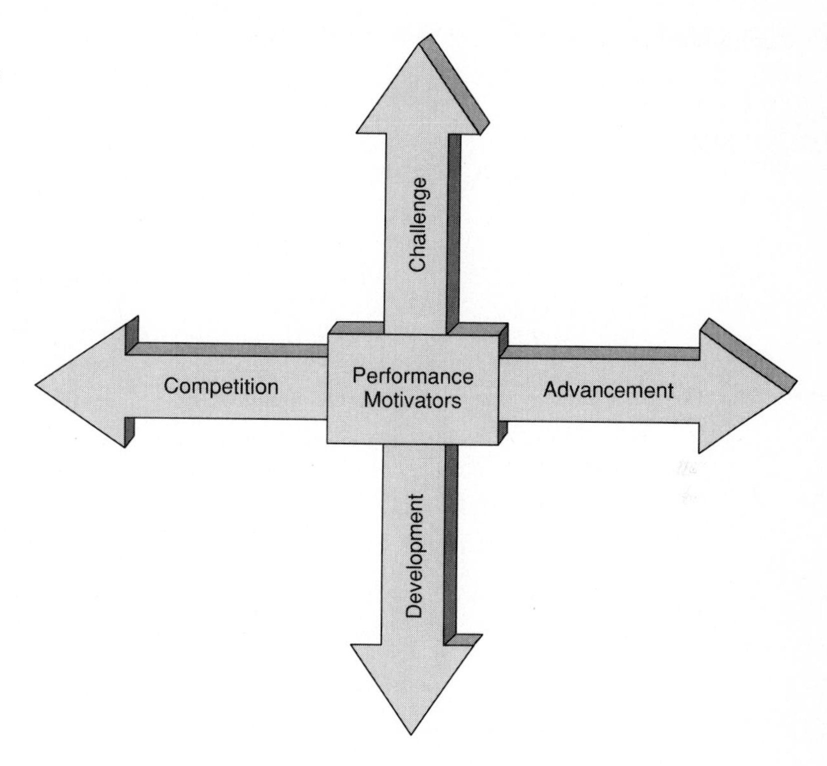

Figure 3–4
These factors help motivate workers to be more productive.

Employees within her unit might compete against each other and the unit as a whole might compete against other assembly units at the company.

EXPECTANCY AND MOTIVATION

Expectancy is an important factor in motivating employees. It has four components: what management expects, what the supervisor expects of employees, what employees expect in return for performance, and what peers expect of each other (Figure 3–5). For employees to stay motivated over the long run, there must be consistency between what they expect for performing well and what they actually get. If the incentives established by management and/or the supervisor come to be viewed as empty promises, their motivational value will be negated. Correspondingly, if the performance expected is so high as to be unattainable, employees will expect to fail and failure will become a self-fulfilling prophecy.

Another side of expectancy that can be a motivator is the expectations of fellow workers. If team members internalize the organization's goals, expectations of performance will manifest themselves through peer pressure. The expectancy of peers can be a powerful motivator.

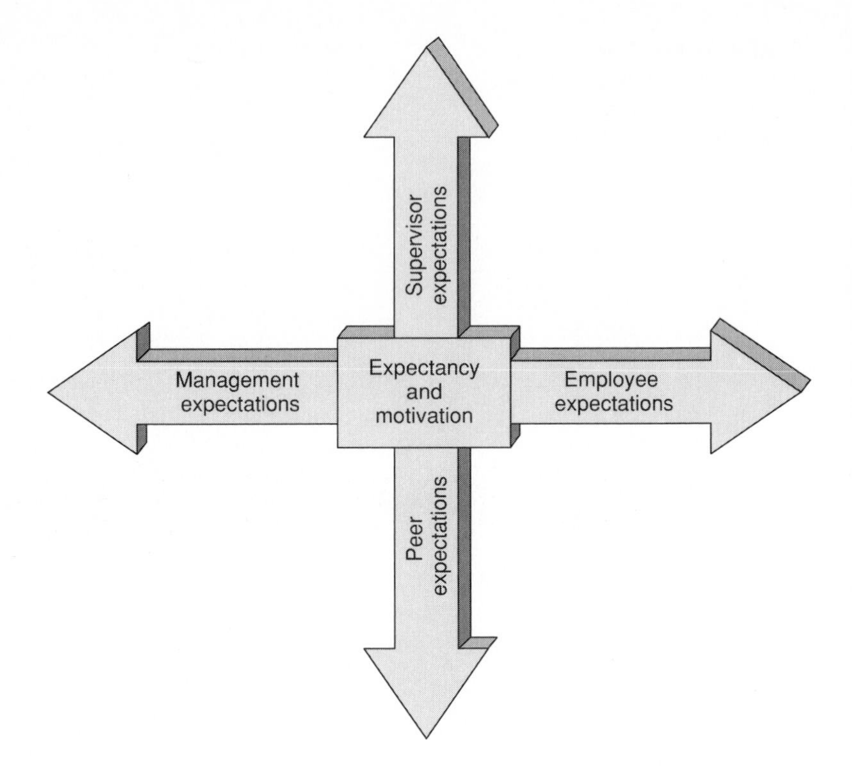

Figure 3–5
Expectancy can play a critical role in motivating employees and keeping them motivated.

ACHIEVEMENT AND MOTIVATION

Achievement motivates some people and not others. The key for supervisors lies in being able to: (1) identify achievement-oriented people; (2) take advantage of this characteristic for motivating them; and (3) keep highly motivated achievers from dampening the morale of employees for whom achievement is less of a motivator.

Recognizing Achievement-Oriented Employees

People who are motivated by achievement are not difficult to pick out of a crowd. They tend to be task-oriented, independent, need continual reinforcement, and focus intently on evaluations of their performance. They usually accumulate physical evidence of their achievements such as trophies, plaques, certificates of accomplishment, and other memorabilia (Figure 3–6).

Using Achievement to Motivate Employees

Perhaps the easiest way to motivate achievement-oriented employees is by sharing the organization's goals with them so that each goal accomplished becomes a personal achievement for them. Another effective strategy is to make achievement-oriented employees responsible for specific goals that they can call their own. In addition, it is

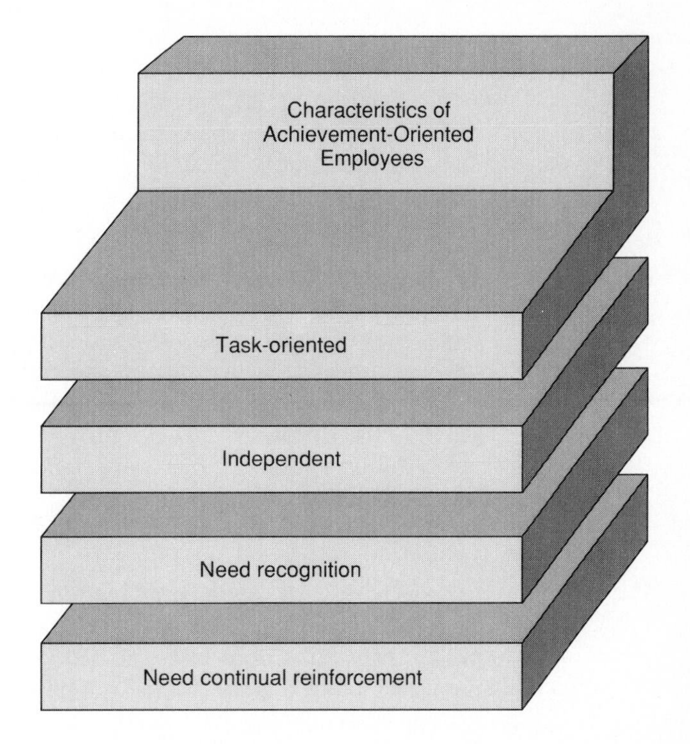

Figure 3–6
Achievement-oriented employees can be recognized by these characteristics.

also necessary to provide appropriate recognition when they achieve goals. Part of what motivates achievement-oriented people is the recognition achievement brings them.

One of the drawbacks of challenging achievement-oriented employees is that it can lead to problems with other workers who are motivated more by affiliation than achievement. Employees with an affiliation orientation are team players who are motivated by the social interaction among peers. They view their achievements in terms of those of the team and do not try to stand out in a crowd.

Affiliation-oriented employees sometimes resent achievement-oriented employees and criticize them for not being team players. Supervisors can minimize the potential for friction by taking the precautions set forth in the following paragraphs.

Talking with both types of employees one-on-one and explaining the strengths and contributions of each will help. Supervisors can explain how achievement- and affiliation-oriented people can complement each other in ways that result in higher productivity for all. Where possible, supervisors should group affiliation types with affiliation types and achievement types with achievement types. Let the achievement types compete with each other.

Where work is done in teams, supervisors should make sure that there is an appropriate balance between affiliation- and achievement-oriented workers on the teams. An achievement-oriented worker can be a "spark plug" for the team if handled properly.

JOB DESIGN AND MOTIVATION

Job design can be an important factor in motivating employees and keeping them motivated over time. There are three basic approaches to job design: people-oriented, task-oriented, and balanced.

Task-Oriented Job Design

In the past, job design was typically **task-oriented.** This means that job processes were set up in such a way as to get the most out of the floor space available, machines used, tools required, and humans involved. Time and motion studies were used to continually improve processes by identifying and eliminating wasted motion.

Task-oriented job designs resulted in such early industrial innovations as the assembly line. On paper, task-oriented job designs look good. However, in reality they rarely live up to expectations over extended periods of time. This is due primarily to the human factor.

Task-oriented job designs break down because they do not give sufficient consideration to the psychological and physiological needs of the people doing the work. This leads to boring, unchallenging, monotonous work. Some of this type of work can now be accomplished by industrial robots, but much of it cannot. As a result, the long-term productivity gains from task-oriented job designs are limited. Knowledge of this phenomenon led to the development of people-oriented job design.

People-Oriented Job Design

In **people-oriented job design** the psychological and physiological needs of human workers are the foremost consideration. The philosophy of this approach is that, in the long run, workers who are simultaneously comfortable and challenged will be more productive even if the tasks they perform are not perfectly laid out.

The science of ergonomics grew out of the people-oriented school of thought. It is the study of how human workers interact with technology and how they react in their environment. The buzzword "user friendly" is part of the language of ergonomics.

Taken to an extreme, even the people-oriented approach can have a detrimental effect on productivity. People can actually become so comfortable that they lose the positive edge that is necessary for high productivity. This realization led to the development of a balanced orientation in job design.

Balanced Orientation in Job Design

The **balanced job design** seeks to strike the optimum balance between the other two orientations (Figure 3–7). The key element of the balanced approach is the maximum participation of people who do a job in the design of the job. They are encouraged to weigh both task and people considerations in trying to arrive at an optimum job design that incorporates the best of both concepts.

There are are several advantages of the balanced approach. The most important of these include the following: (1) it brings the firsthand knowledge of the people who

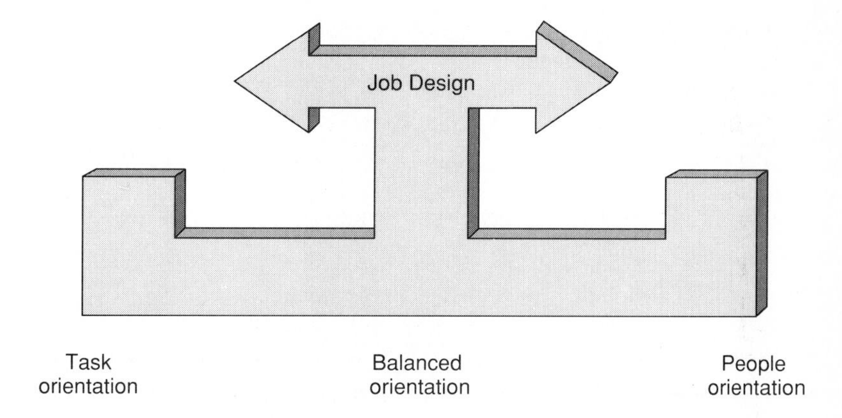

Figure 3–7
Three approaches to job design shown as a continuum.

actually do the job to bear on designing the job; (2) it focuses attention on the work instead of the people who are doing the work, which is less threatening for employees; and (3) it promotes ownership of the job design by workers. The balanced approach has the most potential for motivating employees.

How to Use Job Design to Motivate

The first step in using job design as a motivator is to adopt the balanced approach. In applying this approach, supervisors may find the following rules of thumb helpful (Figure 3–8):

- Design the job so that workers see the big picture rather than isolated, seemingly independent pieces of the job. One of the many problems with the old assembly line approach is that workers cannot see where their part of the job fits into the overall process or how their task contributes to the final product.
- Design the job so that workers must apply a variety of skills rather than performing one monotonous task that never changes.
- Design the job so that workers can be as autonomous as possible. The more decisions about the work that can be made by the people doing it, the better.
- Design the job so that workers are allowed to come into contact with users of their product as frequently as possible. On occasion, when a salesperson meets with a customer, arrange for a worker to participate in the meeting. Feedback straight from the user can be a powerful motivator.

To understand how these strategies can be applied, consider the example of Defense Manufacturing Company, Inc. This company manufactures high-density printed circuit boards used in military applications. Initially, the traditional assembly line approach was used. One worker would mount a given component or perform some other discrete task and pass the board to the next worker. With this approach the company experienced the types of problems that have come to be associated with assembly lines.

Figure 3–8
Tips for using job design as a motivator.

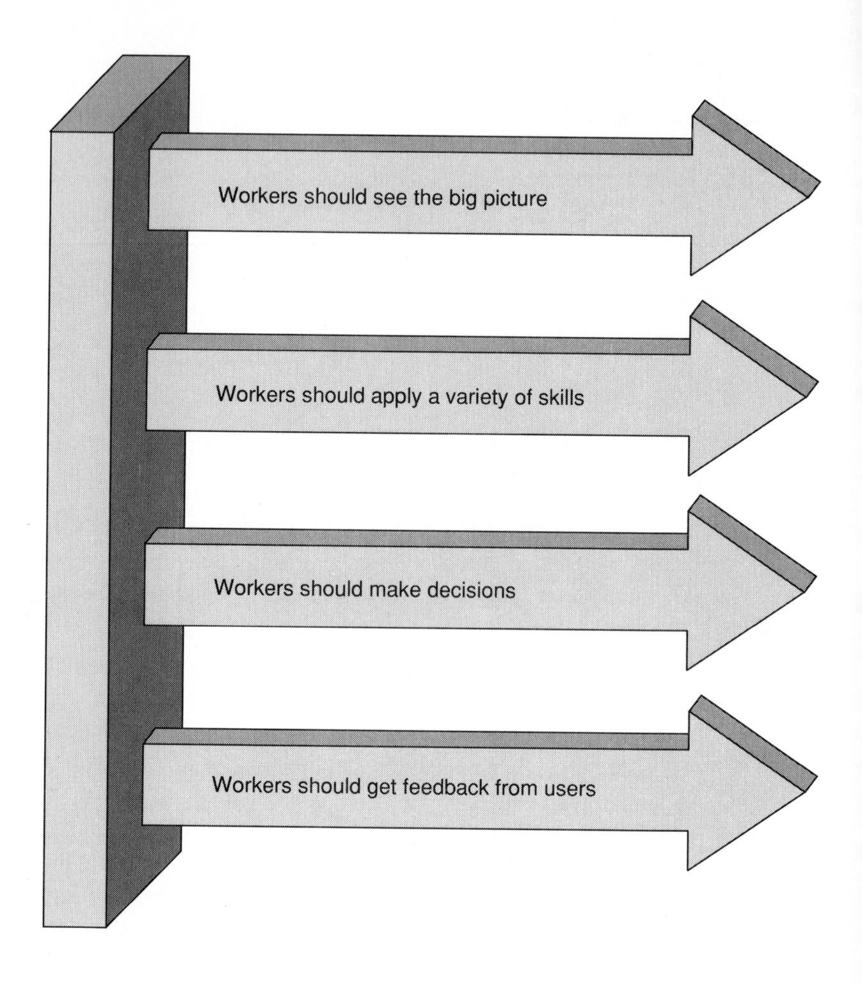

Workers complained of boredom and monotony, which led to a lack of interest on their part. Eventually, both productivity and quality began to suffer.

In an attempt to solve these problems, workers were asked to make recommendations. As a result, the assembly line was broken down and replaced with manufacturing cells staffed by small teams of workers. Each team is responsible for producing an entire circuit board. The manufacturing cell approach allows team members to see a part through from start to finish, requires them to use a variety of different skills, and allows them to make decisions about the work they do.

In addition to setting up the cells, a new policy was enacted allowing workers to occasionally accompany sales personnel on customer visits to get firsthand feedback. The result of these changes have been higher productivity, improved quality, lower absentee rates, higher morale, and lower turnover rates.

COMPETITION AND MOTIVATION

To a degree, we are all competitive. Children's competitive instinct is nurtured through play, reinforced by sports and school activities. Supervisors can use the adults' competitive instinct when trying to motivate employees, but competition on the job should be carefully organized, closely monitored, and strictly controlled. Competition that is allowed to get out of hand can do more harm than good. Competition that is not controlled can lead to cheating and hard feelings among fellow workers.

Competition can be organized between teams, shifts, divisions, or even plants. It can focus on a number of different productivity-related factors or combinations of factors such as production quotas, absentee rates, safety records, quality, waste rates, uptime on machines, and a variety of other factors (Figure 3–9).

What follows are some tips that will help supervisors use competition in a positive way while ensuring that it does not get out of hand.

- Involve the employees who will compete in planning programs of competition.
- Where possible encourage competition among groups rather than individuals, while simultaneously promoting individual initiative within groups.

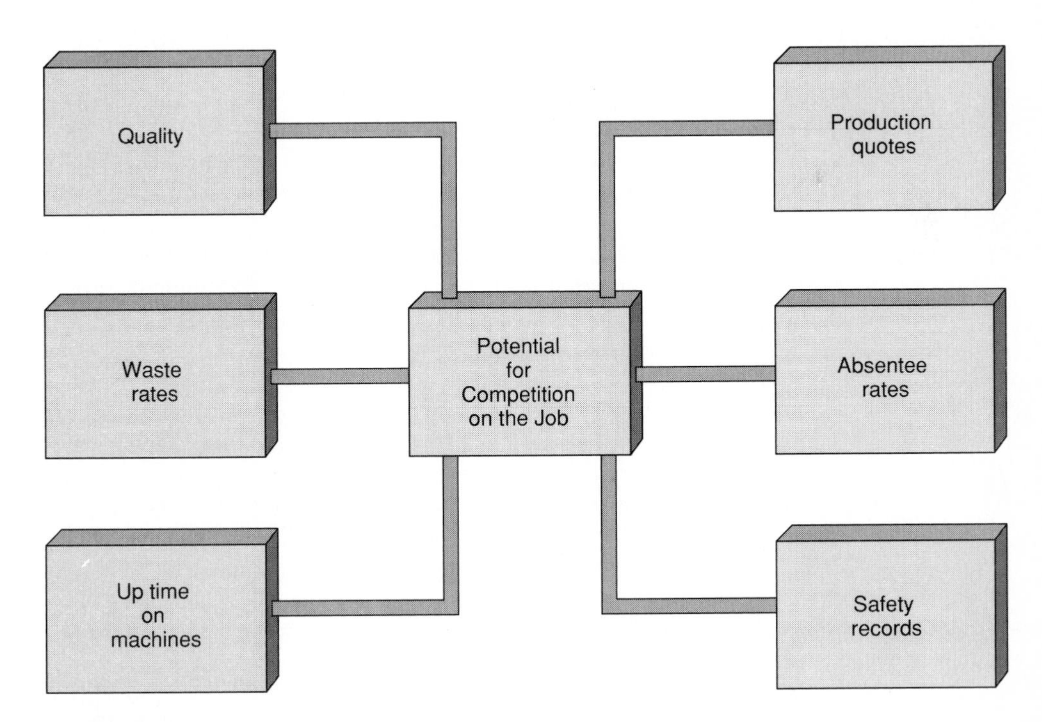

Figure 3–9
If used properly competition can be a positive motivator.

- Make sure the competition is fair by dividing work equally, making sure that resources available to competing teams are equitably distributed and that human talent is as appropriately spread among the teams as possible.

COMMUNICATION AND MOTIVATION

Just as people have a natural instinct for competition, they also have a natural desire to be informed. They want to know how they are doing as individuals, how the team is doing, how their division is doing, and how the company is doing. Providing up-to-date, accurate information on a continual basis is an excellent way to motivate employees and keep them motivated. In fact, without effective communication, all the other strategies for motivating employees break down.

Communication involves more than just conveying information. It also involves giving instructions, listening, persuading, inspiring, and understanding. For the supervisor, communication can be viewed as preventive maintenance. In order to use communication to prevent problems and to keep small problems from becoming big ones, supervisors must be tuned into what is being said by employees, verbally and nonverbally, as well as what is going unsaid. This requires supervisors to have the following types of abilities:

- Ability to understand nonverbal communication
- Ability to empathize with employees and see things through their eyes
- Ability to read between the lines and determine what the real problem is
- Ability to keep an open mind and truly listen to what employees are saying

In an interview in *Inc.* magazine, management consultant Warren Blaisdell stresses the importance of communication as a motivator during difficult times.[6] According to Blaisdell, the most important time to communicate openly and frequently with employees is during hard times. He says,

> You really have to make time, even when other demands may seem greater. The most important time to communicate, to spend an hour or so with your employees, is when you're having problems. . . . People sense problems anyway, especially in small companies. They know what's going on, and the more you try to hide, the more susceptible the environment will be to rumors and misleading information.[7]

PROMOTIONS AND MOTIVATION

Like so many factors relating to motivation, **promotions** can have positive or negative effects depending on how they are handled. The two basic approaches to promotions are promoting from within and promoting from outside. Of the two, promotion from within is the approach most likely to be a motivator if used properly. What follows are some demotivators that should be avoided when promoting from within:

- *Don't* promote solely on the basis of seniority. A person with ten years' experience might, in reality, have one year of experience that has been repeated ten times. Seniority is a legitimate factor to consider in promotions, but should not be the only factor. If a senior employee is promoted over a less senior but more skilled worker, morale will suffer.
- *Don't* promote on the basis of popularity. Personal popularity is no guarantee of success in a new position. It is not uncommon to find that an employee who is well liked as a person does not work well when placed in a supervisory role over former co-workers.
- *Don't* promote on the basis of friendship. Of all the *don'ts*, this is the most critical. Promotions viewed by other workers as being influenced by friendship are doomed to failure from the outset.

A good rule of thumb is to promote from within whenever possible, but to base promotions on qualifications rather than popularity, friendship, or solely on seniority. This will ensure that promotions motivate, rather than demoralize.

NEW EMPLOYEES AND MOTIVATION

New employees represent a special challenge for supervisors. It is important to help new workers make a positive start. To get new employees motivated, supervisors should take personal charge of their introduction to the new job (Figure 3–10).

This should begin with an orientation that is handled personally by the supervisor rather than a subordinate. This will serve two purposes. First, it shows the new employee he or she is important enough to rate the supervisor's personal attention. Second, it shows existing employees that the new employee is important enough to rate the supervisor's personal attention.

The next step involves letting new employees know what is expected of them. This will establish the company's corporate culture in the minds of new employees. **Corporate culture** is the name given to the overall collection of the company's expectations of employees. If the corporate culture encompasses such philosophies as an emphasis on quality, the customer is always right, or service first, new employees should know it from the start.

Supervisors should carefully assign new employees' first task. Success is a powerful motivator. It will breed further success. Failure is an equally powerful demoralizer. The key is to make a new employee's first assignment a confidence builder. In this way, initial success will breed additional success.

Supervisors should spend more than the usual amount of time with the new employee. A one-on-one, face-to-face conference at the end of each day can be an effective motivating strategy. During the conference, supervisors can identify problems the employee is having and deal with them right away, give the employee feedback on performance, and provide encouragement.

Encouragement is particularly important in motivating new employees. In the words of Randolph Barker and Sandra Barker, "Supervisors who use encouragement

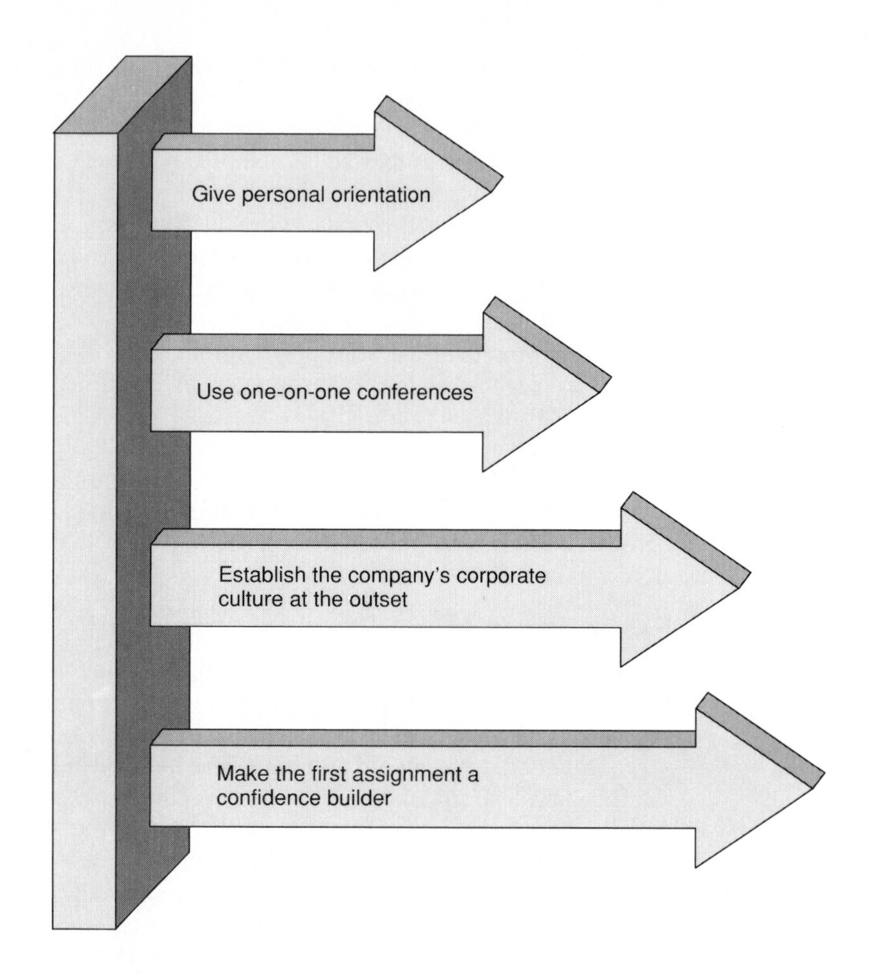

Figure 3–10
Strategies for motivating the new employee.

stimulate their employees' internal motivation to produce. By recognizing employees' efforts, unique contributions, strengths and improvements, supervisors promote self-appreciation and self-confidence in their workers."[8]

Barker and Barker recommend eight strategies for encouraging employees. These strategies are summarized as follows:[9]

- *Focus on efforts* rather than the person. Compliment the effort, not the person. This will keep employees from tying their personal worth to their performance. Remember, even the best employees can have a bad day. This doesn't make them bad people.
- *Build on strengths and assets* to help employees build self-esteem. Most employees will consistently perform well on some aspect of the job. Build on this strength to help overcome and correct weaknesses in other aspects of the job.
- *Recognize improvement* as a way to reinforce an employee's self-esteem. Success breeds success. By acknowledging improvement, supervisors will encourage additional improvement.

- *Break down complex tasks into simpler ones.* It is discouraging to employees to face difficult, complex tasks. When possible, supervisors should break tasks down into a series of tasks that successively increase in difficulty. This will encourage employees by allowing them to experience success.
- *Focus on specific contributions.* Giving specific examples when offering encouragement lets employees know exactly what they did right. Relating the behavior to the accomplishment of one or more of the organization's goals lets the employee know the specific contribution he or she has made.
- *Separate the behavior from the person.* It can be harmful to an employee's self-esteem to tie his or her worth to everyday job performance. The appropriate message when an employee does not meet work standards is, "You are good; it is your performance that was bad." This will make it easier for employees to accept constructive criticism intended to improve their job performance.
- *Focus on mistakes as learning activities.* Employees will make mistakes, particularly those who are willing to risk attempting to improve performance. When an employee makes a mistake say, "All right, this didn't work. What can we learn from the experience?" This communicates the message that mistakes can be opportunities for improvement rather than failure. This will encourage employees to take the calculated risks necessary for improved performance.
- *Promote problem solving* by encouraging self-sufficiency. For example, when an employee comes to you with a problem try saying, "Think about it for a while and bring me a suggested solution. If you get stuck we'll put our heads together." This will show employees you have confidence in them but that you are willing to collaborate if necessary. This approach will promote problem solving.

Encouragement can be an invaluable strategy for supervisors trying to help employees motivate themselves. In the words of Barker and Barker, "Encouragement is a valuable management skill contributing to employee motivation, responsibility, and productivity. Used properly by supervisors, encouragement represents one of the most powerful inducers of intrinsic rewards."[10]

PROBLEM EMPLOYEES AND MOTIVATION

Problem employees have always posed a challenge for supervisors. The drug user, slow achiever, sick leave abuser, grouch, know-it-all, goldbrick, constant complainer, and gossip can strain a supervisor's patience to the breaking point (Figure 3–11). However, firing a problem employee is not an automatic option. Employees, no matter how many problems they cause, have rights. In today's workplace it is difficult to fire an employee in a way that does not leave the company susceptible to charges of wrongful discharge.

As a result, the modern supervisor is well advised to make a concerted effort to help problem employees become contributing members of the team. This issue is treated at length in Chapter 13. Here we discuss the first and most critical step in the process of motivating problem employees: determining what is causing the behavior.

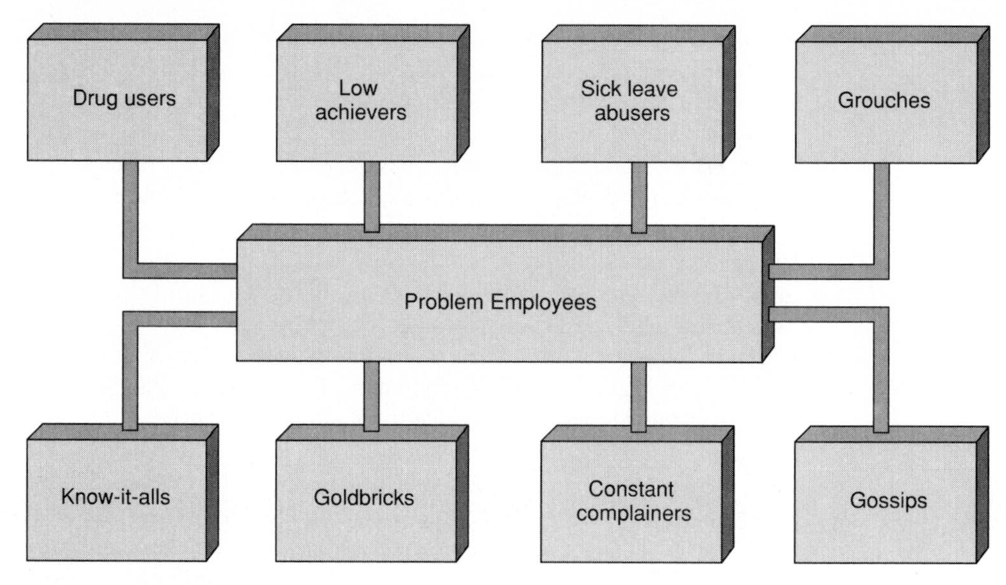

Figure 3–11
Some of the better known types of problem employees.

Determining the Cause of Problem Behavior

Getting to the root causes of unacceptable behavior will not be possible in all cases. In addition, supervisors should resist the temptation to engage in "armchair psychology." However, if supervisors apply the following strategies, they may occasionally succeed in turning a problem employee into an asset (Figure 3–12):

- *Do not jump to conclusions.* Keep an open mind until the causes of the behavior have been identified.
- *Be patient.* It may take time to learn what is causing the negative behavior.
- *Spend as much one-on-one time with problem employees as possible.* Let them do the talking. Listen, make mental notes, and look for clues. To draw the employee out, avoid judgmental comments, gestures, or facial expressions.
- If the cause of the negative behavior can be identified, *tailor your motivational techniques as dictated by the cause.*

EMPOWERMENT AS A MOTIVATIONAL STRATEGY

According to Peter Kizilos "powerlessness is the root cause of many of the problems that management is concerned with in the workplace today."[11] The solution to feelings of powerlessness is the empowerment of employees. Empowerment is not just another name for participatory management, as is sometimes assumed. Participatory manage-

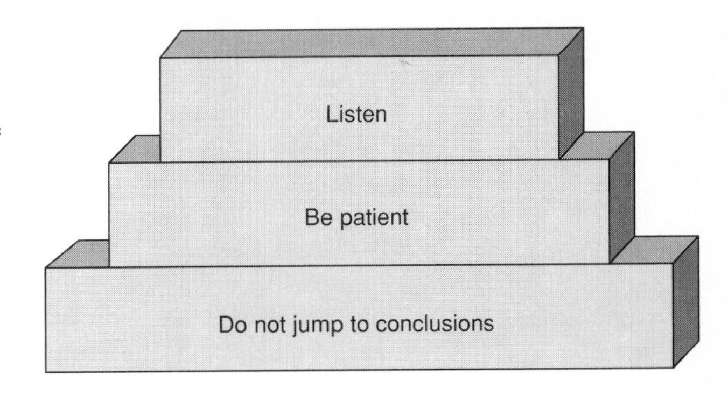

Figure 3–12
To motivate problem employees supervisors must identify the causes of problem behavior.

ment is about managers and supervisors asking for their employees' help. **Empowerment** is about getting employees to help themselves. An empowered employee feels ownership of her or his job. Empowered employees can make decisions, take risks in an attempt to improve performance, speak out when they disagree, and finally, they can make mistakes.

The growing interest in empowering employees stems from the need to continually enhance competitiveness. According to Kizilos,

> There are both economic and psychological reasons why empowerment has become a hot topic in today's business world. In the new era of stiff global competition, American corporations need to respond much more quickly to market forces. . . . As a result, the pressure on employees to produce or perish is increasing. Companies are looking for ways to improve their own bottom lines and help employees cope with the added stress.[12]

Empowerment of employees will work only if management truly commits to it. If fundamental changes in management are not made, employees will not be empowered and they will know it. According to Kizilos, too many companies are attempting, only half-heartedly, to empower employees.[13] They are not making the fundamental changes in organizational structure or management style needed. Supervisors should keep this in mind as they work with higher management to implement policies and practices that support and promote empowerment.

Supervisors and managers interested in using empowerment to motivate employees to continually improve should begin with a thorough examination of organizational structure and management practices. The following questions represent a start:

- *How many layers of bureaucracy are there between workers and decision makers?* Empowered employees need to be able to put new ideas to the test without filling out forms and waiting for them to make their way up the line and back down again. Does the organizational structure allow for this?
- *Does the employee evaluation system allow for risk-taking and mistakes?* Employees will simply stick with tried and approved work procedures if they know that mistakes

will be reflected in their next evaluation. This does not mean that empowerment must accept carelessness. Rather, it means that calculated risks are encouraged, but there are no negative consequences when they do not work out as planned.

- *Are employees penalized for speaking out against policies and practices they feel impede quality and/or productivity?* If management truly wants empowered employees, it must establish a formal mechanism for encouraging them to speak out. Then management must listen.

According to Kizilos, it can take companies eight to ten years to overcome all of the internal characteristics that mitigate against empowerment.[14] He says,

> but altering people's orientation toward empowerment—turning a bureaucratic goldbricker into a risk-seeking, initiative-taking, problem-solving go-getter—is no simple training task. . . . Organizational inertia tends to discourage people from acting in an empowered fashion. Individual fiefdoms carefully nurtured over time don't suddenly crumble overnight.[15]

Supervising Empowered Employees

Supervisors of empowered employees spend more time coordinating and facilitating than giving orders and controlling. Experienced supervisors must let go of some of their traditional controls if employees are to be fully empowered. This is not to say that empowered employees are an out-of-control mob or a haphazard work team, although this is a concern often felt and occasionally expressed by traditional managers and supervisors. On the contrary, empowered employees, like all employees, must work within limits and controls.

According to Kizilos, "Part of a manager's job is to say what is negotiable and what is not. Part of the skill of being able to manage a group of people is knowing what the boundaries are and being clear about those to people."[16] In this statement lies the key to successfully supervising empowered workers. Define clearly for them the limits that must guide and control their work performance (i.e., budgets, deadlines, market data, etc.). These are the non-negotiables. With the limits clearly defined, communicate the message that within the limits employees are in charge of their jobs and the team is in charge of the unit's workload. This means they can make decisions, try new work methods, and suggest improvements. They are also free to consult the supervisor for help at any point in the process.

For their part, supervisors monitor work performance continually and measure it against all applicable controls. If a unit's workload is beginning to go over budget or fall behind schedule, the supervisor works out appropriate solutions with the employees.

Kizilos relates an example of how supervisors or managers can work with empowered employees.[17] A computer company that subscribed to empowerment of employees in both theory and practice was developing a new product with a five-year development cycle. If the product could be brought on-line in time, the computer company would gain an edge on the competitors. After a couple of years in development problems occurred that resulted in the development team asking for more time than was originally projected.

The development team was composed of high-achieving, empowered employees who were determined not to release the product until they got it right no matter how behind schedule that turned out to be. If this had been a traditional situation, management might have called the team together and given them the "your jobs are on the line here, do it or else" rationale. Instead, their supervisor sat down with team members and explained the realities of the situation. He showed them his market data and compared their position as a company with that of the competition. It clearly showed that if they did not meet the deadline, the company was out of business. This approach did not threaten or patronize the team. Rather, it challenged them. The deadline became their deadline, not the company's.

Supervisors who wish to motivate through empowerment must learn to strike a proper balance between employee autonomy and appropriate control. Being a supervisor of empowered employees is akin to being a coach. According to Kizilos, "the coaching skills . . . bear a striking resemblance to parenting skills. These skills involve using control properly and appropriately, handling bad ideas."[18] Supervisors of empowered employees will be presented on occasion with a bad idea.[19] The key here is to protect the work team from a bad idea without turning off the employee who thought it up. One way is to tell the employee to try it and learn from the results. Another is to discuss the idea openly and share information the employee may not have foreseen. There is nothing wrong with turning down bad ideas as long as it is done in a way that does not disempower employees.

MOTIVATING PART-TIME WORKERS

Part-time employees represent approximately 25 percent of the total workforce in this country.[20] The reasons part-time employees work and their needs with regard to work often differ from those of their full-time counterparts. Consequently, what it takes to motivate part-time employees often differs too.

Daniel Kopp and Lois Shufeldt developed a six-phase program called P-TIMER that can be used by supervisors to motivate part-time employees.[21] The P-TIMER program consists of six strategies, which are summarized here:

- *Positive reinforcement* to show part-time employees that they are full members of the team and that their contributions are appreciated.
- *Team-building* efforts that include part-time employees and let them know they are members of the team.
- *Information sharing* that allows part-time employees to know what is going on. Better informed workers, part-time or full-time, are better prepared to do their jobs.
- *Money* paid at a rate that says, "you are important" to part-time employees.
- *Expectations* explained fully so that part-time employees understand what is expected of them.

- *Feedback* that lets part-time employees know how they are doing. Praise and recognition can be as effective in helping part-time employees motivate themselves as they are with full-time employees.

With part-time employees representing one-fourth of the workforce now and that number likely to increase over time, modern supervisors will need to be able to motivate part-timers. The six strategies set forth in the P-TIMER program can help.

INCENTIVE PROGRAMS AND MOTIVATION

Tim Puffer explains what is taking place in the modern workplace with regard to incentive programs:

> Most companies have no problem developing incentive programs for their sales forces. . . . But with the shift toward a service economy, a growing number of companies are becoming just as concerned with non-sales performance issues such as productivity and customer service.[22]

Increasingly, companies faced with intense competition in the marketplace are developing incentive programs to motivate employees to higher levels of productivity and quality. For this reason it is important for modern supervisors to know how to construct effective incentive programs. In order to do this, supervisors need to know which incentives will work best.

According to Puffer, what motivates today's workers may not be what motivated workers in the past.[23] This is because today's worker is more likely to be a minority, woman, single parent, or older worker and the needs of such workers may differ from those of their more traditional counterparts.

According to Don Roux, a sales promotion and marketing consultant, pay raises and promotions are part of the picture, but "today's workers are more concerned with personal recognition than corporate recognition (i.e., the gold watch, holiday turkey, or jewelry bearing the company logo)."[24] Jennifer Hurwitz, a consultant who designs corporate incentive programs, says, "People also want to be involved in decision-making. . . . And they want individualized personal recognition from their peers."[25]

Puffer suggests the following seven strategies that can help ensure the effectiveness of incentive programs:[26]

- *Define objectives.* Begin by deciding what is supposed to be accomplished by the incentive program. Higher quality? Higher productivity? Improved safety? Examples of incentive program objectives might be:
  - To increase productivity by 22 percent.
  - To decrease the reject rate by 25 percent.
  - To decrease the accident rate by 50 percent.
- *Lead by example.* For incentive programs to work, supervisors and higher managers must set a positive example of modeling the type of behavior they want the incentives

to reinforce. For example, if an objective is to decrease absenteeism and tardiness, supervisors must set an example of attendance and punctuality.

- *Develop specific criteria.* On what basis will the incentives be awarded? This question should be answered during the development of the program. Specific criteria define the type of behavior and level of performance that is to be rewarded as well as guidelines for measuring success. For example, an objective might be to improve quality. Measurable criteria might be the reject rate, percentage of waste, and rework rate.
- *Make rewards meaningful.* For an incentive program to be effective, the rewards must be meaningful to the recipients. Giving an employee a reward that he or she does not value will not produce the desired results. To determine what types of rewards will be meaningful, it is necessary to involve employees.
- *Only the employees who will participate in an incentive program know what incentives will motivate them.* In addition, employees must feel it is *their* program. This means that employees should be involved in the planning, implementation, and evaluation steps.
- *Keep communications clear.* It is important for employees to fully understand the incentive program and all of its aspects. Communicate with employees about the program, ask for continual feedback, listen to the feedback, and act on it.
- *Reward teams.* Rewarding teams can be more effective than rewarding individuals. This is because work in the modern industrial setting is more likely to be accomplished by a team than an individual. When this is the case, other team members might resent the recognition given to an individual member. Such a situation can cause the incentive program to backfire.

SUMMARY

1. All people have two separate but overlapping sets of needs with regard to their work. The first set consists of general needs and includes survival, safety/security, social, esteem, and self-actualization needs. The second set consists of more specific needs and includes financial reward, personal satisfaction, and societal contribution.
2. The work ethic is a philosophy that can be summarized as believing that work is intrinsically good and intrinsically rewarding.
3. Job satisfaction is an issue when trying to motivate people. Factors affecting job satisfaction fall into one of two categories; morale maintenance factors and motivating factors. Maintenance factors include wages, benefits, working conditions, co-worker relationships, and proximity. Motivators include challenge, competition, advancement, and development.
4. Expectancy is an important factor in motivating employees. There are four components: what management expects; what the supervisor expects; what peers expect; and what employees expect in return for performance.
5. Achievement is a motivating factor for some people, but not others. Supervisors can recognize achievement-oriented employees because they tend to be task-oriented,

independent, need continual reinforcement, and tend to accumulate physical evidence of their achievement.

6. Job design can be an important factor in motivating employees. There are three basic approaches to job design: people-oriented; task-oriented; and a balanced approach.
7. Competition, if used properly, can be a motivator. Involve employees in planning the competition; arrange competition among groups rather than individuals; and make sure the competition is fair.
8. Communication can be an effective motivator if supervisors are good communicators. In order to be good communicators, supervisors must have the ability to understand nonverbal communication, empathize, read between the lines, listen, and keep an open mind.
9. Promotions can motivate if handled properly. Some *don'ts* to keep in mind about promotions are: don't promote solely on the basis of seniority; don't promote on the basis of popularity; and don't promote on the basis of friendship.
10. To motivate new employees have personal orientations and one-on-one conferences. Establish the corporate culture at the outset and make the first assignment a confidence builder.
11. To motivate problem employees, listen, be patient, and do not jump to conclusions.

KEY TERMS AND PHRASES

Achievement
Annual motivators
Balanced job design
Communication
Competition
Corporate culture
Empowerment
Esteem needs
Expectancy
Financial reward
Frequent motivators
Job design
Job satisfaction
Morale maintenance
Motivation
Occasional motivators
People-oriented job design
Performance motivators
Personal satisfaction
Problem employees
Promotions
Safety/security needs
Self-actualization needs
Social needs
Societal contribution
Survival needs
Task-oriented job design
Work ethic

CASE STUDY: Motivating with Innovative Compensation[27]

The following real-world case study contains a motivation-related problem of the type supervisors may face in the modern workplace. Read the case study carefully and answer the accompanying questions.

Jeffrey Banks learned the importance of competitiveness, team building, and motivation in the NFL. The former middle linebacker has used what he learned to become a leader in the outdoor advertising industry. Because competition in this industry is so intense, Banks knows he must attract and keep talented employees. In an effort to motivate his team to higher levels of performance while simultaneously encouraging them to stay with his company, Banks established an incentive program.

The original incentive program included a profit-sharing plan that distributed a portion of company profits among all employees on a pro-rata basis. Banks has since added other components to the plan. Each component is carefully designed to serve two purposes: (1) improve job performance on a continual basis; and (2) keep high-performing employees on the team. Banks describes the payoff of his incentive program as producing a workforce that "understands that we're all in a family working toward a common goal, toward rewards we're all going to share."

1. Are the objectives of the incentive program described in this case study clear?
2. Are the rewards in this incentive program meaningful with regard to long-term motivation?
3. How would you improve this program if given an opportunity?

REVIEW QUESTIONS

1. What are the five general areas of need all people have?
2. What are the three specific work-related needs all people share?
3. Define the term *work ethic*.
4. Give three reasons why the work ethic is not as strong as it could be in this country.
5. Briefly explain how supervisors can help enhance the work ethic.
6. What is job satisfaction and how does it affect motivation?
7. List the job satisfaction factors that maintain morale.
8. List the job satisfaction factors that can motivate people to higher levels of performance.
9. Explain the term *expectancy* as it relates to motivation.
10. How can a supervisor recognize an achievement-oriented employee?
11. How can a supervisor use achievement as a motivator?
12. How can a supervisor prevent friction between achievement- and affiliation-oriented employees?
13. Explain the following terms: *people-oriented job design; task-oriented job design; balanced job design.*
14. Explain just one strategy for using job design as a motivator.
15. Explain briefly how supervisors can ensure that competition is a motivator rather than a demotivator.
16. What abilities must a supervisor have in order to be a good communicator?
17. Explain briefly how to avoid making promotions a demotivator.
18. Explain briefly how to motivate a new employee.
19. Explain briefly how to deal with problem employees.

20. List and briefly explain four strategies supervisors can use to make their praise of employees more meaningful.
21. Define and differentiate among the following types of motivators: frequent, occasional, and annual.
22. List and briefly explain four strategies supervisors can use for encouraging employees in ways that will help motivate them.

SIMULATION ACTIVITIES

The following activities are provided to give you opportunities to develop supervisory skills by applying the material presented in this chapter and previous chapters in solving simulated supervisory problems.

1. Amy Parker is hired as the new supervisor for Petroleum Products Company's secretarial pool and given only one assignment: Improve productivity and do it fast! At first Amy does not understand why her unit's reputation for work is so bad. Every member of the secretarial pool is well trained and experienced. The company has invested in the latest word processing systems. Office procedures appear to be well designed and everyone follows them. In spite of all this, the pool appears to work in slow motion and the stack of backlogged work grows higher every day. Amy decides the problem is the unit's work ethic. It does not have one. How can Amy Parker help improve the work ethic of her unit?
2. Rasheed Press is a large printing company that produces a wide range of printed materials from business cards to books. It recently won a large contract that will test its production capabilities. Howard Cromwell, the printing supervisor, plans to divide the work evenly between his two printing teams and have them compete. In this way he hopes to increase production. If you were Howard, how would you proceed?

ENDNOTES

1. Schnake, M. E., *Human Relations* (Columbus, OH: Merrill, 1990), pp. 124–125.
2. Luke, R. A., Jr. "Meaningful Praise Makes a Difference," *Supervisory Management,* February 1991, p. 3.
3. Ibid.
4. Stein, F. J. "A Matrix Approach to Help Identify Employee Motivators," *Supervisory Management,* January 1991, p. 9.
5. Yankelovich, D. and Immerwahr, J. *Action Guide to Motivating People* (Waterford, CT: Prentice-Hall, 1987), p. 7.
6. Lyons, N. J. and Posner, B. G. "Hard Times," *Inc.,* November 1990, p. 84.
7. Ibid.
8. Barker, R. T. and Barker, S. B. "Encouragement: An Essential Skill," *Supervisory Management,* June 1990, p. 5.
9. Ibid.
10. Ibid.

11. Kizilos, P. "Crazy About Empowerment?" *Training,* December 1990, pp. 47–56.
12. Ibid.
13. Ibid., p. 49.
14. Ibid., p. 50.
15. Ibid., p. 51.
16. Ibid., p. 51.
17. Ibid., p. 52.
18. Ibid., p. 54.
19. Ibid., p. 55.
20. Kopp, D. G. and Shufeldt, L. M. "Motivating the Part-Time Worker," *Supervisory Management,* January 1990, p. 4.
21. Ibid.
22. Puffer, T. "Eight Ways to Construct Effective Service Reward Systems," *Reward & Recognition* Supplement, *Training,* August 1990, pp. 8–12.
23. Ibid.
24. Ibid., p. 9.
25. Ibid., p. 10.
26. Ibid., p. 11.
27. Fraser, J. A. "State of the Art," *Inc.,* November 1990, pp. 68–76.

CHAPTER FOUR

Communication and the Supervisor

CHAPTER OBJECTIVES

After studying this chapter, you will be able to define or explain the following topics:

- Communication
- Classes of Communication
- Communication as a Process
- Inhibitors of Communication
- Communication Networks
- How to Communicate Better by Listening
- How to Communicate Better Nonverbally
- How to Communicate in Writing
- How to Communicate Corrective Feedback
- Six Steps to Improved Communication
- Electronic Communication

CHAPTER OUTLINE

- Communication Defined
- Classes of Communication
- Communication as a Process
- Inhibitors of Communication
- Communication Networks
- Communicating by Listening
- Communicating Nonverbally
- Communicating Verbally
- Communicating in Writing
- Communicating Corrective Feedback
- Steps to Improved Communication
- Electronic Communication
- Summary
- Key Terms and Phrases
- Case Study Application Problem
- Review Questions
- Simulation Activities
- Endnotes

Of all the skills needed by the modern supervisor, communication skills are the most important. All the other components of supervision presented in this book depend either directly or indirectly on effective communication. It is fundamental to leadership, motivation, problem solving, management, training, discipline, labor relations, ethics, and all other areas of concern to the modern supervisor. This chapter will help supervisors become more effective communicators. Note that although communication is important enough to warrant its own chapter, in practice it must be fully integrated into all the other functions supervisors perform.

COMMUNICATION DEFINED

Inexperienced supervisors often make the mistake of confusing *telling* with *communicating.* When a problem develops, they are likely to explain, "But I told him what to do." Inexperienced supervisors occasionally confuse *hearing* with *listening.* They are likely to say, "This isn't what I told you to do. You had to hear me. I was standing right next to you!"

In both cases the supervisor has confused telling and hearing with communicating. This point can be illustrated by a quotation attributed to many different politicians: "I know you believe you understand what you think I said, but I am not sure you realize that what you heard is not what I meant."

This is an amusing quote, but it does make a point. What you say is not necessarily what the other person hears and what the other person hears is not necessarily what you intended to say. The key word is *understand.* This word is the key to communication. Communication may involve telling, but it is not *just* telling. It may involve hearing, but it is not *just* hearing. For the purpose of this book, **communication** is defined as follows:

> ***Communication is the transfer of information that is understood from one source to another.***

A message can be sent by one person and received by another, but until the message is fully understood by both, there is no communication (Figure 4–1). This applies to spoken, written, and nonverbal messages.

Communication versus Effective Communication

When information conveyed is understood, there is communication. However, understanding by itself does not necessarily make effective communication. **Effective communication** occurs when the information that is understood is acted on in the desired manner.

For example, a supervisor might ask her team members to arrive at work fifteen minutes early for the next week to ensure that an important order goes out on schedule.

Figure 4–1
Unless the message is understood, there is no communication.

Each team member verifies that he or she understands both the facts and the reasons in the message. However, without informing the supervisor, two team members decide they are not going to comply.

This is an example of communication that is not effective. The two nonconforming employees understood the message, but decided against complying with it. The supervisor in this case failed to achieve acceptance of the message.

Effective communication is a higher level of communication. It implies not just understanding, but understanding and acceptance (Figure 4–2). This means that effective communication may require persuasion, motivation, monitoring, and leadership.

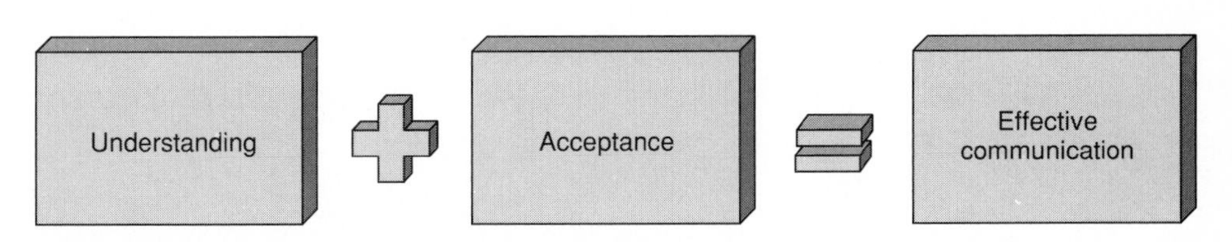

Figure 4–2
Effective communication requires both understanding and acceptance.

Communication Levels

Communication can take place on several levels in a company. These levels are as follows (Figure 4–3):

- One-on-one level
- Team or unit level
- Company level
- Community level

Although supervisors are actively engaged primarily in the first two levels, they may be involved at all levels, at least indirectly. Consequently, modern supervisors should be familiar with all four levels of communication.

One-on-one level communication is just what the name implies: one person communicating with one other person. This might involve face-to-face conversation, a telephone call, or even a simple gesture or facial expression.

Team or unit level communication is communication within a peer group. The primary difference between one-on-one and team communication is that, with the

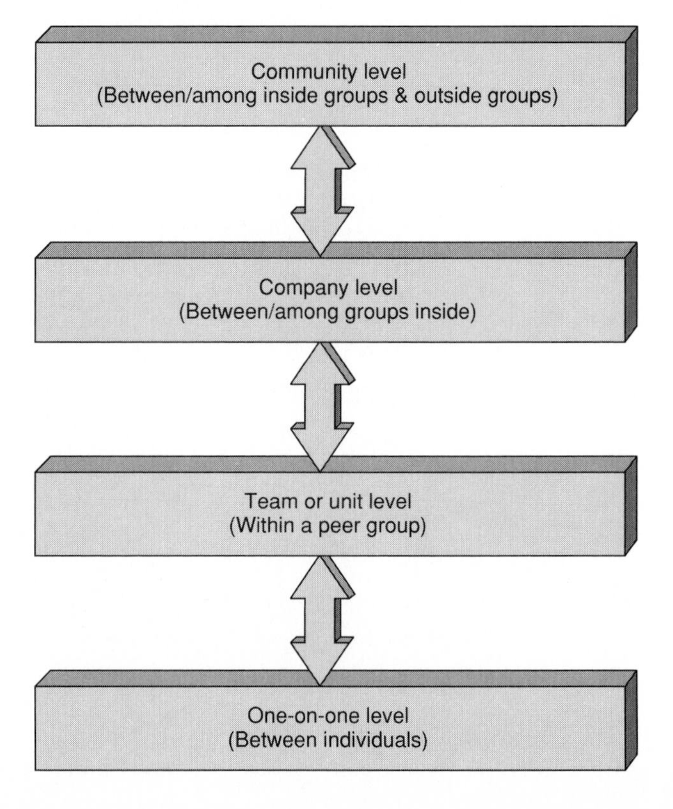

Figure 4–3
In an industrial firm, communication can take place on four different levels.

latter, all team members are involved in the process at once. A team meeting called to solve a problem or set goals would be an opportunity for team level communication.

Company level communication is communication among groups. A meeting involving the sales department, design department, and a production team would represent an opportunity for company level communication.

Community level communication occurs among groups inside a company and groups outside the company. Perhaps the most common examples of community level communication are the company's sales force with potential clients and the company's purchasing department with vendors. An example that is becoming increasingly commonplace and increasingly important is a company's communication with members of the press and electronic media.

CLASSES OF COMMUNICATION

W. B. Rossnagel has identified ten **classes of communication** replies that can help supervisors improve their communication with employees.[1] Each class represents a type of verbal reply or response. Rossnagel recommends that supervisors learn the types of responses associated with each class so they can be conscious of avoiding them and the communication problems they can cause. The classes of communication are summarized as follows:

- *Class I*: Statements you make about something you have witnessed and hold to be *unquestionably* correct. Such statements leave no room for discussion.
- *Class II*: Statements made by a second party to a third party about what you have said. The accuracy of such statements tends to diminish as they pass from person to person.
- *Class III*: Statements passed on by the third party to others about what he or she heard from the second party.
- *Class IV*: Statements that are answers given to questions other than the ones you were asked. Such responses give the appearance of evasiveness or that you don't know the answers.
- *Class V*: Statements with built-in semantics or room for interpretation that may not tell the whole story. For example, say you were asked to accomplish a specific task and you got it done by asking a fellow worker to do it for you. If your boss asks, "Did you get that job done?" and you respond, "I sure did!" you have given a Class V response.
- *Class VI*: A guess by a person who does not want to admit he or she does not have the answer.
- *Class VII*: Exaggerations. Rossnagel gives the example of using a word such as *frequently* when *occasionally* would be more appropriate.
- *Class VIII*: Cover-up statements made to buy time. For example, if your boss asks, "Did you get the information I asked for yet?" and you respond, "Yes, but I'm right in the middle of something. Let me call you back in an hour," when you really don't have the information, you've made a Class VIII statement.

Differences in Meaning

Differences in meaning represent a common problem in communication. We all have different backgrounds and levels of education. We might even come from different cultures and have different national origins. As a result, words, gestures, and facial expressions can have altogether different meanings to different people. This is why modern supervisors must invest the time to know their team members.

Insufficient Trust

Insufficient trust can inhibit effective communication. If receivers do not trust senders, they may be overly sensitive and guarded. They might concentrate so hard on reading between the lines for the "hidden agenda" that they miss the message. This is why trust building between supervisors and employees is so important. It is well worth all the time and effort required.

Information Overload

Information overload is more of an inhibitor than it has ever been. Computers, modems, satellite communication, facsimile machines, electronic mail, and the many other technological devices developed to promote and enhance communication can actually cause a breakdown in communication.

Because of advances in communication technology and the rapid and continual proliferation of information, we often find ourselves with more information than we can deal with effectively. This is **information overload.** Supervisors can guard against information overload by screening, organizing, summarizing, and simplifying the information they convey to employees.

Interference

Interference is any external distraction that inhibits effective communication. It might be something as simple as background noise or as complex as atmospheric interference with satellite communications. Regardless of its source, interference either distorts or completely blocks the message. This is why supervisors must be attentive to the environment in which they plan to communicate.

Condescending Tones

A condescending tone when conveying information can inhibit effective communication. People do not like to be talked down to. Problems in this regard typically result from the tone rather than the content of the message. Supervisors should never talk down to employees.

Listening Problems

Listening problems are one of the most serious inhibitors of effective communication. Problems can result when the sender does not listen to the receiver and vice-versa. An entire section is devoted to listening later in this chapter.

Premature Judgments

Premature judgments by either the sender or the receiver can inhibit effective communication. This is primarily because they interfere with listening. When we make a quick judgment, we are prone to stop listening at that point. You cannot make premature judgments and maintain an open mind. Therefore it is important for supervisors to listen nonjudgmentally when talking with employees.

Inaccurate Assumptions

Our perceptions are influenced by our assumptions. Consequently, **inaccurate assumptions** can lead to inaccurate perceptions, as in the following example. John Andrews, a technician, has been taking an inordinate amount of time off from work lately. His supervisor, Joe Little, assumes John is goldbricking. As a result, whenever John makes a suggestion in a team meeting, Joe assumes he is just lazy and suggesting the easy solution.

It turns out this is an inaccurate assumption. John is actually a highly motivated, highly skilled worker. His excessive time off is the result of a problem he is having at home, a problem he is too embarrassed to discuss. In this case, because of an inaccurate assumption, the supervisor is missing opportunities to take advantage of the suggestions of a highly motivated, highly skilled worker. In addition, he is overlooking a need for building trust. Perhaps if John Andrews trusted his supervisor more, he would be less embarrassed to discuss his personal problem with him.

COMMUNICATION NETWORKS

In a modern company, much of the communication that takes place goes through networks, either formal or informal (Figure 4–6). A **network** is a group of senders linked by some means with a group of receivers. A formal network might consist of all

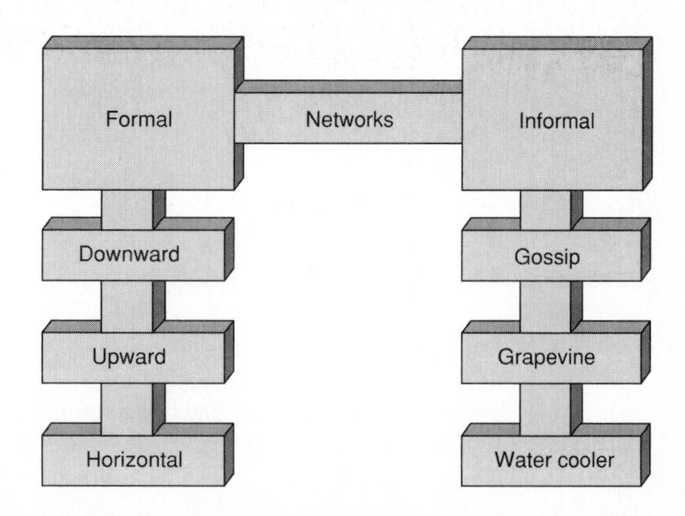

Figure 4–6
Supervisors should be aware of both types of networks.

supervisors in a company linked to each other and higher management by an electronic mail system. Any person in the system can be a sender or receiver. The networking mechanism is the electronic mail system.

An informal network would be what is alternately referred to as the gossip circle, water cooler crowd, or grapevine. In this case, all participants in the network can be senders and receivers. The network itself is one-on-one conversation passed along from person to person. Formal networks are used for communicating official company messages. Informal networks are used to convey unofficial and often inaccurate messages.

COMMUNICATING BY LISTENING

Perhaps the most important communication skill of supervisors is listening. It is also the one people are least likely to have. Are you a good listener? Consider the following questions:

1. When in a group of people, are you more likely to talk or listen?
2. When talking with someone, do you often interrupt before he or she completes the statement?
3. When talking with someone, do you find yourself tuning out and thinking ahead to your response?
4. When talking with someone, could you paraphrase what he or she said and say it back?
5. In a conversation, do you tend to state your opinion before the other party has made his or her case?
6. When talking with someone, do you give your full attention or continue with other tasks simultaneously?
7. When you do not understand, do you ask for clarification?
8. In meetings, do you tend to daydream or stray from the subject?
9. When talking with someone, do you fidget and sneak glances at your watch?
10. Do you ever finish statements for people who do not move the conversation along fast enough?

A skilled listener will respond to these questions as follows:

1. Listen	6. Full attention
2. No	7. Yes
3. No	8. No
4. Yes	9. No
5. No	10. No

A better way to find out if you are a good listener is to ask. Ask a friend, your spouse, a fellow worker, and an employee you can trust to give an objective answer. Do not be overly concerned if you find you are not a good listener. Listening is a skill and, like all skills, it can be developed. In order to become a good listener you need to know: (1)

what listening is; (2) barriers that inhibit listening; and (3) strategies that promote effective listening.

What Is Listening?

Hearing is a natural process, but listening is not. A person with highly sensitive hearing abilities can be a poor listener. Conversely, a person with impaired hearing can be an excellent listener. Hearing is the physiological decoding of sound waves, but listening involves *perception*. Listening can be defined in numerous ways. In this book we define **listening** as follows:

> ***Listening is receiving the message, correctly decoding it, and accurately perceiving what is meant.***

Inhibitors of Effective Listening

Listening breaks down when the receiver does not accurately perceive the message. Several inhibitors can cause this to happen. These inhibitors include the following (Figure 4–7):

- Lack of concentration
- Preconceived ideas
- Thinking ahead
- Interruptions
- Tuning out

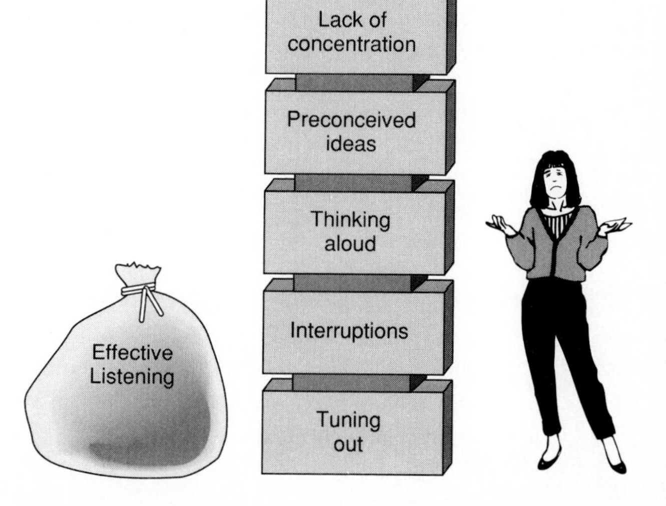

Figure 4–7
These factors can form an impenetrable barrier to effective listening.

To perceive the message accurately, listeners must *concentrate on what is being said, how it is being said, and in what tone.* Part of effective listening is properly reading nonverbal cues (covered in the next section).

Concentration requires the listener to eliminate as many extraneous distractions as possible and to mentally shut out the rest. J. Lamar Roberts, chief executive officer and president of a successful and dynamic community bank, requires his supervisors to keep their desks clear of all projects except the one being worked on at the moment. In this way, when an employee enters the office, this project can be easily pushed aside, leaving a clean desk. Work left on the desk can distract the supervisor and make the employee feel like an intruder.

Supervisors who cling to preconceived notions cannot listen effectively. **Preconceived ideas** can cause supervisors to make premature judgments that turn out to be wrong. Be patient, wait, and listen.

Supervisors who jump ahead to where they think the conversation is going may get there and find they are all alone. *Thinking ahead* is typically a response to being hurried, but supervisors will find it takes less time to hear an employee out than it does to start over after jumping ahead to the wrong conclusion.

Interruptions not only inhibit effective listening, they frustrate and often confuse the speaker. If clarification is needed during a conversation, make a mental note and wait for the speaker to reach an interim stopping point. Mental notes are preferable to written notes. Writing notes can distract the speaker and/or cause the listener to miss a critical point. If you find it necessary to make written notes, keep them short and to a minimum.

Tuning out inhibits effective listening. Some people become skilled at using body language to make it appear they are listening while their mind is focusing on other areas of concern. Supervisors should avoid the temptation to engage in such ploys. A skilled speaker may ask you to repeat what he or she just said.

Supervisors can become effective listeners by applying several simple strategies. These strategies are presented in Figure 4–8. You might want to commit them to memory or keep this list handy and refer to it frequently until you can remember the strategies without referring to the checklist.

Total Quality Newsletter relates the story of how Milliken Company turned its managers into good listeners with very positive results.[3] In 1979 CEO Roger Milliken had called a meeting of his company's 300 managers. One risk-taking manager stuck his neck out and told Milliken that of the 300 managers in attendance, only five were effective listeners. The CEO responded by asking all managers to take a pledge to become good listeners.

Many of the managers thought they were good listeners, but Milliken convinced them they needed to improve. According to Newt Hardie, Milliken's vice president for quality, "That was a moment of revelation, and the beginning of important cultural change for the company. . . . If you don't listen well, you can't involve people. . . . If you don't listen, they know you think they're unimportant."[4]

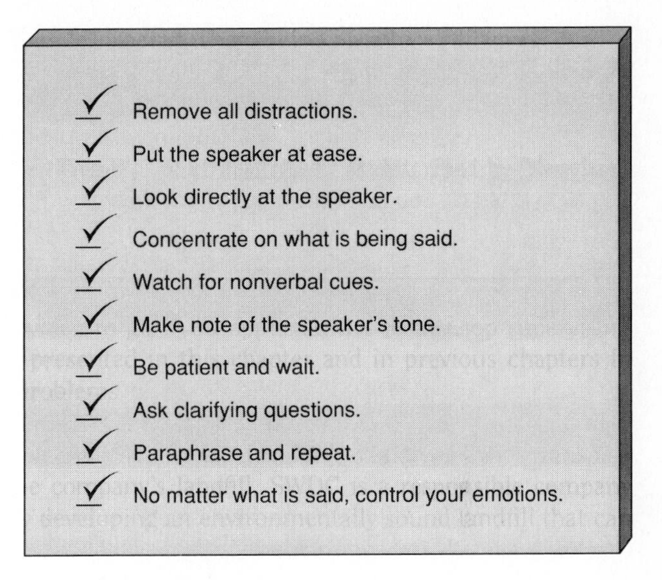

Figure 4–8
Checklist of strategies for effective listening.

COMMUNICATING NONVERBALLY

Nonverbal messages represent one of the least understood, but most powerful modes of communication. Often nonverbal messages are more honest and telling than verbal messages provided the receiver is attentive and able to read nonverbal cues.

It has become popular to call nonverbal communication body language. However, body language is only part of nonverbal communication. There are actually three components (Figure 4–9):

- Body factors
- Voice factors
- Proximity factors

Communications consultant Roger Ailes explains the importance of nonverbal communication as follows:

> You've got just seven seconds to make the right first impression. . . . you broadcast verbal and nonverbal signals that determine how others see you. . . . And whether people realize it or not, they respond immediately to your facial expressions, gestures, stance, and energy, and they instinctively size up your motives and attitudes.[5]

Body Factors

A person's posture, body poses, facial expressions, gestures, and dress can convey a message. Even such extras as makeup or the lack of it, well-groomed or unkempt hair,

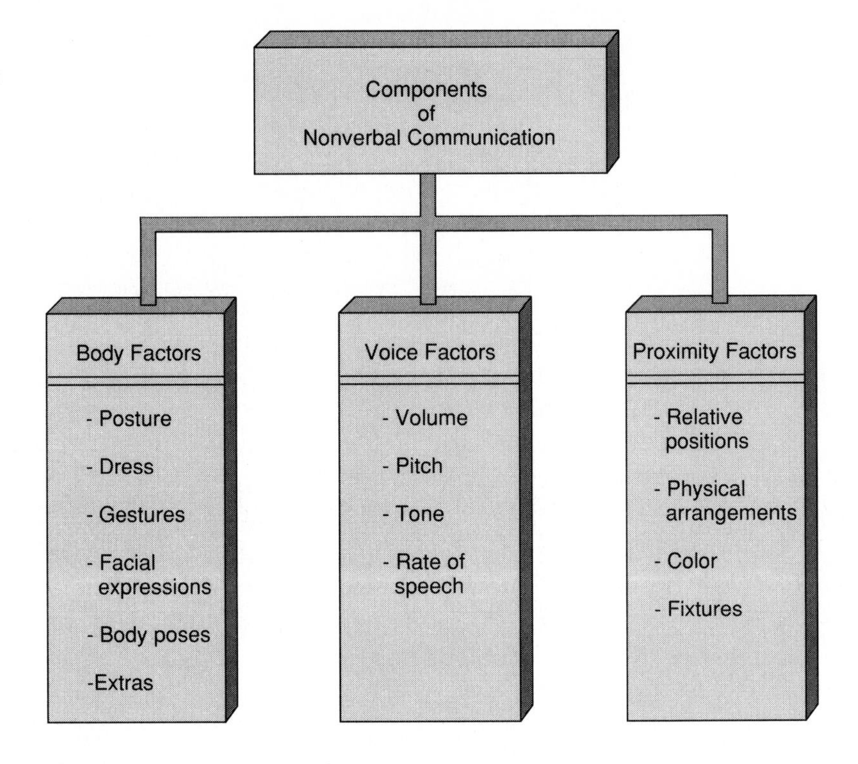

Figure 4–9
Nonverbal communication encompasses more than just body language.

and shined or scruffy shoes can convey a message. Supervisors should be attentive to these **body factors** and how they add to or distract from the verbal message.

One of the keys to understanding nonverbal cues lies in the concept of **congruence.** Are the spoken message and the nonverbal message congruent? They should be. To illustrate this point, consider the hypothetical example of Chem-Tech Company. An important element of the company's corporate culture is attractive, conservative dress. This is especially important for Chem-Tech's sales force. For the men, white shirts, dark suits, and shined shoes are the expected norm.

John McNamara is an effective sales representative, but lately he has taken to flashy dressing. He wears loud sports coats, open-neck print shirts, and casual shoes. When questioned by his supervisor, John said he understands the dress code and agrees with it. This is **incongruence.** His verbal message says one thing, but his nonverbal message says another. This is an exaggerated example. Incongruence is not always so obvious. A simple facial expression or a casual gesture can be an indicator of incongruence.

When verbal and nonverbal messages are not congruent, supervisors should take the time to dig a little deeper. An effective way to deal with incongruence is to gently confront it. A simple statement such as, "Cindy, your words agree, but your eyes disagree" can help draw an employee out so the supervisor gets the real message.

Voice Factors

Voice factors are also an important part of nonverbal communication. In addition to listening to the words, supervisors should listen for such factors as volume, tone, pitch, and rate of speech. These factors can portray anger, fear, impatience, unsureness, interest, acceptance, confidence, and a variety of other messages.

As with body factors, it is important to look for congruence. It is also advisable to look for groups of nonverbal cues. Supervisors can be misled by attaching too much meaning to isolated nonverbal cues. A single cue taken out of context has little meaning. But as a part of a group of cues, it can take on significant meaning.

For example, if you look through the office window and see a person leaning over a desk pounding his fist on it, it would be tempting to interpret this as a gesture of anger. But what kind of look does he have on his face? Is his facial expression congruent with desk-pounding anger? Or could he simply be trying to knock loose a desk drawer that has become stuck? On the other hand, if you saw him pounding the desk with a frown on his face and heard him yelling in an agitated tone, your assumption of anger would be well based. He might just be angry because his desk drawer is stuck. But, nonetheless, he would still be angry.

Proximity Factors

Proximity factors range from where you position yourself when talking with an employee to how your office is arranged, to the color of the walls, to the types of fixtures and decorations. A supervisor who sits next to an employee conveys a different message than one who sits across a desk from the employee. A supervisor who goes to the trouble to make his or her office a comfortable place to visit is sending a message that invites communication. A supervisor who maintains a stark, impersonal office sends the opposite message.

Supervisors who want to send the nonverbal message that employees are welcome to stop and talk should consider the following guidelines:

- Have comfortable chairs available for visitors.
- Arrange chairs so you can sit beside visitors rather than behind the desk.
- Choose soft, soothing colors rather than harsh, stark, or overly bright or busy colors.
- If possible, have refreshments such as coffee, soda, and snacks available for visitors.

COMMUNICATING VERBALLY

Verbal communication ranks close to listening in its importance to the modern supervisor. Supervisors can improve their verbal communication skills by being attentive to the following factors (Figure 4–10):

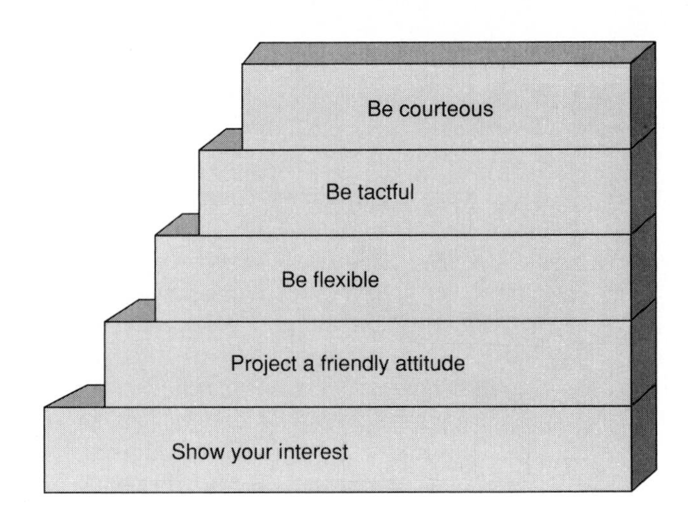

Figure 4–10
Steps to improving your verbal communication skills.

Interest

When speaking with employees, show an interest in your topic. Show that you are sincerely interested in communicating your message to them. Also show interest in the receivers of the message. Look them in the eye and, when in a group, spread your eye contact evenly among all receivers.

Attitude

A positive, friendly attitude will enhance verbal communication. A caustic, superior, condescending, or disinterested attitude will shut off communication. So will an argumentative attitude. Be patient, be friendly, and smile.

Flexibility

Be flexible. Flexibility can enhance verbal communication. For example, if a supervisor calls his team together to explain a new company policy, but finds they are uniformly focused on a problem that is disrupting their work schedule, he must be flexible enough to put his message aside for the time being and deal with the problem. Until the employees work through what is on their minds, they will not make good listeners.

Tact

Tact is an important ingredient in verbal communication, particularly when delivering a sensitive or potentially controversial message. Tact has been called the ability to hammer in the nail without breaking the board. The key to tactful verbal communication lies in thinking before talking.

Courtesy

Courtesy promotes effective verbal communication. Being courteous means showing appropriate concern for the needs of the receiver. For example, calling a team meeting ten minutes before quitting time is discourteous and will inhibit communication. Most workers have after-work obligations. A meeting called ten minutes before the end of the workday is likely to conflict with these obligations. Courtesy also means not monopolizing. When communicating verbally, give the receiver ample opportunities to ask questions, to seek clarification, and to state his or her point of view.

In addition to applying the preceding factors, supervisors should learn to be skilled questioners. Knowing how and when to question is an important verbal communication skill. It is how supervisors can find out what employees really think and feel. What follows are some general rules of questioning professional counselors use to draw out the feelings and thoughts of their clients. Modern supervisors can apply these rules to enhance their verbal communication with employees (Figure 4–11).

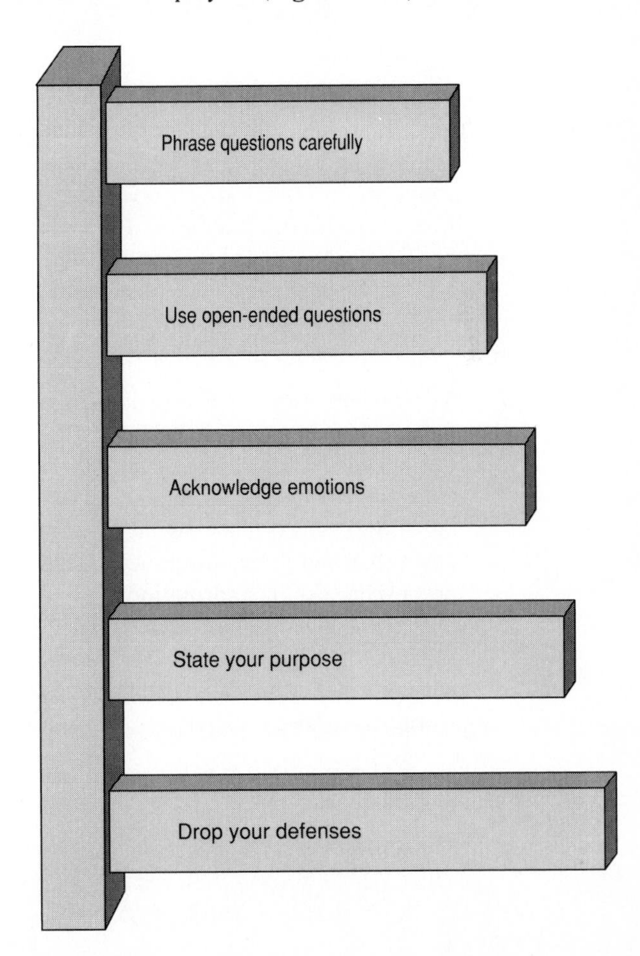

Figure 4–11
Strategies for effective questioning.

Drop Your Defenses

Human interaction is emotional interaction. There is no such thing as fully objective discourse between people. We all have our public and private faces and it is rare when what we say matches completely what we feel. People learn early in life to build walls and put up defenses. In order to communicate effectively, it is necessary to get behind the walls and break through the defenses. A strategy counselors use is to drop their defenses first. When employees see you open up, they will be more likely to follow suit and respond more honestly to your questions.

State Your Purpose

The silent question people often ask themselves when you ask a question is, "Why is he asking that? What does he really want?" You will learn more from your questions by stating your purpose at the outset. This will allow the receiver to focus on your question rather than on a perceived hidden meaning.

Acknowledge Emotions

Avoid what counselors call the "elephant in the living room" syndrome when questioning employees. Human emotions can be difficult to deal with. As a result, some people respond by simply ignoring them. This is like walking around an elephant in the living room and pretending you don't see it. Ignoring the emotions of people you question may cause them to close up. If a person shows anger, you might respond by saying, "I see I've made you angry," or "You seem to feel strongly about this." Such nonjudgmental acknowledgments will usually draw a person out.

Use Open-Ended Questions and Phrase Questions Carefully

To learn the most from your questions, make them open-ended. This allows the person being questioned to do most of the talking and lets you do most of the listening. Counselors feel they learn more when listening than when talking. Closed-ended questions force restricted or limited responses. For example, the question, "Can we meet our deadline?" will probably elicit a yes or no response. However, the question, "What do you think about this deadline?" gives the responder room to offer opinions and other potentially useful information.

Practice using these questioning techniques at home, at work, and even in social settings. It will take practice to internalize them so that they become natural. However, with practice supervisors can become skilled questioners and, as a result, more effective communicators.

COMMUNICATING IN WRITING

The ability to communicate effectively in writing is important for modern supervisors. The types of writing required of supervisors can be mastered, like any other skill, with the appropriate mixture of coaching, practice, and genuine effort to improve. This section provides the coaching. Prospective supervisors must provide the practice and effort.

Several rules of thumb that can enhance the effectiveness of your written communication follow (Figure 4–12):

1. Plan before you write.
2. Be brief.
3. Be direct.
4. Be accurate.
5. Practice self-editing.

Planning

One of the reasons some people have trouble writing effectively is that they start writing before deciding what they want to say. This is like getting in a car and driving before deciding where to go. The route is sure to be confusing, as will the message if you write without first planning.

Planning a memorandum, letter, report, notification, or any other written document is a simple process. It is a matter of deciding who you are writing to, why, and what you want to say. Figure 4–13 is a planning sheet that will help supervisors plan before they write.

Taking the time to complete a planning sheet such as the one in Figure 4–13 can help ensure that the intended message is communicated. Once you have become used to filling out a planning sheet, you will be able to complete this step mentally.

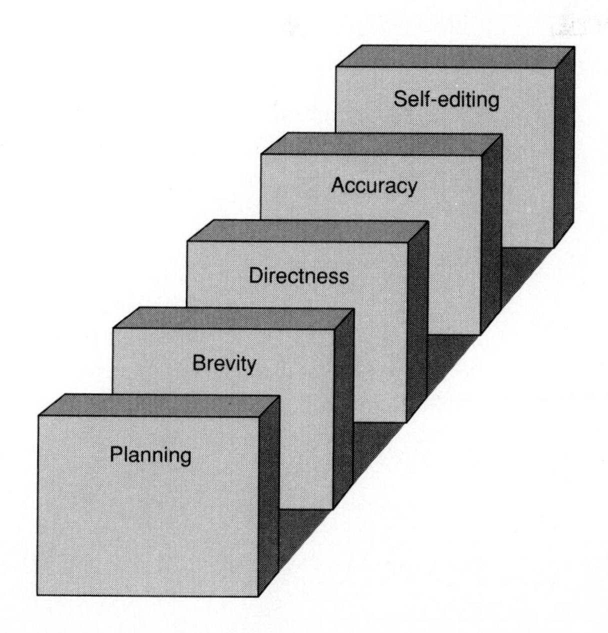

Figure 4–12
Enhancing the effectiveness of your written communication.

Figure 4–13
Plan before you write. It will enhance your effectiveness.

Planning Sheet

1. I am writing to...
____
____

2. My purpose in writing is to...
____
____

3. I want to make the following points...
a.______________________
b.______________________
c.______________________
d.______________________
e.______________________

4. I want the reader to do the following...
a.______________________
b.______________________
c.______________________
d.______________________
e.______________________

Brevity

One of the by-products of modern technology is the potential for information overload. This conflicts with another by-product, the tendency toward shortened attention spans. Modern computer and telecommunication technology have conditioned us to expect instant information with little or no effort on our part.

The various forms of written communication can run counter to our expectation of instant information with no effort. Reading takes time and effort. Keep **brevity** in mind when writing. In as few words as possible, explain your purpose, state your points, and tell readers what you want them to do.

Directness

Directness is an extension of brevity. It means getting to the point without beating around the bush. This is especially important when the message is one readers will not particularly like. No purpose is served by obscuring the message.

Accuracy

Accuracy is important in written communication. Be exact. Avoid such terms as *some time ago, approximately,* and *as soon as possible.* Take the time to identify specific dates, numbers, quantities, and so on. Then double-check to make sure they are accurate.

Self-Editing

A first-draft writer is rare. But people who *send* their first draft are very common. This can cause you to overlook errors that are embarrassing and obscure or confuse the message.

Practice **self-editing.** In your first draft, concentrate on *what* you are saying. In the second draft, concentrate on *how* you say it. These are two different processes that should not be mixed. Even professional writers cannot effectively edit for content and for grammar, sentence structure, and spelling simultaneously.

Figure 4–14 is an example of a memorandum written by a supervisor who did not follow the rules of thumb listed in Figure 4–12. Figure 4–15 is an example of a simpler memorandum from another supervisor. Both are trying to convey the same message.

The notice in Figure 4–14 was not planned. It rambles, goes on too long, is not direct, contains obscure, inaccurate terms and has not been self-edited. Since it is not dated, there is no way for readers to determine when the meeting will be held. The notice in Figure 4–15 was well planned. The plan for this memorandum would read as follows:

1. Writing to all members of Fabrication Team B.
2. The purpose is to notify them of a meeting to set production goals for next year.
3. Make the following points:
   a. Date: April 10, 1990
   b. Time: 8:00 a.m.
   c. Place: Conference Room B

Figure 4–14
Sample team notice.

Notice to All Members
of Fabrication Team A

As you know, a week or two ago we talked about getting production goals for next year and how important it is to set goals if you ever want to improve and I know you do just as I do. the time has come to decide what our goals will be next year as the new production year begins in a month or so. I thought we should get together and talk about what we want our goals to be so let's meet in the Conference Room A. Please be sure to arrive on time. I think it might take all day to complete the goal setting process so maybe you should bring a lunch with you on this day so we don't have to interrupt our thoughts to go to lunch. The meeting will start at 8:00 a.m. It will be helpful if you will have already decided on your proposed goals when you get there so we can get a head start on the process. the meeting will be next Tuesday. I am looking forward to meeting with you so please arrive on time.

John J. Supervisor

Posted April 6, 1990

Notice to All Members
of Fabrication Team B

The current production year ends June 30, 1990. It is time to set our production goals for next year. Please plan to meet with me:

Date:	**April 10, 1990**
Time:	**8:00 a.m.**
Place:	**Conference Room B**

Special Instructions:
1. Bring your proposed goals with you.
2. Bring a "brown bag" lunch so we can work through if necessary.
3. Please be prompt. Arrive on time!

Thank you.

Jo Anne P. Supervisor

Figure 4–15
Sample team notice.

4. I want the readers to:
   a. Bring their proposed goals.
   b. Bring a brown bag lunch.
   c. Arrive on time.

The example in Figure 4–15 followed this plan, is brief, direct, accurate, and obviously was self-edited before it was distributed. By taking the time to go through this process, the supervisor who wrote the notification in Figure 4–15 has produced a written document that is easy to understand and will communicate the message.

COMMUNICATING CORRECTIVE FEEDBACK

In dealing with employees it is important for supervisors to give **corrective feedback.** However, in order to be effective, corrective feedback must be communicated properly. Robert Luke offers the following guidelines supervisors can use for enhancing the effectiveness of their corrective feedback.[6]

- *Be positive.* To be corrective, feedback must be accepted and acted on by the employee. This is more likely to happen if it is delivered in a positive manner.
- *Be prepared.* Focus all feedback specifically on the behavior. Do not discuss personality traits. Give specific examples of the behavior you would like to see corrected.

- *Be realistic.* Make sure the behaviors you want to change are controlled by the employee. Don't expect an employee to correct a behavior he or she cannot control.
- *Don't be completely negative.* Find something positive to say. Give the employee the necessary corrective feedback, but don't focus wholly on the negative.

Luke also recommends the following options for giving corrective feedback:[7]

- *Tell-ask-listen.* Tell the employee about the behavior, ask for his or her input, and listen to that input. According to Luke, "The session will have been a success if the employee leaves feeling the issues are important, is appreciative of your input, and is committed to correcting the problems."[8]
- *Listen-ask-tell.* Listen first. It may be necessary to ask an open-ended, general question such as, "How are things going with your job?" to get the ball rolling. Once the employee starts talking, listen. If the employee adequately covers the area in which corrective action is needed, reinforce his or her comments. Ask the employee what he or she thinks can be done to improve. If the employee does not appear to be fully aware of the problem, you may give your corrective feedback (tell). According to Luke, "By communicating, you resolve more problems and eliminate the possibility of hard feelings and mixed messages."[9]

STEPS TO IMPROVED COMMUNICATION

Consider the words of Kim McKinnon, manager of personnel development at the Santa Barbara Research Center: "Effective communication is something that all supervisors should master. If each supervisor worked at having some basic skills, interpersonal communication would flow more smoothly among supervisors and their employees."[10]

McKinnon recommends the following strategies supervisors can use for improving their communications skills:[11]

- *Obtain new information.* Make an effort to stay up to date with new information relating to the workplace. You cannot communicate what you don't know.
- *Prioritize and determine time constraints.* Communicating does not mean simply passing on everything you learn to your employees. Such an approach might overload them and thereby inhibit communication. Analyze your information and decide what your employees need to know. Then prioritize it from *urgent* to *when time permits* and share the information accordingly.
- *Decide whom to inform.* Once your information has been prioritized, decide who needs to have it. Employees have enough to keep up with without receiving information they don't need. Correspondingly, don't withhold information that employees can use. Achieving the right balance will enhance your communication.
- *Determine how to communicate.* There are a variety of different ways to communicate (i.e., verbal, written, one-on-one, in groups, etc.). A combination of methods will probably be most effective. The next section in this chapter deals with this issue in more depth.

- *Communicate the information.* Don't just tell your employees what you want them to know or write them a memorandum. Follow up. Ask questions to determine if they have really gotten the message. Encourage them to ask you questions for clarification. Agree on the next steps (i.e., what they should do with the information).
- *Check accuracy and get feedback.* Check to see that your communication was accurate. Is the employee undertaking the next steps as agreed? Get feedback from employees to ensure that their understanding has not changed and that progress is being made.

Selecting the Appropriate Communication Method

One of the six steps to improved communication recommended in the previous section was "determine how to communicate." Since most workplace communication is either verbal or written, supervisors need to know when each method is the most and least effective. According to D. A. Level, written communication is most effective for communicating general information and for information requiring action on the part of the employee.[12] For example, general information such as new company policies or announcements of activities that carry dates, times, places, or other specific information are appropriate for communicating in writing. A message that says, "Please bring your automobile registration to work no later than noon Friday if you want to have a parking sticker," is appropriately communicated in writing. Verbal communication is appropriate when reprimanding employees or attempting to resolve conflict between employees. In these cases, verbal interaction in private is the best approach.

Written communication is least effective in the following instances:

- To communicate a message requiring immediate action on the part of employees. The more appropriate approach in such a case is to communicate the message verbally and follow it up in writing.
- To commend an employee for doing a good job. This should be done verbally and publicly, then followed up in writing.
- To reprimand an employee for poor performance. This message can be communicated more effectively if given verbally in private. This is particularly true for occasional offenses.
- To resolve conflict among employees about work-related problems. The necessary communication in such instances is more effectively given verbally and in private.

Verbal communication is least effective in the following instances:

- To communicate a message requiring future action on the part of employees. Such messages are more effectively communicated when given verbally and followed up in writing.
- To communicate general information such as company policies, personnel information, directives, or orders.
- To communicate work progress to an immediate supervisor or higher manager.
- To promote a safety campaign.

By following the guidelines set forth in this section, supervisors can choose a method of communication that is appropriate to the situation at hand, thereby enhancing the effectiveness of their communication.

ELECTRONIC COMMUNICATION[13]

In the age of high technology, **electronic communication** has become an important means of sending written messages. Electronic mail or E-mail consists of written messages transferred electronically from computer to computer. Electronic communication is doing for written communication what the telephone did for verbal communication.

- *Messages can be transmitted rapidly.* For example, consider the time it might take to transmit a letter to a supplier in another city. Even if overnight express services are used, it will take as much as twenty-four hours. The same letter sent electronically would be received in a matter of seconds.
- *Messages can be transmitted simultaneously to more than one person.* This is particularly helpful when the same notification must be sent to a large number of people. The sender inputs the message and enters the codes of all who are to receive it. The message is sent instantly.
- *Messages can be printed if a hard copy is needed.* A hard copy is a paper printout of the information shown on the screen of a computer terminal or personal computer. Electronic messages can be printed if a hard copy is needed.
- *Messages can be stored for future reference.* The computer can serve as an electronic filing cabinet for storing messages that have been sent if they must be referred to later.
- *Messages can be prompted and acknowledged electronically.* Recipients of electronic messages can acknowledge their receipt by simply pressing a key. This allows the sender to know not just that the message was received, but when. Recipients can also be prompted. This means a message, light, or some other type of visual prompt can inform the receiver that a message is waiting. Typically, the prompt cannot be cleared until the message is accessed.

Just as there are advantages with electronic communication, there are also potential disadvantages. The first disadvantage is one inherent in any form of written communication: inability to transmit body language, voice tone, facial expressions, and eye contact. All of these supplementers of verbal communication are missing with electronic communication.

Another disadvantage is the potential for overuse of electronic communication. Because of the ease of sending a written message by simply pressing a key, users may send more messages than they really need to, send frivolous messages, or send messages that could be more appropriately delivered verbally.

Figure 4–16 is an example of a main menu for an electronic mail system. Notice that it allows users to view a listing of their messages, read specific messages, print

Figure 4–16
Hypothetical main menu for an E-mail system.

Electronic Mail System MAIN MENU	
Function	Key
LIST of your messages	L
READ your messages	R
PRINT your messages	P
CREATE a message	C
HELP	H
EXIT	E

specific messages, create a new message, access instructions on how to use the system (HELP), and exit the main menu. The READ option would take users into a submenu that allows them to read, acknowledge, erase, and/or store messages. The PRINT option allows users to print a specific message without having to first access and read it. The CREATE option allows users to write and send a message and store it if they wish to.

SUMMARY

1. Communication is the transfer of information that is understood from one source to another. Effective communication goes beyond just understanding. It means the message is also acted on in the desired manner. Effective communication requires persuasion, motivation, monitoring, and leadership.
2. Communication involves four components: sender, receiver, medium, and message.
3. Communication is inhibited by: differences in meaning, insufficient trust, information overload, interference, condescending tone, listening problems, premature judgments, and inaccurate assumptions.
4. In an industrial firm communication takes place through networks. A network is any group of senders linked together by some means with a group of receivers. Networks may be formal or informal.
5. Listening is the supervisor's most important communication skill. Listening is receiving the message, correctly decoding it, and accurately perceiving what it means.
6. Nonverbal communication is often more honest and telling than any other form. Supervisors should be aware of body, voice, and proximity factors. It is also important to watch for congruence between verbal and nonverbal cues.
7. Verbal communication ranks close to listening in importance. Supervisors can enhance the effectiveness of their verbal communication by showing interest, projecting a friendly attitude, being flexible, using tact, and being courteous.

8. Supervisors can improve their written communication by planning before writing, being brief, being direct, being accurate, and practicing self-editing.
9. There are ten classes of responses that impede communication. These classes represent various types of evasiveness and vagueness that supervisors should avoid and should encourage their team members to avoid.
10. To enhance the effectiveness of their corrective feedback, supervisors should be positive, be prepared, be realistic, and avoid being completely negative.
11. Electronic communication offers advantages and disadvantages. Advantages include: they can be transmitted rapidly, they can be transmitted simultaneously to a large number of people, and they can be prompted, acknowledged, printed, and stored. Disadvantages include the potential for overuse and the inability to send nonverbal cues electronically.

KEY TERMS AND PHRASES

Accuracy
Body factors
Brevity
Classes of communication
Communication
Community level communication
Company level communication
Congruence
Corrective feedback
Directness
Effective communication
Electronic communication
Grapevine
Inaccurate assumptions
Incongruence
Information overload
Interference
Listening
Medium
Message
Network
Nonverbal communication
One-on-one level communication
Planning
Preconceived ideas
Proximity factors
Receiver
Self-editing
Sender
Tact
Team or unit level communication
Voice factors
Verbal communication
Written communication

CASE STUDY: TGIF and Communication[14]

The following real-world case study concerns a communication-related problem of the type that might be encountered by modern supervisors. Read the case study and answer the accompanying questions.

MeraBank was experiencing a variety of problems: high turnover, employee conflict, low morale, and low productivity, to name just a few. Clearly something had to be done fast, but what? Executives at MeraBank were anxious to find solutions so they decided to ask all employees, "What do you think will improve conditions in the organization?" But they didn't just ask, they threw a party.

In order to solicit employees' input more effectively, MeraBank's management team hosted a TGIF (Thank God It's Friday) party on company time and invited all employees. The cost of admission was one idea for improving conditions at work. Ideas could be serious, frivolous, or in between as long as they provided some degree of insight into how the company could improve conditions.

When the suggestions had been summarized and analyzed they showed a clear pattern. Employees felt that the company needed to improve communication at all levels. Some of the suggestions submitted included the following:

- Publish a newsletter for employees as a way to keep them informed.
- Involve employees in more activities and decisions relating to their jobs.
- Provide communications training for all employees.
- Consider communications skills on employee evaluations.

This approach to improving communication proved very effective. Two years after the TGIF party idea was implemented, conditions at MeraBank had improved markedly. Absenteeism was down and morale was up. Communication and teamwork had become the norm.

1. What level or levels of communication do you think took place at the TGIF party?
2. Evaluate the use of a party as a means to improve communication, especially with regard to overcoming typical inhibitors of communication.
3. If you were a supervisor at MeraBank, how would you build on the TGIF party idea to further improve communication?

REVIEW QUESTIONS

1. Define the term *communication.*
2. Distinguish between communication and effective communication.
3. What are the four levels of communication in a company?
4. Explain the communication process.
5. What are the three basic categories of communication media?
6. List five communication inhibitors.
7. What is a network? Give two examples.
8. Define the term *listening.*
9. List five inhibitors of effective listening.
10. What are the three components of nonverbal communication?
11. Explain the concept of congruence.
12. List five factors that can improve a supervisor's verbal communication.
13. List five rules of thumb that can improve a supervisor's verbal communication.
14. List five rules of thumb that can improve a supervisor's written communication.

15. Explain the importance of self-editing.
16. Explain how to be more effective in communicating corrective feedback.
17. What are the advantages and disadvantages of electronic communication?

SIMULATION ACTIVITIES

The following activities are provided to give you opportunities to develop supervisory skills by applying the material presented in this chapter and in previous chapters in solving simulated supervisory problems.

1. John Borg is a supervisor with the Citrus Bottling Company. His unit is responsible for bottling a minimum of 1,500 containers of grapefruit juice, orange juice, and various other juice mixtures. John sees himself as a man on the fast track. He intends to break all productivity records at Citrus Bottling Company and, in so doing, ensure a promotion for himself and the top producers in his unit. He knows all his team members will have to accept the challenge if he is to succeed. John knows what he wants, but does not know how to proceed. If he came to you for help in outlining a strategy, what would you advise?
2. Jo Ann Schembera is a new supervisor for the Pro Sports Athletic Wear Company. Pro Sports is a leading producer of uniforms for athletic teams. Since she was hired from outside, Jo Ann will have to win the trust and confidence of hew new team members. She wants to begin by having a face-to-face, one-on-one conference with each team member. Jo Ann wants to get to know her team members and she wants them to know her. She became a supervisor because of her technical skills, but has never had to be a good communicator. What advice can you give to Jo Ann?
3. Mike Durham never had to do much writing as a technician for Phenolic Plastics Corporation. But now, as a new supervisor, he will have to write letters and memos frequently. Mike wants to send a memo to his boss summarizing the situation he has inherited and what he intends to do about it. Mike has come to you for help. Find a partner to play the role of Mike. Have Mike explain his problems and what he intends to do about them. Using this information, develop a planning sheet that will help Mike write an effective memo.

ENDNOTES

1. Rossnagel, W. B. "The Ten Classes of Communication," *Supervisory Management,* April 1990, p. 4.
2. Ibid., p. 4.
3. "Listening is the Key to Winning Workers' Hearts and Minds," *Total Quality Newsletter,* December 1990, p. 6.
4. Ibid.
5. Ailes, R. "The Seven-Second Solution," *Management Digest,* January 1990, p. 3.
6. Luke, R. A. "How to Give Corrective Feedback to Employees," *Supervisory Management,* March 1990, p. 7.

7. Ibid.
8. Ibid.
9. Ibid.
10. McKinnon, K. "Six Steps to Improved Communication," *Supervisory Management,* February 1990, p. 9.
11. Ibid.
12. Level, D. A. "Communication and Situation," *Journal of Business Communication,* 1972, Number 9, pp. 19–25.
13. Huseman, R. C. and Miles, E. W. "Organizational Communication in the Information Age: Implications of Computer-Based Systems," *Journal of Management,* 1988, Volume 14, Number 2, pp. 181–204.
14. Roarty, C. J. "A Party Crashes the Communication Barrier," *Personnel Administrator,* 1989, Volume 32, Number 11, pp. 66–69.

CHAPTER FIVE

Ethics and the Supervisor

CHAPTER OBJECTIVES

After studying this chapter, you will be able to define or explain the following topics:

- Ethics
- Ethical Behavior in Organizations
- The Supervisor's Role in Ethics
- The Company's Role in Ethics
- How to Handle Ethical Dilemmas

CHAPTER OUTLINE

- An Ethical Dilemma
- Ethics Defined
- Ethical Behavior in Organizations
- The Supervisor's Role in Ethics
- The Company's Role in Ethics
- Handling Ethical Dilemmas
- Summary
- Key Terms and Phrases
- Case Study Application Problem
- Review Questions
- Simulation Activities
- Endnotes

There is almost universal agreement that the business practices of industrial firms should be above reproach with regard to ethical standards. Few people are willing to defend unethical behavior and, for the most part, industry in the United States operates within the scope of accepted legal and ethical standards.

According to Peter Drucker, "Business Ethics is rapidly becoming the 'in' subject. . . . there are countless seminars on it, speeches, articles, conferences and books, not to mention the many earnest attempts to write 'business ethics' into the law."[1]

However, unethical behavior does occur frequently enough that modern supervisors should be aware of the types of ethical dilemmas they will occasionally face. They should also know how to deal with these dilemmas. This chapter is designed to help prepare the supervisor to deal successfully and effectively with ethics on the job.

AN ETHICAL DILEMMA

According to Edward Stead, Dan Worrell, and Jean Stead,

> Managing ethical behavior is one of the most pervasive and complex problems facing business organizations today. Employees' decisions to behave ethically or unethically are influenced by a myriad of individual and situational factors. Background, personality, decision history, managerial philosophy, and reinforcement are but a few of the factors which have been identified by researchers as determinants of employees' behavior when faced with ethical dilemmas.[2]

Tom Richards is supervisor of the metal cutting shop for Alloy Tech Corporation. He is an effective supervisor and the metal cutting shop is an efficient, productive operation. But Tom has a problem. Alloy Tech is losing market share to Morton Metal, Inc., its chief competitor down the street. His manager has made it clear to Tom that the metal cutting shop is to blame and he expects the situation to be corrected quickly.

Tom knows the source of his problem, and well he should; he trained him. Jake Bronkowski is the secret to Morton Metal's recent success in the marketplace. He is the most talented CNC machinist Tom Richards has ever known. He represents the quintessential example of the student who surpasses the teacher. Not only does he know process planning, resources planning, setup, tooling, and programming, he can retrofit manual machines for CNC operation, troubleshoot inoperable machines, and repair both the electronic and mechanical components of machines that are down. In short, Jake Bronkowski is a one-person machine shop.

Since he was lured away by Morton Metals just six months ago, Alloy Tech has seen its market share decline steadily and that of Morton Metals increase correspondingly. In desperation, Tom has talked with Jake in an attempt to get him back. This is where Tom's problem became an ethical dilemma. Jake Bronkowski is willing to return to Alloy Tech. In fact, he would like to. However, he knows his value on the open market. He knows that several companies would like to hire him. As a result, Jake has made some demands that Tom feels are not just exorbitant, but unethical.

Jake wants a wage and benefits package that exceeds by far the salary schedule approved by Alloy Tech's management. He wants his own office and a four-day work week. These demands are clearly out of line, but Tom is desperate to grant them. In fact, Tom had already received his boss' approval to meet all of Jake's demands and put him to work when Jake made his final demand. It is this demand that is causing the crisis of conscience that Tom is currently experiencing.

Jake's final demand is that Tom move Alice McCormick, a machine operator at Alloy Tech, out of the metal cutting shop. Alice was a secretary who became a machine operator through the company's upward mobility program. She went to night school, completed the requisite training, and started work as a trainee in the metal cutting shop. Alice loves her job and has realized a significant pay increase with it. At this point in her development, Alice is just an average machine operator but she works hard and has the potential to develop into a good technician over time.

Jake and Alice had had a personal relationship when Jake worked at Alloy Tech. But the relationship did not last and Jake is still bitter. As a result, Tom Richards faces an ethical dilemma not unlike those faced by supervisors every day somewhere in this country. He is caught between the demands of his company to protect the bottom line and the demands of morality to behave ethically.

What should Tom Richards do? Is there an acceptable compromise here? Does the end justify the means in these cases? These are the types of questions modern supervisors will face on the job as they deal with such concerns as privacy, hiring, firing, promotions, performance appraisal, working conditions, product quality, production quotas, and other people-oriented concerns.

ETHICS DEFINED

There are many definitions of the term *ethics*. However, no one definition has emerged as universally accepted. According to Paul Taylor, the concept can be defined as "inquiry into the nature and grounds of morality where morality is taken to mean moral judgments, standards, and rules of conduct."[3] According to Peter Arlow and Thomas Ulrich, ethical dilemmas in the workplace are more complex than ethical situations in general.[4] They involve societal expectations, competition, and social responsibility as well as the potential consequences of an employee's behavior to customers, fellow workers, competitors, and the public. The result of the often conflicting and contradictory interests of workers, customers, competitors, and the general public is a natural tendency for ethical dilemmas to occur frequently in the workplace.

Any time ethics is the topic of discussion, such terms as *conscience, morality*, and *legality* will be frequently heard. Although these terms are closely associated with ethics, they do not by themselves define it. For the purposes of this book, **ethics** will be defined as follows:

Ethics is the study of morality.

Morality refers to the **values** that are subscribed to and fostered by society in general and individuals within society. Ethics attempts to apply reason in determining rules of human conduct that translate morality into everyday behavior. **Ethical behavior** is that which falls within the limits prescribed by morality.

How, then, does the supervisor know if someone's behavior is ethical? Ethical questions are rarely black and white. They typically fall into a **gray area** between the two extremes of the continuum. And this gray area is often clouded further by personal experience, self-interest, point of view, and external pressure.

Guidelines for Determining Ethical Behavior

Before presenting the guidelines supervisors can use to sort out matters that fall into the gray area between clearly right and clearly wrong, it is necessary to distinguish between what is **legal** and what is *ethical.* They are not necessarily the same thing.

In fact, it is not uncommon for people caught in the practice of questionable behavior to use the "I didn't do anything illegal" defense. A person's behavior can be well within the scope of the law and still be unethical. The following guidelines for determining ethical behavior assume the behavior in question is legal (Figure 5–1):

- Apply the **morning-after test.** If you make this choice, how will you feel about it tomorrow morning?
- Apply the **front page test.** Make a decision that would not embarrass you if printed as a story on the front page of your hometown newspaper.
- Apply the **mirror test.** If you make this decision, how will you feel about yourself when you look in the mirror?
- Apply the **role reversal test.** Trade places with the people affected by your decision and view the decision through their eyes.
- Apply the **common sense test.** Listen to what your instincts and common sense are telling you. If it feels wrong, it probably is.

Kenneth Blanchard and Norman Vincent Peale suggest their own test for determining the ethical choice in a given situation.[5] Their test consists of the following three questions:

1. Is it legal?
2. Is it balanced?
3. How will it make me feel about myself?

Naturally, if a potential course of action is not legal, no further consideration of it is in order. If an action is not legal, it is also not ethical. A course of action that is balanced will be fair to all involved. This means supervisors and their team members have responsibilities that extend well beyond the walls of their unit, organization, and company. A course of action that will leave you feeling good about yourself is one that is in keeping with your own moral structure.

Figure 5–1
Guidelines supervisors can use to determine what is ethical.

Guidelines for Ethical Choices
1. Apply the morning after test
2. Apply the front page test
3. Apply the mirror test
4. Apply the role reversal test
5. Apply the common sense test

Blanchard and Peale also discuss what they call "The Five P's of Ethical Power."[6] Their "Five P's" can be summarized as follows:

- *Purpose:* Individuals see themselves as ethical people who let their conscience be their guide and in all cases want to feel good about themselves.
- *Pride:* Individuals apply internal guidelines and have sufficient self-esteem to make decisions that may not be popular with others.
- *Patience:* Individuals believe right will prevail in the long run and they are willing to wait when necessary.
- *Persistence:* Individuals are willing to stay with an ethical course of action once it has been chosen and see it through to a positive conclusion.
- *Perspective:* Individuals take the time to reflect and are guided by their own internal barometer when making ethical decisions.

These tests and guidelines will help supervisors and their team members make ethical choices in the workplace. In addition to internalizing the guidelines themselves, supervisors may want to share them with the employees they supervise.

ETHICAL BEHAVIOR IN ORGANIZATIONS

Research by L. K. Trevino suggests that ethical behavior in organizations is influenced by both individual and social factors.[7] Trevino identified three personality measures that can influence an employee's ethical behavior: (1) ego strength; (2) Machiavellianism; and (3) locus of control.

An employee's **ego strength** is his or her ability to undertake self-directed tasks and to cope with tense situations. A measure of a worker's **Machiavellianism** is the extent to which he or she will attempt to deceive and confuse others. **Locus of control** refers to workers' perspective concerning who or what controls their behavior. Employees with an internal locus of control feel they control their own behavior. Employees with an external locus of control feel their behavior is controlled by external factors (i.e., rules, regulations, their supervisor, etc.).

J. F. Preble and P. Miesing suggest that social factors also influence ethical behavior in organizations.[8] These factors include gender, role differences, religion, age, work experience, nationality, and the influence of other people who are significant in an individual's life. F. Luthans and R. Kreitner suggest that people learn appropriate behavior by observing the behavior of significant role models (i.e., parents, teachers, public officials, etc).[9] Since supervisors represent perhaps the most significant role model for their team members, it is critical that they exhibit ethical behavior that is beyond reproach in all situations.

THE SUPERVISOR'S ROLE IN ETHICS

Using the guidelines set forth in the previous section, supervisors should be able to make responsible decisions concerning ethical choices. Unfortunately, deciding what is ethical is much easier than actually doing what is ethical. In this regard, trying to practice ethics is like trying to diet. It is not so much a matter of knowing you should cut down on eating, it is a matter of following through and actually doing it.

It is this fact that defines the role of supervisors with regard to ethics. Supervisors have a three-part role. First, they are responsible for setting an example of ethical behavior. Second, they are responsible for helping employees make the right decision when facing ethical questions. Finally, supervisors are responsible for helping employees follow through and actually undertake the ethical option once the appropriate choice has been identified. In carrying out their roles, supervisors can adopt one of the following approaches (Figure 5–2):

- Best ratio approach
- Black and white approach
- Full potential approach

Best Ratio Approach

The **best ratio approach** is the pragmatic approach. Its philosophy is that people are basically good and under the right circumstances will behave ethically. However, under certain conditions they can be driven to unethical behavior. Therefore, the supervisor should do everything possible to create conditions that promote ethical behavior and try to maintain the best possible ratio of good choices to bad. When hard decisions must be made, supervisors should make the choice that will do the most good for the most people. This is sometimes referred to as *situational ethics.*

Black and White Approach

In the **black and white approach** right is right, wrong is wrong, and conditions are irrelevant. The supervisor's job is to make ethical decisions and carry them out. It is also to help employees choose the ethical route regardless of circumstances. When difficult decisions must be made, supervisors should make fair and impartial choices regardless of the outcome.

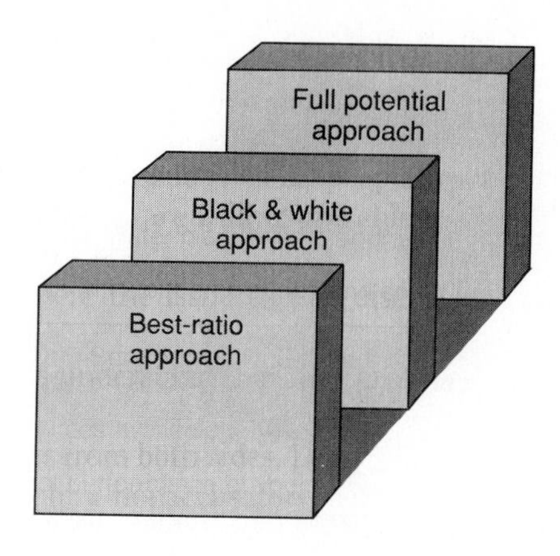

Figure 5–2
Three basic approaches supervisors can take to handle ethical problems.

Full Potential Approach

Supervisors make decisions under the **full potential approach** based on how they will affect the ability of those involved to achieve their full potential. The underlying philosophy is that people are responsible for realizing their full potential within the confines of morality. Choices that can achieve this goal without infringing on the rights of others are considered ethical.

Decisions might differ, depending on the approach selected. For example, consider the ethical dilemma presented at the beginning of this chapter. If the supervisor, Tom Richards, applied the best ratio approach, he might opt to go along with Jake Bronkowski's demand to move Alice McCormick out of the metal cutting shop. He might justify this choice based on what he perceives as the best decision for the most people. On the other hand, if he took the black and white approach, he could not justify giving in to Jake's demand.

THE COMPANY'S ROLE IN ETHICS

Industrial firms have a critical role to play in promoting ethical behavior among their employees. Supervisors cannot expect employees to behave ethically in a vacuum. A company's role in ethics can be summarized as: (1) creating an internal environment that promotes, expects, and rewards ethical behavior; and (2) setting an example of ethical behavior in all external dealings (Figure 5–3).

Creating an Ethical Environment

A company creates an **ethical environment** by establishing policies and practices that ensure all employees are treated ethically and then enforcing these policies. Do employees have the right of due process? Do employees have access to an objective grievance procedure? Are there appropriate health and safety measures to protect employees? Are hiring practices fair and impartial? Are promotion practices fair and objective? Are

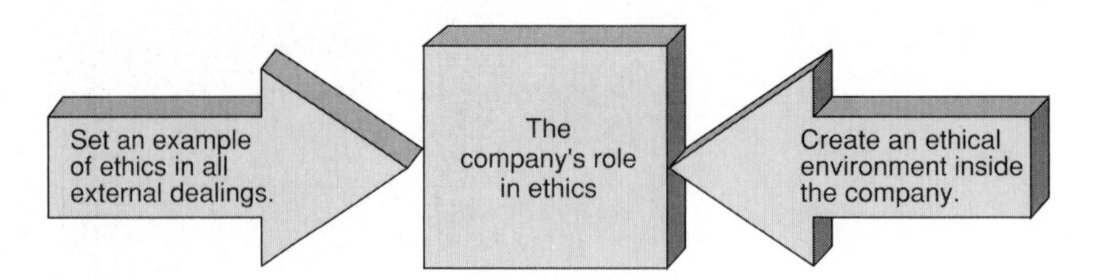

Figure 5–3
Supervisors cannot practice ethics in a vacuum. The company has a critical role to play.

employees protected from harrassment based on race, sex, or other reasons? A company that establishes an environment that promotes, expects, and rewards ethical behavior can answer yes to all these questions.

One effective way to create an ethical environment is to develop an **ethics philosophy** and specific guidelines that operationalize the philosophy. These should be put in writing and shared with all employees. Martin Marietta Corporation of Orlando, Florida has a *Code of Ethics and Standards of Conduct* that is shared with all employees. It begins with the following statement of philosophy:

> Martin Marietta Corporation will conduct its business in strict compliance with applicable laws, rules, regulations, and corporate operating unit policies, procedures and guidelines, with honesty and integrity, and with a strong commitment to the highest standards of ethics. We have a duty to conduct our business affairs within both the letter and the spirit of the law.[10]

Such a statement sets the tone for all employees at Martin Marietta. It lets them know that higher management not only supports ethical behavior, but expects it. This approach makes it less difficult for supervisors when they find themselves caught in the middle between the pressures of business and the maintenance of ethical behavior in their units.

In addition to its corporate ethics philosophy, Martin Marietta also publishes a more specific *Credo* and *Code of Conduct*. The Martin Marietta credo is summarized as follows:

> In our daily activities we bear important obligations to our country, our customers, our owners, our communities, and to one another. We carry out these obligations guided by certain unifying principles:
>
> - Our foundation is INTEGRITY
> - Our strength is our PEOPLE
> - Our style is TEAMWORK
> - Our goal is EXCELLENCE[11]

This **ethics credo** shows employees that they have obligations that extend well beyond their work units and that how they perform their work can have an impact, negative or positive, on fellow employees, the company, customers, and the country. Key concepts set forth in the credo are *integrity, people, teamwork,* and *excellence.* Supervisors who stress, promote, and model these concepts will make a major contribution to ethical behavior in the workplace.

Although the emphasis on ethics in the workplace is relatively new, the concept is not. According to David Shanks, Robert Wood Johnson, who built Johnson & Johnson into a major international corporation, developed an ethics credo for his company as early as the mid-1940s.[12] Johnson's credo read:

- To customers and users: quality and service at reasonable prices.
- To suppliers: a fair opportunity.
- To employees: respect, equal opportunity, and a sense of job security.
- To communities: a civic responsibility.
- To the environment: protection.
- To shareholders: a fair return.[13]

Written philosophies and guidelines such as those developed by Martin Marietta Corporation and Johnson & Johnson are the first step in creating an ethical environment in the workplace. Supervisors can play a key role in promoting ethical behavior on the job by encouraging higher management to develop written ethics philosophies/credos/guidelines and then by modeling the behavior they encourage.

Setting an Ethical Example

Companies that take the "Do as I say, not as I do" approach to ethics will not succeed. Employees must be able to trust their company to conduct all external and internal dealings in an ethical manner. Companies that do not pay their bills on time; companies that pollute; companies that do not live up to advertised quality standards; companies that do not stand behind their guarantees; and companies that are not good neighbors in their communities are not setting a good ethical example. Such companies can expect employees to mimic their unethical behavior.

In addition to creating an ethical internal environment and handling external dealings in an ethical manner, companies must support supervisors who make ethically correct decisions; not just when such decisions are profitable, but in all cases. For example, in the ethical dilemma presented at the beginning of this chapter, say the supervisor decided the ethical choice was to deny Jake Bronkowski's demand to move Alice McCormick. Management gave the supervisor permission to meet Jake's demand, and the sooner the better. This is obviously the profitable choice. But is it the ethical choice? If Tom Richards does not think so, will Alloy Tech stand behind him? If not, everything else the company does to promote ethics will break down.

HANDLING ETHICAL DILEMMAS

No person will serve long in a supervisory capacity without confronting an ethical dilemma. How, then should supervisors proceed when they confront such a problem? There are three steps (Figure 5–4):

- Apply the various guidelines for determining what is ethical that were presented earlier in this chapter.
- Select one of the three basic approaches to handling ethical questions.
- Proceed in accordance with the approach selected and with consistency.

Apply the Guidelines

In this step, supervisors apply as many of the tests set forth in Figure 5–1 as necessary to determine what the ethically correct decision is. In applying these guidelines, supervisors should attempt to block out all mitigating circumstances and other factors that tend to cloud the issue. At this point, the goal is only to identify the ethical choice. Deciding whether or not to implement it comes in the next step.

Select the Approach

Supervisors have three basic approaches to use when deciding how to proceed. As shown in Figure 5–2, they are the best ratio, black and white, and full potential approaches. These approaches and their ramifications can be debated ad infinitum; no one can tell a supervisor which approach to take. It is a matter of personal choice. Factors that will affect the ultimate decision include the supervisor's own personal makeup, the company's expectations, and the degree of company support.

Proceed with the Decision

The approach selected will dictate how the supervisor should proceed. Two things are important in this final step. The first is to proceed in strict accordance with the approach selected. The second is to proceed consistently. **Consistency** is critical when handling ethical dilemmas. Fairness is a large part of ethics and consistency is a large part of fairness. The grapevine will ensure that all employees know how a supervisor handles a given ethical dilemma. Some will agree and some will disagree regardless of the decision. Such is the nature of human interaction. However, regardless of the

Figure 5–4
Handling ethical dilemmas.

Steps for Handling Ethical Dilemmas

1. Apply the guidelines.
2. Select the approach.
3. Proceed accordingly and consistently.

differing perceptions of the problem, employees will respect the supervisor for being consistent. Conversely, even if the decision is universally popular, supervisors may lose respect if the decision is not consistent with past decisions.

SUMMARY

1. Ethics is the study of morality. Morality refers to the values that are subscribed to and fostered by society. Ethics attempts to apply reason in determining rules of human conduct that translate morality into everyday behavior.
2. Ethical behavior is that which falls within the limits prescribed by morality.
3. *Legal* and *ethical* are not the same thing. If something is illegal it is also unethical. However, just because something is legal does not mean it is ethical. An act can be legal but unethical.
4. To determine if a choice is ethical, supervisors can apply the following tests: morning after, front page, mirror, role reversal, and common sense.
5. Supervisors have a three-pronged role with regard to ethics. Supervisors are responsible for setting an ethical example, helping employees identify the ethical choices when facing ethical questions, and helping employees follow through and actually undertake the ethical option.
6. Supervisors have three approaches available to them in handling ethical dilemmas: the best ratio, black and white, and full potential approaches.
7. The company's role in ethics is to create an ethical environment and to set an ethical example. An effective way is to develop a written ethics philosophy and share it with all employees.
8. Three personality characteristics that can influence an employee's ethical behavior are ego strength, Machiavelliansim, and locus of control.
9. Supervisors facing ethical dilemmas should apply the tests for determining what is ethical, select one of the three basic approaches, and proceed consistently.

KEY TERMS AND PHRASES

Best ratio approach
Black and white approach
Common sense test
Consistency
Ego strength
Ethical environment
Ethical behavior
Ethics
Ethics credo
Ethics philosophy
Front page test
Full potential approach
Gray area
Legal
Locus of control
Machiavellianism
Mirror test
Morality
Morning-after test
Role reversal test
Values

CASE STUDY: Ethics at Martin Marietta

The following real-world case study contains an ethics-related situation of the type supervisors may confront on the job. Read the case study carefully and answer the accompanying questions.

Martin Marietta Corporation is one of this country's largest government contractors in the area of missile manufacturing. As a producer of military hardware, Martin Marietta receives its funds from public tax dollars. Consequently, it is critical that the company and its employees operate in accordance with the highest standards of ethical behavior.

To ensure that this happens, Martin Marietta established a written *Code of Ethics and Standards of Conduct* manual that is shared with all employees. Employees must acknowledge receipt of this copy in writing. The manual begins with a statement of the company's philosophy regarding ethical behavior in the workplace:

> Martin Marietta Corporation will conduct its business in strict compliance with applicable laws, rules, regulations, and corporate operating unit policies, procedures and guidelines, with honesty and integrity, and with a strong commitment to the highest standards of ethics. We have a duty to conduct our business affairs within both the letter and the spirit of the law.[14]

The booklet goes on to give employees specific guidance in such areas as bidding, negotiation, performance of contracts, conflicts of interest, acceptance of gifts, political contributions, and other pertinent areas of concern. It also explains compliance and disciplinary measures.

1. If you were a supervisor at Martin Marietta, how would having such a booklet help you?
2. What problems might you encounter as a supervisor if your company had no such guidelines or ethics?
3. How would you use a booklet such as the one produced by Martin Marietta to promote ethical behavior among members of your work team?

REVIEW QUESTIONS

1. Define the term morality.
2. Define the term ethics.
3. Briefly explain each of the following ethics tests: morning-after test, front page test, mirror test, role reversal test, common sense test.
4. What is the supervisor's role with regard to ethics?
5. Briefly explain the following approaches to handling ethical behavior: best ratio approach, black and white approach, full potential approach.
6. Briefly explain a company's role with regard to ethics.

7. Explain how a supervisor should proceed when facing an ethical dilemma.
8. Write a brief ethics philosophy for a plastics recycling company.
9. List the individual and situational factors that might influence an employee's ethical behavior.
10. List and briefly describe the "Five P's" of ethical power as described by Blanchard and Peale.

SIMULATION ACTIVITIES

The following activities are provided to give you opportunities to develop supervisory skills by applying the material presented in this chapter and in previous chapters in solving simulated supervisory problems.

1. The Solid Waste Disposal Company (SWDC) has three teams of drivers who transport compacted solid waste to the company's landfill. SWDC is a responsible company that has spent a lot of money developing an environmentally sound landfill that can accept all of the various types of solid waste the company accepts. Its one disadvantage is its distance from the compacting plant. Drivers complain that the 100 mile round trip required for each load limits their ability to earn the financial incentives available to drivers who exceed their weekly tonnage quotas.

   These financial incentives are at the heart of Don Morgan's ethical dilemma. Don supervises the three teams of drivers for SWDC. One of his best, most loyal, most experienced drivers needs the financial incentives badly. The driver, Tim McGhee, has a child in the hospital. The child's treatment is stretching Tim's finances to the breaking point. Don has learned that in order to maximize the incentive money received, Tim McGhee is dumping every third load in the company's old landfill. This landfill, which has been closed for three years, is not rated to accept some of the potentially hazardous material SWDC accepts. How should Don Morgan handle this dilemma?
2. Leisure Wear Corporation and Modern Apparel, Inc. are major textile producers and competitors. Recently Modern Apparel has been winning almost every time it competes against Leisure Wear. The grapevine has it that this is due to a revolutionary breakthrough in the formula Modern Apparel uses to produce a particular synthetic material. The new formula is allowing Modern Apparel to produce its fabric at half the cost of Leisure Wear's competing fabric.

   Nancy Davies, supervisor of the research and development unit for Leisure Wear, has been pulling her hair out trying to duplicate Modern Apparel's breakthrough, but to no avail. The pressure on her is tremendous. Top management is pushing her hard to come up with the formula. Friends in other sections of the plant have intimated they may be laid off if things do not turn around soon. If this happens, the first employee to be let go will be Nancy's niece, a new employee in Leisure Wear's mailroom. What makes this even more difficult for Nancy is that Leisure Wear is located in a one-company town. Prospects of finding other jobs will not be good for employees who are laid off.

Nancy may have found a solution to her problem. A disgruntled employee in the research and development unit at Modern Apparel has called asking for a job. He has promised that, if hired, he will bring the new formula with him. It happens that Nancy has an opening in her unit. How should she handle this problem?

ENDNOTES

1. Drucker, P. F. "What Is Business Ethics?" *Across the Board,* October 1981, pp. 22–32.
2. Stead, E. W., Worrell, D. L., and Stead, J. G. "An Integrative Model for Understanding and Managing Ethical Behavior in Business Organizations," *Journal of Business Ethics,* 1990, Number 9, p. 233.
3. Taylor, P. *Principles of Ethics: An Introduction* (Encino, CA: Dickson Publishing Company, 1975).
4. Arlow, P. and Ulrich, T. A. "Business Ethics, Social Responsibility, and Business Students: An Empirical Comparison of Clark's Study," *Akron Business and Economic Review,* 1980, Number 3, pp. 17–23.
5. Blanchard, K. and Peale, N. V. *The Power of Ethical Management* (New York: Ballantine Books, 1988), pp. 10–17.
6. Ibid., p. 79.
7. Trevino, L. K. "Ethical Decision Making in Organizations: A Person-Situation Interactionist Model," *Academy of Management Review,* 1986, Volume 11, Number 3, pp. 601–617.
8. Preble, J. F. and Miesing, P. "Do Adult MBA and Undergraduate Business Students Have Different Business Philosophies?" Proceedings of the National Meeting of the American Institute for the Decision Sciences, November 1984, pp. 346–348.
9. Luthans, F. and Kreitner, R. *Organizational Behavior Modification and Beyond: An Operant and Social Learning Approach* (Glenview, IL: Scott Foresman, 1985).
10. Martin Marietta Corporation. *Code of Ethics and Standards of Conduct* (Orlando, FL: Martin Marietta Corporate Ethics Office, 1990), inside front cover.
11. Ibid., p. 2.
12. Shanks, D. C., "The Role of Leadership in Strategy Development," *Journal of Business Strategy,* January/February 1989, p. 32.
13. Ibid., p. 33.
14. Martin Marietta Corporation. *Code of Ethics and Standards of Conduct,* p. 2.

PART TWO

The Supervisor as a Manager

CHAPTER SIX

Management and the Supervisor

CHAPTER OBJECTIVES

After studying this chapter, you will be able to define or explain the following topics:

- The Term *Management*
- Key Concepts in the Definition of Management
- The Term *Manager*
- The Primary Management Functions
- The Two Aspects of Management
- The Basic Principles of Management
- Basic Approaches to Management
- The Differences Between Supervision and Management
- How to Manage Conflict
- Organizational Culture as it Relates to Management
- The Concept of Managing for Competitiveness
- A Checklist for Managers

CHAPTER OUTLINE

- What Is Management?
- Key Concepts in the Definition of Management
- What Is a Manager?
- Primary Functions of Management
- Two Aspects of Management
- Basic Principles of Management
- Basic Approaches to Management
- Differences Between Supervisors and Managers
- Managing Conflict
- Management and Organizational Culture
- Managing for Competitiveness
- Checklist for Managers
- Summary
- Key Terms and Phrases
- Case Study Application Problem
- Review Questions
- Simulation Activities
- Endnotes

Supervisors represent their company's first level of management. They are part of the management team. Therefore, supervisors must be knowledgeable about management as a process and where they fit into the process. This chapter helps prospective and practicing supervisors understand the management process, primary management functions, basic principles of management, and basic approaches to management. It also explains where supervisors fit into the management process as well as differences between supervision and management.

WHAT IS MANAGEMENT?

The term **management** refers to the people in a company who are formally responsible for managing the company. When used in this way, the term encompasses supervisors, middle managers, managers, and executive managers. To **manage** means to create and maintain a work environment that promotes the efficient and effective accomplishment of organizational goals. The key concepts in this definition of management as a process are *work environment, goals, efficiency,* and *effectiveness* (Figure 6–1).

KEY CONCEPTS IN THE DEFINITION OF MANAGEMENT

The four concepts contained in the definition of management are part of the key to understanding how and where supervisors fit in. Work environment, goals, efficiency, and effectiveness are fundamental to the process of management.

Work Environment

A **work environment** is a system. Like any system it has several components. The components of the work environment are people, facilities, and resources (Figure 6–2).

Figure 6–1
Key components in the definition of management as a process.

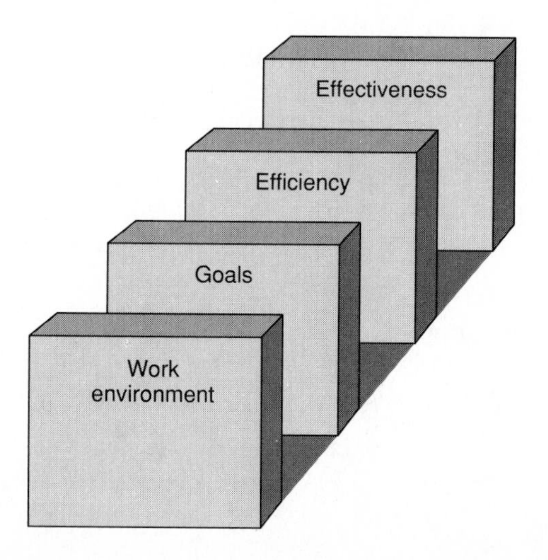

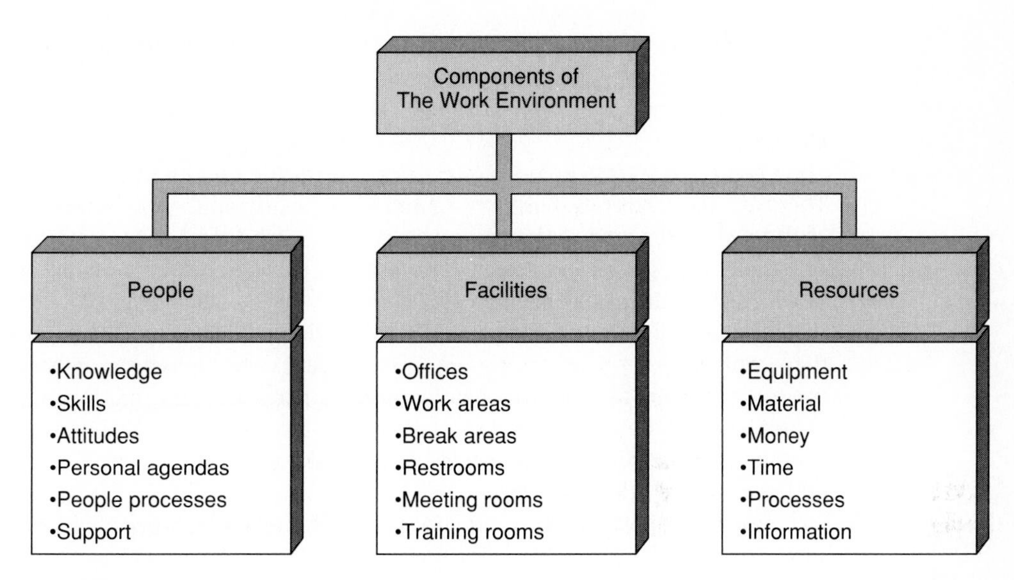

Figure 6–2
The work environment consists of three major components.

The people component consists of the knowledge, skills, attitudes, and personal agendas of people who work in the environment. It also includes the processes through which they interact and the amount of support people get from co-workers, supervisors, and managers.

Knowledgeable, skilled workers at all levels contribute to a positive work environment because they are better able to perform their tasks; support fellow workers in performing theirs; meet quality standards, work quotas, and deadlines; and comply with the expectations of fellow workers, supervisors, and managers.

An example of this phenomenon can be seen in athletic teams. Teams that are able to win consistently are fun to play on. You have probably heard the phrase "Everybody loves a winner, but nobody loves a loser." It also applies in a work setting. The better people are at performing their jobs and the better their fellow workers are, the more enjoyable work is.

People with positive attitudes also contribute to a positive work environment. Nobody likes to spend eight hours or more a day in a setting where people are continually bickering among themselves, complaining, or whining. This type of environment will wear down even the most dedicated workers. More importantly, negative attitudes are contagious.

Fortunately, so are positive attitudes. It is invigorating and stimulating to associate with positive people who take a can-do approach to work. Resources invested in developing positive attitudes among employees will be resources invested well.

The personal agendas of people include their goals, hopes, dreams, and ambitions. When the personal agendas of workers are consistent with the goals of the organization, they contribute to a positive work environment. The obverse is also true. This is why it is so important for supervisors to get to know their employees and spend time helping them understand the relationship between their personal goals and those of the organization.

People processes can add to or detract from the quality of the work environment. These are the processes, both formal and informal, through which human interaction is channeled. For example, the processes used for hiring, firing, promoting, and handling grievances are people processes. The process through which employees suggest improvements is also a people process.

When people processes are perceived as being fair, objective, open, and supportive, they contribute to a positive work environment. This is why it is so important for supervisors to be attentive to how employees perceive the organization's people processes. One of the easiest traps to fall into as a supervisor is responding to employee complaints with such statements as "They make us do it this way." Supervisors must understand that *they* means *we*.

When a people process is not contributing to a positive work environment, supervisors should work with higher management to correct the situation. Although a process may be administratively expedient, it cannot be justified unless it also contributes to a positive work environment.

The level of support afforded workers by co-workers, supervisors, and managers can also contribute to a positive work environment. It is important for employees to feel they have the support necessary to do their jobs. It is also important for employees to feel that they will be supported when the unavoidable personal problems we all experience intrude in their lives. How do co-workers, supervisors, and managers respond when an employee experiences such personal problems as a death, a divorce, or a sick child?

Supervisors should never underestimate the importance of supporting employees in good times and bad. Not only will it contribute to a positive work environment, it will build strong ties of loyalty. Conversely, few mistakes management can make will so quickly and effectively dampen morale as failing to support employees. This is one of the reasons so many companies in this country are either providing on-site child care or are now considering it.

Facilities are also an important part of the work environment. Consider all the time, money, and effort we put into making our homes comfortable and well suited to our personal tastes. We all like to feel comfortable and *at home.* This is also true at work. Employees spend too much time at work to be uncomfortable with the offices, work areas, break areas, restrooms, meeting rooms, training rooms, and other areas. Supervisors should look at facilities critically and listen to how employees perceive their facilities. If facilities are not contributing to a positive work environment, supervisors should work with higher management to improve them. Figure 6–3 is a checklist that will help supervisors evaluate the facility available to their employees. After completing the checklist yourself, ask several employees to complete it and compare your response with theirs.

Figure 6–3
Facility assessment checklist for supervisors.

Directions for Supervisors:
In completing this assessment checklist reverse roles and be an employee. Go to each area and rate it as an employee who spends time there. How would you rate it if you worked there?

0 = Does not contribute to a positive work environment
5 = Contributes greatly to a positive work environment

	0 – 5
1. Office space	
2. Work areas	
3. Break areas	
4. Restrooms	
5. Meeting rooms	
6. Training rooms	

Comments: ______________________________

Resources are also an important component of the work environment. When employees have the resources they need to do a good job, a positive work environment results. The obverse is also true. Resources typically needed in a work setting include equipment, materials, money, time, technical processes, and information.

Old and outdated equipment, material shortages, funding shortages, insufficient time, inadequate technical processes, and insufficient information will dampen the morale of dedicated workers trying to do their jobs. One of the key management roles of supervisors is ensuring that their employees have the resources needed to do their jobs. Figure 6–4 is an assessment tool supervisors can use to determine if resources are contributing to a positive work environment.

Goals

The term *goals* refers to both the organization's goals and those of individual employees. A **goal** is a desired result toward which resources are directed. The most successful management team will be the one that is able to ensure that these two sets of goals are complementary.

Supervisors play two key roles here. The first is in matching the right people with the right jobs during the hiring process. The second is in staying abreast of the personal goals of employees and helping them understand how they relate to the organization's goals.

Figure 6–4
Assessment tool for determining employee perceptions of resources provided.

Resource Assessment Instrument

Directions:
Indicate on the scale how frequently each resource is available to you.

0 = never 1 = seldom 2 = occasionally
3 = usually 4 = almost always 5 = always

1. Equipment 0 – 5
2. Material 0 – 5
3. Money 0 – 5
4. Time 0 – 5
5. Processes 0 – 5
6. Information 0 – 5

Comments:

Efficiency and Effectiveness

The most competitive companies are those that produce their products and provide their services efficiently and effectively. Efficiency and effectiveness are critical concerns of management. The terms are sometimes used interchangeably. They should not be. Efficiency and effectiveness are related, but they are not the same thing.

Efficiency means doing work better. *Better* can mean different things such as faster or less expensive. Doing a job efficiently means doing it with a minimum of wasted resources. Those resources may be time, money, material, effort, or any other resource required to do a job. Efficiency is sometimes expressed as a ratio of **output** to **input** (Figure 6–5). The more output per unit of input, the more efficient the operation.

Effectiveness means setting the right goals and achieving them (Figure 6–6). It takes both to be effective. Setting the right goals means setting goals that will result in making the organization more productive, more competitive, and more profitable.

It is not uncommon for market conditions or other external factors to influence managers to set shortsighted goals that actually do more harm than good in the long run. Even when such goals are accomplished, the manager who set them must be considered ineffective because he or she set the wrong goals.

Efficiency and effectiveness are usually considered simultaneously because to know one without knowing the other is to know only half the story. An organization can be efficient without being effective. This happens when an organization efficiently pursues the wrong course of action. This is like being in a race and running faster than all other competitors, but in the wrong direction.

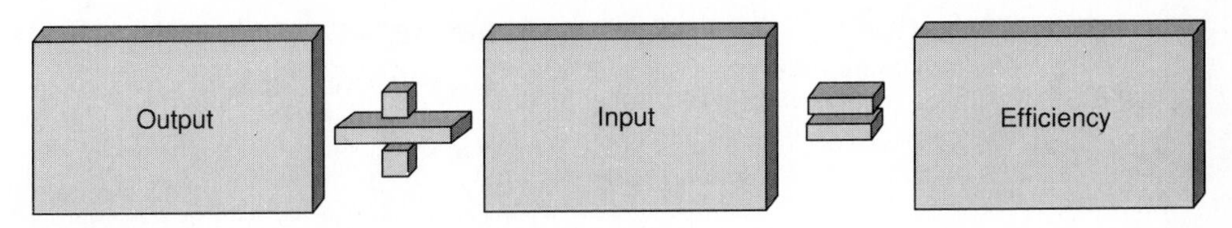

Figure 6–5
Efficiency can be expressed as a ratio of output to input.

Figure 6–6
Effective managers set the right goals and achieve them.

Conversely, an organization can be effective without being efficient. This happens when managers set the right goals and accomplish them, but in the process expend more than the allocated resources.

This is why it is important for supervisors always to consider efficiency and effectiveness simultaneously. By doing so they build in checks and balances that ensure reliable information upon which to base decisions.

WHAT IS A MANAGER?

The term **manager** refers to a job classification, not a job title. People whose positions are classified as management can have a variety of different titles. Typical titles in an industrial setting include chief executive officer, president, vice president, general manager, superintendent, executive director, director, and foreman.

In addition to having different titles, managers also have different levels (Figure 6–7). Supervisors are the first level of management. The next levels are mid-manager, manager, and executive manager. Regardless of the title or level all managers share two common characteristics: **authority** and **responsibility** (Figure 6–8).

Managers have the authority to set goals for the organization, make decisions relating to the accomplishment of the goals, and set resources in motion to accomplish the goals.

Correspondingly, managers are responsible for the goals they set, the decisions they make, and the resources they put in motion to accomplish the goals. Another term for responsibility is **accountability.** Managers are held accountable not just for their personal performance, but for the performance of all people who report to them directly

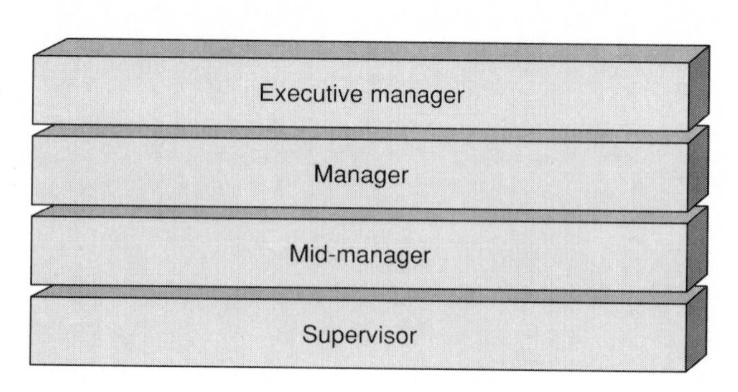

Figure 6–7
Levels of managers.

and indirectly. Managers are accountable to all persons inside and outside the company who might be affected by their performance on the job (Figure 6–9).

The list of people to whom managers are accountable is comprehensive. Even a cursory examination of the list can be an eye-opening experience, one managers should undertake frequently as a reminder of their responsibilities.

Characteristics of a Good Manager

The characteristics that make a person a good manager can be identified and described. According to Yoshio Hatakeyama, president of the Japanese Management Association, effective managers must have the following characteristics:[1]

- Have ideas of their own and mobilize people to accomplish them.
- Behave like the owner and operator of their department.
- Have a clear grasp of the demands of the overall organization and the problems within their department.
- Be able to convince others of the lucidity of their ideas and act autonomously to realize those ideas.
- Be able to influence and mobilize workers, colleagues, superiors, and outsiders.
- Make things happen according to what they believe.
- Want to be judged by achievement, not credentials.
- Achieve at least one significant and lasting accomplishment for the good of the organization in each successive management position.

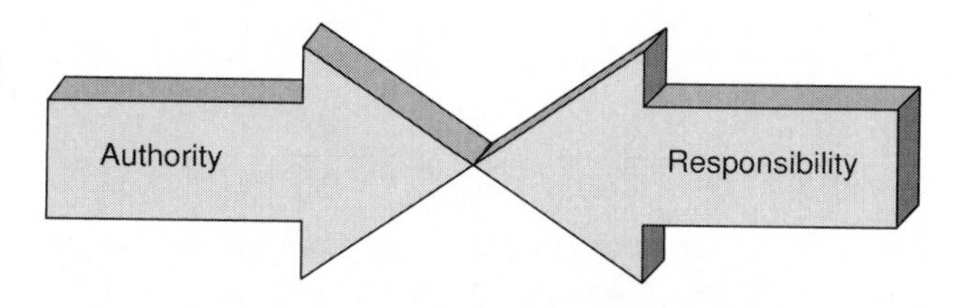

Figure 6–8
All managers, regardless of title and level, share these two characteristics.

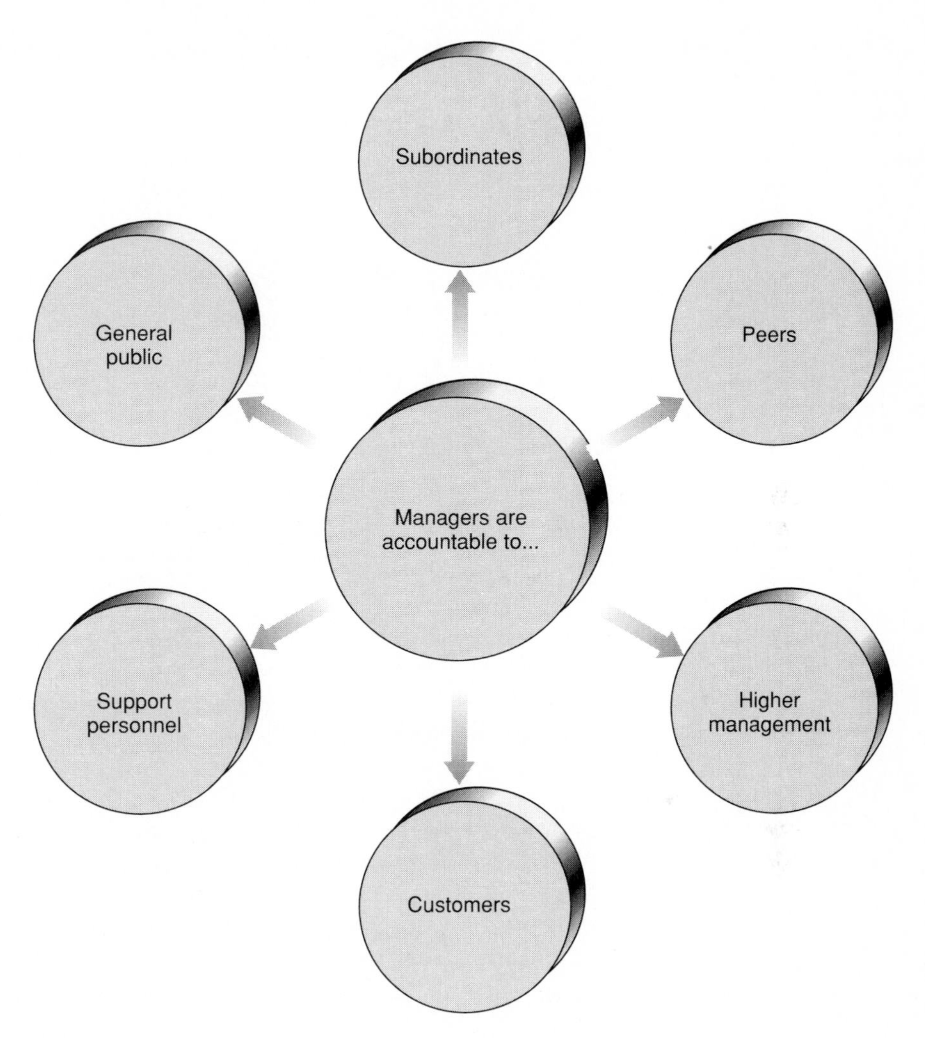

Figure 6–9
People to whom managers are responsible and accountable.

PRIMARY FUNCTIONS OF MANAGEMENT

Regardless of the level of management, all managers are responsible for performing five basic functions (Figure 6–10): planning, organizing, staffing, facilitating, and monitoring. These primary functions of management are the subjects of the remaining chapters in this section of this book. What follows here are brief descriptions of these fundamental management functions. This section also examines a new function, issues management.

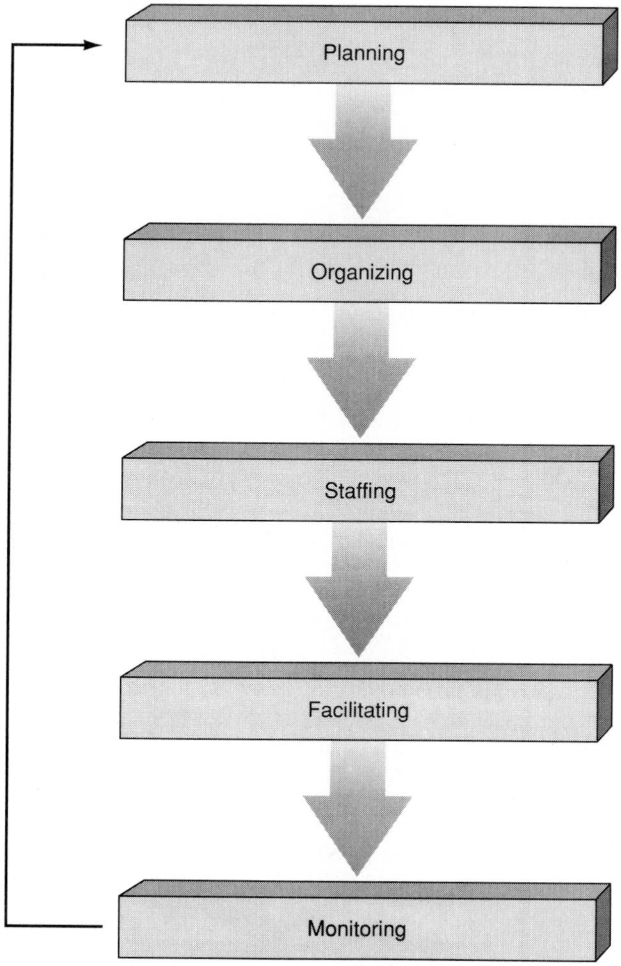

Figure 6–10
Basic management functions.

Planning

Organizing

Staffing

Facilitating

Monitoring

Planning

Planning is the process through which goals are set, timetables are established, and action plans are developed for meeting the goals. The goals, objectives, strategies, timetables, and schedules contained in a management plan form the basis for another important component of planning, budgeting.

Organizing

Organizing is the process through which managers marshal the resources needed to accomplish goals set in the planning component. These resources include the people, materials, technology, time, funding, and information needed to do the job. Organizing can also involve structuring jobs and assigning work.

Staffing

Staffing is the process through which people needed to do jobs are put to work. It involves developing job descriptions, deciding on qualifications, securing applicants, screening, interviewing, hiring, orienting, and training.

Facilitating

Facilitating is a catch-all term that encompasses all the tasks managers perform to keep people on the optimum course toward accomplishing the organization's goals. It involves motivation, leadership, communication, counseling, and problem solving.

Monitoring

Monitoring is the continual assessment of progress toward accomplishing goals. During the planning process, operational schedules and timetables were developed. By monitoring, managers know if they are on, ahead, or behind these schedules and timetables. They also know if resources are being used efficiently and effectively. By monitoring, supervisors know when facilitation is necessary.

Issues Management

Mark A. Newman coined the term *issues management.*[2] An issue is a trend that might become harmful to a company. Issues management involves responding in a way that prevents the trend from harming the company. It has five steps:

1. *Issue identification* involves staying abreast of local, national, and international issues and examining them for potential impact now or in the future.
2. *Issue analysis* involves analyzing and prioritizing issues.
3. *Policy formulation* involves developing a plan for dealing with the issue.
4. *Policy implementation* involves implementing the plan.
5. *Evaluation* involves reviewing results and adjusting accordingly.

The goal of this process is to identify issues in the early stages of development while a variety of options still exist. An example of an opportunity for issues management came during Iraq's invasion of Kuwait in July and August, 1990. Managers of oil-dependent companies in the United States might have applied issues management in an attempt to minimize the impact any interruption in oil supplies might have had for them.

TWO ASPECTS OF MANAGEMENT

According to Yoshio Hatakeyama, there are two aspects to management, *occupational* and *human.*[3] The occupational aspect can be divided into two distinct components, *maintenance management* and *structural innovation.* Maintenance management consists of performing the routine, day-to-day aspects of the job. Structural innovation involves continually finding new ways to improve productivity and quality by reforming existing concepts and methods. Hatakeyama stresses that managers must accomplish both.

According to Hatakeyama, "The basis of the human aspect of the manager's job is the realization that, in effect, he controls the lives of workers. The manager must reflect on this fact with humility, give . . . scrupulous attention to . . . workers, and never lead them astray."[4] He stresses that the foundation of the human aspect of management is *trust.* Only when workers trust a manager can he or she motivate them and only when workers can be motivated can they be developed. The progression is as follows: (1) build trust, (2) motivate, and (3) develop.

Hatakeyama theorizes about a manager's overriding objectives: (1) get the work he or she is responsible for done efficiently and effectively, and (2) stimulate workers while continually enhancing their abilities. An effective manager, says Hatakeyama, will "lead his section in a cycle of growth in which the work and the workers not only coexist, but actually influence and stimulate one another."[5]

BASIC PRINCIPLES OF MANAGEMENT

Principles are practical guidelines upon which managers can base actions and behavior (Figure 6–11). The principles listed in this figure were originally developed in the early 1900s and, for the most part, they still apply. These principles and the extent of their applicability in the modern workplace are explained in the following subsections.

Figure 6–11
These basic principles apply to all levels of management.

Basic Principles of Management

1. Divide work to allow for appropriate specialization.
2. Give managers the authority needed to carry out their responsibilities.
3. Managers are responsible for both morale and discipline.
4. Managers may have many subordinates, but no employee should have more than one boss.
5. An organization should have only one set of overall goals. Goals of sub-units should be tied directly to these.
6. Employees should put the organization's goals ahead of their personal goals.
7. Reward for work should reflect both the level of contribution and efficiency/effectiveness.
8. Organizations should have a well defined governance structure.
9. All policies and practices of an organization should be fair, impartial, and equitable.
10. An organization's management should encourage initiative.

Division of Work

Of all the traditional principles of management, **division of work** is the only one that can no longer be applied in all cases. Work in a modern industrial setting should still be specialized. For example, the typical industrial firm still needs managerial, administrative support, and production or process personnel. Work within these categories is divided even further.

However, in the same settings where work has traditionally been specialized, the emphasis has begun to shift more toward flexibility. Assembly lines and transfer lines where workers perform highly specialized tasks are beginning to be replaced with flexible assembly cells where workers perform a variety of tasks.

This is due in part to the fact that we have learned that human workers do not do boring, repetitive tasks very well. It is also due to the fact that managers have come to realize that workers perform better when they can see the overall picture and where their work fits into it rather than just a highly specialized, isolated piece of the picture. Finally, it is also due to the fact that managers learned the value of flexible work in decreasing absenteeism.

Authority/Responsibility

Managers at all levels must have the authority necessary to carry out the responsibilities for which they are held accountable. This means managers must have the authority to set goals, make decisions relative to these goals, and put resources into action in ways they feel will most efficiently and effectively accomplish the goals.

Morale/Discipline

Managers are responsible for the **morale** of the workers over whom they have authority. At the same time, managers are responsible for enforcing the organization's policies, procedures, and practices. Managers are also responsible for administering **disciplinary** procedures when necessary.

Only One Boss

Workers at all levels should report to only one supervisor. This does not mean that employees cannot move laterally as well as up and down in the organization in collecting input or communicating ideas. However, once input has been collected and ideas have been communicated, orders should come from just one supervisor and workers should be accountable to just one supervisor. At best, other approaches cause confusion. At worst, they cause conflict.

Uniform Goals

The purpose of setting goals is to give direction. This principle implies that all resources should be pointed in the same direction. The only way to ensure that this happens is to have uniform goals. This means an organization should have one master set of goals.

Any goals set by subunits of the organization must grow from and be consistent with the master set.

Organization's Needs First

All workers in an organization have their own personal goals, dreams, and ambitions. The organization has its goals. This principle means that all workers should perform their jobs with the goals of the organization foremost in their minds. Ideally, the two sets of goals will be consistent and complementary. However, when they conflict, the employee should put the organization's goals first.

In recent years, managers have come to realize that this "company first" philosophy, though still valid, can be taken too far. Some workers will occasionally commit their efforts so exclusively to accomplishing the organization's goals that they neglect other important aspects of their lives. The results of such behavior are predictable. They include family problems, divorce, exhaustion, burnout, nervous breakdowns, and other problems that will turn an organization's best performers into nonperformers.

Modern managers put a great deal of effort into employing people whose goals are consistent with those of the organization, then encouraging them to achieve a healthy balance between the work and nonwork aspects of their lives.

Reward for Work

In any organization some positions are more critical than others, a function of how much they contribute to the accomplishment of the organization's goals. This is also a function of the law of supply and demand. This principle of management requires that salary levels reflect the criticality of the position.

It is also true that two people in equally critical positions, for example, two engineers, will not necessarily perform with equal effectiveness. In fact, they probably will not. This principle requires that provisions be made for rewarding people for performance. This is sometimes accomplished through incentive pay programs. It can also be accomplished by having discretionary ranges within otherwise standardized salary schedules.

Governance Structure

It is important for all personnel in an organization to know how the organization is structured so they will know where they fit into the organization, where their subunit fits in, whom they report to, and who reports to whom throughout the organization. A well-defined *governance structure* promotes communication and eliminates confusion.

Equitable Policies and Practices

Experience has shown that fairness and objectivity in dealing with people are good business. They engender loyalty and enhance morale. As a result, turnover decreases and productivity increases. For this reason managers should ensure that an organization's policies and practices are fair, impartial, and equitable.

Encouraging Initiative

The people most familiar with problems are the people involved in them. Consequently, the people most likely to be able to solve work-related problems are the people doing the work. For this reason it is important to encourage individual **initiative** at all levels. When initiative is encouraged, people will find better ways to do their jobs. When people continually try to find better ways to do their jobs, the organization's level of productivity is continually improved.

BASIC APPROACHES TO MANAGEMENT

Figure 6–12 depicts the five basic approaches to management. These approaches are sometimes called by different names, but regardless of the names, the approaches are widely accepted. This section also discusses what has come to be known as the Japanese approach to management and several other approaches.

Scientific Approach

The fundamental concepts undergirding **scientific management** are *evaluation* and *analysis*. Process and performance criteria are established at all levels and for all components of the organization, then evaluated continually. The information collected is analyzed. Based on this analysis, decisions are made and action is taken. This has traditionally been the most widely used approach.

Figure 6–12
Basic approaches to management.

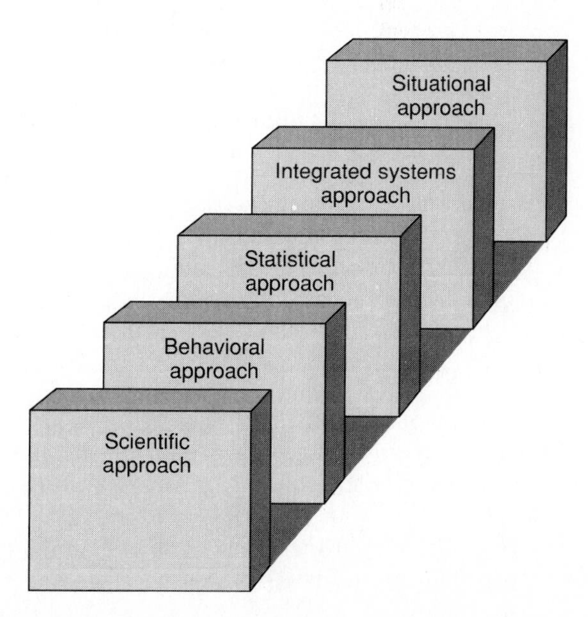

Behavioral Approach

The **behavioral management** approach focuses on people as opposed to tasks. Managers attempt to win the full cooperation of employees in achieving the organization's goals. This approach is based on the assumption that managers can understand human behavior well enough to know how to win people's complete cooperation.

Statistical Approach

Statistical management focuses on the application of mathematical and scientific principles to decision making. In this approach decisions are based on numbers and statistics. Problems are reduced to mathematical equations and solutions are based on quantifiable data.

Integrated Systems Approach

The fundamental principle of **integrated systems management** is that any organization is a system made up of interdependent subsystems so that a decision made in one subsystem may affect all subsystems. This approach requires managers to take a more holistic view when making decisions.

Situational Approach

The underlying principle of the **situational management** approach is that all the other approaches have merit depending on the situation. At times the scientific approach works best; at other times the behavioral, statistical, or integrated systems approach might work better. In the situational approach managers apply the methods that are appropriate to the situation. As circumstances change, so does the approach.

Which of these approaches is best? No one can say because the same approach might work for one manager and fail for another. However, a general rule of thumb applies here: Even in a highly technical work setting, an organization's most important asset is still its people. This makes such concepts as participation and consensus building vitally important. For this reason many modern managers combine the behavioral approach with one of the others in an attempt to create a *balanced* approach. The difference between a balanced approach and the situational approach is that the balanced approach always includes elements of the behavioral approach while the situational approach may or may not, depending on the situation.

The Japanese Approach

A great deal of attention has been given to what has come to be known as the *Japanese management style.* James R. Lincoln, writing in *California Management Review,*[6] lists several characteristics of Japanese management, including the following:

- Long-term employment
- Age/seniority preferences

- Cohesive work groups
- Close supervisor-subordinate contact
- Quality circle participation
- Employee welfare services

According to Lincoln, these and other characteristics of the Japanese approach to management breed high levels of commitment and positive work attitudes among workers. His research suggests this approach would bring similar results in the United States.

Other Approaches to Management

W. Steven Brown divides all management styles into four categories: autocratic, bureaucratic, democratic, and idiosyncratic.[7] *Autocratic managers* are clearly in charge. They make the decisions. They give the orders. *Bureaucratic managers* manage strictly by the book. The book in their case, is the company policy manual. *Democratic managers* solicit feedback from all people involved in or affected by a decision. This is a form of participatory management. *Idiosyncratic managers* are attuned to employees' individual differences and use these differences when working with them. Brown stresses the importance of applying an appropriate combination of these styles as dictated by circumstances and individuals.

Regardless of the management approach, according to Alex Mironoff, many corporations need to "de-Stalinize."[8] Mironoff claims that organizations are almost always designed on a totalitarian model. He says, "the totalitarian model works hours a day to sap our best energies."[9] Mironoff makes the point that organizations can no longer survive and succeed in an intensely competitive international marketplace if they are centrally controlled and patterned after "every . . . totalitarian system from feudalism to Stalinism."[10]

He claims that the reason quality circles and other attempts at improving quality and productivity so often fail in U.S. industrial firms is that management is not willing to let go of control and harness the power of free enterprise from within. The approach to management recommended by Mironoff is one that ensures that even the lowest paid wage earner is as personally interested in the bottom line as is the chief executive officer. This is accomplished by applying the theory that "People who have something at stake and something to gain will outperform those who don't."[11] This concept is even more valid when employers are given the freedom (empowered) they need to improve their performance.

To give all employees a stake in the company's success, Mironoff recommends such incentives as employee stock ownership, performance bonuses, or profit sharing. An example of using this approach to management is Springfield Remanufacturing Corporation (SRC) of Springfield, Missouri.[12]

CEO John Stack and his employees worked together to buy out a struggling subsidiary of International Harvester and created Springfield Remanufacturing. Stack owns 19 percent of the company's stock and the employees own the rest. As a result, 475 employees and Stack all have a common interest in the company's success. Policies and training programs all promote success. The key to this alignment of interests is employee ownership.

DIFFERENCES BETWEEN SUPERVISORS AND MANAGERS

Supervisors are managers in that they represent the first level of management. As first-level managers, supervisors' jobs are similar to higher-level managers in certain ways and different in other ways. The best way to understand this is to examine the skills needed to perform the duties of each. The duties and responsibilities of managers at *all* levels require skills that can be divided into three broad categories (Figure 6–13):

- Technical skills
- Paper skills
- People skills

Technical Skills

Technical skills are necessary to perform a specific job needed to produce a product or provide a service. Technical skills typically include job-specific knowledge; the ability to use specialized tools, machines, equipment, and systems; and the ability to operate and control processes.

Paper Skills

Paper skills are those necessary to perform the paperwork aspects of a job. They include planning, organizing, budgeting, record keeping, correspondence, reporting, and monitoring.

People Skills

People skills are needed to influence human behavior in a positive way. They include the ability to motivate, inspire, and communicate effectively.

All levels of managers need a certain amount of skill in each of these areas. The differences between levels of management are in the amount of skill needed in each

Figure 6–13
The skills needed by managers at all levels can be divided into these three categories.

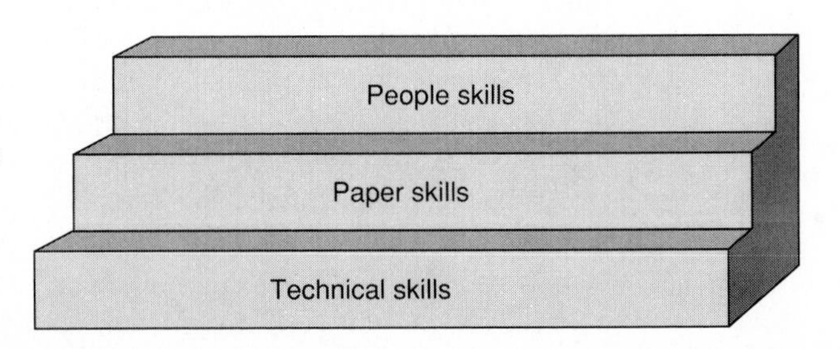

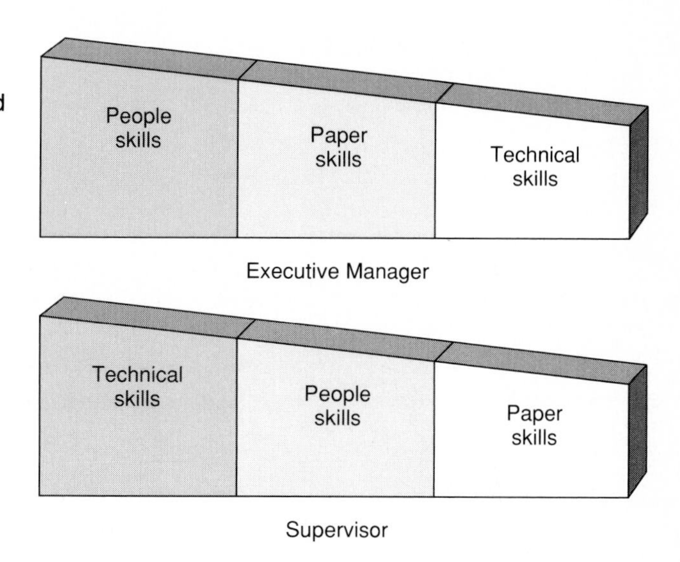

Figure 6–14
Comparison of skills needed by supervisors and executive managers.

area. For example, supervisors need more technical skills than do executive managers. Conversely, executive managers need more people skills than do supervisors. Figure 6–14 depicts this comparison graphically.

Supervisors typically prioritize these three categories of skills as follows:

1. Technical skills
2. People skills
3. Paper skills

Compare this with the prioritization by executive managers:

1. People skills
2. Paper skills
3. Technical skills

MANAGING CONFLICT

According to K. W. Thomas and W. H. Schmidt, mid-managers spend approximately 26 percent of their time dealing with conflict between individuals and groups.[13] A rule of thumb to remember is that the lower the level of management, the more time spent handling conflict. This means that supervisors are likely to spend a substantial amount of time in conflict resolution.

Conflict can be defined as what occurs when one person (or group) thinks another has or is about to frustrate an important concern.[14] Conflict is a process that consists of the following stages: (1) latent conflict, (2) perceived conflict, (3) felt conflict, (4) manifest conflict, and (5) conflict aftermath (Figure 6–15).[15]

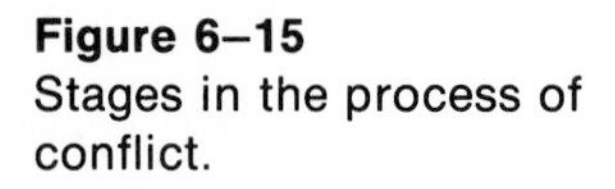

Figure 6–15
Stages in the process of conflict.

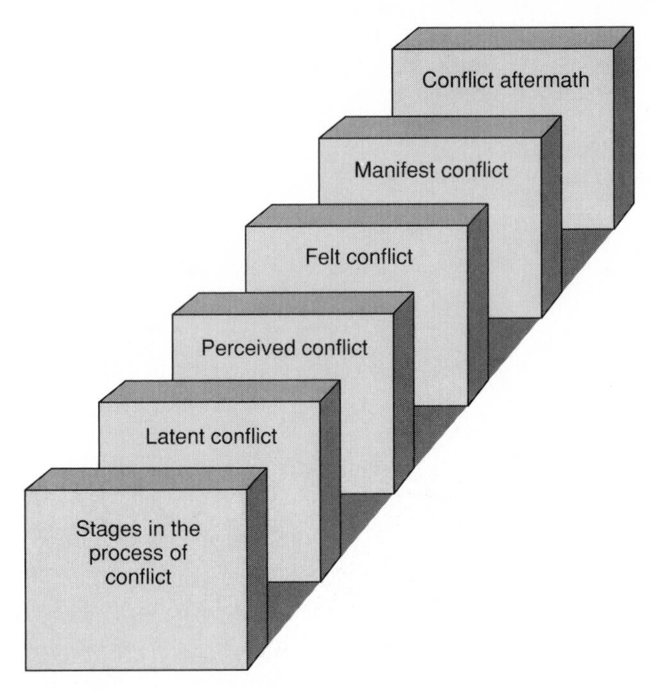

Latent conflict means that the potential for conflict exists but actual conflict has not yet occurred. *Perceived conflict* is the stage in the process at which one or both parties becomes aware of a problem. *Felt conflict* is the stage when one or more parties not only perceive a problem, but feel strongly enough to attempt to do something about it. During *manifest conflict* induced behaviors can be observed. At this stage the conflict either worsens or it is resolved. *Conflict aftermath* is the stage that takes place when the conflict is not resolved. When this is the case, conflict behaviors tend to increase and become more aggressive in nature or they are suppressed until they reach an explosion point.

Clearly, supervisors must know how to resolve conflict effectively. The conflict they must deal with is likely to fall into one of three categories: (1) interindividual, (2) intragroup, and (3) intergroup. Interindividual conflict is between two individuals or among more than two, but does not involve groups. Intragroup conflict is between members of the same group. Intergroup conflict is between or among two or more groups.

Why Conflict Occurs

There are several predictable sources of conflict with which supervisors should be familiar. According to Mel Schnake, these sources are: (1) limited resources, (2) incompatible goals, (3) role ambiguity, (4) different values, (5) different perceptions, and (6) communication problems (Figure 6–16).[16]

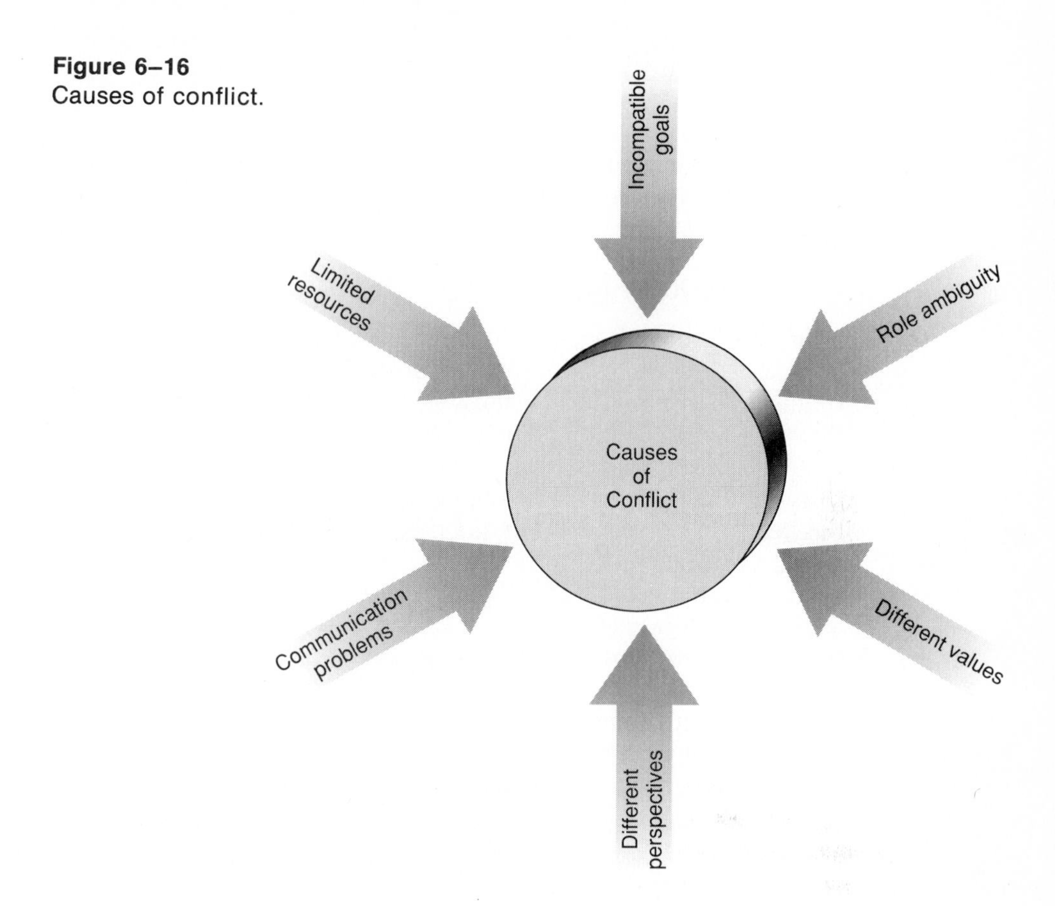

Figure 6–16
Causes of conflict.

Limited resources often lead to conflict in the workplace. It is not uncommon for an organization to have fewer resources (i.e., funds, supplies, personnel, time, equipment, etc.) than might be needed to complete a job. When this happens, who gets the resources and in what amounts? *Incompatible goals* often lead to conflict and incompatibility of goals is inherent in the workplace. For example, conflicts between engineering and manufacturing are common in modern industry. The goal of engineering is to design a product that meets the customer's needs. The goal of manufacturing is to produce a high-quality product as inexpensively as possible. In an attempt to satisfy the customer, engineering might create a design that is difficult to manufacture economically. The result? Conflict.

Role ambiguity can also lead to conflict by blurring "turf lines." This makes it difficult to know who is responsible and who has authority. *Different values* can lead to conflict. For example, if one group values job security and another values maximum profits, the potential for conflict exists. *Different perceptions* can led to conflict. How people perceive a given situation depends on their background, values, beliefs, and individual circumstances. Since these factors are sure to differ among both individuals and groups, particularly in an increasingly diverse workplace, perceptional problems are not uncommon.

The final predictable cause of conflict is *communication.* Effective communication is difficult at best. Improving the communication skills of employees at all levels is an ongoing goal of management. Knowing that communication will never be perfect, communication-based conflict should be expected.

How People React to Conflict

In order to deal with conflict in an effective manner, supervisors need to understand how people react to conflict. According to Thomas, the ways in which people react to conflict can be summarized as follows: (1) competing, (2) collaborating, (3) compromising, (4) avoiding, and (5) accommodating.[17]

A typical reaction to conflict is *competition* wherein one party attempts to win while making the other lose. The opposite reaction to conflict is *accommodation.* In this reaction, one person puts the needs of the other person first. *Compromise* is a reaction in which the two opposing sides attempt to work out a solution that helps both to the extent possible. *Collaboration* involves both sides working together to find an acceptable solution for both. *Avoidance* involves shrinking away from conflict. This reaction is seen in people who are not comfortable facing up to conflict and dealing with it.

In some situations a certain reaction to conflict is more appropriate than another. It is important for supervisors who are responsible for resolving conflict to understand what is and is not an appropriate reaction to conflict. Thomas has summarized the various situations in which specific reactions to conflict are appropriate. Samples of this summary follow:[18]

- *Competing* is appropriate when quick action is vital, or when important but potentially unpopular actions must be taken.
- *Collaborating* is appropriate when the objective is to learn or to work through feelings that are interfering with interpersonal relationships.
- *Compromising* is appropriate when the goal is to achieve temporary settlements to complex issues or to arrive at expedient solutions under time pressure.
- *Avoiding* is appropriate when you perceive no chance of satisfying your concerns or to let people cool down and have time to regain a positive perspective.
- *Accommodating* is appropriate when you are outmatched and losing anyway or when harmony and stability are more important than the issue at hand.

How Conflict Should Be Handled

Supervisors have two responsibilities regarding conflicts: (1) *conflict resolution* and (2) *conflict stimulation.* Where conflict is present, supervisors need to resolve it in ways that serve the long-term best interests of the organization. This will keep conflict from becoming a detriment to performance. Where conflict does not exist, supervisors need to stimulate it to keep the organization from becoming stale and stagnant.[19] Both of these concepts, taken together, are known as **conflict management**.

D. Tjosvold set forth the following guidelines that can be used by supervisors in attempting to resolve conflict:[20]

- Determine how important the issue is to all people involved.
- Determine if all people involved are willing and able to discuss the issue in a positive manner.
- Select a private place where the issue can be discussed confidentially by everyone involved.
- Make sure that both sides understand that they are responsible for both the problem and the solution.
- Solicit opening comments from both sides. Let them express their concerns, feelings, ideas, and thoughts, but in a nonaccusatory manner.
- Guide participants toward a clear and specific definition of the problem.
- Encourage participants to propose solutions. Examine the problem from a variety of different perspectives and discuss any and all solutions proposed.
- Evaluate the costs versus the gains (cost/benefit analysis) of all proposed solutions and discuss them openly. Choose the best solution.
- Reflect on the issue and discuss the conflict resolution process. Encourage participants to express their opinions as to how the process might be improved.

How and When Conflict Should Be Stimulated

Occasionally an organization will have too little conflict. Such organizations tend to be those in which employees have become overly comfortable and management has effectively suppressed free thinking, innovation, and creativity. When this occurs, stagnation generally results. Stagnant organizations need to be shaken up before they die. Supervisors can do this by stimulating *positive conflict* or conflict that is aimed at revitalizing the organization. S. P. Robbins suggests that a yes response to any of the following questions suggests a need for conflict stimulation by the supervisor/manager:[21]

- Are you surrounded by employees who always agree with you and tell you only what you want to hear?
- Are your employees afraid to admit that they need help or that they've made mistakes?
- Do decision-makers focus more on reaching agreement than on arriving at the best decision?
- Do managers and supervisors focus more on getting along with others than on accomplishing objectives?
- Do managers and supervisors place more emphasis on not hurting feelings than on making quality decisions?
- Do managers place more emphasis on being popular than on high job performance and competence?
- Are employees highly resistant to change?
- Is the turnover rate unusually low?
- Do employees, supervisors, and managers avoid proposing new ideas?

Each time one of these questions is answered in the affirmative, it is an indication that conflict may need to be stimulated. It may be possible to have a vital, energetic, developing, improving, organization without conflict, but this isn't likely to happen. Innovation, creativity, and the change inherent in continual improvement typically breed conflict. Therefore, the absence of conflict can also be an indication of the absence of vitality. Since this is the case, supervisors need to know how to stimulate positive conflict.

According to Robbins, the various techniques for stimulating conflict fall into three categories: (1) improving communication, (2) altering organizational structure, and (3) changing behavior.[22]

- *Improving communication* will ensure a free flow of ideas at all levels. Open communication will introduce a daily *agitation factor* that will ensure against stagnation while at the same time providing a mechanism for effectively dealing with the resultant conflict.
- *Altering organizational structure* in ways that involve employees in making decisions that affect them and that empower them will help prevent stagnation. Employees in organizations that are structured to give them a voice will use that voice. The result will be positive conflict.
- *Changing behavior* may be necessary, particularly in organizations that have traditionally suppressed and discouraged conflict rather than dealing with it. Supervisors and managers who find themselves in such situations may find the following procedure helpful: (1) identify the types of behaviors you want employees to exhibit; (2) communicate with employees so that they understand what is expected; (3) reinforce the desired behavior; and (4) handle conflict as it emerges using the procedures set forth in the previous section.

MANAGEMENT AND ORGANIZATIONAL CULTURE[23]

Among the responsibilities of management that supervisors help carry out are establishing an organizational culture and instilling it in all employees. The term *organizational culture* is another way of describing those values and beliefs the organization holds dear. Collectively, these values and beliefs help guide the behavior of all employees not just by rule but by assumption. For example, if quality is part of an organization's culture, it can be assumed that behavior detrimental to quality would be viewed as unacceptable even if there were no written policy specifically prohibiting the behavior. An organization's culture lets employees, customers, and vendors know what is important to the organization and how things are expected to be done.

R. Mitchel and J. H. Dobrinski relate how new CEO Jack Welch changed the organizational culture at General Electric (GE).[24] The culture Welch inherited at GE was stiff, formal, and bureaucratic. It encouraged formal relationships between managers and employees, the avoidance of risk taking, conformance, and stability. The market in which GE had to compete in 1981 (when Welch took over) was not kind to companies that encouraged bureaucracy at the expense of innovation and productivity.

Recognizing this, Welch moved immediately to change the organizational culture at GE. The new culture is more open, less formal, and more encouraging of innovative risk taking. According to Welch, "You want to open up the place so people can flower and grow, expand, hit the home run. When you're tight-bound, controlled, checked, nitpicked, you kill it."[25]

Supervisors have an important role to play in passing on the organization's culture to employees. V. Sathe sets forth the following five strategies supervisors and managers can use to ensure that the organization's culture is instilled and perpetuated.[26]

- *Hire employees who will fit in with the organization's culture.* Supervisors should play an active role in the hiring process (see Chapter 9). By doing so, they can prevent the hiring of an applicant who would not accept the organization's culture. Once an applicant is hired, instilling the organization's culture should be a major element of his or her orientation.
- *Remove employees who refuse to accept the organization's culture.* Even one team member whose behavior consistently runs counter to the organization's culture can disrupt the team and undermine management's efforts to promote teamwork and interorganizational camaraderie. Such team members should be removed in accordance with the procedures set forth in Chapter 9.
- *Monitor and model behavior.* Supervisors and managers should continually monitor the behavior of employees while simultaneously modeling behavior that is in accordance with the organization's culture. It is important to redirect inappropriate behavior before it becomes habitual. In doing so it is equally important to be able to say, "Do as I do." Supervisors and managers who don't model the desired behavior will likely be ineffective in bringing it about.
- *Justify the desired behavior.* Employees are more likely to practice appropriate behavior if they understand why it is important. Supervisors and managers should take the time to justify desired behavior. Employees who understand the *why* behind an organization's culture will be more likely to understand and appreciate the culture. This will help employees internalize the culture, accept it, and make it their own.
- *Communicate continually.* Circumstances change with time. As they change, employees have to adapt. Successfully adapting and changing requires continual communication. What do these new circumstances mean with regard to the organization's culture? Is the culture still valid? What is the appropriate behavior now? Talking, listening, observing, and asking will keep the wheels of change oiled.

MANAGING FOR COMPETITIVENESS

The concept of managing for competitiveness, or MFC, was developed by the Institute for Corporate Competitiveness, a private institute based in Northwest Florida that is dedicated to the improvement of productivity, quality, and competitiveness in business and industry. Unlike other management theories such as management by objectives (MBO) or total quality management (TQM), the sole focus of MFC is competitiveness and how organizations can continually improve it. The basic principles of MFC are summarized as follows:[27]

- *MFC requires an executive-level commitment to the highest ethical standards.* Organizations practicing MFC should work with employees to develop an ethics philosophy and guidelines for operationalizing the philosophy. Both should be stated in writing and shared with all employees. Managers at all levels should model the desired behavior.
- *MFC requires an executive-level commitment to building and maintaining a world-class workforce.* Organizations practicing MFC should continually train and retrain employees at all levels. Training should be based on careful and continual analysis of employee needs and should be viewed as a normal part of doing business. Organizations practicing MFC increase the level of training during difficult economic times and emphasize the teamwork approach at all times.
- *MFC requires an organizationwide effort to continually improve productivity.* Organizations practicing MFC must make individual employees responsible for their own productivity and empower them to continually improve it. Suggestions for improvements are not just encouraged, they are expected and they are acted on.
- *MFC requires an organizationwide effort to continually improve quality.* Organizations practicing MFC must also make individual employees responsible for quality and empower them to continually improve it. Suggestions for improvements are not just encouraged, they are expected and they are acted on.
- *MFC requires an organizationwide effort to continually decrease overhead costs.* Organizations practicing MFC must involve all employees at all levels in finding ways to continually reduce fixed costs. This might mean eliminating layers of bureaucracy, reducing inventory, or any other strategy that reduces fixed expenses. Suggestions from employees are not just encouraged, they are expected.
- *MFC requires an organizationwide effort to continually improve long-term profitability.* Organizations practicing MFC must take the long view. To do so, it may be necessary to educate owners and stockholders. Focusing too intently on the next quarterly dividend can force a company to make decisions that are harmful in the long run.
- *MFC requires an organization to make individual employees profit-conscious CEOs of their respective jobs.* Organizations practicing MFC must establish appropriate incentive programs that allow employee work teams to benefit financially from their improved productivity, cost savings, and enhanced quality.
- *MFC requires an organization to commit to becoming the world champion in its field.* Organizations practicing MFC must be committed to adhering to all of the preceding principles and to undertaking any other efforts necessary to achieve and maintain world-champion status.

CHECKLIST FOR MANAGERS

Keeping up with all the strategies managers and supervisors can use on a daily basis to enhance communication and stimulate improvement can be difficult for even the most experienced managers and supervisors. Bill Rossnagel recommends the use of a written checklist that is posted where it can be easily seen.[28] The checklist contains questions arranged under the following open-ended statements:

- Have you remembered to:
- Have you considered:
- Have you tried:
- Have you listed:
- Have you communicated with:
- Have you issued:
- Have you verified:

The specific statements that complete these open-ended statements are written in list form. The actual checklist statements under each open-ended statement will vary from situation to situation and should be developed by the individual manager/supervisor. Here are some samples by Rossnagel:

- *Have you remembered to:*
  1. Make your *unannounced monthly inspection* of your facilities?
  2. Indicate to all key employees the importance of using calendars and daily phone logs for maximum time management?[29]
- *Have you listed:*
  1. Emergency phone numbers on the front door of any plant without a guard?
  2. Key company numbers?[30]

SUMMARY

1. Management means the creation and maintenance of a work environment that promotes the efficient and effective accomplishment of organizational goals. Key concepts in this definition are work environment, goals, efficiency, and effectiveness.
2. Efficiency means doing work better or with a minimum of waste. Effectiveness means setting the right goals and achieving them.
3. Managers have the authority to set goals for an organization and the responsibility to achieve them. Managers are held accountable.
4. The primary functions of management are planning, organizing, staffing, facilitating, and monitoring.
5. Some basic principles of management include the division of work, giving authority along with responsibility, responsibility for morale and discipline, structuring so that employees have only one boss, maintenance of a uniform set of goals, placing the organizational goals first, rewarding work, establishing a clear governance structure, establishing equitable policies, and practicing and encouraging initiative.
6. The five basic approaches to management are scientific, behavioral, statistical, integrated systems, and situational.
7. Supervisors differ from managers in the level of technical, people, and paper skills they need. Supervisors rank these categories of skills as follows: (1) technical, (2) people, and (3) paper.

KEY TERMS AND PHRASES

Accountability
Authority
Behavioral management
Conflict management
Discipline
Division of work
Effectiveness
Efficiency
Equitable policies
Facilitating
Facilities
Goal
Governance structure
Initiative
Input
Integrated systems management
Manage
Management
Manager
Monitoring
Morale
Organizing
Output
Paper skills
People skills
People processes
Planning
Resources
Responsibility
Scientific management
Situational management
Staffing
Statistical management
Technical skills
Work environment

CASE STUDY: Special Treatment for Management?[31]

The following real-world case study illustrates a management problem. Read the case study carefully and answer the accompanying questions.

In May 1990, Roger Smith, chairman of the board of General Motors, was approaching retirement after ten years at the helm. During his tenure, General Motors had had several good years. However, 1990 saw the automotive giant in the midst of an aggressive cost-cutting program. To bring costs down, General Motors was preparing to ask the United Auto Workers union to accept a cost-sharing plan for medical care. In addition, the company had frozen merit raises for middle managers. Over a three-year period, General Motors' cost-sharing strategies had succeeded in reducing corporate-wide costs by $12.3 billion.

Therefore, many were shocked to learn that Smith and General Motors' top management team planned to ask shareholders to approve an expensive change to the pension plan for executive managers. The costly change would benefit over 3,000 top-level managers. Roger Smith alone would see his annual retirement pension increase from $700,000 to $1.25 million.

General Motors based the proposal on the fact that the company's pension plan for executives ranked near the bottom when compared with other major corporations in the United States. The General Motors management team was concerned that such a low ranking might make it difficult to attract high-quality executive talent.

1. Do you agree with General Motors' proposal to increase executive pensions? Why or why not?
2. Analyze this decision in terms of the following principles of management: morale/discipline; organization's needs first; equitable policies and practices.
3. Suppose you are a supervisor at General Motors and Roger Smith asks all supervisors to respond to the question, "If you were me, what would you do?" What would you tell him?

REVIEW QUESTIONS

1. Define the term *management.*
2. Briefly explain the impact of the work environment on workers.
3. Explain the difference between efficiency and effectiveness.
4. What is a goal?
5. What are people processes?
6. What role do facilities play in the work environment?
7. What role do resources play in the work environment?
8. Define the term *manager.*
9. Explain the term *accountability.*
10. Briefly explain the following functions of management: planning; staffing; facilitating.
11. Briefly explain the following principles of management: division of work; uniform goals; only one boss; company first.
12. Contrast and compare the scientific and behavioral approaches to management.
13. Contrast and compare the statistical and integrated systems approaches.
14. Explain the situational approach to management.
15. Explain the differences between a supervisor and a manager.

SIMULATION ACTIVITIES

The following activities are provided to give you opportunities to apply what you learned in this chapter and previous chapters to the solution of simulated supervisory problems.

1. Identify a manager and a supervisor in a local industrial company. Interview them about the management approach they use. Contrast their approaches with the Japanese style described by James R. Lincoln in *California Management Review.*
2. You are the supervisor of two teams of machine operators that produce different parts of the same product. Team A works the first shift, Team B the second. Both teams are in the middle of rush orders that are behind schedule and now a conflict has developed between the lead operator in Team A and her counterpart in Team B. The Team B leader claims his members get a late start on their work every day because Team A members break down and clean their machines at the end of their shift. Team A's leader counters that not to do so would cause her team members to lose thirty minutes a day of productive time at the beginning of their shift. How do you plan to resolve this conflict? Explain every step you plan to take.

ENDNOTES

1. Hatakeyama, Y. *Manager Revolution! A Guide to Survival in Today's Changing Workplace* (Cambridge, MA: Productivity Press, 1985), pp. 2–16.
2. Newman, M. A. "The Issue Here is . . . ," *Sky*, July 1990, pp. 84–89.
3. Hatakeyama, Y. *Manager Revolution!* pp. 17–27.
4. Ibid., p. 19.
5. Ibid., p. 23.
6. Lincoln, J. R. "Employee Work Attitudes and Management Practice in the U.S. and Japan: Evidence from a Large Comparative Survey," *California Management Review,* Fall 1989, pp. 89–106.
7. Brown, W. S. "Management by Ones: One-on-One, One-by-One," *Supervisory Management,* May 1990, pp. 3–7.
8. Mironoff, A. "De-Stalinizing the Corporation," *Training,* August 1990, p. 30.
9. Ibid.
10. Ibid., p. 32.
11. Ibid., p. 33.
12. Ibid.
13. Thomas K. W. and Schmidt, W. H. "A Survey of Managerial Interests with Respect to Conflict," *Academy of Management Journal,* 1976, Number 19, pp. 316–318.
14. Thomas, K. W. "Conflict and Conflict Management," in M. D. Dunnette, Ed., *Handbook of Industrial and Organizational Psychology* (New York: Wiley & Son, 1976), p. 163.
15. Pondy, L. R. "Organizational Conflict: Concepts and Models," *Administrative Science Quarterly,* 1967, Number 12, pp. 296–320.
16. Schnake, M. E., *Human Relations* (Columbus, OH: Merrill, 1990), pp. 362–364.
17. Thomas, K. W. "Conflict and Conflict Management," p. 163.
18. Thomas, K. W. "Toward Multi-Dimensional Values in Technology: The Example of Conflict Behaviors," *Academy of Management Review,* 1977, Number 2, p. 487.
19. Robbins, S. P. "Conflict Management and Conflict Resolution Are Not Synonymous Terms," *California Management Review,* 1978, Number 21, pp. 67–75.
20. Tjosvold, D. "Making Conflict Productive," *Personnel Administrator,* 1984, Number 29, pp. 121–130.
21. Robbins, S. P. "Conflict Management and Conflict Resolution," pp. 67–75.
22. Ibid.
23. Schein, E. H. "Coming to a New Awareness of Organizational Culture," *Sloan Management Review,* 1984, Number 25, pp. 3–16.
24. Mitchel, R. and Dobrinski, J. H. "Jack Welch: How Good a Manager?" *Business Week,* December 14, 1987, pp. 92–103.
25. Ibid.
26. Sathe, V. "Implications of a Corporate Culture: A Manager's Guide to Action," *Organizational Dynamics,* 1983, Number 12, p. 18.
27. Institute for Management and Supervision; *Mission Statement and Principles,* January 1990, p. 1.

28. Rossnagel, B. "A Checklist to Take You Through Hectic Days," *Supervisory Management,* February 1991, p. 12.
29. Ibid.
30. Ibid.
31. Treece, J. B. "GM: Wrong Move. Wrong Time. Wrong Reason," *Business Week,* May 28, 1990, p. 32.

CHAPTER SEVEN

Planning, Scheduling, and the Supervisor

CHAPTER OBJECTIVES

After studying this chapter, you will be able to define or explain the following topics:

- Planning
- The Supervisor's Role in Planning
- The Difference Between Goals and Plans
- The Various Types of Plans
- The Planning Process
- Scheduling Techniques
- How Policies Relate to Planning
- How to Plan a Plan

CHAPTER OUTLINE

- Planning Defined
- Supervisor's Role in Planning
- Difference Between Goals and Plans
- Types of Plans
- The Planning Process
- Scheduling Techniques
- Rules of Thumb for Scheduling
- PERT and CPM
- Gantt Charts
- Computer-Aided Process Planning
- Policies and Planning
- Planning the Plan
- Other Planning Concerns
- Summary
- Key Terms and Phrases
- Case Study Application Problem
- Review Questions
- Simulation Activities
- Endnotes

One of the primary management functions of supervisors is planning. Recall that a fundamental principle of management is that an organization should ensure uniformity in planning by maintaining one master set of organizational goals. All subunit planning should be an extension of these master goals.

It is important for supervisors to be skilled planners. Therefore, they must invest the time necessary to develop, monitor, adjust, and carry out plans. This chapter will help prospective and practicing supervisors develop planning skills and an understanding of the importance of planning.

PLANNING DEFINED

Some people think of planning as goal setting. Others think of it as developing strategies. Still others think of it as establishing benchmarks against which performance can be monitored. The truth is, planning is all of the above. It is a matter of deciding where you are going, how you will get there, and when you will arrive (Figure 7–1). Therefore, **planning** can be described as follows:

> *Planning is the process of setting goals, developing strategies for accomplishing them, and establishing benchmarks for monitoring progress.*

SUPERVISOR'S ROLE IN PLANNING

Supervisors have a key role to play in planning. Most companies are made up of subunits and each subunit typically has a supervisor. Supervisors are responsible for the planning for their subunits. Higher management is responsible for setting the goals and

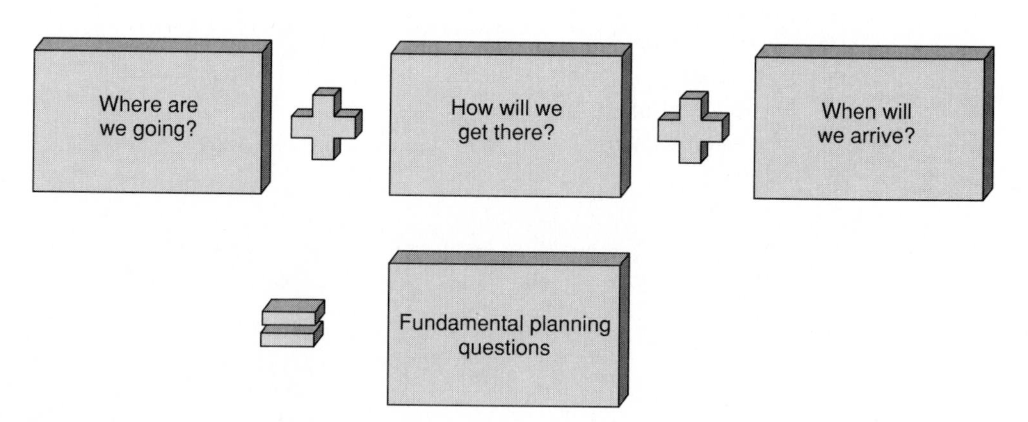

Figure 7–1
These questions are fundamental to the planning process.

Figure 7–2
The supervisor's plans operationalize those of higher management.

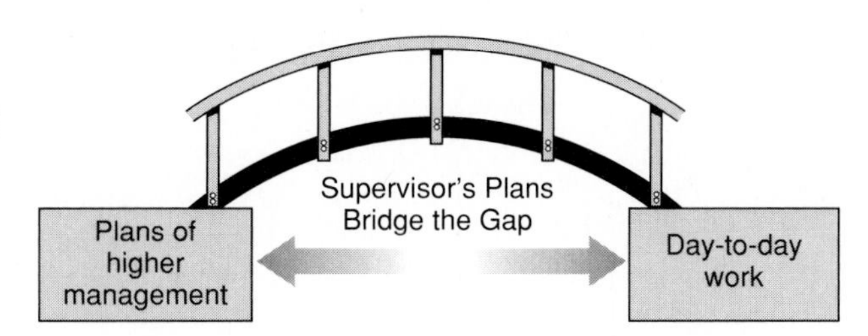

developing the broad plans of a company; supervisors are responsible for developing plans that translate the plans of higher management into everyday action (Figure 7–2). They are responsible for monitoring day-to-day progress and adjusting their plans as appropriate. They are also responsible for including the people they supervise in the planning process.

Overall company plans that include subunit plans developed by supervisors tend to be more effective plans. There are two reasons for this. First, supervisors who participate in the planning process are more likely to internalize the plans as their own. Second, this approach ensures the input of the people who know the most about getting the job done on a daily basis. Supervisors and the people they supervise know more than anyone about the technology, processes, materials, and other resources that will be focused on carrying out plans.

Because of the nature of the process, higher management develops long-range plans. Supervisors develop short-range plans, providing guidance for periods ranging from one day to one year. Making daily, weekly, and monthly plans takes up most of the time supervisors devote to planning (Figure 7–3).

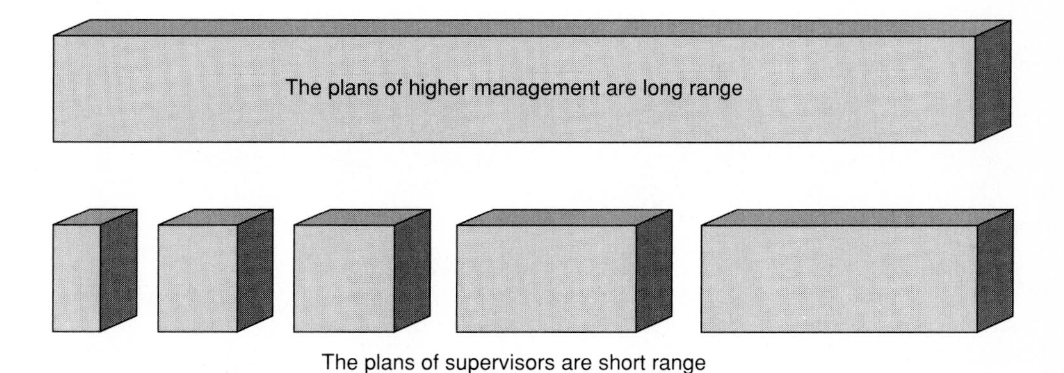

Figure 7–3
The short-range plans of supervisors implement the long-range plans of higher management.

DIFFERENCE BETWEEN GOALS AND PLANS

It is important for prospective and practicing supervisors to understand the various terms and concepts associated with planning. These terms include *goals, plans, objectives, policies, regulations, schedules,* and *procedures* (Figure 7–4).

Goals and **objectives** are similar. In fact, the terms can be used interchangeably, and frequently are. Those who distinguish between the two argue that goals are stated in broader terms than objectives; objectives are more specifically stated than goals.

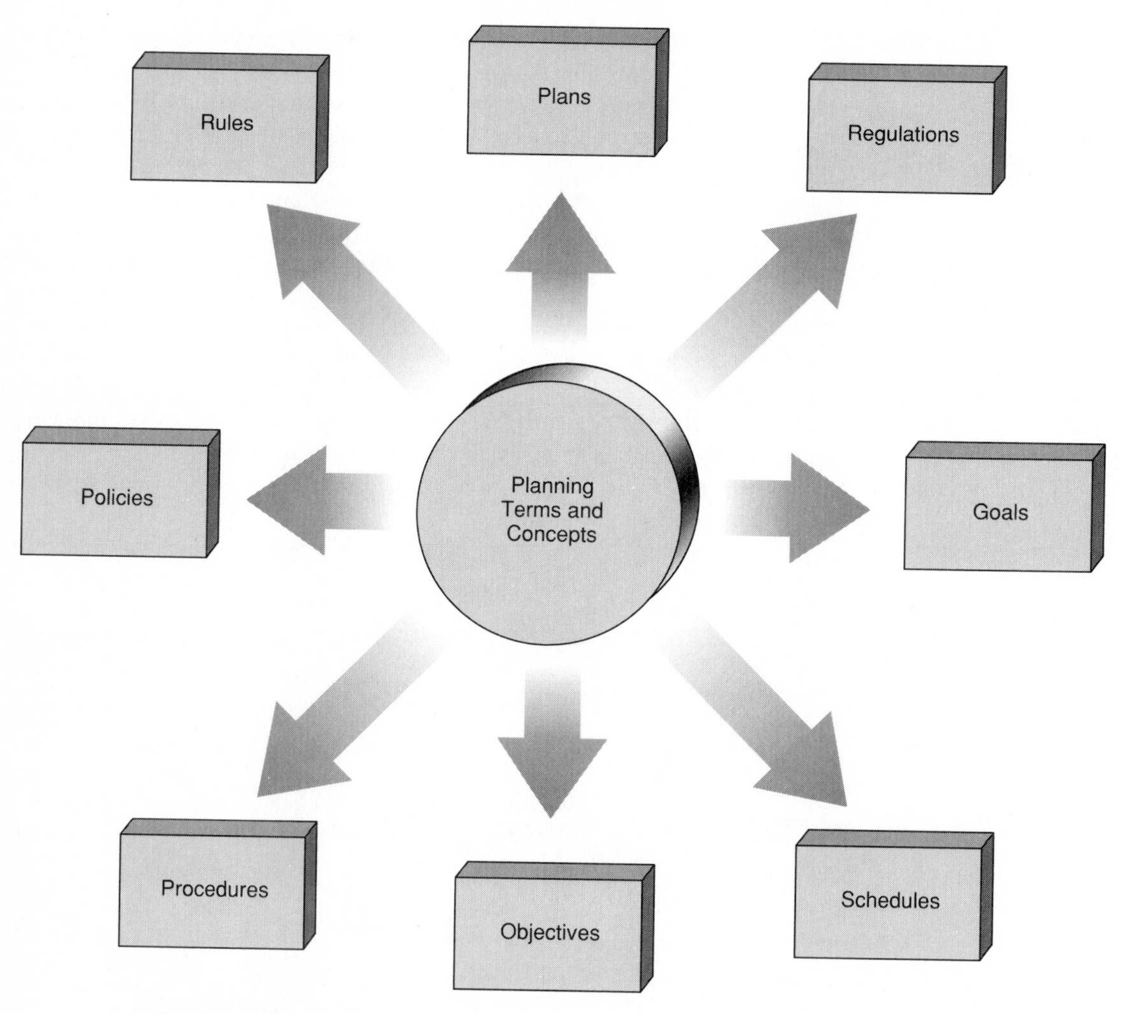

Figure 7–4
These terms are widely used in planning.

Consequently, objectives can be used to more clearly delineate the desired outcomes set forth as goals. For the purpose of this chapter, we will view goals and objectives as being the same. Both define desired future outcomes. They answer the question "Where are we going?" Goals or objectives represent one component of a plan. A **plan** is simply the written product of the planning process.

Policies are guidelines that establish the broad parameters within which resources are focused on accomplishing goals. Rules and **regulations** translate policies into language that defines the day-to-day do's and don'ts that control workers' behavior. **Procedures** are the methods used to implement plans on a day-to-day basis. **Schedules** establish the timetables within which goals will be accomplished (Figure 7–5).

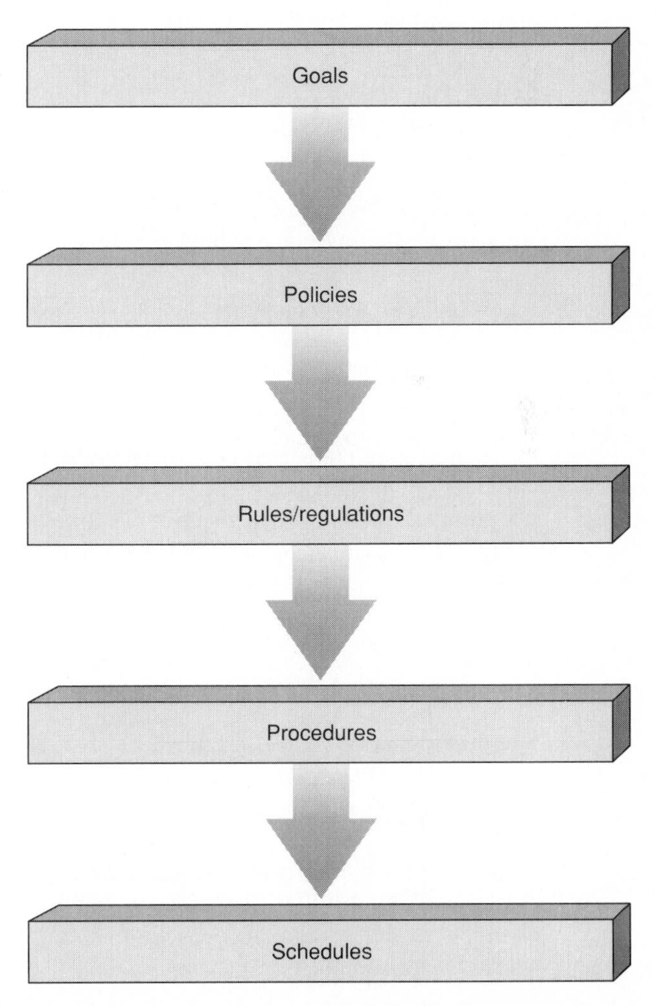

Figure 7–5
Components of a plan.

TYPES OF PLANS

Management consultant Peter Drucker has this to say about planning:

> Management has no choice but to anticipate the future, to attempt to mold it, and to balance short-range and long-range goals. It is not given to mortals to do well any of these things. But lacking divine guidance, management must make sure that these difficult responsibilities are not overlooked or neglected, but taken care of as well as is humanly possible.[1]

These responsibilities are taken care of through planning.

There are a number of different types of plans with which prospective and practicing supervisors should be familiar. When classified according to duration, these are long-range and short-range plans; when classified according to purpose, standing and single-use plans; and when classified according to level, strategic and operational plans.

Long-Range and Short-Range Plans

Whether a plan is considered long-range or short-range is not a function of the span of time associated with it. Rather, the distinction is a function of the time span over which the plan is effective.[2] According to Drucker,

> Long-range planning should prevent managers from uncritically extending present trends into the future, from assuming that today's products, services, markets, and technologies will be the products, services, markets, and technologies of tomorrow, and above all, from dedicating their resources and energies to the defense of yesterday.[3]

Long-range plans are typically thought of as being effective from one to five years, and possibly longer. Long-range plans are the responsibility of higher management.

Short-range plans are effective for time periods of less than one year and perhaps for periods as brief as one day. They are typically weekly, monthly, and quarterly plans. Short-range plans are the responsibility of supervisors.

Standing and Single-Use Plans

Standing plans cover those aspects of the job that do not change or at most change infrequently. These plans cover such issues as personnel practices, purchasing procedures, equity practices, and safety procedures. **Single-use plans** are throwaway plans. They are developed, used once over a short period, and thrown away, or revised to become a new plan. For example, a schedule established for a specific job would be a single-use plan. A budget for an individual project would be a single-use plan. Standing plans are typically developed by higher management. Single-use plans are typically developed by supervisors.

Strategic and Operational Plans

Drucker makes the following statement about **strategic planning:**

> Planning what is our business, planning what will it be, and planning what should it be have to be integrated. . . . The skill we need is not long-range planning. It is strategic decision-making, or perhaps strategic planning.[4]

Long-range plans are typically strategic plans, sometimes called **master plans.** These are plans developed by higher management to chart a broad course for the organization. **Operational plans** are short-term plans developed by supervisors to carry out the organization's strategic plans. An operational plan grows out of a strategic plan and should be tied directly to specific goals in the strategic plan.

It is important for modern supervisors to understand strategic planning. Drucker's thoughts in this regard can be summarized as follows:[5]

- *Strategic planning is not a bag of tricks or a group of technologies.* Rather, strategic planning involves analytical thinking and an effective commitment of resources to action.
- *Strategic planning is not forecasting.* Humans cannot predict the future. This is why strategic planning is necessary. It is about creating the future, not predicting it.
- *Strategic planning is not about future decisions but about the futurity of present decisions.* Strategic planning is about preparing an organization for an uncertain, unpredictable future.
- *Strategic planning is not an attempt to eliminate risk.* Strategic planning is about choosing rationally among risk-taking options rather than forging ahead blindly or randomly.

Drucker concludes his discussion of strategic planning with the following definition:

> the continuous process of making present entrepreneurial (risk-taking) decisions systematically and with the greatest knowledge of their futurity; organizing systematically the efforts needed to carry out these decisions; and measuring these decisions against the expectations through organized, systematic feedback.[6]

THE PLANNING PROCESS

Planning is a process. A plan is the product that results from the planning process. This section focuses on the various steps in the planning process. There are different opinions on the number of steps. For the purposes of this book, the planning process recommended for supervisors has been divided into the following ten steps (Figure 7–6):

1. Review the plans of higher management to determine the direction for your unit.
2. Share the plans of higher management with your employees and solicit their input.
3. Set goals for your unit that are logical extensions of those of higher management.
4. Develop an overall plan for your unit.
5. Develop more specific plans for each of your subunits.
6. Develop strategies with measurable performance criteria.
7. Establish timetables and schedules.
8. Assign responsibility/make assignments.
9. Review the finished plan with everyone involved in carrying it out.
10. Monitor continually and make adjustments where and when appropriate.

Review Plans of Higher Management

Supervisors develop plans to carry out the plans of higher management. Every organization or unit within a company is responsible for carrying out a portion of a company's master plan. Before beginning to plan at this level, supervisors should review the plans of higher management. These plans will give supervisors the direction they need to develop effective plans for their units. Every component of a supervisor's plan should make a specific measurable contribution to fulfilling the master plan developed by higher management.

Share the Master Plan with Employees

It is important for the employees who will do the work in carrying out your part of the master plan to see the big picture. The master plan will show your employees where the company is heading, where the employees fit into the process, and how their efforts contribute to fulfilling the company's goals.

Figure 7–6
Steps in the planning process for supervisors.

Planning Process for Supervisors

1. Review the plans of higher management.
2. Solicit input from your employees.
3. Set goals for your unit.
4. Develop an overall plan for your unit.
5. Develop more specific plans for your subunits.
6. Establish benchmarks or quotas.
7. Establish timetables/schedules.
8. Assign responsibility.
9. Review the finished plan with everyone included in carrying it out.
10. Monitor continually and adjust.

Remember, the goals explain where you are and the procedures explain how you will get there.

Develop More Specific Plans for Subunits

In this step you divide the goals and corresponding procedures as they relate to subunits or to specific functions within the unit. For example, if you are the machine shop supervisor, some of the goals/procedures may relate specifically to the sawing function while others relate to milling. At this point, strategies are developed for achieving the goals. Strategies are specific action statements. Performance criteria are also developed. They operationalize the strategies and, in every case possible, should include numbers or other measurable elements. If your goals were stated properly they contain numbers. These numbers should be broken down and spread over the procedures assigned to each subunit. For example, if a goal is to increase output by five subassemblies per unit, what part will machine operators play? How about sawing machine operators? The procedures should clearly state what part of each function will contribute to meeting this goal.

Establish Timetables/Schedules

If your goals were properly stated, they contain a time element: so many units per day, week, month, etc. In this step you establish timetables or schedules. For example, if your unit has 90 days to produce 180 parts you should establish a schedule to produce at least two parts per day. At 30 days, your unit should have produced 60 parts, at 60 days 120 parts, and at 90 days 180 parts. Schedules and timetables make it easy to monitor progress.

Assign Responsibility/Make Assignments

It is critical that all people who have to carry out a plan know what their responsibilities are. This means specific assignments should be made and monitored.

Review the Finished Plan

Once assignments have been made it is important to bring the team together to review the finished plan. All people who will have a role in carrying out the plan should review it and familiarize themselves with benchmarks, quotas, timetables, schedules, and assignments. At this point supervisors ensure that all team members understand and accept the plan.

Monitor and Adjust

Once a plan has been reviewed, it is implemented. Seldom will everything fall into place exactly as planned. There are usually too many variables that are beyond the scope of a supervisor's control. Consequently, it is important for supervisors to monitor progress and make adjustments wherever and whenever they are needed.

This is an important point. Supervisors and the plans they develop must be flexible. Supervisors should avoid falling into the trap of sticking with a plan that is not working. This happens when supervisors become so attached to their plans that they come to value them more than what the plans were supposed to achieve. Plans only have value to the extent that they work. If a plan is not working, supervisors should adjust and move on.

SCHEDULING TECHNIQUES

Scheduling involves scheduling work to be done, people to do the work, and the materials and equipment needed. These are not separate tasks. Rather, they must be integrated to ensure a well-coordinated, steady flow of work. Scheduling is a fundamental part of the planning responsibilities of supervisors. It is both an art and a science. The scientific aspects of scheduling can be learned. The most widely used scientific scheduling techniques are covered in this section. The art of scheduling comes from experience.

Here we discuss general rules that should be observed when scheduling and how to apply such techniques as routing charts, PERT, CPM, Gantt charts, and CAPP (Figure 7–9).

RULES OF THUMB FOR SCHEDULING

Regardless of the scheduling technique used, some general rules of thumb should be observed. These grow out of the art of scheduling and apply across the board.

Figure 7–9
Some of the more widely used scheduling techniques and systems.

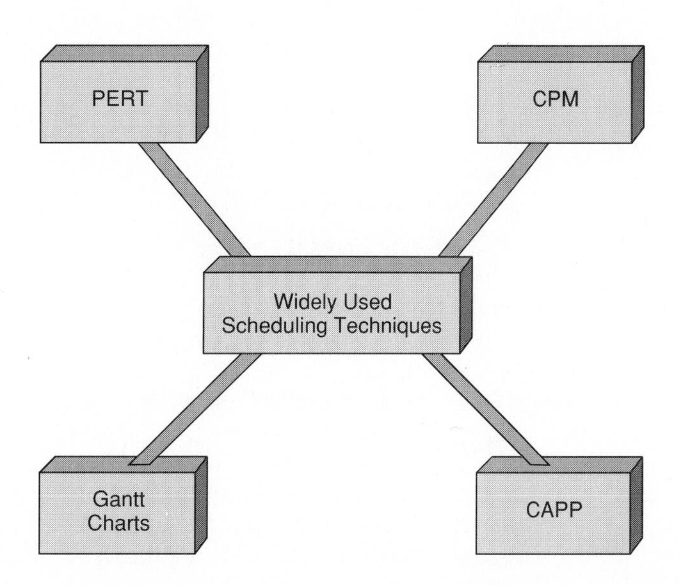

Schedule for Less than Capacity

Wise supervisors hold a certain amount of capacity in reserve. This allows them to respond to emergencies, unexpected absences, breakdowns, and other unforeseen circumstances. When you schedule at 100 percent of capacity everything has to fall in place exactly as planned or your plans go awry. Experienced supervisors know that things seldom work out exactly as planned. For this reason, it is prudent to build a buffer into your schedule.

Prepare for Slack Periods

When you schedule at less than capacity, there will be times when workers will be idle, particularly in those occasional periods when reality matches planning. Wise supervisors prepare for this by scheduling in backlogged work, routine maintenance, and other types of work that are sometimes put off. This will keep workers from becoming bored or stretching their work to cover the allotted time.

Avoid Scheduling Overtime

Loading overtime in at the front end of a project is not a wise practice. Supervisors should avoid scheduling overtime unless a project cannot be completed on time without it.

There is an ongoing debate as to whether or not productivity falls off during overtime hours. Data to prove either side of the debate do not exist. However, it is known that fatigue has a negative impact on productivity. Common sense suggests that employees who have already worked a full shift are bound to experience fatigue during overtime hours.

Schedule Accurately

Supervisors will always feel the pressure to hurry. It goes with the job. This is because customers typically want their orders yesterday. It is also because a better delivery is sometimes more important than cost in winning a bid or getting an order. This is the nature of business and it is not likely to change.

This kind of pressure can lead inexperienced supervisors to promise more than they can deliver. Avoid this trap. It is much better to give higher management a completion date that is later than they want than to give an unrealistic date that cannot be met.

The obverse is also true. Supervisors should avoid the practice of **padding** a schedule and then delivering early. If this happens often, your schedules will lose credibility with higher management and they will begin to promise early delivery dates.

Wise supervisors schedule accurately and stick to their schedules. They also work with the sales force, either directly through a higher manager, to ensure that sales personnel are not making unrealistic promises. Sales people need to be kept up to date about work in progress, back orders, and other factors that will have an impact on completion dates for future projects. A positive working relationship between those who

Figure 7–10
Sample routing ticket.

MIL-TECH MANUFACTURING

Item ______ Job ______ Date ______

Operation	Date Needed	Assignment
•Clean	1-5	Alexander
•Drill	1-5	Jones, drill #5
•Deburr	1-6	Jones, machine #5
•Prep	1-6	Garcia
•Paint	1-6	Garcia, #22
•Pack and ship	1-7	My-Ling

Comments:

sell a company's products/services and those who provide them is essential to accurate scheduling.

Routing Tickets[7]

Typically the first step in scheduling, whether using manual or computerized techniques, is to develop a *routing ticket.* Each time a new order is received a routing ticket is completed. A routing ticket summarizes all the major operations to be performed, the dates when they must be completed, and assigns the people, equipment, and operations necessary to complete the job. Figure 7–10 is an example of a routing ticket. The greatest benefit of a routing ticket is that it lets workers know when to expect work, when it should be completed, and where it goes next. The disadvantage of routing tickets is that they are isolated. They do not show an overall picture of the total workload in a unit.

PERT AND CPM

Program Evaluation and Review Technique, or **PERT**, was developed in the late 1950s through a partnership of the U.S. Navy's Special Projects Office and a private management consulting firm. The original purpose of PERT was the planning, scheduling, and controlling of the Polaris submarine and its missile system.

The value of PERT was recognized quickly after this initial project and it began to be used by other governmental agencies. It is now widely used in government, business, and industry. PERT is particularly effective for sequencing and coordinating projects that include a large number of activities that must be completed within a specified time. This is an accurate description of most projects supervisors are responsible for.

CPM is the acronym for Critical Path Method. Like PERT, CPM was developed in the late 1950s. It is used to identify the various paths that lead from the beginning of a project to its completion. The longest of these paths (in time) is the **critical path.** Once the critical path for a project has been identified, supervisors know where they need to focus their attention and resources.

The PERT Process

PERT consists of seven steps. CPM comes into play in Step 6. These steps are as follows:

1. Identify the project goal.
2. Identify the events.
3. Sequence the events.
4. Identify the activities.
5. Estimate the time for each activity.
6. Set up the network.
7. Implement the network and adjust as needed.

The PERT process is best illustrated by example. Consider the fictitious example of Michael Bates, shop floor supervisor for Container Corporation of America. CCA manufactures containers that are used for shipping non-nuclear armaments for the U.S. Department of Defense. The current contract stipulates completion of the containers in 75 days. Michael Bates used PERT as his scheduling technique. Figure 7–11 is his PERT planning worksheet.

The first step was for Bates to identify the project goal. In this case the goal was to complete all containers within 75 days. The second step was to identify the events. Each PERT **event** represents a specific point in time or benchmark in the overall project. Bates identified 10 events.

Next, Bates had to sequence the events. In any project, some events must precede others, some can occur simultaneously, and some must follow others. Column C of Figure 7–11 shows the planned **sequence** of events. Notice that Events 3, 4, and 5 must be preceded by Events 1 and 2, but they can be accomplished simultaneously.

The fourth step was to identify the activities. **Activities** are tasks that must be completed before an event can occur. For example, in Column D of Figure 7–11, Activity 2-3 is all the tasks that must be accomplished before Event 3 can occur.

Fifth, estimate the time for each activity. With PERT this is done by using an **optimistic time,** a **pessimistic time,** and a **most likely time.** The equation used for calculating the estimated time is:

$$\frac{\text{optimistic time} + 4\ (\text{most likely time}) + \text{pessimistic time}}{6}$$

Container Corporation of America
PERT Worksheet

Project name/number Non-Nuclear Munitions Containers (Job 90-816)

Project goal Complete all containers in 75 days

A	B	C	D	E
Event number	Event	Sequence of events	Related activities	Estimated time (days)
1	Begin project	None	None	0
2	Receive materials	1	1-2	0
3	Cut top components	1,2	2-3	7
4	Cut bottom components	1,2	2-4	13
5	Cut side components	1,2	2-5	5
6	Assemble containers	1,2,3,4,5	3-6	12
7	Pressure test containers	1,2,3,4,5,6	6-7	9
8	Paint containers	1,2,3,4,5,6,7	7-8	4
9	Stack containers for delivery	1,2,3,4,5,6,7,8	8-9	1
10	Complete project	1,2,3,4,5,6,7,8,9	9-10	3

Figure 7–11
Sample PERT worksheet.

For example, assume your time estimates for an activity are as follows:

optimistic time = 5
most likely time = 9
pessimistic time = 12

The formula for determining the PERT estimated time would be applied as follows:

$$\frac{5 + 4\,(9) + 12}{6} = 8.8 \text{ days}$$

The PERT estimate in this example is 8.8 days. Michael Bates applied this process in determining the estimated times shown in Column E of Figure 7–11.

Finally, set up the network. A **network** is a graphic representation of a group of events and activities. A sequence of activities that connects one event with another event is called a **path**. Identifying the critical path in a project is one of the primary functions of PERT.

Setting Up the Network

The network allows supervisors to schedule a series of related and dependent activities that proceed to a final goal. Networks can be particularly useful when several different

Figure 7–12
Sample network diagram.

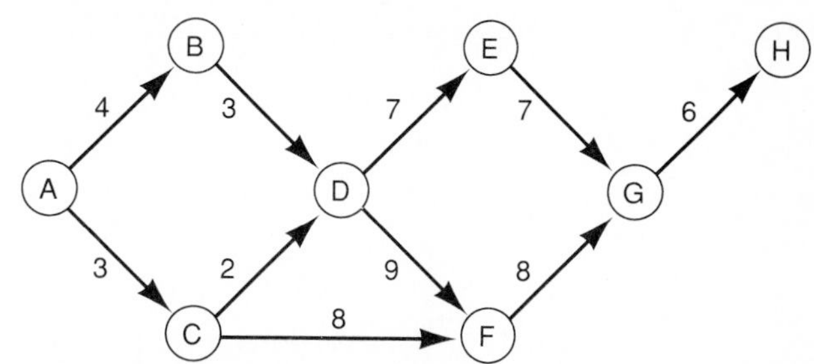

operations must come together at critical times. A network diagram has three basic components: (1) tasks or completed segments, (2) activities, and (3) relationships.[8] Tasks are represented by circles and activities by arrows. These symbols, together with the time associated with each activity as indicated by numbers placed near the center of an arrow, represent relationships. Each number indicates time in a predetermined unit (i.e., hours, days, weeks, etc.). These times are computed as shown in the previous section. Figure 7–12 is an example of a network diagram that illustrates these components.

A network diagram such as the one shown in Figure 7–12 is developed in stages. The first stage involves developing the PERT planning worksheet as explained in this chapter.

Using this information, the circles and arrows are drawn and times are labeled appropriately. With this done, the diagram is examined to identify the shortest route from beginning to completion. This sequence of activities is the critical path.

The final step is to implement the network and adjust as necessary. This step and all the others have been simplified by the advent of the personal computer. Software is now readily available that makes using PERT more practical and less time consuming.

GANTT CHARTS

Another traditional scheduling method available to supervisors is the **Gantt chart,** named after its originator, Henry Laurence Gantt, an industrial engineer. Although the Gantt chart was developed in the early 1900s it is still widely used today.

Gantt charts give supervisors a way to extract information specific to their units from route sheets and to display it graphically. This makes the information easier to understand and communicate. The underlying purpose of a Gantt chart is to allow supervisors to increase the utilization rate for machines and equipment in their units.

Developing Gantt Charts

The best way to illustrate the use of Gantt charts is by example. Consider the example of Emile Enfante, machining supervisor for Austin Metal Works, Inc. Enfante has three production orders (90-11, 90-12, and 90-13) he must fill. The order sheet (Figure 7–13) gave the machines to be used, the required sequence, and the estimated time for each machine (in hours).

Figure 7–13
Sample order sheet.

Austin Metal Works, Inc.
Machine Order Sheet

	Order number(s)								
	90–11			90–12			90–13		
Sequence	1	2	3	1	2	3	1	2	3
Machine	A	B	C	C	A	B	B	A	C
Time	2	6	4	4	8	4	6	4	2

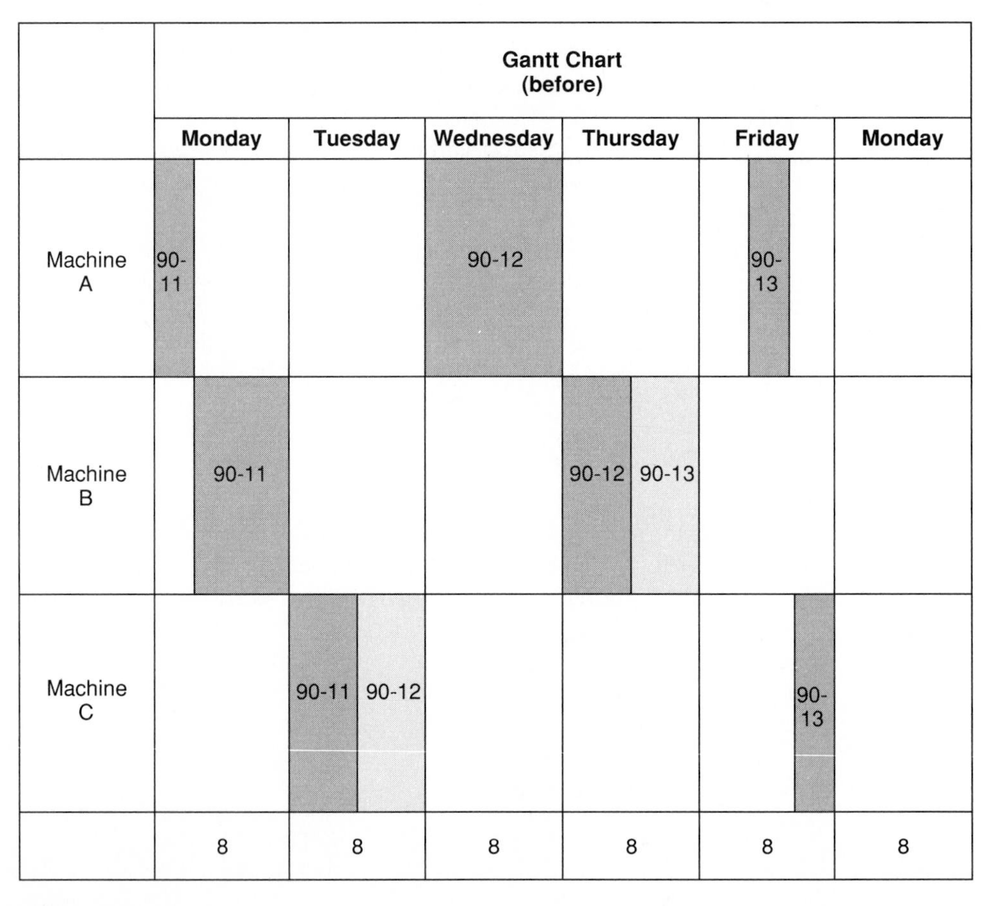

Figure 7–14
Sample Gantt chart before condensing the work.

To make this information easier to understand, Enfante decided to develop two Gantt charts. The first was a graphic representation of the orders as received and depicted on the order sheet (Figure 7–14). This chart was developed so that Enfante could visualize the work to be done and identify ways to better utilize his machines. It lays out the work in line sequence according to the way the orders were received (i.e., 90-11 first, 90-12 second, and 90-13 third).

You can see from this chart that Enfante would experience a great deal of idle time on all three machines if he processed the orders as received. Rather than do this, he developed the Gantt chart in Figure 7–15. In this chart he adhered to the required sequence within orders, but scheduled the work without regard for when he received the order. As you can see he made much better use of his machines and he will finish all three orders twelve hours earlier than he would have. This is the value of Gantt charts.

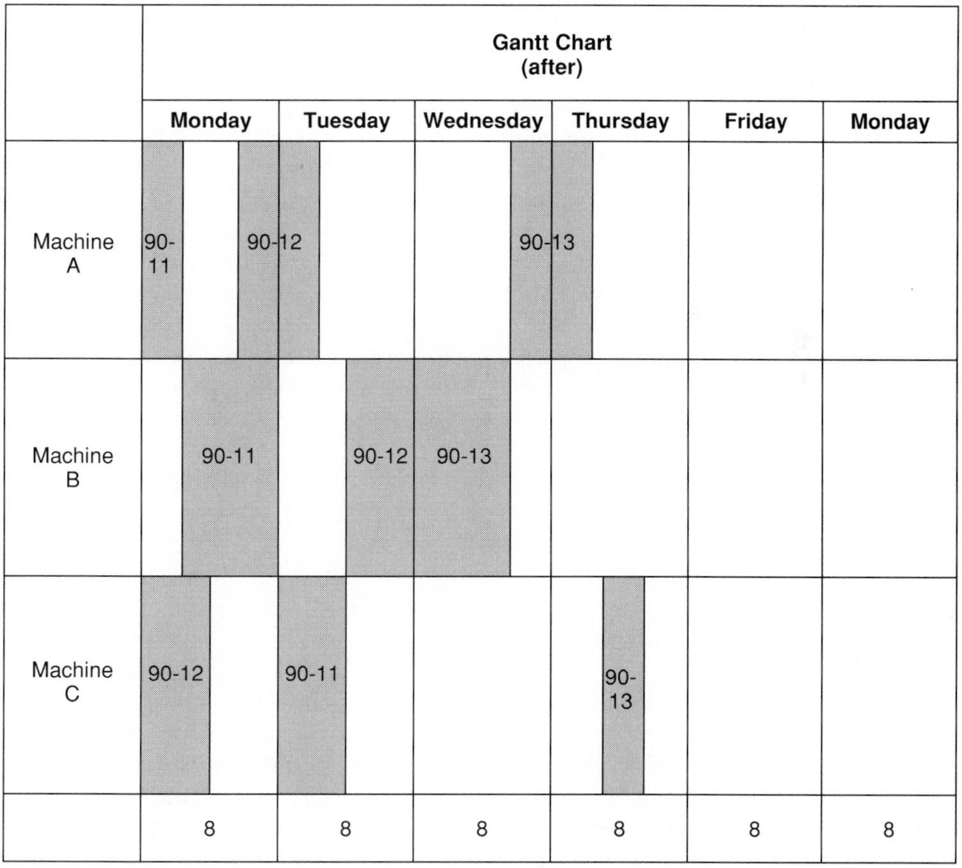

	Gantt Chart (after)					
	Monday	**Tuesday**	**Wednesday**	**Thursday**	**Friday**	**Monday**
Machine A	90-11 90-12		90-13			
Machine B	90-11	90-12	90-13			
Machine C	90-12	90-11		90-13		
	8	8	8	8	8	8

Figure 7–15
Sample Gantt chart after condensing the work.

COMPUTER-AIDED PROCESS PLANNING

A significant technological development in planning for the manufacture of discrete parts is computer-aided process planning or **CAPP**. Once a part has been designed, the various processes needed to produce it can be sequenced in a variety of ways. Typically there are so many potential sequences that prior to CAPP, process planning was really a game of chance.

With CAPP, supervisors can quickly and easily determine the optimum sequence of operations for producing a given part. A CAPP system is used to collect and store all known information about a given company's manufacturing capability, general manufacturing processes, and design principles. This information is used for determining the optimum sequence of processes for producing the part in question. A CAPP-based plan specifies the machines to be used, the tooling required, the sequence of operations, speed and feed settings for the machines, and other planning data.

Policies and Planning

Policies are guidelines established by higher management to ensure that all employees are working toward the same goals and that this is being done in accordance with the company's corporate philosophy. Policies help translate the company's philosophy or operating principles into day-to-day practice. It is important for supervisors to know and understand company policies.

Some policies are *must* statements that give supervisors no discretion. Others are more broadly stated so that supervisors can exercise judgment. An example of a policy that a supervisor would be required to follow to the letter is as follows:

> This company supports equal access and equal opportunity in spirit and in practice. There will be no discrimination based on race, gender, age, religion, marital status, national origin, or any other factor unrelated to job performance.

This policy leaves the supervisor no room for making judgments or applying discretion. It must be followed precisely as stated. An example of a policy that allows supervisory discretion is as follows:

> Overtime work should be held to a minimum and used only when absolutely necessary to meet deadlines.

This policy allows supervisors to exercise discretion in deciding if overtime work will be assigned in their units. Clearly the company discourages overtime as a matter of policy. However, this policy is stated in a way that shows higher management understands overtime may occasionally be necessary. The person in the best position to know when overtime work is called for is the supervisor.

Do Supervisors Set Policy?

Policies are set by higher management. Management may solicit input from supervisors when setting policy or they may not. In any case, policy setting is not within the supervisor's realm of responsibility. However, supervisors do carry out policy. It is at the operational level that policies take on their true meaning.

To illustrate this concept let's return to the example of the nondiscrimination policy. The company's nondiscrimination policy leaves no room for interpretation. Therefore, when hiring or promoting, supervisors cannot allow age, race, gender, marital status, or other irrelevant factors to affect their decisions.

To see how a supervisor might operationalize this policy consider the example of Humberto Vega, documentation supervisor for MI-TEK Engineering. Four of his CAD system operators have applied for a CAD technician position Vega intends to fill. The technician slot involves dome design and would be a promotion for all four applicants. Three of the applicants are men and one is a woman, the only woman in the documentation department. Humberto Vega has interviewed all four applicants and feels the best qualified is the woman, My Ling Kusaki. However, during the interviews the male candidates advised Vega against promoting Kusaki. They claimed the other men in the department would resent it. One candidate made the statement, "She will probably get married and leave anyway." Another candidate commented, "How do you know she won't file pregnancy leave as soon as you promote her?"

Vega dismissed these comments and offered the promotion to My Ling Kusaki, thereby implementing his company's equal access/equal opportunity policy. With the task accomplished, though, Vega clearly has some additional work to do in the area of equity.

Written and Unwritten Policies

Not all policies are written. In fact, some companies operate within a framework defined by policies that have developed over time, but have never been written. Should all policies be written? There are those who argue convincingly for written policies and those who argue equally convincingly that not all policies can or should be written.

Certainly, from the supervisor's perspective, written policies are desirable. It is much easier to follow guidelines that can be referred to when questions arise. Also, written policies are easier to communicate to employees, especially new employees.

The need for written policies has become more and more evident in recent years. Litigation, particularly that relating to personnel matters, has been less difficult for companies with written policies. Written policies also offer supervisors a form of protection. Supervisors who are sued for their actions on the job may be less susceptible with regard to personal liability if they were following clearly written company policy.

Can Supervisors Change Policy?

Supervisors cannot directly change policy. Policies are set by higher management and supervisors are expected to carry them out. If a given policy impedes a unit's ability to accomplish its goals, the supervisor should make this fact known to higher management. In this way supervisors may be able to convince management to change a policy, but supervisors themselves should never unilaterally change a policy.

Higher management establishes policies that are in the best interests of the company as a whole. Such policies will occasionally inconvenience individual units or saddle them with what may appear to be an unfair burden. However, supervisors should not automatically assume that policies that inconvenience their units are bad.

When given the opportunity to view the company as a whole, supervisors may agree that the questionable policy makes more sense. However, after having taken a holistic view, if a policy still does not make sense, supervisors should try to influence change.

Handling Unpopular Policies

You will not be a supervisor long before it will be necessary for you to carry out an unpopular company policy. This is one of many such instances of when supervisors must be leaders. The worst mistake a supervisor can make in these cases is to say, "I hate to do this, but company policy says I have to."

Regardless of how you feel about a policy personally, as a supervisor you are responsible for giving it your full support in carrying it out. If you do not agree with a policy, let higher management know in private. In public, support the policy until management changes it. Do not fall into the trap of blaming the company. Remember, as a supervisor, you are the company. This is one of the most difficult lessons new supervisors have to learn. But it is important that it be learned. Supervisors who take credit for popular policies and blame the company for unpopular policies will eventually lose credibility. When supervisors administer policy properly, their actions and examples become the everyday manifestation of policy in the eyes of employees.

PLANNING THE PLAN

At this point you have studied the supervisor's role in planning, the planning process itself, scheduling techniques, and several other related topics. But how do you pull all this together? How can you remember everything that should be included in a plan? Figure 7–16 is a planning checklist that can be a helpful tool in developing short-term operational plans. It is a composite of the planning process reduced to six steps. Once you have developed several plans, these steps will become second nature. Until then, the checklist will help you pull the various steps together and ensure that all elements are included.

OTHER PLANNING CONCERNS

In addition to enhancing the efficiency and effectiveness of their organizations through planning, supervisors can make better use of their time through planning. This section explains how supervisors can save time by planning organizational meetings and by planning their own time.

Planning Meetings

Meetings are a necessary part of the supervisor's life. Meetings are a handy way to communicate with all employees at once, conduct roundtable discussions, and give

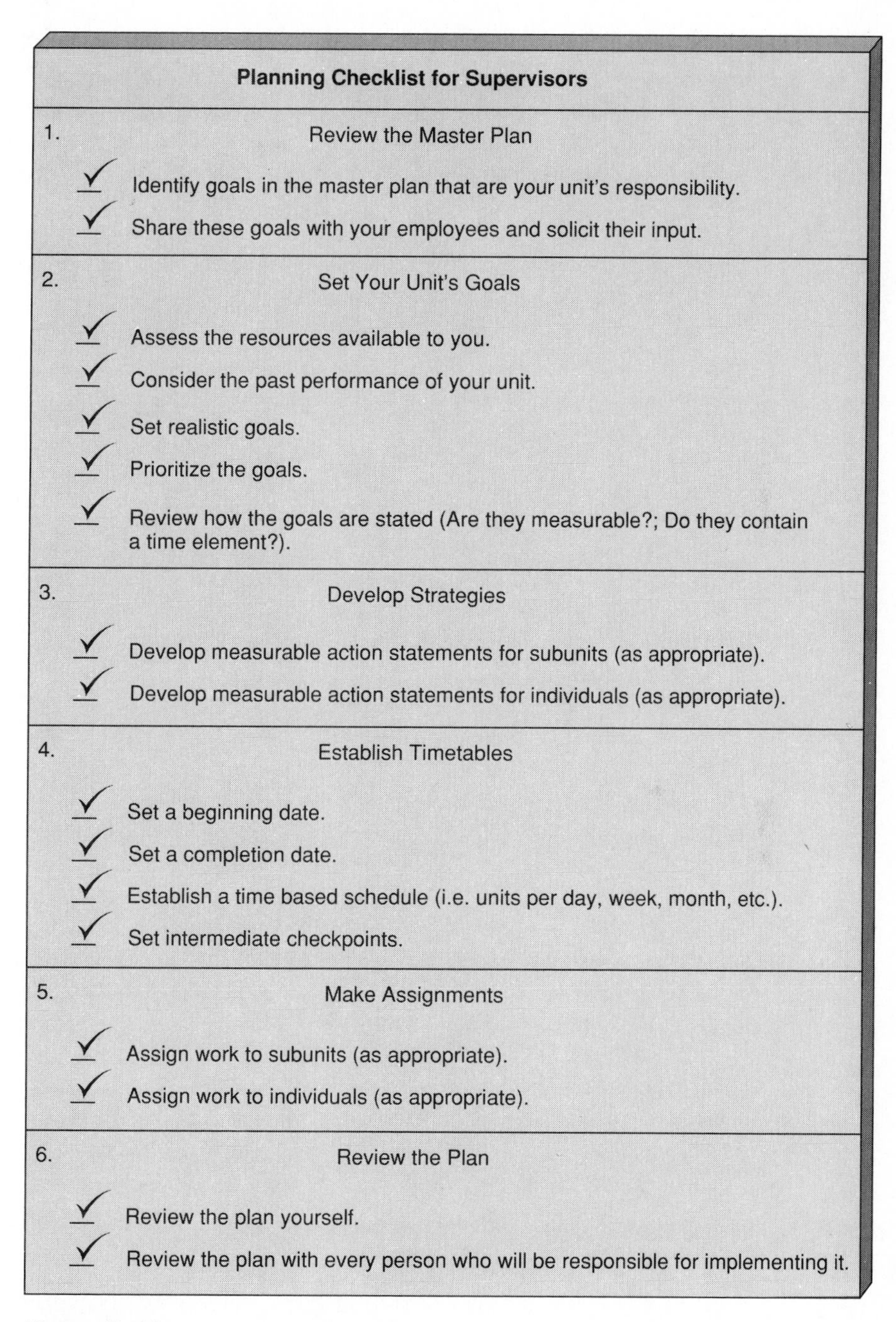

Planning Checklist for Supervisors

1. Review the Master Plan
   - ✓ Identify goals in the master plan that are your unit's responsibility.
   - ✓ Share these goals with your employees and solicit their input.
2. Set Your Unit's Goals
   - ✓ Assess the resources available to you.
   - ✓ Consider the past performance of your unit.
   - ✓ Set realistic goals.
   - ✓ Prioritize the goals.
   - ✓ Review how the goals are stated (Are they measurable?; Do they contain a time element?).
3. Develop Strategies
   - ✓ Develop measurable action statements for subunits (as appropriate).
   - ✓ Develop measurable action statements for individuals (as appropriate).
4. Establish Timetables
   - ✓ Set a beginning date.
   - ✓ Set a completion date.
   - ✓ Establish a time based schedule (i.e. units per day, week, month, etc.).
   - ✓ Set intermediate checkpoints.
5. Make Assignments
   - ✓ Assign work to subunits (as appropriate).
   - ✓ Assign work to individuals (as appropriate).
6. Review the Plan
   - ✓ Review the plan yourself.
   - ✓ Review the plan with every person who will be responsible for implementing it.

Figure 7–16
Supervisor's planning checklist.

employees opportunities to participate in decision making. Unfortunately, they can be time consuming if not carefully planned in advance.

Gilda Dangot-Simpkin[9] lists ten planning strategies for better meeting management:

1. Always have an agenda.
2. Formulate issues to be discussed as specific questions.
3. Include only employees whose input is essential to the discussion.
4. Circulate the agenda beforehand.
5. Assign time frames for each item on the agenda.
6. Resolve each item on the agenda.
7. Have minutes taken to record the actions taken and follow-up action that needs to be taken.
8. Follow up on assignments.
9. Ensure that everyone has opportunities for input.
10. Enforce the agenda and its time frames.

These ten strategies can improve the effectiveness and efficiency of meetings and, concurrently, save valuable time. Notice that the first five are planning strategies. A rule of thumb to remember is, "It takes less time to plan a meeting than it does to conduct an unplanned meeting."

Planning Your Time

A supervisor's time is always pressed. The first step in effectively managing time is proper planning. Merrill Douglas advocates the "Oats formula" for planning personal time.[10]

The "O" is for *objectives.* Supervisors should begin each day by deciding what they hope to accomplish. The "A" is for *activities.* Activities are what you have to do to accomplish your objectives. The "T" is for *time estimates.* It is important to estimate realistically both the time the activities will require and the time available. The "S" is for *schedules.* This involves deciding when you will conduct your activities and scheduling the time available to you.

This simple approach to planning your time can ensure better use of it. Merrill makes the point that planning time causes supervisors to focus on what is important while ignoring what is not. It also identifies potential trouble areas in advance.

SUMMARY

1. Planning is the process of setting goals, developing strategies for accomplishing them, and establishing benchmarks for the monitoring process.
2. Supervisors are responsible for developing operational plans that translate the

strategic plans of higher management into everyday action. Supervisors develop short-range plans that are based on the long-range plans of higher management.

3. Goals and objectives define desired future outcomes. They answer the question "Where are we going?" Policies are broad guidelines. Rules and regulations translate policies into language that contains the everyday behavior of workers. Procedures are the methods used to carry out plans on a day-to-day basis. Schedules establish timetables within which goals are to be accomplished. Plans contain all the information relevant to accomplishing goals.
4. Long-range plans cover periods of more than a year and are the responsibility of higher management. Short-range plans cover periods of a year or less and are the responsibility of supervisors. Long-range plans are sometimes called strategic plans. Short-range plans are sometimes called operational plans. Standing plans cover those aspects of the job that do not change. Single-use plans typically cover one project.
5. There are ten steps in the planning process: (1) review management's plans, (2) solicit the input of employees, (3) set goals, (4) develop an overall plan, (5) develop specific plans for subunits, (6) develop strategies, (7) set up schedules, (8) assign responsibility, (9) review the finished plan, and (10) monitor and adjust as necessary.
6. When setting goals, observe the following rules of thumb: (1) examine the master plans, (2) solicit input from employees, (3) assess the resources available to you, (4) consider past performance, (5) set realistic goals, (6) prioritize the goals, and (7) make sure the goals are stated properly.
7. When scheduling, observe the following rules of thumb: (1) schedule for less than capacity so you can accommodate unforeseen circumstances, (2) prepare for slack periods, (3) avoid scheduling overtime, and (4) schedule accurately.
8. PERT is a widely used scheduling technique that consists of the following seven steps: (1) identify the project goal, (2) identify the events, (3) sequence the events, (4) identify the activities, (5) estimate the time for each activity, (6) set up the network, and (7) implement the network and adjust as needed.
9. CPM is a scheduling technique in which the longest or critical path from the beginning to the end of a project is identified. Once the critical path has been identified, attention and resources are focused on it to ensure a project stays on schedule.
10. Gantt charts are used as scheduling tools for making better use of machines, equipment, and systems. They help supervisors schedule work according to the availability of machines as opposed to the order in which work requests are received.
11. Policies are broad guidelines established by higher management to ensure that all employees are working toward the same goals. Supervisors can have input into policies, but they do not set policy, nor should a supervisor change a policy. The actions of a supervisor should be seen by employees as representing company policy.

KEY TERMS AND PHRASES

Activity
CAPP
CPM
Critical path
Event
Gantt chart
Goal
Long-range plan
Master plan
Measurable goals
Most likely time
Network
Objective
Operational plan
Optimistic time
Padding
Path
PERT
Pessimistic time
Plan
Planning
Policy
Procedure
Realistic goals
Regulation
Schedule
Sequence
Short-range plan
Single-use plan
Standing plan
Strategic planning
Time element

CASE STUDY: Planning for Competitive Manufacturing at IBM[11]

The following real-world case study illustrates a planning problem. Read the case study carefully and answer the accompanying questions.

In the early 1980s IBM faced intense competition from both foreign and domestic sources. To maintain its position as a leading producer of computer systems, IBM needed to ensure that all of its various operations were achieving optimum levels of efficiency and effectiveness.

IBM officials knew that their manufacturing operations, which comprised 49 percent of corporate assets, would need special attention. They decided to develop a strategic plan for manufacturing competitiveness that focused on seven key areas: low cost, reduced inventories, total quality, automation, organization, integration of manufacturing systems, and affordability. The manufacturing plan included the following strategies:

- Become the low-cost producer in IBM's various markets
- Reduce inventory through standardization and adoption of the just-in-time delivery concept
- Increase quality through adoption of the zero-defect concept
- Enhance the design-production interface through automation
- Create a production management center as part of the IBM organization
- Fully integrate manufacturing systems

- Enhance affordability and competitiveness by evaluating manufacturing performance using competitors' performance as the criteria

1. Assume you are a manufacturing supervisor with IBM. What impact will this new strategic plan have on you?
2. Outline the steps you would follow in developing a plan for operationalizing IBM's strategic plan.
3. One goal of the strategic plan is full integration of manufacturing systems. As a manufacturing supervisor responsible for operationalizing this goal, what inhibitors might you encounter?

REVIEW QUESTIONS

1. Define the term *planning.*
2. Briefly explain the supervisor's role in planning.
3. Explain the difference between a goal and a plan.
4. What is the purpose of a policy?
5. What is the purpose of a procedure?
6. Distinguish between strategic plans and operational plans.
7. Distinguish between standing plans and single-use plans.
8. Using brief statements, list the ten steps in the planning process.
9. Why is it important for supervisors to review the plans of higher management?
10. Why is it important for supervisors to solicit the input of their employees when setting goals?
11. What are realistic goals and why are they important?
12. Explain the time element concept in goal setting.
13. Why is it important to schedule for less than capacity?
14. How can supervisors prepare for slack periods?
15. Explain why it is important for supervisors to plan accurately.
16. What is PERT?
17. What is CPM?
18. What is the purpose of a Gantt chart?
19. What is CAPP?
20. What is a policy?
21. How do policies relate to planning?
22. Where do supervisors fit into the policy-making process?
23. Explain briefly how supervisors should handle unpopular policies.

SIMULATION ACTIVITIES

The following activities are provided to give you opportunities to apply what you learned in this and previous chapters to the solution of supervisory problems. The problems may be undertaken by individuals or in groups.

1. Select an operation with which you are familiar (i.e., changing a tire, preparing a meal, tuning an automobile engine, etc.). Prepare a PERT planning worksheet similar to the one in Figure 7–11 for your operation.
2. Using the PERT planning worksheet developed in Activity 1, develop a network diagram and identify the critical path.

ENDNOTES

1. Drucker, P. *Management* (New York: Harper & Row, 1985), p. 121.
2. Ibid., p. 122.
3. Ibid.
4. Ibid.
5. Ibid., pp. 123–126.
6. Ibid., p. 125.
7. Small Business Administration. *Business Basics: Inventory and Scheduling Techniques,* Self-Instructional Booklet Number 1017, pp. 41–42.
8. Ibid., p. 43.
9. Dangot-Simpkin, G. "Saving $2000 Daily by Better Meeting Management," *Supervisory Management,* August 1990, pp. 1–2.
10. Douglas, M. "The Oats Formula: How to Better Plan Your Time," *Supervisory Management,* February 1990, pp. 10–11.
11. Wheelwright, S. C. and Hayes, R. H. "Competing Through Manufacturing," *Harvard Business Review,* January/February 1985, p. 39.

CHAPTER EIGHT

Organization and the Supervisor

CHAPTER OBJECTIVES

After studying this chapter, you will be able to explain or define the following topics:

- Organization
- Organizing
- The Rationale for Organizing
- Formal and Informal Organization
- Organizing as a Process
- Types of Organizational Structures
- Uses and Types of Organizational Charts
- Concept of Accountability
- Concept of Span of Authority
- Concept of Delegation
- How to Organize Yourself

Figure 8–2
The types of questions that should be asked when organizing.

Organizing Questions

1. What is the goal?
2. What activities must be accomplished?
3. How can these activities be grouped?
4. Who should be responsible for what?
5. Who should report to whom?
6. How does it all fit together?
7. How can everything be coordinated?

chains, delegating authority and assigning responsibility, and setting in place procedures for coordinating all organizational efforts. Robert Albanese defines *organizing* as:

> all the managerial work involved in creating an organizational structure and design. It involves such work as identifying activities that will be considered as jobs, deciding on the best way to group jobs together, determining who should report to whom, delegating authority and responsibility to members of the organization, and figuring out the best way to coordinate work.[1]

Figure 8–2 is a list of the types of questions that might be asked when organizing a unit, a project, or an activity.

RATIONALE FOR ORGANIZING

Possibly without realizing it, we live very organized lives. Think about it for a moment. Our time is organized into seconds, minutes, hours, days, and weeks. Our weeks are organized into workdays and weekends. Our days are organized according to work hours and free time. We typically have our workday mornings organized into a routine, as are the workdays themselves.

There is a reason for all this organization: Organization leads to the efficient and effective channeling of resources toward the accomplishment of goals. Your morning routine is organized to get you out of bed, get your morning responsibilities taken care of, and get you to work on time. To understand the need for organization, think of those times when your morning routine is disrupted for some reason. If one member of the family does not take care of his or her responsibility or deviates from the schedule, everything else is thrown off track.

This same concept applies at work. People work better when they agree on what is to be done by whom and when. Establishing organization answers those questions and ensures that people are working together toward the same goals rather than pulling in different directions or working at cross-purposes.

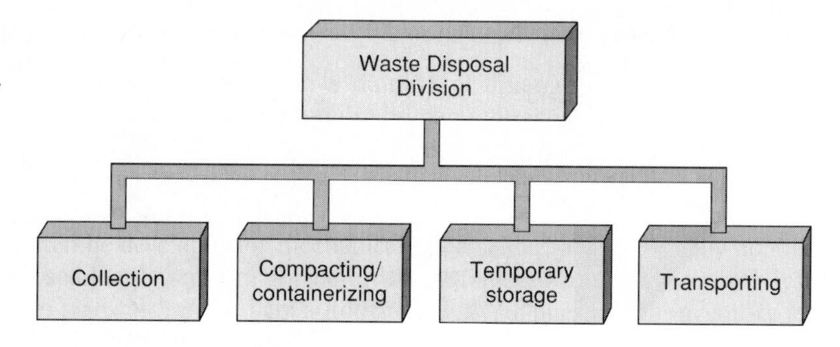

Figure 8–3
This division within a company has four major formal components.

FORMAL VERSUS INFORMAL ORGANIZATION[2]

In any organization there are formal and informal components. The **formal organization** component encompasses the planned, approved, and officially communicated structure, that which appears on an organizational chart. The organizational chart in Figure 8–3 depicts the formal components of the waste disposal division of a large industrial firm.

The formal components illustrate how an organization is supposed to work. For example, in the organization shown in Figure 8–3, one component is supposed to collect waste material. Another is supposed to compact it or put it in containers. A third is supposed to place the material in temporary storage. Finally, another component is supposed to transport the waste material to the disposal site.

Informal organization components are relationships that evolve for expedience, convenience, necessity, or other reasons. At times informal components evolve out of necessity because a formal component is faulty. This could mean that it was not properly structured or it could mean that its people do a less than adequate job. Informal components may also be based on personalities, friendships, or shortcuts. Occasionally people will find what they feel are better ways of accomplishing work than those that are formally established.

Although this chapter is devoted to discussion of the formal components of an organization, supervisors should be aware that informal components exist. Informal components in an organization are not necessarily bad. In fact, they can be good. However, in any case, they bear watching. An informal network can be an indication of a need to revise a formal network. It can also mean that workers are bypassing critical steps in a process. For this reason, modern supervisors should be attuned to the informal components of their organizations.

THE PROCESS OF ORGANIZING

Organizing goes hand-in-hand with planning. Supervisors organize their units to efficiently and effectively carry out their plans. Just as planning is a systematic process, so is organizing. It consists of the following steps (Figure 8–4):

Figure 8–4
Organizing is a step-by-step process that should be undertaken systematically.

Organizing Process

1. Clearly identify the goal(s).
2. Identify all tasks that must be accomplished in order to achieve the goal(s).
3. Divide the tasks into jobs. A job is a group of related activities that can be reasonably accomplished by one person.
4. Group the jobs together in a logical manner based on the nature of the jobs and how they interact with other jobs.
5. Establish formal relationships for people within groups.
6. Establish formal relationships between and among groups.

1. Clearly identify the goal(s).
2. Identify all tasks that must be accomplished in order to achieve the goal(s).
3. Divide the tasks into jobs. A job is a group of related activities that can be reasonably accomplished by one person.
4. Group the jobs together in a logical manner based on the nature of the jobs and how they interact with other jobs.
5. Establish formal relationships for people within groups.
6. Establish formal relationships between and among groups.

To understand how the process of organizing works, consider the hypothetical example of Carlos Alvarez, a supervisor for Crestview Textiles Company. Crestview Textiles is going to have an open house on Monday. It is Friday and the employees who will conduct a major house cleaning on Saturday were selected by drawing names out of a hat earlier in the week. Carlos Alvarez was the only supervisor whose name was drawn. Consequently, he was put in charge of the cleanup.

Alvarez is a good organizer. He has called the cleaning team together and is about to distribute copies of his organizational plan for the cleanup. It contains the following information:

Goal
To clean up the plant and grounds of Crestview Textiles Company in preparation for the upcoming open house.

Tasks That Must Be Accomplished
1. Clean the inside of the plant facility including the following:
   a. Sweep/vacuum all floors as appropriate.
   b. Mop floors where appropriate.
   c. Apply touch-up paint to walls and ceiling as needed.

   d. Wash all windows.
   e. Clean all restroom facilities.
2. Clean the outside of the plant facility including the following:
   a. Mow the lawn.
   b. Trim and edge around the sidewalks.
   c. Trim plants and bushes as needed.
   d. Apply touch-up paint where needed.
   e. Wash all windows.

Division of Tasks into Jobs
1. Indoor tasks are divided into the following jobs:
   a. Sweepers/vacuumers
   b. Floor moppers
   c. Painters
   d. Window washers
   e. Restroom cleaners
2. Outdoor tasks are divided into the following jobs:
   a. Lawn mowers
   b. Trimmers/edgers
   c. Painters
   d. Window washers

Grouping of Jobs
All jobs will be grouped together as indoor or outdoor jobs.

Relationships within Groups
One person will be appointed as the indoor supervisor. One person will be designated as the lead person for each job (i.e., lead painter, lead restroom cleaner, etc.). Lead workers will report to the indoor and outdoor supervisor respectively. The supervisors will report to the overall cleanup coordinator, Carlos Alvarez.

Relationships Between and Among Groups
Communication between job groups will be handled by the lead workers in each group. Communication between the indoor and outdoor teams will be handled by the indoor and outdoor supervisors.

Carlos Alvarez completed the organization process by developing an organization chart to share with all members of the cleanup teams. That chart is shown in Figure 8–5.

ORGANIZATIONAL STRUCTURES[3]

There are several different ways to structure an organization. The most widely used organizational structures are line, line and staff, division or product, geographic, market, centralized or decentralized, and ad hoc (Figure 8–6).

Figure 8–5
Organizational chart for the cleanup teams at Crestview Textiles.

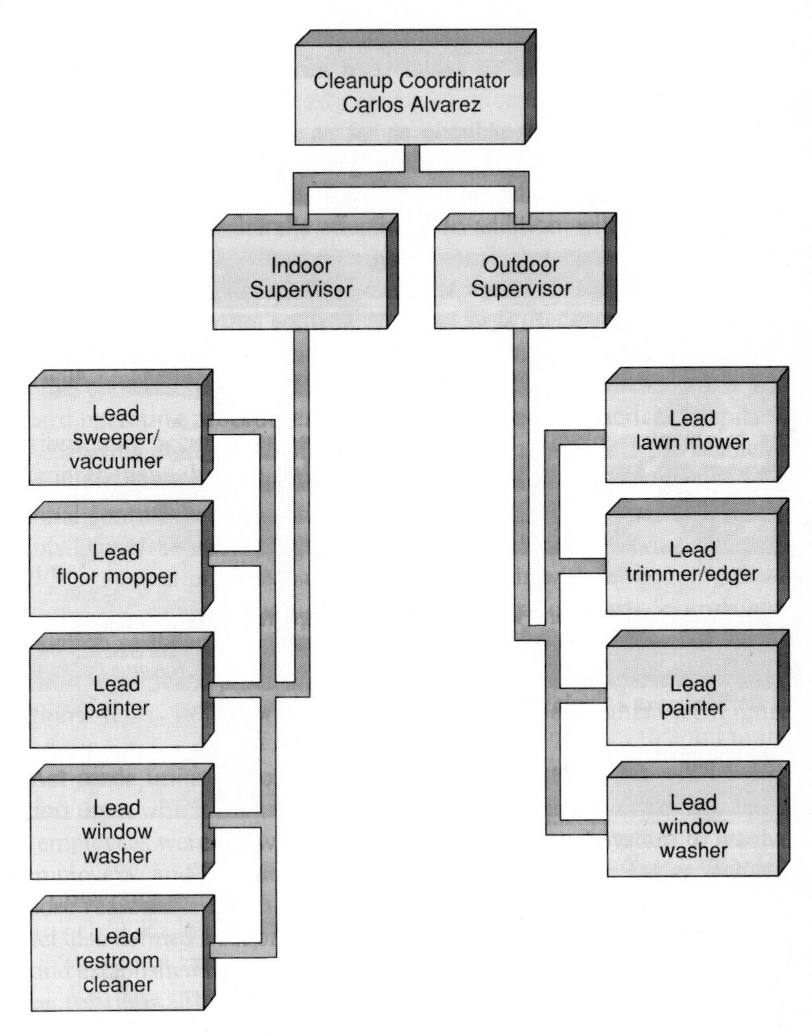

Line Structure

A **line organization** is structured so that related subunits of the organization are arranged under one responsible position. The person filling this position has authority over and is accountable for these subunits. This position might be an executive level manager, manager, mid-manager, or supervisor. Figure 8–7 is an organizational chart for a line organization. In this figure, the machine shop supverisor has responsibility for subunits or teams of machinists.

Line and Staff Structure

Line and staff organizations are line organizations with the addition of staff positions. A **staff position** is one that supports a line position or positions. **Line positions** are those

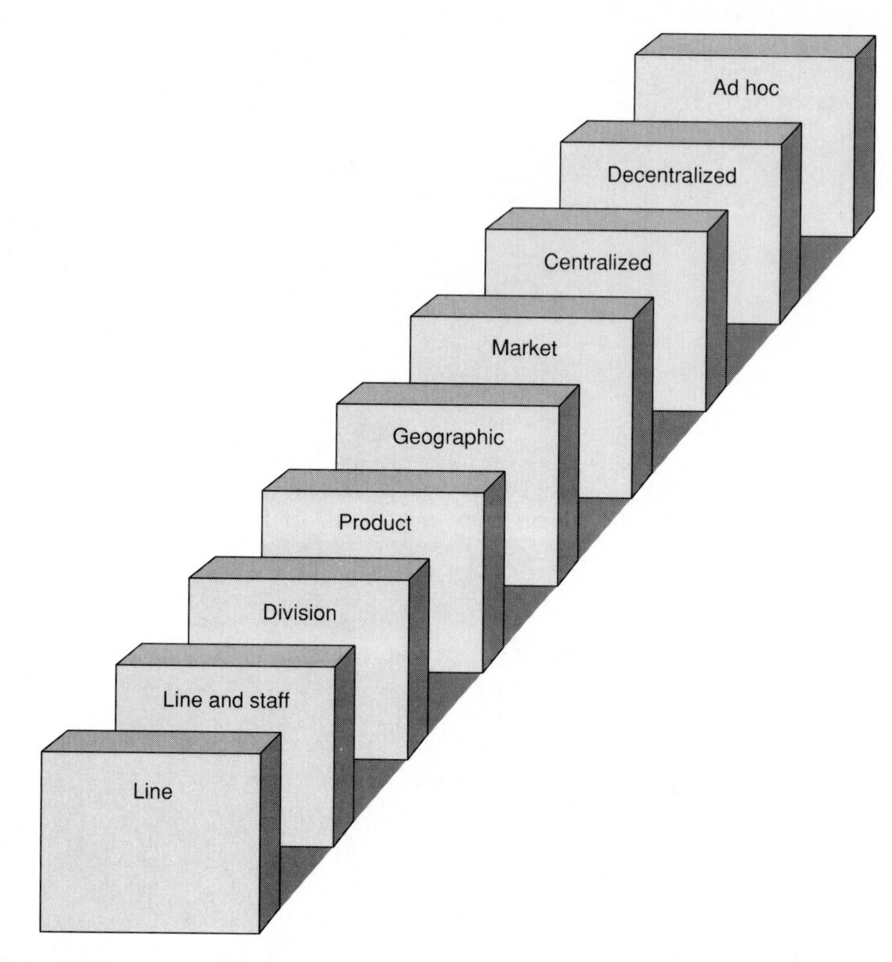

Figure 8–6
Widely used organizational structures.

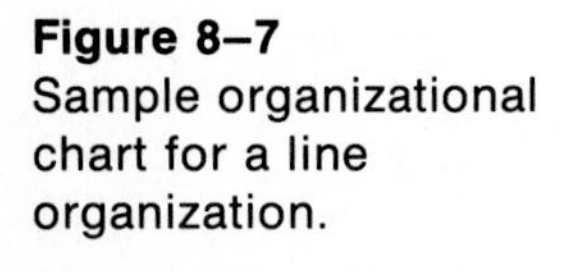

Figure 8–7
Sample organizational chart for a line organization.

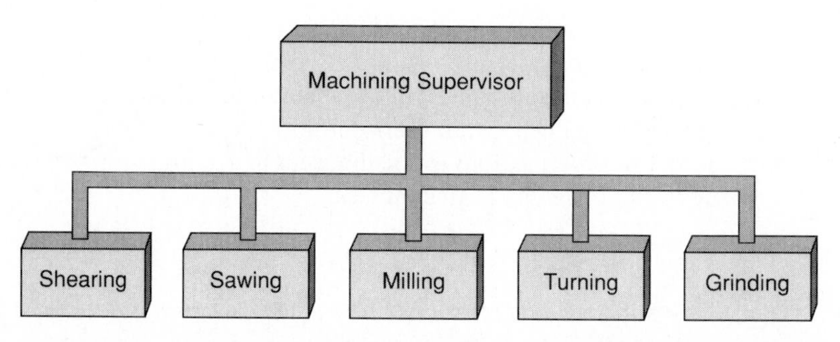

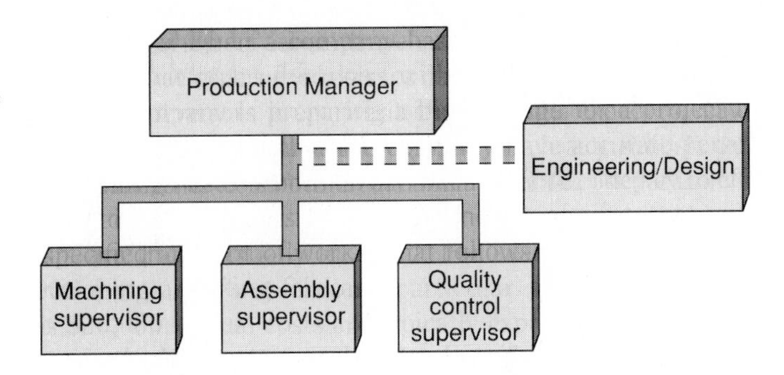

Figure 8–8
Sample organizational chart for a line and staff organization.

that are involved in the activities necessary to produce products or deliver services. Staff positions assist by providing advice, decision-making assistance, and administrative help. For example, a company's management information system (MIS) department is typically a staff unit. Staff positions do not have line authority or responsibility. On an organizational chart, the lines drawn to staff positions are typically dotted lines (Figure 8–8).

Division or Product Structure

Another way to structure an organization is by **division** or **product**. This approach is frequently used by companies that have a diverse product line. For example, if a textile manufacturer produces children's wear, athletic wear, and beach apparel, the company might be separated into divisions that mirror these product lines. Each division has a functional head with responsibility for various subunits. A division-based organization is depicted in Figure 8–9.

Geographical Structure

A company that has plants or offices in several different locations might be structured according to **geographic** location. For example, a company that has plants or offices on the East Coast, on the West Coast, and in the Midwest might have eastern, western, and central divisions. Each division has a functional head. The people filling these positions report to another functional head (Figure 8–10).

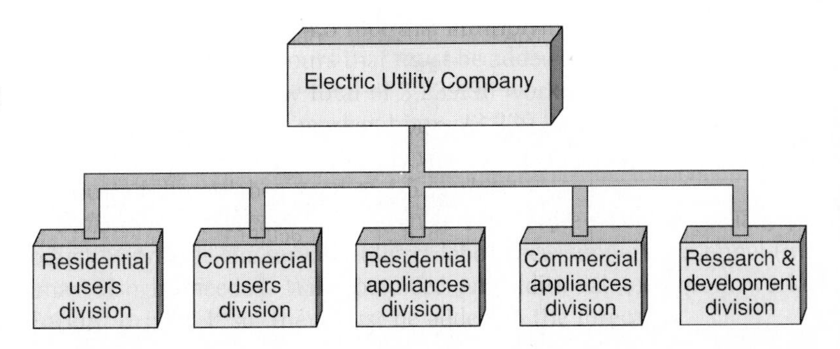

Figure 8–9
Sample organizational chart for a division-based organization.

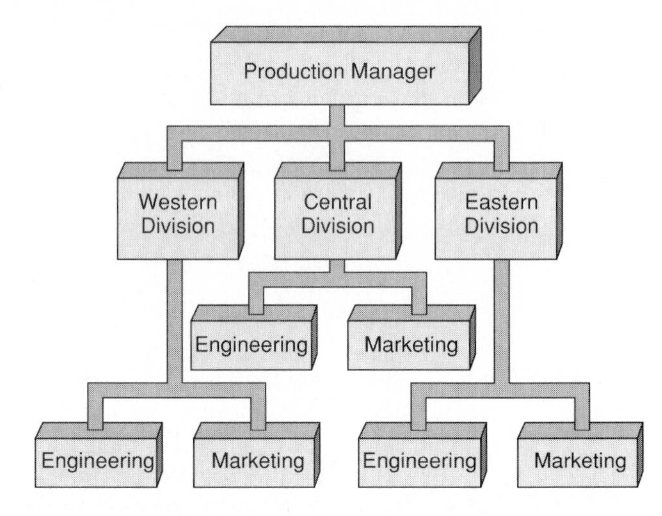

Figure 8–10
Sample organizational chart for a geographically based organization.

Market Structure

An organization may be structured according to the **markets** it serves. This is similar to structuring by product line. If a manufacturer produces both civilian and military aircraft, it has two different markets. Such a company might be structured as two divisions, military products and civilian products.

Centralized and Decentralized Structures

Centralized organizations concentrate selected functions in one location. For example, a given company might have three different plants. Each plant is responsible for producing some part of an overall product. However, all major functional components of the company such as higher management, engineering, accounting, purchasing, human resources, and MIS are located in one of the facilities. They are centralized; they provide their services to all three locations. If this organization were **decentralized**, each of the three plants would have its own functional components such as engineering, accounting, purchasing, human resources, and MIS.

There is a longstanding debate as to which approach is better, the centralized or decentralized structure. The centralized approach tends to create more levels of management and put more layers between those who do the work and those who make major decisions about the work. In other words, centralization tends to create a larger bureaucracy. On the other hand, the centralized approach can cut down on duplication, thereby cutting overhead costs.

The decentralized approach makes communication, participatory management, and decision making easier and potentially more effective. However, it can lead to higher overhead costs. A sensible way to deal with this issue is to examine each case on its own merits and do what is best in each case. There are times when centralization works more effectively. Conversely, there are times when decentralization works better.

Ad Hoc Structure

This approach involves structuring a unit on a temporary basis by borrowing positions from standing components of an organization. This ad hoc structuring approach is used to fulfill a specific need. When this has been done, the structure is disbanded and the personnel return to their normal places in the larger organization.

Ad hoc units are typically formed to complete rush orders, solve one-time problems, or meet temporary needs. One of the problems sometimes experienced with this approach is confusion of authority. People temporarily assigned to the ad hoc unit report to the head of the unit, but they also still report to the functional head of their parent unit. If this issue is not dealt with carefully, it can result in divided loyalties as ad hoc personnel try to report to two different supervisors.

ORGANIZATIONAL CHARTS[4]

Organizational charts are graphic illustrations or blueprints of a company's organizational structure. They serve an important purpose: communication. An organizational chart illustrates the various subunits that make up the company, where each subunit fits in, how subunits relate to each other, and who reports to whom.

Typically organizational charts are arranged vertically from the top down (Figures 8–5 and 8–7 through 8–10). Departments, units, functions, or positions are enclosed in boxes and connected by vertical and horizontal lines. Line functions are connected using dotted lines and are typically drawn outside the main flow of the chart. When space allows boxes containing equal levels of authority are drawn at the same level in the chart.

Although organizational charts are typically drawn as just described, they can take on a variety of shapes. Figure 8–11 is an example of an organizational chart that uses circles arranged in pods. The same chart could have been laid out using boxes in a vertical, top-to-bottom format.

The format used in developing organizational charts is less important that ensuring a clear, simple, easily understood, and up-to-date chart. Organizational structures often change. As this happens, organizational charts should be revised and distributed to all workers.

Benefits of Organizational Charts

Properly developed organizational charts can be of benefit to supervisors. As a communication tool, an organizational chart can do the following:

- Force supervisors and managers to give careful, continual thought to their company's organizational structure.
- Show employees the big picture and where they fit into it.
- Give people from outside a snapshot of the company.
- Communicate the formal lines of authority and responsibility to all employees as well as outsiders.

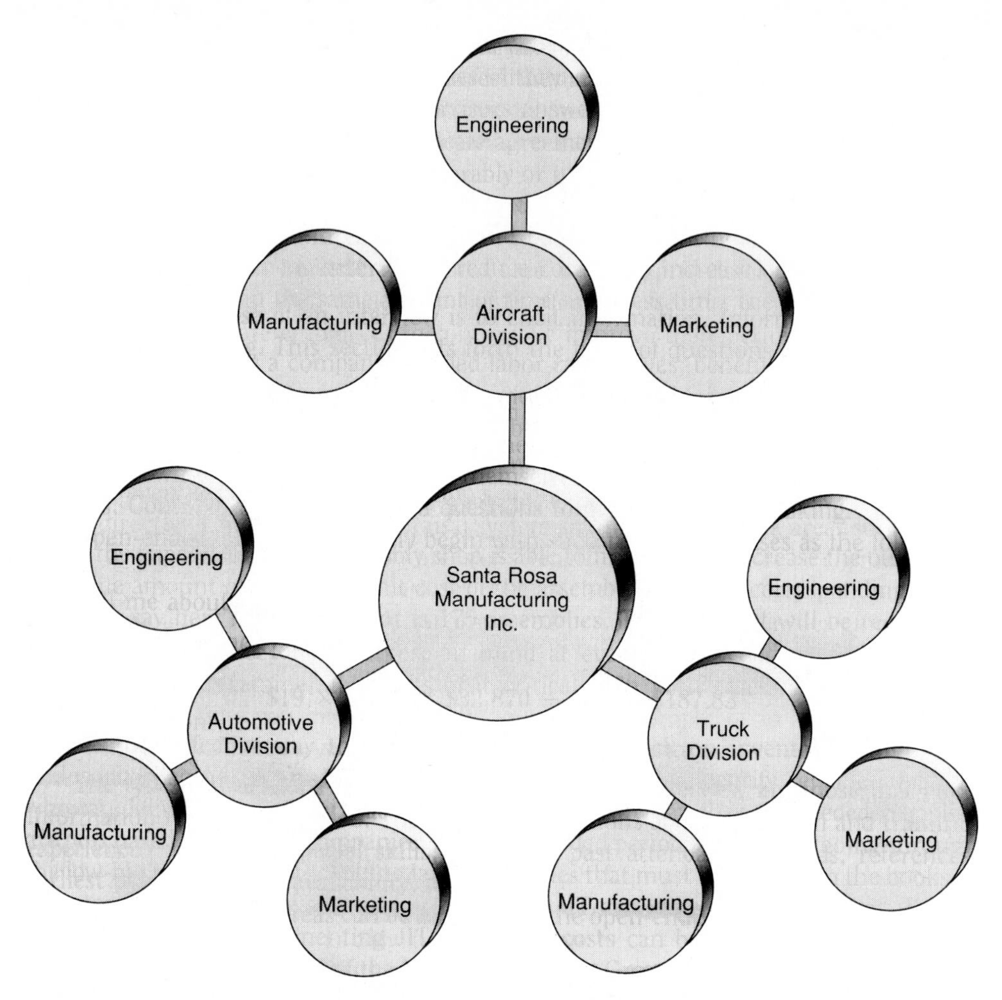

Figure 8–11
Organizational charts can be presented in a variety of different formats.

Weaknesses of Organizational Charts

Even well-developed, up-to-date organizational charts have their limitations. The major weaknesses of organizational charts can be summarized as follows:

- They do not show the informal relationships that inevitably develop in a company.
- They do not convey the personalities, strengths, or weaknesses of the units or people in the boxes.
- Lines connecting boxes can convey an overly simplified image of the actual channels they represent.
- They become outdated quickly and can be time consuming to keep updated.

ORGANIZATION AND ACCOUNTABILITY

One of the benefits of organizational charts is that they show the formal *chain of command* for the organization. This means they communicate the following accountability-related information:

1. Who is responsible for what?
2. Who is accountable to whom?
3. Who can delegate to whom?
4. Who has authority for what?

John Roach, while serving as CEO of Tandy Corporation, said his employees were accountable to him for three things: "Performance. Performance. Performance."[5] John J. Byrne of Geico expressed a similar sentiment about accountability when he said of his employee, "I don't care if he shines his shoes with a brick, so long as he gets results."[6]

Modern supervisors must have a clear understanding of their responsibilities and the authority they have in carrying out these responsibilities. In addition, supervisors should understand the critical difference between responsibility and authority.

It is possible to be given responsibility but not the authority necessary to carry it out. This should not happen, but it can and it does. In such cases, supervisors find themselves held accountable for activities over which they have little or no control.

Supervisors who understand the difference between responsibility and authority will be better prepared to avoid such situations and to speak up when they find themselves about to be placed in one. **Responsibility** is defined by the concept of *accountability*. You are responsible for those duties, decisions, and activities for which you will be held accountable. **Authority** is the formal power afforded you by virtue of the position you hold (Figure 8–12).

Figure 8–12
Responsibility and authority must go hand in hand.

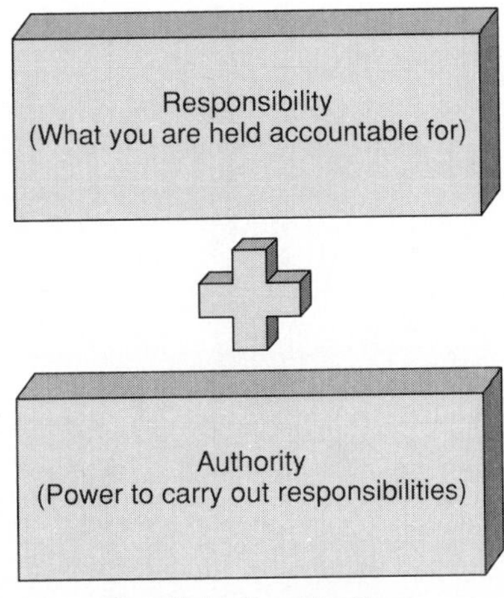

The Right Combination

Formal Authority

Formal authority is that depicted on an organizational chart. It is given by the company to each successive level of management. For example, executive level managers have more authority than managers, who have more than mid-managers, who have more than supervisors, who have more than their employees.

The process of passing authority down from the top of an organization is known as **delegation** (Figure 8–13). Each successive level has as much authority as is delegated by the next higher level.

A supervisor's responsibilities and corresponding authority should be spelled out in writing. In addition, supervisors who are at the same level on an organizational chart should have the same amount of authority, even though their actual duties may be different. For example, if the quality control supervisor and the assembly supervisor are at the same level, they should have the same amount of authority.

Informal Authority

Formal authority, by itself, may not always be sufficient. A weak supervisor who, in the eyes of employees, cannot handle the job may not be able to exercise his or her authority fully and effectively. Authority can be given, but it does not amount to much unless employees submit to it.

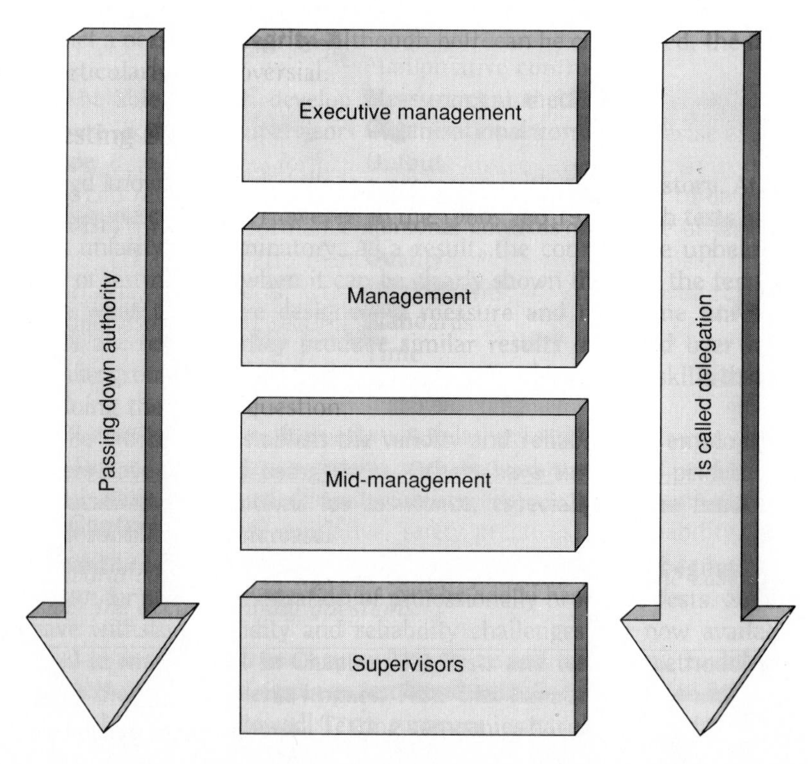

Figure 8–13
The concept of delegation.

Several factors can enhance a supervisor's **informal authority** or ability to get employees to accept their formal authority:

1. Supervisors' personal influence in the organization (the "who you know" factor).
2. Supervisors' power of persuasion and personal charisma.
3. Supervisors' advanced level of job-related skills.
4. Supervisors' length and depth of experience.
5. Supervisors' willingness to work harder and longer than their employees.
6. Supervisors' ability to earn the trust of their employees.

Supervisors should never be satisfied with having just the formal authority that goes with the job. They should also work to reinforce it by developing informal authority. This concept is illustrated by the example of Myra Schulze, the new meter repair shop supervisor for Escambia Gas Company. Schulze was hired from outside and does not yet know her employees very well.

She has been the meter repair shop supervisor for two weeks and things are not going well. Work is stacking up and employees seem to be in no hurry to get it done. She doesn't understand why they don't respond to her directions. After all, she thinks, "I am the boss here."

To make matters worse, it is obvious to Schulze that her employees look to a fellow employee, Doug Maris, for leadership. Unsure of what to do, Schulze consults with one of the company's more senior supervisors. Together they work out the following strategies:

1. Schulze is to display her considerable job skills by repairing one or two of the more difficult to repair meters in the shop.
2. Schulze is to carefully insert information about her many years of experience into casual shop conversation.
3. Schulze is to make sure she is the first person in the shop to arrive in the morning and the last to leave for the next month.
4. Schulze is to be more persuasive and less directive in giving instructions and assignments.
5. Schulze is to be patient. It takes time to establish informal authority.

In this example, Schulze received good advice. A fact of life all supervisors must face is that people respond better to informal than formal authority. It is natural for people to resist formal authority until supervisors are able to prove they have earned it and deserve it.

SPAN OF AUTHORITY[7]

Few people in any company have complete authority. Even chief executive officers typically have boards of directors to whom they are responsible. Board members have stockholders to whom they must be accountable. And, in the broadest sense, stockhold-

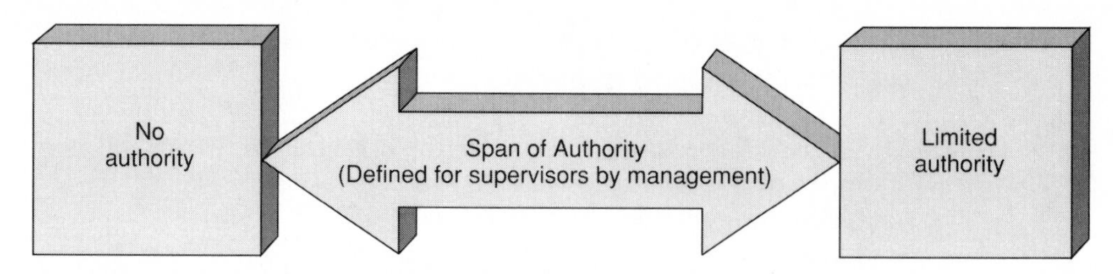

Figure 8–14
A supervisor's span of authority is defined by higher management.

ers are responsible to society. Therefore, supervisors will have either **limited authority** or no authority depending on the area of responsibility. The limits of a supervisor's authority define his or her **span of authority** (Figure 8–14).

The limits of a supervisor's authority should be clearly defined. Can the supervisor order supplies on his or her own signature? Can a supervisor hire and fire on his or her own authority? Can a supervisor decide how best to schedule and assign work in his or her unit? In most companies, the answer to the first two questions would be no. Conversely, the answer to the third would be yes.

Most supervisors will have a finite number of duties within the span of authority that can be undertaken on the basis of their signature alone. Most will require a second signature or approval from a high manager. Where supervisors typically have the most autonomy is in the day-to-day scheduling and carrying out of the work of their unit. For most other matters supervisors' authority is typically limited to recommending.

ORGANIZATION AND DELEGATION

Delegation has already been defined as the passing of authority down the line from higher in the organization to lower. Delegation is an important concept for supervisors to understand, particularly as it relates to responsibility.

Responsibility in an organization should be relatively fixed and clearly assigned. It is set forth on organizational charts and in job descriptions. *Responsibility cannot be delegated* without revising a job description. Duties, tasks, assignments, and the authority needed to carry them out can be delegated, but responsibility for these things cannot.

A supervisor who is responsible for conducting a weekly tool and supply inventory for her unit can delegate that job to an employee. However, she is still responsible for ensuring that the inventory is done and done properly. If it is not done or if there are problems with it, the supervisor will be the one called to account by higher management. This is an important concept for supervisors to understand.

It is illustrated in the following example of Mohammed Saad, documentation supervisor for Fairfield Engineering, Inc. In order to secure a larger design contract, Fairfield's management had to promise an earlier completion date than could be achieved without occasional overtime.

Last night had been one of those occasions when several CAD technicians had had to work late to complete several critical engineering change orders. The change orders were important enough to warrant Mohammed Saad's personal attention, but he had tickets on the fifty-yard line for a football game he really wanted to see.

Instead of working overtime with his team, he had delegated supervision of the change orders to one of his more experienced CAD technicians and gone to the football game. Upon reporting for work this morning Saad learned two things: (1) the change orders had not been completed the night before; and (2) his boss and several other higher level managers wanted to see him right away.

Notice in this fictitious example that higher management did not want to talk with the technician to whom Saad had delegated the change orders. They wanted to talk with Saad. Even though he delegated the assignment and gave his most experienced technician the authority to use his best judgment if design questions arose, Saad was still responsible for ensuring that the project was completed. He could not delegate away his responsibility.

This does not mean that Saad had to miss the football game. However, it does mean that he should have at least called in to determine how the work was progressing. Had he done this, he would have known that problems had caused the project to bog down and he needed to get back to the office.

Why and When Should You Delegate?

The reason for delegating is to enhance your efficiency and effectiveness. The time to delegate is when you have more tasks to complete than you can get done in a timely manner. It is particularly important to delegate routine, time-consuming duties that do not require critical decisions. For example, if you are working on your unit's mid-year budget report but also need to prepare several routine in-house supply requisitions, ask an employee to prepare the requisitions. If an employee has come to you with a problem that needs your attention now, but it is time to check actual production levels against planned levels, delegate the production check to an employee.

How to Delegate Effectively[8]

Delegating does not mean passing off work you don't want to do to a subordinate. If delegation is to make supervisors more efficient and effective it must be handled properly. Here are some tips that will help you become an effective delegator:

1. Delegate only to enhance your effectiveness, not to get rid of work you don't like to do.
2. Prioritize the work that needs to be done in terms of its complexity, sensitivity, and importance. Keep the most complex, most sensitive, most important work for yourself and delegate the lesser work.
3. Do not delegate highly sensitive or confidential work.
4. When delegating work, give complete instructions, definite timetables, and a clear

picture of your expectations. Explain what decisions the subordinate can make alone and what should be cleared with you.
5. When delegating work, explain why the subordinate was chosen and the importance of the task. Employees will resent having work delegated to them if they feel the supervisor is passing off his or her dirty work.
6. Use delegated work as a training opportunity for subordinates and make handling of delegated work part of performance evaluations.
7. When delegating work, let the subordinate's fellow employees know that he or she is doing the work for you and that they should cooperate.
8. When delegating work, monitor and follow up. Do not just assume that because the work was delegated it can be considered done.
9. Use delegation to enhance the job satisfaction of subordinates who might need a break in their routine or who enjoy variety.
10. Use delegation as a form of positive feedback that tells employees "I can count on you" or "I trust your work and your judgment."

Potential Problems with Delegation

If it is not handled properly, delegation can cause more problems than it solves. Here are some practices supervisors should avoid when delegating work:

1. Continually delegating dirty work (i.e., boring, unchallenging, distasteful work).
2. Always delegating work to the same subordinate instead of spreading it around.
3. Delegating work to a subordinate who is already overburdened with his or her own work.
4. Delegating work and then continually looking over the subordinate's shoulder to see how he or she is doing it.
5. Delegating work and not monitoring or following up at all.

Other Aspects of Delegation

We typically think of delegation as delegating to subordinates. This is the traditional approach to delegation, but there are others. Joan Pastor and Risa Gechtman recommend that supervisors also delegate sideways and upward.[9]

Sideways delegation involves delegating to fellow supervisors. Under the best of circumstances this is a strategy that must be handled carefully. How you ask or what you have to offer in return can make the difference in whether or not your peer accepts the work.

In organizations where a teamwork ethic prevails, simply asking for help in a positive way may be enough. If this is not sufficient the "help me this time and I'll help you next time" approach may work. Be sure to follow through and return the favor without waiting to be asked. This will enhance your credibility among fellow supervisors. Credibility is what makes sideways delegation work.

Upward delegation involves returning work that has been delegated to you to the delegator. It is a way of ensuring that you do not receive more delegated work than you can do. Supervisors who do a good job of completing work delegated to them are sure to be rewarded with yet more work. There is a great deal of truth in the adage that work gravitates to those who get it done.

At some point such supervisors are sure to have more work delegated to them than they can accomplish. When this happens, upward delegation may be the answer. The key is to make a list of tasks delegated to you and keep it current. Then when the boss wants to delegate more than you can handle, pull out the list and ask, "Which of these other projects can I put off to do this new one?" This gives the responsibility back to the boss.

ORGANIZE YOURSELF

Companies organize to promote efficiency and effectiveness in achieving their goals. It is also important for supervisors to organize themselves for efficiency and effectiveness. Consider the following comments made by practicing supervisors:

> "At work, I'm always in a hurry. It seems as if there is never enough time."
>
> "I have a stack of paperwork at the bottom of my in-basket that never makes it to the top. It's been there so long I don't remember what it is."
>
> "My desk is always covered with paper. My calendar is buried so deep I rarely see it."
>
> "It seems that all I do is move from one crisis to the next. I don't ever feel I'm in control."
>
> "I have telephone messages I haven't returned on my desk, in my pockets, in desk drawers, and taped to the telephone. I never seem to be able to find the time to return calls."
>
> "There are stick-up reminders all over my office. I have so many stuck up here and there that I don't even read them anymore."

The supervisors who made these comments need to get themselves organized. This section provides practical strategies for getting yourself organized and staying organized. Not only will being organized improve your personal efficiency and effectiveness, it will set an example for the employees you supervise or will supervise in the future.

Strategies for Getting Organized

Getting yourself organized is a matter of establishing the right habits. Organization, like carelessness, is habitual. The following strategies will help practicing and prospective supervisors in any setting organize themselves (Figure 8–15).

1. *Act, don't react.* Many supervisors let the events of the day control them. Such supervisors are continually reacting to one crisis after another. They never seem to get ahead or gain control. The key to turning this around is to stop reacting and start acting. In this case, acting means planning. A little time spent planning your day can

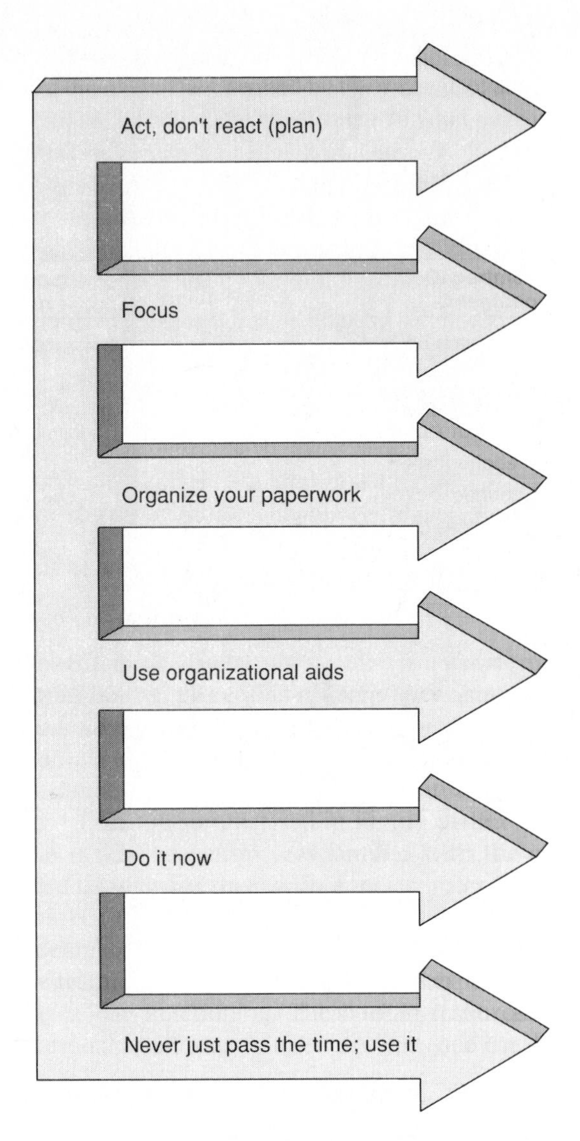

Figure 8–15
Strategies for organizing yourself.

cut down on the amount of wasted time. Make a habit of ending each work day by planning the next.

2. *Focus.* Most supervisors have to juggle several projects at one time. As a result they sometimes find themselves devoting a few hurried and disjointed minutes to each project, but never quite finishing any of them. The solution to this problem is to **focus.** Learn to prioritize the various projects you are juggling, select the most important one, and devote an amount of uninterrupted time to focusing on it. You may have to move on to another project before completing the first. This is fine as

long as you focus on one project at a time and stick with each until real progress has been made toward completing it.

3. *Organize your paperwork.* Cluttered desks and overflowing in-baskets are sure signs of supervisors who need to organize their paperwork. Establish a paperwork system and stick to it. Most paperwork can be divided into several categories such as reading material, signature material, and "to do" projects. "To do" projects can be divided into those that must be dealt with immediately, those that have regular or low priority, and those that are pending (Figure 8–16). By placing paperwork in files that correspond with these categories and arranging the work files according to priority, supervisors can get their paperwork organized and keep it organized. It is a good idea to review the files as one of your last tasks at the close of each workday. You might even pull the files you plan to work on the next day so they are ready when you arrive.
4. *Use organizational aids.* Supervisors should make wise use of **organizational aids** available to them. This means using a calendar (either traditional or on a computer); a watch; a firm-backed notebook that contains a notepad, pen, and calculator; and a directory for important telephone numbers.
5. *Do it now.* Modern supervisors should adopt a do-it-now approach to both paperwork and returning telephone calls. Often the most difficult part of a project is getting started. There is a natural tendency for people to procrastinate, especially when facing projects they do not want to do or don't think they will enjoy.

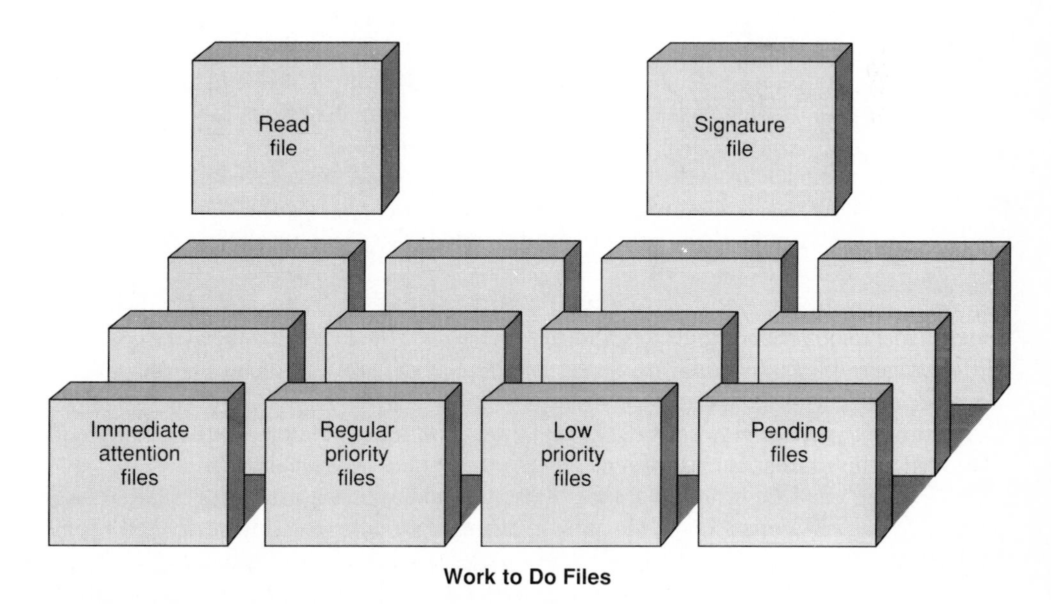

Figure 8–16
Paperwork organization system for supervisors.

An effective way to overcome the inertial phenomenon is to adopt the do-it-now approach. Don't worry about a project, don't even think about it; just start it. Once you start, inertia will work in your favor. This approach is particularly important when returning telephone calls. Promptly and consistently returning calls is a great way to build credibility while simultaneously holding down the amount of backlogged work you have to deal with.

6. *Never just pass time; use it!* Time is a valuable resource in the modern workplace. Like any resource it should be used efficiently and effectively. Never just pass the time of day; use it. One of the advantages of the paperwork organization system illustrated in Figure 8–16 is that it allows you to make productive and wise use of even short periods of time that otherwise might be wasted. For example, if you find yourself with brief snatches of time on your hands, you can take out one of your work files and use the time productively. Experienced supervisors learn to complete most if not all of their routine paperwork during these occasional lulls.

SUMMARY

1. An organization is a group of people structured to effectively and efficiently accomplish its goals.
2. Organizing is the act of creating organization. It involves structuring, grouping, establishing reporting chains, assigning responsibility, delegating authority, and establishing procedures.
3. The rationale for organizing is that it leads to more efficient and effective channeling of resources toward the accomplishment of goals.
4. Formal components of an organization are the planned, approved, and officially sanctioned components. They appear on organizational charts. Informal components are relationships among people in an organization that evolve for expedience, convenience, necessity, or for other reasons.
5. The process of organizing consists of six steps: (a) identify goals; (b) identify tasks; (c) divide tasks into jobs; (d) group the jobs; (e) establish formal relationships for individuals within groups; and (f) establish relationships between and among groups.
6. The most widely used organizational structures are line, line and staff, division/product, geographic, market, centralized/decentralized, and ad hoc.
7. Organizational charts are graphic illustrations or blueprints of a company's formal organizational structure.
8. Authority is the formal power afforded people by virtue of their position. Responsibility encompasses what you are held accountable for.
9. The limits of a supervisor's authority define his or her span of authority.
10. Delegation is the passing of authority down the line from higher in the organization to lower. Authority can be delegated, but responsibility cannot.
11. Supervisors should organize themselves. Six strategies for organizing are: (a) act, don't react; (b) focus; (c) organize your paperwork; (d) use organization aids; (e) do it now; and (f) never just pass time, use it.

KEY TERMS AND PHRASES

Ad hoc organization
Authority
Centralized organization
Decentralized organization
Delegation
Division/product organization
Focus
Formal authority
Formal organization
Geographic organization
Hierarchy
Informal authority
Informal organization
Limited authority
Line organization
Line position
Line and staff organization
Market organization
Organization
Organizational aids
Organizational chart
Organizing
Responsibility
Span of authority
Staff position

CASE STUDY: Japanese Organization Strategies[10]

The following real-world case study illustrates a management problem. Read the case study carefully and answer the accompanying questions.

W. G. Ouchi calls companies that use the Japanese approach to organization Type J companies. Such companies have the following characteristics:

- Expectation of lifetime employment for employees
- Slow promotion associated with lifetime employment
- Consensual decision-making involving all employees affected by the decision
- Collective responsibility so that no individual employee is responsible for a decision, good or bad
- Implicit rather than formal controls growing out of common philosophies and shared attitudes
- Less specialization, allowing workers to learn a variety of jobs
- Concern for all aspects of an employee's life (i.e., family, children, social life, etc.)

In Ouchi's opinion, American companies can successfully adopt the Type J approach.

1. Do you agree that American companies can successfully adopt the Type J approach?
2. What do you think will happen if a company adapts some but not all of the Type J characteristics?
3. What advantages and disadvantages might result from the adoption of a lifetime employment policy in an American company?

REVIEW QUESTIONS

1. What is an organization?
2. Define the term *organizing*.
3. Explain the rationale for organizing.
4. Compare and contrast formal and informal organizations.
5. List the steps in the organization process.
6. Explain the following types of organizations: line; line and staff; division/product; ad hoc.
7. Compare and contrast centralized and decentralized organizations.
8. Explain the purpose of an organizational chart.
9. List four benefits of organizational charts.
10. List three weaknesses of organizational charts.
11. Explain the difference between responsibility and authority.
12. Compare and contrast formal and informal authority.
13. List three strategies supervisors can use to enhance their informal authority.
14. Explain the concept of span of authority.
15. Explain why you agree or disagree with the following statement: Supervisors should delegate some of their responsibility to subordinates.
16. Explain the concept of delegation.
17. What is the rationale for delegating?
18. List four strategies for effective delegation.
19. List three potential problems with delegation.
20. List six strategies supervisors can use to organize themselves.

SIMULATION ACTIVITIES

The following activities are provided to give you opportunities to apply what you learned in this chapter and previous chapters in solving simulated supervisor problems.

1. Your company has built a new facility across town. As senior supervisor of the manufacturing division you have been selected to supervise the transfer of all machines, equipment, tools, and supplies in the manufacturing division. You will be responsible for dismantling, crating, shipping, uncrating, reconnecting, and start-up testing of all machines, equipment, and systems. Show on paper how you will organize the effort. Complete all six steps in the organization process.
2. Visit an industrial firm in your community and obtain an organizational chart for the company or a major subunit of it. Analyze the chart in terms of the following questions:
   - What type of organizational structure(s) does the chart depict?
   - Does the chart show clearly who is responsible for what?
   - Is the chart current? When was it last updated?
   - How would you change the chart to improve it?

ENDNOTES

1. Albanese, R. *Management* (Cincinnati: South-Western Publishing Company, 1988), p. 275.
2. Ibid., pp. 277–278.
3. Mintzberg, H. *The Structuring of Organizations* (Englewood Cliffs, NJ: Prentice-Hall, 1979), pp. 32–39.
4. Ibid., p. 37.
5. Kalbacker, W. "Computer-Age Trail Boss," *Success,* September 1983, p. 15.
6. Sherman, S. P. "Muddling to Victory at Geico," *Fortune,* September 5, 1983, p. 66.
7. Van Fleet, D. D. "Span of Management: Research and Issues," *Academy of Management Journal,* 1983, Volume 26, pp. 546–552.
8. Jenks, J. and Kelly, J. "When a Manager Is Duty Bound Not to Pass the Buck," *The Wall Street Journal,* July 1, 1985, p. 12.
9. Pastor, J. and Gechtman, R. "Delegating Up, Down, and Sideways," *Supervisory Management,* August 1990, p. 9.
10. Ouchi, W. G. *Theory Z: How American Business Can Meet the Japanese Challenge* Addison-Wesley, (Reading MA: 1981).

CHAPTER NINE
Staffing and the Supervisor

CHAPTER OBJECTIVES

After studying this chapter, you will be able to explain or define the following topics:

- Staffing
- An Overview of the Staffing Process
- Legal Considerations Relating to Staffing
- Forecasting and Job Analysis Techniques
- The Recruiting Component of Staffing
- The Interviewing Component of Staffing
- The Employment Test Issue
- The Selection Component of Staffing
- The Orientation Component of Staffing
- Other Staffing Concerns

CHAPTER OUTLINE

- Staffing Defined
- Overview of the Staffing Process
- Legal Considerations of Staffing
- Forecasting Staffing Needs
- Interviewing
- The Employment Testing Issue
- Selection
- Orientation
- Other Staffing Concerns
- Summary
- Key Terms and Phrases
- Case Study Application Problem
- Review Questions
- Simulation Activities
- Endnotes

In the modern workplace, human workers are viewed as valuable resources rather than simply as **labor** as was often the case in the past. This attitude has resulted in many modern companies changing what used to be called **personnel** to **human resources.** Such companies see people as critical resources who have a vested interest in the organization's success and, as a result, should be full participants in the overall enterprise.

Staffing is a key aspect of human resources development and management. Supervisors have a key role to play in the staffing process. This chapter should help supervisors understand their role in staffing and help them become proficient in all aspects of the staffing process.

STAFFING DEFINED[1]

Staffing, the key component in a company's overall resources development and management effort, is the process through which people are recruited, interviewed, se-

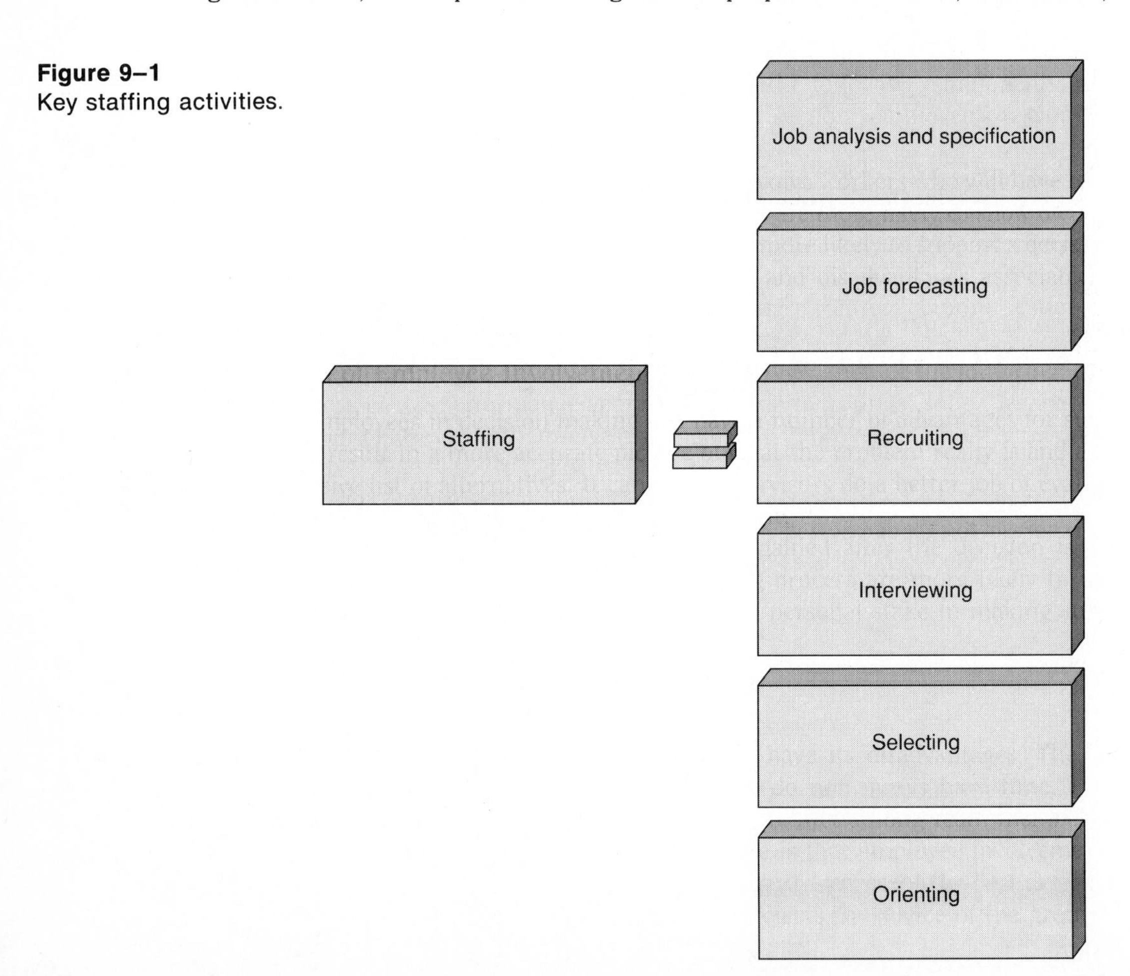

Figure 9–1
Key staffing activities.

lected, and oriented (Figure 9–1). Effective staffing results in filling an accurately forecasted number of positions with the best qualified people when they are needed. Staffing also involves job analysis and job forecasting.

OVERVIEW OF THE STAFFING PROCESS[2]

Staffing is a process consisting of six steps (Figure 9–2), which sometimes are given different names. In this book the six steps in the staffing process are named as follows:

1. Analyzing and specifying
2. Forecasting
3. Recruiting
4. Interviewing
5. Selecting
6. Orienting

Analyzing and Specifying

This step involves **analyzing** the needs of the company to determine what types of job positions are required and in what numbers. Typically these needs are displayed on an organizational chart (see Chapter 8). There must be clear-cut specifications that delineate the job requirements and minimum qualifications. These specifications are set forth in a job description (Figure 9–3). Supervisors typically are not responsible for this task.

Figure 9–2
The staffing process.

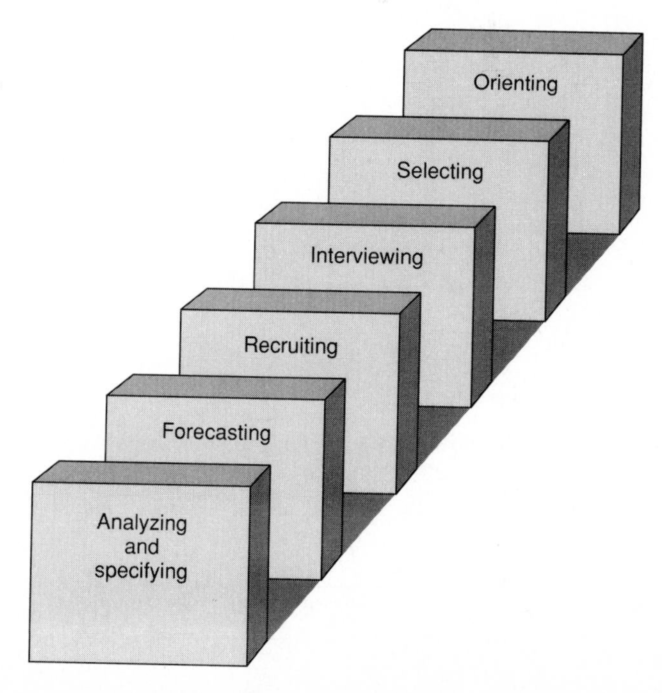

CNC Machine Set-Up and Operator (Level III)

Statement of Purpose

The purpose of this position is to provide a Level III machine set-up and operator who can set up metal fabricating machines to shape and mold metal parts as required by drawings, blueprints, or other source documents/verbal instructions. Works under and reports to the machine shop foreman. Work is performed in a shop environment with noise, sharp objects and tools. Employee is subject to cuts, bruises, and burns in day-to-day work operation. Required to wear protective clothing, to include safety eyeglasses and safety shoes.

Duties and Responsibilities

Sets up and operates machines, such as brakes, rolls, shears, saws, and heavy-duty presses, etc. as specified by blueprints, layouts, and templates.

Selects position, clamps dies, blades, cutters and fixtures, etc. into machine.

Uses rule square, shims, templates, built-in gauges and hand tools, etc.

Positions by direction or at clamp stops, guides and turntables.

Turns handwheels to set pressure and depths of ram stroke.

Adjusts rolls and speed of machines.

Locates and marks bending or cutting lines and reference points onto work piece using rule, compass, straight edge, templates, etc.

Positions work piece manually/machine against stops and guides or aligns layout marks with dies or cutting blades.

Starts machine, repositions work piece and may change dies for multiple or successive passes.

Inspects work.

Reads complex blueprints.

Has ability to make minor programs and changes in programs for changes in blueprints.

Sharpens drill, rod, and mills.

Reads and interprets complex blueprints/drawings.

Required to do machine maintenance.

Capable of supervising or assisting of training level I or Level II CNC machine operators.

Interfaces with engineers and programmers.

THIS LIST OF DUTIES AND RESPONSIBILITIES IS NOT INTENDED TO BE ALL INCLUSIVE AND MAY BE EXPANDED TO INCLUDE OTHER DUTIES OR RESPONSIBILITIES THAT MANAGEMENT MAY DEEM NECESSARY FROM TIME TO TIME.

Qualifications

Graduate from a technical school, or equivalent of an apprenticeship preferred. Good interpersonal skills required.

Human Resources Manager	Manager/Supervisor

Figure 9–3
Sample job description.

Forecasting

Forecasting involves predicting the number of employees that will be needed in each job classification in order to achieve the organization's goals. Supervisors play a key role in this step since they are in the best position to know how many employees will be needed. This step is covered in more detail later in this chapter.

Recruiting

Recruiting involves undertaking those activities necessary to bring in applicants, such as advertising job openings and screening applications, which may come from inside the organization or from outside. These tasks are usually accomplished by the human resources or personnel department. Supervisors do not play a major role, if any role at all, in this step.

However, it is not uncommon for a supervisor to need to fill a vacancy on short notice. How can the process be speeded up in these occasional emergency situations? R. W. Wendover recommends the following strategies:[3]

- Keep an updated list of all positions for which you are responsible. Note the positions you have had to fill most frequently.
- Determine how long the high turnover positions remain filled on the average and the reasons employees give for quitting (is there something you can do in response to these reasons?).
- Determine where these high turnover positions typically have been filled from (inside the company or outside?).
- Using this information, develop a chart that forecasts likely turnover in your unit for the next twelve months.
- Decide in advance where you plan to draw from (the applicant pool). For example, will it be local vocational schools, community colleges, or the Private Industry Council (PIC)? Keep an updated directory of telephone contacts in these agencies for quick reference. Stay in touch with your contacts and nurture your relationship with them so that when needed on short notice they will respond.

Interviewing

Once the applicants have been narrowed down to those who appear to be qualified, interviews are arranged, usually by the human resources department. A human resources specialist may even participate in the interviews to ensure that all applicable equal access–equal opportunity mandates are observed. However, the key role in this step is played by the supervisor. Supervisors typically ask the job-related questions that help determine which applicant is the best qualified. This step is covered in more detail later in this chapter.

Selecting

When filling a job opening there are numerous selection factors to consider. Which applicant has the best technical skills? Which applicant appears to have the most growth potential? Which applicant appears to have the best employability skills (i.e., dependability, loyalty, personability)? Which applicant will best fit into the organizational unit? Supervisors play a key role in answering these types of questions. Human resources personnel help ensure that all factors considered are fair, equitable, and legal. This step is covered in more detail later in this chapter.

Orienting

Too often the staffing process stops with the selection step. Once selected, new employees are put to work and expected to fend for themselves. This inevitably leads to problems that could have been prevented had the new employees been given an orientation.

Orienting new employees involves making them familiar with their new employer, their fellow employees, their new job, and what is expected of them. It also involves clearing up unknowns that might inhibit their productivity.[4] The purpose of orienting employees is to give them the information they need to succeed in their new jobs and to help them break down the barriers that are inherent in being new. This step is covered in more depth later in this chapter.

LEGAL CONSIDERATIONS OF STAFFING

Staffing has evolved over the years into a complex and potentially dangerous activity for companies. Gone are the days when supervisors could hire whomever they wanted without concern for legal, ethical, or equity considerations. Society now insists that staffing practices be fair and equitable, as they should be. To ensure this is the case, numerous laws relating to staffing have been enacted over the years. Some of these are summarized in the checklist in Figure 9–4.

From this list you can see that the legal considerations of staffing have become too complex for supervisors alone to keep up with. This is one of the reasons the human resources department has become so important. Human resources specialists are responsible for staying up to date with regard to the legal aspects of staffing and for developing in-house procedures that translate applicable laws into everyday practice.

However, even with the assistance provided by human resources personnel, supervisors are still actively involved in the staffing process. Consequently, this section discusses the legal considerations with which supervisors should be familiar. More specifically, supervisors should be familiar with legal issues relating to equal employment opportunity, employee compensation, health and safety, and industrial relations (Figure 9–5).

Equal Employment Opportunity[5]

Staffing policies, procedures, and practices must ensure equal employment opportunity for all people. More specifically, they cannot discriminate based on race, gender, age, national origin, handicapping conditions, religion, or marital status in any aspect of the staffing process, or in guiding any employment decision. The equal employment oppor-

Federal Legislation Relating to Staffing (1963–Present)

- Equal Pay Act (1963)
- Civil Rights Act, Title VII (1964)
- Age Discrimination in Employment Act (1967, 1978)
- Occupational Safety and Health Act (1970)
- Equal Employment Opportunity Act (1972)
- Vocational Rehabilitation Act (1973)
- Employee Retirement Income Security Act (1974)
- Freedom of Information Act (1974)
- Privacy Act (1974)
- Vietnam Era Veterans Readjustment Act (1974)
- Minimum Wage Law (1977)
- Pregnancy Discrimination Act (1978)
- Civil Service Act, Title VII (1978)

Figure 9–4
Summary of legislation relating to staffing.

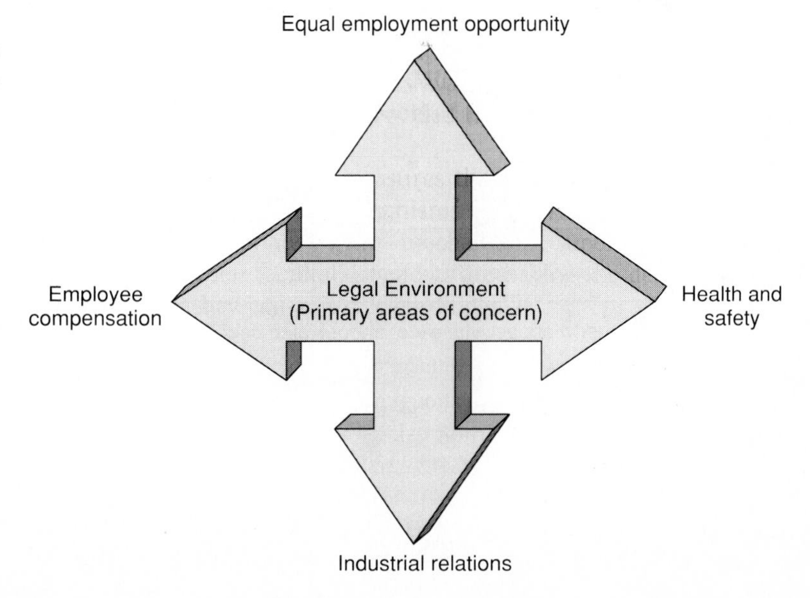

Figure 9–5
Primary areas of concern for supervisors about the legal aspects of staffing.

tunity provisions of staffing grow out of two pieces of federal legislation: (1) The **Civil Rights Act** of 1964 (Title VII); and (2) the **Equal Employment Opportunity Act** of 1972.

Since these legislative initiatives were enacted the courts have heard thousands of cases relating to discrimination in the workplace. A percentage of these worked their way up through the system to the U.S. Supreme Court. Two concepts have evolved from these cases for challenging a company's staffing practices: (1) disparate impact and (2) disparate treatment (Figure 9–6). **Disparate impact** involves discrimination that affects a *protected class* of people (i.e., minorities, women, the handicapped). **Disparate treatment** involves discrimination against individuals.

In a case in which disparate impact is alleged, the courts apply a three-step process. These steps can be summarized as follows:

Step One

The employee who feels wronged (complainant) files a suit alleging discrimination that has resulted in disparate impact against a protected class.

Step Two

The employer is required to show how the staffing practice in question relates to the job. The concept of *job relatedness* is critical. For example, an elderly person cannot be required to pass a heavy-lifting test unless it can be shown that heavy lifting is an integral part of the job.

Figure 9–6
Concepts used to challenge staffing practices.

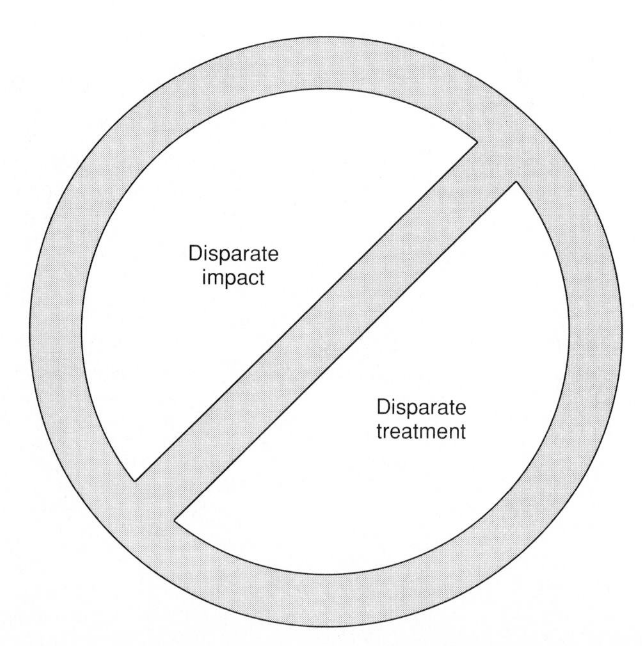

Step Three

The employee is required to show that the employer's justification in Step Two is unacceptable *or* that the employer could have implemented a practice that would be likely to have less of a negative impact.

Cases alleging disparate treatment focus on the *intent* of the employer. The key question is, "Did the employer implement a staffing practice with the intent to discriminate?" The burden of proof shifts from the complainant to the employer if the following conditions exist:

1. The complainant is a member of a protected class.
2. The complainant applied for the job, was qualified for the promotion, was eligible for the raise, etc.
3. The complainant did not get the job, promotion, raise, etc.
4. The employer continued the job search, gave another person the promotion, gave another person the raise, etc.

At this point, the burden shifts to the employer to provide nondiscriminatory reasons for its staffing or employment-related decision. The court rules on the viability of these reasons. The development of the disparate impact and disparate treatment concepts have given any person (or group) who feels discriminated against in the workplace an avenue for resolving the situation.

The handbook used by human resources personnel for ensuring compliance with equal opportunity laws is *Uniform Guidelines on Employee Selection Procedures,* published by the Equal Employment Opportunity Commission (EEOC). Prospective and practicing supervisors can benefit from familiarizing themselves with this document. It is available in most college and university libraries.

In addition to the Civil Rights Act of 1964 (Title VII), as amended in 1972, several other federal laws prohibit discrimination against specific protected classes of people. The **Age Discrimination in Employment Act** of 1967 (amended in 1978) singled out people between the ages of 40 and 69 for special protection in the workplace. The Vocational Rehabilitation Act of 1973 made people with mental and/or physical handicaps a protected class. The U.S. Supreme Court has ruled that employees who have a contagious disease may be considered handicapped and, therefore, are protected under the Vocational Rehabilitation Act. For example, employees with AIDS are covered by the act and are therefore protected from discrimination. The Vietnam Era Veterans Readjustment Act of 1974 extended special protection in the workplace to Vietnam veterans.

Some states have established their own equal employment opportunity agencies. These agencies may supplement and extend the efforts of the EEOC, but they cannot supersede them. The EEOC is traditionally deferred to by the U.S. Supreme Court in equal employment questions.

Compensation and Benefits

The United States has a long history of enacting laws to regulate **compensation and benefits** (Figure 9–7). One of the principal laws that still governs the behavior of

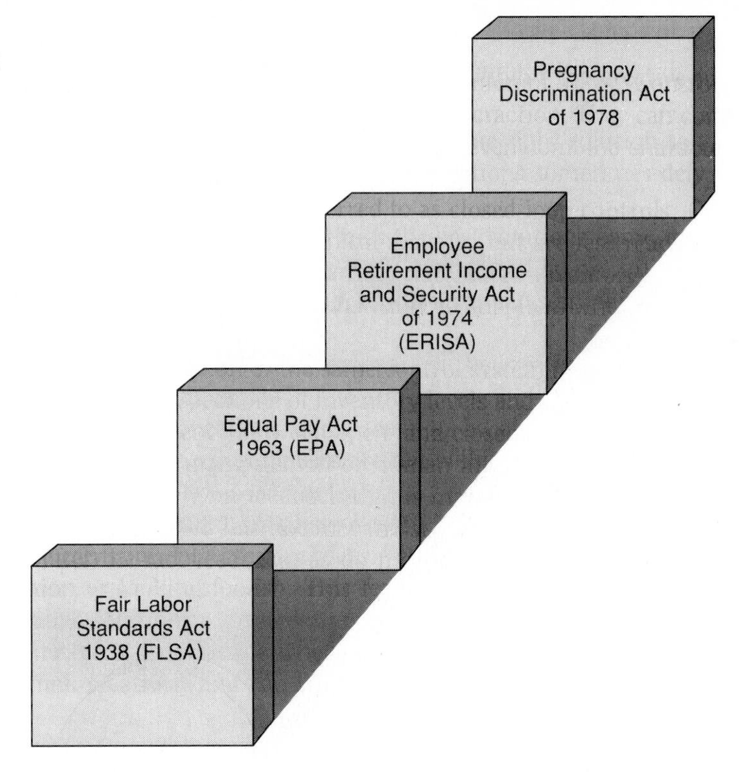

Figure 9–7
Compensation and benefit laws.

employers in this regard was passed in 1938: the **Fair Labor Standards Act** (FLSA). This act established minimum wages, maximum hours, overtime guidelines, and child labor standards. Enforcement of the FLSA is the responsibility of the U.S. Department of Labor.

Two key concepts in the FLSA are those of exempt and nonexempt employees. *Nonexempt* employees are those to whom all the provisions of the law apply. *Exempt* employees are those in executive, administrative, and professional positions to whom certain provisions in the law do not apply. One provision that does not apply to exempt employees is the overtime pay provision. This is why companies are not obligated to pay overtime pay to their salaried managers and professionals.

The **Equal Pay Act** (EPA) of 1963 was enacted to ensure that men and women receive equal pay for equal work. For example, male and female machine operators doing the same job must receive the same pay under the provisions of this act.[6] To determine if two jobs constitute equal work and therefore must receive equal pay the following criteria are used (Figure 9–8):

1. Does the work require **equal skill**?
2. Does the work require **equal effort**?

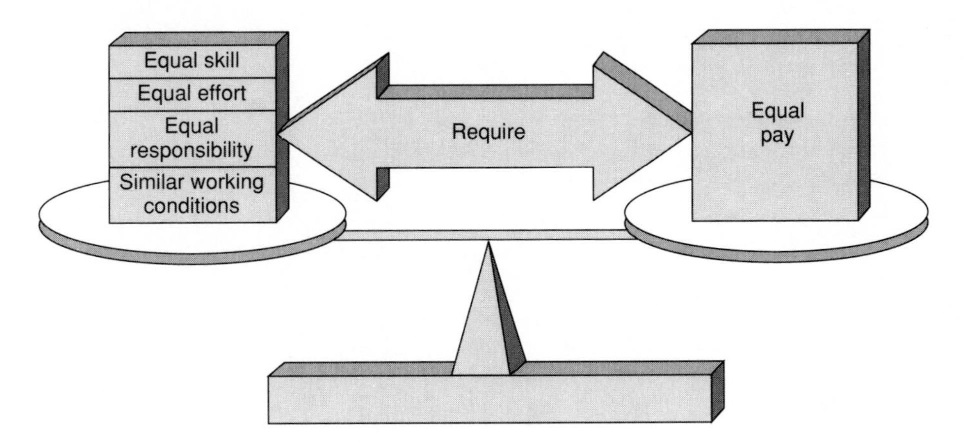

Figure 9–8
Equal pay provisions.

3. Does the work involve **equal responsibility**?
4. Is the work performed under **similar working conditions**?

The EPA is administered by the EEOC. It has come into play frequently in recent years in suits in which women allege that they are paid less than men to do the same job.

In a few exceptional cases equal pay for equal work is not required. Most such cases fall into the following three categories (Figure 9–9):

1. **Seniority**-based pay systems
2. **Merit** pay systems
3. **Incentive**-based systems in which unequal pay is based on either quality or quantity

Other systems that can be shown to discriminate on any basis other than gender may be allowed under the EPA. However, such programs will be carefully scrutinized.

In addition to the FLSA and the EPA, several other laws govern compensation and benefit plans. Prominent among these are the **Employee Retirement Income and Security Act** (ERISA) of 1974, the **Pregnancy Discrimination Act** of 1978, and the **Federal Insurance Contribution Act** (FICA) of 1978. ERISA protects employee retirement pensions from misuse or poor management. It ensures employee pensions against the potential bankruptcy of employers.

The Pregnancy Discrimination Act protects pregnant workers from discrimination by requiring employees to treat pregnancy as they would a disability. This allows pregnant workers to draw the same benefits they would receive for a disability resulting from an accident. FICA requires that employers pay at least half of an employee's Social Security contribution for each pay period.

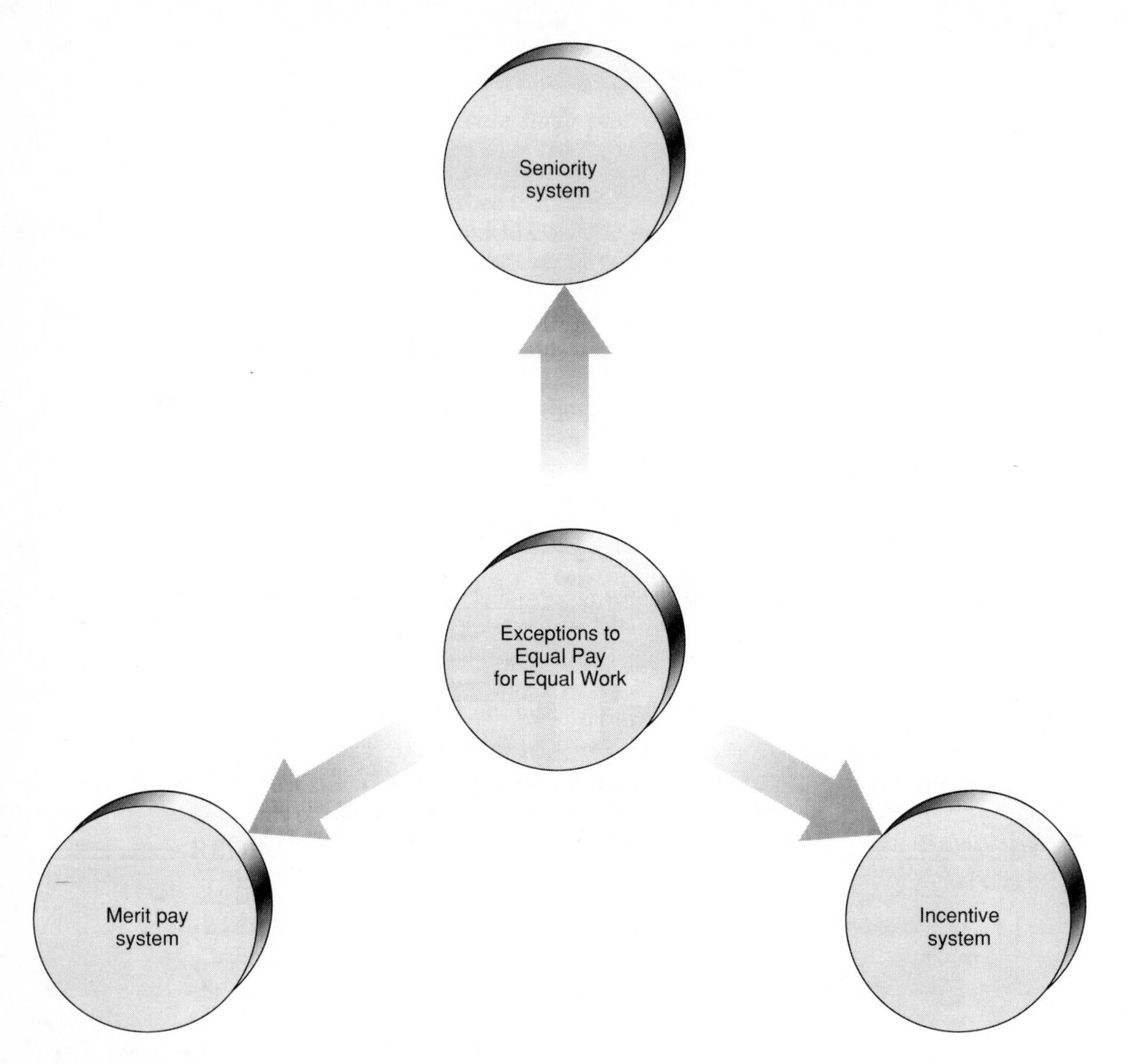

Figure 9–9
Cases in which equal pay may not necessarily apply.

Health and Safety

Individual states enact worker's compensation laws to protect employees who are temporarily or permanently disabled on the job. Typical legislation requires employers to pay medical costs and premiums that are used to pay injured workers a percentage of their normal pay.

Perhaps the most comprehensive and far-reaching piece of health and safety legislation is the *Occupational Safety and Health Act* or OSHA of 1970. OSHA is federal legislation that requires employers to open their doors to federal health and safety inspectors.

The purpose of the act is "to assure as far as possible every working man and woman in the nation safe and healthful working conditions and to preserve our human resource."[7] Section 5(a) of the act reads as follows:

> Each employer
> 1. shall furnish to each of his employees employment and a place of employment which are free from recognized hazards that are causing or are likely to cause death or serious physical harm to his employees;
> 2. shall comply with occupational safety and health standards promulgated under the Act.[8]

OSHA inspections may occur unannounced at any time. Employers are also required to keep comprehensive health, safety, and accident records and file periodic reports.

Industrial Relations[9]

Industrial relations laws were born during the worst years of the Depression. Consequently, the earliest industrial relations legislation was tilted heavily in favor of organized labor. The most significant piece of legislation passed at this time was the **National Labor Relations Act** of 1935, sometimes referred to as the **Wagner Act** (Figure 9–10).

The Wagner Act made unionization of employees legal. The act is credited with being the foundation upon which the union movement in the United States was built. For the first time employees were allowed to organize, use the court system to resolve differences with employers, and bargain collectively for higher wages, better working conditions, and more reasonable hours.

The Wagner Act also defined a set of unfair labor practices employers are prohibited from engaging in and established the **National Labor Relations Board** (NLRB) to oversee employer-employee relations. The NLRB is empowered to conduct union elections, schedule hearings, conduct investigations into alleged unfair labor practices, and issue injunctions when unfair practices are discovered.

In 1947 the **Taft-Hartley Amendment** was attached to the Wagner Act. Taft-Hartley was put in place to balance what had come to be seen, at least by employers, as labor relations legislation slanted too heavily in favor of labor. It identified unfair labor practices on the part of unions, protected the free speech of employers, and provided a mechanism for employees to vote out or decertify a union.

The Labor-Management Reporting and Disclosure Act—also called the **Landrum-Griffin Act**—was enacted in 1959. The purpose of the act was to rid unions of corruption. It established a bill of rights for individual union members, enacted controls on union dues, and put in place reporting requirements with which unions must comply. These and other laws and guidelines with applications in the workplace are covered in greater detail in Chapter 21.

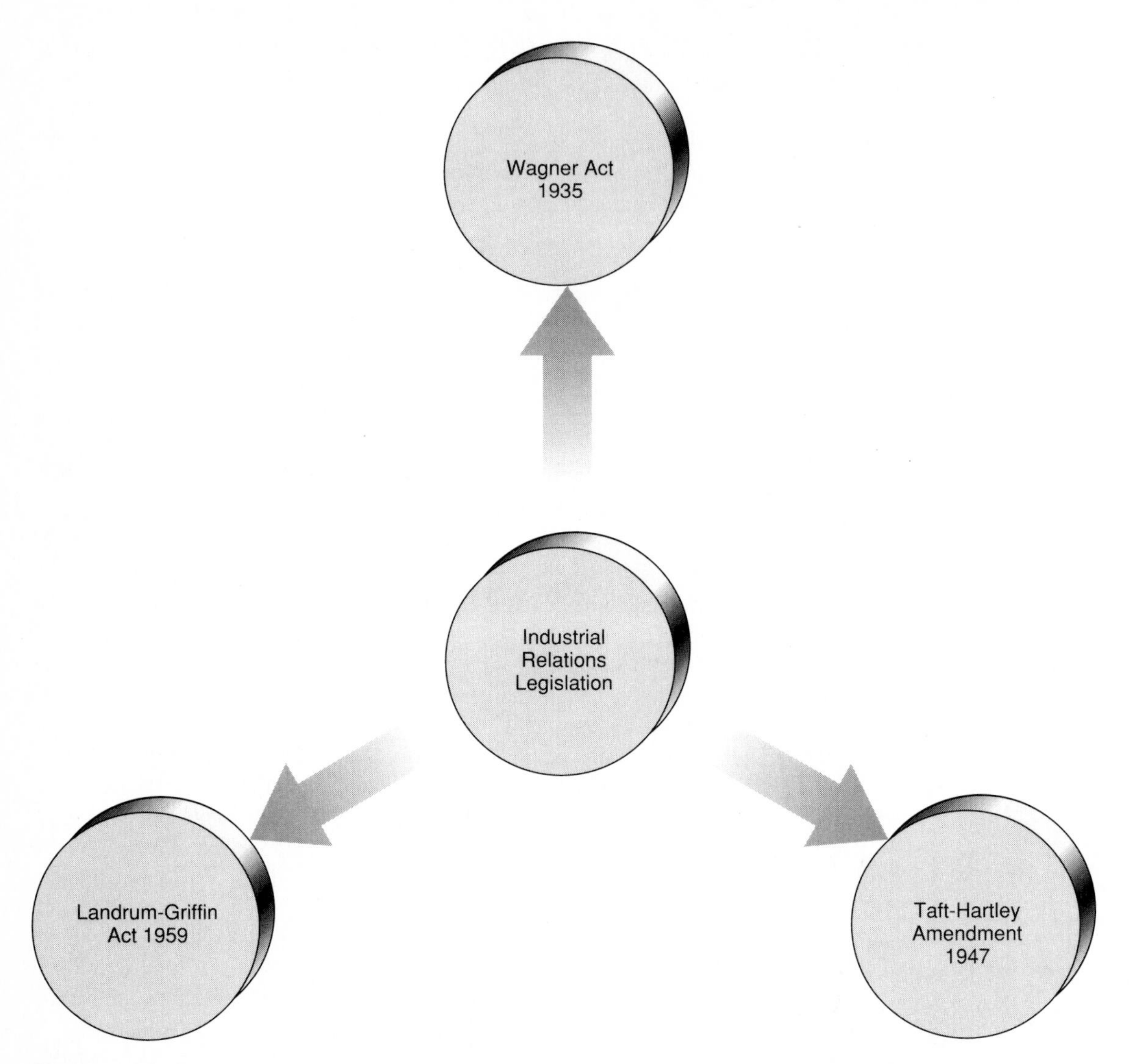

Figure 9–10
Industrial relations legislation.

FORECASTING STAFFING NEEDS

Forecasting staffing needs is a major responsibility of supervisors. Forecasting is the process through which supervisors determine their personnel needs for a given period or for a specific project. There are two occasions when supervisors are called on to forecast staffing needs.

The first occasion is when a company has a given amount of work and needs to know how many of what type of workers will be needed to complete it. The second occasion is when a company is preparing a bid package for a project it would like to have. In both cases, it is critical that the company have accurate personnel forecasts.

Forecasting staffing needs is both an art and a science. The key to effective forecasting is the ability to accurately estimate the amount of time that will be needed to accomplish a specified amount of work. What follows is an annotated list of the steps involved in developing a staffing forecast:

1. *Clearly identify the amount of work to be done during the period in question.* You will typically be given a specified amount of work to be completed within a specified amount of time (i.e., produce 250 front panels in 60 days).
2. *Compute the direct labor needs.* This step involves both the art and the science of forecasting. The art is in determining the total number of worker hours needed to complete the work. Experienced supervisors have an advantage here. They can look back at previous contracts and base their estimates on these.

   The cost accounting department might be able to help, if the product is one the company has produced before, by providing an hours per part multiplier. If the product is new to your company, be sure to allow for preparation time (i.e., writing CNC programs, tooling up, etc.).

   Divide the estimated total worker hours by 8 to determine how many worker days will be needed. For example, if you estimated 3,200 worker hours, dividing by 8 will convert this to 400 worker days.

   Divide the worker days figure by the number of workdays in the period of time in question. Continuing our example, if there are 90 workdays in the period in question, divide the 400 worker days by 90. This computation reveals the need for 4.44 workers.

   Complete this step by adjusting your calculations to accommodate **absenteeism.** Attendance records for your unit will provide a weekly or monthly absentee rate per employee. Multiply that rate times the projected number of employees (4.44). For example, if the monthly absentee rate in your unit is one-quarter of a day (2 hours) per employee, multiply 4.44 employees times 2 hours. The result is a loss of 8.88 hours to absenteeism per month.

   To continue our example, using 20 workdays per month as the conversion factor, 90 working days converts to 4.5 months. Multiplying 4.5 months times 8.88 hours per month results in 39.96 hours that must be added to the total worker hours estimate (3,200 + 39.96) for a new total of 3,239.96 worker hours.

   Dividing the adjusted total of worker hours (3,239.96) by 8 results in a new workdays total of 405. Dividing the adjusted workdays total (405) by 90 workdays results in an adjusted direct labor need of 4.5 workers.
3. *Compute the indirect labor needs.* In addition to the people who will actually produce the product or provide the services (**direct labor**) new support personnel (**indirect labor**) may also be needed. Will you need an additional tool room worker? Secretary? Forklift driver? If so, they must be added to the forecast.

4. *Add the direct figures to the indirect figures to get a total labor figure.* Suppose you determined that a part-time tool handler will be needed to work half-time. This indirect position (.5) is added to the adjusted direct figure (4.5) for a total forecast of 5 workers to complete the projected work in a 90-day period.

To see how these steps are actually applied, consider the example of Harrison Washington, senior supervisor for Military Clothing Manufacturers, Inc. MCM has just won a large contract to produce field jackets for the U.S. Army. The contract gives MCM 6 months or 120 workdays to complete the project. Already working at capacity, MCM will need to add personnel. Harrison Washington has been asked to forecast the personnel needed for his unit. Harrison's calculations are as follows:

Step 1: Identify the Work
Produce 10,000 field jackets in 120 workdays.

Step 2: Compute the Direct Labor

Total work hours = 40,000
(4 hours per jacket × 10,000)

Total worker days = 5,000
(40,000 ÷ 8)

Workers needed (nonadjusted figure) = 42
(5,000 ÷ 120)

Total absentee hours = 504
(2 hours per month × 42 × 6 months)

Adjusted total work hours = 40,504
(40,000 + 504)

Adjusted total workdays = 5,063
(40,504 ÷ 8)

Adjusted direct labor need = 42.19
(5,063 ÷ 120)

Step 3: Compute the Indirect Labor
Hire two material runners and one warehouse worker.

Step 4: Add the Direct and Indirect Figures

42.19 direct labor + 3 indirect labor = 45.19

By these calculations, Harrison Washington is forecasting a need for 45 new workers to complete the contract in 120 workdays. These calculations represent the science of forecasting. The art comes into play when taking into account such intangible factors as the expected level of competence for new hires. Will MCM be able to find 45 fully trained, experienced workers or will the company have to provide on-the-job training? Will MCM's current workforce remain stable? How will these factors affect the contract? Such considerations become part of the forecasting process for experienced supervisors.

INTERVIEWING

Interviewing is an important part of the staffing process for supervisors. In this step the supervisor's experience and expertise come into play. The human resources department can be very helpful in making sure that equal opportunity guidelines are adhered to, that open positions are advertised, that applicants are screened, and that interviews are scheduled. But during an interview, it is the supervisor who is in the best position to determine whether or not an applicant is really qualified. Because of this, supervisors should be prepared to play an active role in interviews.

The purpose of an interview is to determine the real qualifications of people who, from their applications, appear to be qualified. The best way to do this is in a face-to-face conversation involving the applicant and one or more persons who can determine from interacting through questions and answers if someone really knows a given job. The interview also gives supervisors an opportunity to explain the realities of the job to applicants and gauge their reactions.

Interviewing is a process that can be learned. To be effective interviewers, supervisors need to know what to look for when conducting an interview, the types of questions to ask, the types of questions that should not be asked, and some general interviewing guidelines. These concepts are discussed in the paragraphs that follow.

General Interviewing Guidelines

Here are some guidelines supervisors should observe when conducting or participating in an interview.

1. *Be relaxed but businesslike.* Being overly formal will make the applicant nervous and nervous applicants are not at their best. You do not want to lose a well-qualified employee simply because he or she became nervous and did not interview well.
2. *Ask open-ended questions* that let the applicant do most of the talking. Instead of asking "Do you know how to run a Model XYZ machine?" say "Tell me about your machine operation experience." This approach will cause the applicant to reveal more information than will using closed-ended questions. Remember, the more an applicant talks, the more you will learn.
3. *Listen intently* to the applicant and do not create distractions. Do not fiddle, fidget, flip through the applicant's resume, or take notes. These things can be annoyingly distracting to an applicant. Sit still, listen, and look the applicant in the eyes.
4. *Prepare* before the interview. Familiarize yourself with the applicant's paperwork. Prepare a list of questions you plan to ask. In today's litigation-prone society, it can be important to ask all applicants the same list of questions.

5. *Do not hurry.* Avoid stealing glimpses at your watch or appearing to be rushed in any way. It is important for applicants to feel they are being given ample time to present their best case.
6. *Remain neutral.* Do not communicate agreement or disagreement verbally or nonverbally. The time to comment, favorably or unfavorably, on an applicant's response is after the interview has concluded and the applicant has departed.

Questions to Ask in an Interview[10]

The primary purpose of an interview is to elicit information. Information is obtained through questioning. This section sets forth the types of questions that will elicit the most information.

As was stated in the previous section, an **open-ended question** is one that does not predetermine the answer or the nature of the answer. Rather, it lets applicants formulate their own answers and define for themselves the breadth and depth of their responses. Counselors use open-ended questions to get their clients talking.

Open-ended questions typically begin with such terms or phrases as the following:

Tell me about . . .

Describe for me . . .

Explain . . .

Expound on . . .

The types of questions that should be asked in an interview are those that elicit information relating to the job. These include questions about education and training, experience, specific or special skills, past jobs, past attendance records, references, earliest possible date of availability, and health.

Questions in these areas can be asked using the open-ended approach. For example:

"Explain your educational background for me."

"Tell me about your work experience."

"Expound on your reasons for leaving your last job."

Besides using open-ended questions to elicit more complete information, ask only questions that elicit information that is directly related to the job. Remember, if a selection is questioned by an unsuccessful applicant, you may be called upon to show how your questions relate to what the courts call a **bona fide occupational qualification.**

Questions to Avoid in an Interview[11]

In today's litigious society, it is almost as important to know what questions to avoid as it is to know which to ask. This is important because a supervisor can unwittingly violate an equity statute simply by asking the wrong type of question. Questions that do not relate to a bona fide occupational qualification should be avoided. Questions in the following areas are prohibited:

- Race, color, or culture
- National origin, ancestry, lineage
- Pregnancy
- Gender
- Personal preferences (sexual or otherwise)
- Religion
- Credit rating
- Age
- Birthplace
- Marital status
- Citizenship
- Children
- Relatives
- Place of residence
- Membership in organizations or clubs
- Criminal record
- Disability
- Military service (type of discharge)
- Physical ability

If physical ability, such as the ability to lift a specified amount of weight, can be proven to be a bona fide occupational qualification, the human resources department can arrange an appropriate physical test. Let the results of such a test speak for themselves. Do not ask questions about physical ability during an interview.

Certain questions about a person's criminal record can be asked, but these are limited. Again, it is best to let the human resources department deal with inquiries about an applicant's criminal record.

Characteristics to Look for in an Interview

Is this person right for the job? This question should be answered by the interviewing process. Although you are limited as to the questions you may ask, you are not limited with regard to listening and observing. This is another reason for asking open-ended questions. They may elicit information that you are not allowed to ask.

As you listen and observe, try to determine if the applicant has a positive attitude. Will this person fit in and be a contributing member of the team? If the position is physically demanding you can observe if the applicant is clearly in poor condition. Did the applicant discuss his or her personal interests or hobbies? How do these relate to the job? What was the applicant's attitude toward previous employers? What does this attitude suggest to you? As you listen and observe, be a detective. Look for spoken and unspoken clues that give you more insight into the suitability of applicants.

Contacting References

Society's propensity for litigation has made contacting **references** more complicated than it used to be. You may find that references are reluctant to give information beyond verifying employment dates. This is because references have been sued by applicants for making negative statements and by employers for being unrealistically positive.

When you contact references, ask whatever you want. If references respond, make a mental note of the information and use it in the selection process. One question is safe for references to answer and will tell you much of what you want to know. That question is, "Is this applicant eligible for reemployment with your company?" Even without further details, knowing the answer to this question can tell you a great deal.

THE EMPLOYMENT TESTING ISSUE

Employment testing as a part of the staffing process is an issue with which the modern supervisor should be familiar. Some companies test applicants and claim it is an effective part of the process; other companies avoid testing. Proponents claim it can be an effective way to predict future behavior on the job. Opponents claim testing can discriminate unfairly and violate the right to privacy.

It is not the purpose of this book to recommend for or against employment testing. Rather, prospective and practicing supervisors should be aware of the issue so that they can be informed participants regardless of whether their company tests or does not test.

The first point to understand is that a distinction must be drawn between tests that attempt to measure knowledge and/or skills that are directly related to the job and tests that attempt to predict a person's integrity. Although both can be challenged, the latter type of testing is particularly controversial.

Job Knowledge/Testing Skills

Testing for job-related knowledge and skills is a procedure with a long history. At one time the practice was quite common. However, in the 1960s and 1970s such tests began to be challenged as unfairly discriminatory. As a result, the courts have upheld the viability of this type of testing only when it can be clearly shown that: (a) the tests are **valid** (they measure what they were designed to measure and not some unrelated factor); (b) the tests are **reliable** (they produce similar results over and over when administered to similar groups); and (c) the tests measure knowledge and skills that are *directly related* to doing the job in question.

Because it can be difficult to establish the validity and reliability of employment tests many employers have stopped using them. Others have turned to professional testing firms or educational institutions for assistance, especially as the functional literacy problem has continued to increase.

Employers in need of workers with at least basic literacy skills are beginning to administer or contract for the administration of professionally produced tests. Several such tests that have withstood validity and reliability challenges are now available. (These are examined in more detail in Chapter 16.) Tests and testing methodologies must themselves pass the test of **adverse impact**. Tests that have an adverse impact on protected classes of people are not allowed. Testing companies have responded to this in a number of ways. One way has been to compare test results among like groups. For example, test results for Hispanics are compared only with those of other Hispanics.

As more companies began to use **integrity tests**, job applicants began to challenge them. At the heart of the controversy over such tests is the nature of the questions asked. Following are examples of the types of questions found on integrity tests:

Cheating a little on your income tax is not really all that wrong. (true or false)

Office workers don't work as hard as shop workers. (true or false)

Have you ever taken something that did not belong to you? (yes or no)

Do you ever borrow things and forget to bring them back? (yes or no)

If you found a wallet in the street and it contained a lot of money, would you return it and the money? (yes or no)

Proponents of integrity tests say they can predict whether employees are likely to lie and steal on the job. Opponents claim they discriminate unfairly, invade the privacy of test-takers, and are not valid predictors of behavior. This is an issue that is likely to continue to be controversial.

The American Bar Association (ABA) sets forth the following parameters concerning employment testing:[12]

- *Lie detector testing.* **Lie detector testing** is allowable in only a limited number of situations (i.e., government work, company security, national security). Even in these cases the use of lie detectors is strictly controlled and tightly regulated.
- *Preemployment screening test.* **Preemployment screening tests** are allowable only if they are job-related. No matter how fair a screening test might appear to be, if it has an adverse impact on the protected group it is not allowable. This is determined by comparing the pass rate of minorities (including women) with that of white males. The pass rate for minorities should be at least 80 percent of that for white males.
- *Drug testing and medical examinations.* **Drug testing** and medical examinations are allowable as long as the employer keeps the results confidential and does not use them to discriminate against protected classes.

SELECTION

Having conducted interviews, how do you select the best applicant? This section provides several pointers that will help supervisors increase their chances of selecting the best applicant, a more difficult task than you might expect. These pointers are as follows:

1. Put aside personal bias.
2. Make a checklist of skills, qualities, and characteristics you are looking for.
3. Require a physical examination.
4. Check references.

Yoshio Hatakeyama, president of the Japan Management Association, lists the following methods of personal selection:[13]

1. By seniority
2. By ability
3. By experience
4. By education
5. By desire

Hatakeyama's thoughts on the relative merits of each selection methods are summarized as follows:[14]

- *Seniority.* The seniority approach is appropriate when there are many qualified candidates for a position, none of whom clearly stands out from the rest. In such a case, the choice will have little or no effect on the quality of work produced and the seniority approach will cause the least amount of friction in the workplace.
- *Ability.* When the person selected must produce immediate results, this approach is most appropriate. Often this is the most important consideration. When it is used, workers with more seniority than the one selected should be given the attention necessary to prevent friction and discontent.
- *Experience.* Choosing on the basis of experience is appropriate when it is important to be able to give the selected employee immediate responsibility with a minimum of supervision. Experience differs from seniority in that it relates to time in a specific job while seniority relates to overall time of employment in the company. The downside of this approach is that often the most experienced workers are also the most set in their ways and, therefore, the least likely to change.
- *Education.* In situations where a new job has been developed at which no one is experienced, it is important to select a person with the potential to grow, learn, and develop quickly. A candidate with the appropriate educational background relating to the job is likely to be able to satisfy this need.
- *Desire.* This is an appropriate approach any time a candidate who has the desire also has the necessary foundation upon which more specific job-related knowledge and skill can be built. Since the two most important criteria for any selection method are that the job in question be performed well and that the abilities of the employee selected be enhanced, desire is an important criterion. A person who really wants a given job is likely to do what is necessary to be successful at it. The downside of this approach is that an applicant may really want a position he or she is not yet ready for. Therefore, the judgment of the responsible manager/supervisor is critical with this approach.

Put Aside Personal Bias

You are not seeking to hire a buddy. You are looking for a person who can do a job. It is important to keep this in mind. All people have personal biases. Consequently, it is easy to fall into the trap of recommending the applicant who is most like you, or whose background is similar to yours, or who has characteristics you are comfortable with but that have little or nothing to do with the job.

Before recommending an applicant for employment, ask yourself if the recommendation is affected at all by **personal bias.** If it is, step back and take a second look. Remember, you will be held accountable if the new employee does not work out.

Make a Checklist

One of the best ways to enhance the effectiveness of the selection process is to sit down and make a comprehensive checklist before asking that a position be advertised. This checklist should contain all the qualities you are looking for in an applicant. Making such a checklist is an effective way to eliminate bias from the process.

Figure 9–11 is an example of a checklist developed as a tool for selecting the best applicant among several for a CNC machinist position. This supervisor is looking for

Selection Checklist

Formal Education

______ Associate degree in machining
______ Technical certificate in machining

Experience

How many months of experience does the applicant have operating the following machines?

- Smith Brown model IV CNC mill . . . . ______
- Any model five-axis CNC mill . . . . ______
- Kysaki-Osaka vertical CNC lathe . . . . ______
- Marcellini model 123 CNC drill . . . . ______
- Experience on other CNC machines as listed below:

  ____
  ____
  ____

Intangibles (Rate each Good, Average, or Poor)

- Positive work ethic. . . . . ______
- Dependability. . . . . ______
- Ability to learn. . . . . ______
- Willingness to change. . . . . ______
- Teamwork ability . . . . ______
- Cooperation . . . . ______
- Health/vitality. . . . . ______

Figure 9–11
Sample selection checklist.

applicants with formal education in machining, experience in operating specific machines, and experience in operating any nonspecified machines. The supervisor also wants an employee with a positive work ethic, the ability to learn, a willingness to change, the ability to be a team player, and who is dependable, cooperative, healthy, and vital.

This supervisor would determine applicants' formal education and experience by examining applications and asking questions during interviews. Getting a fix on the intangibles is not as easy. One of the most effective ways is to contact references and ask. If references will not or cannot provide enough information to allow for an intelligent judgment, you will have to put your detective skills to work.

A person's dependability can be judged by noting when he or she showed up for the interview and by asking the applicant to provide some type of follow-up information by a specified time. Is it provided on time? Is the material complete and properly prepared? These things are evidence of dependability or the lack of it.

An applicant's ability to learn and willingness to change can be determined through questions asked during the interview. What major changes has the applicant confronted in his or her career? A change from manual machine operation to computer control? A change from a civilian to a government specification?

An applicant's cooperation can be judged by observing his or her attitude toward the various steps in the process. Does the applicant cooperate fully in completing the application, getting a physical examination, and scheduling an interview? Is the applicant cooperative during the interview? Did the applicant willingly complete tests that were part of the process?

The physical examination will provide information on the applicant's health. His or her vitality will be apparent in the interview and on the plant tour if a tour is part of the process.

The applicant's ability to be a team player may be the most difficult of the characteristics in Figure 9–11 to determine. In describing their hobbies, applicants will occasionally provide clues. Do they enjoy solitary activities or those that involve other people? During the interview do they say "I" or "We" more frequently? Can they give tangible examples of their teamwork skills?

The only characteristic set forth in Figure 9–11 that has not been addressed is the work ethic. The work ethic is a person's beliefs about work itself. A person with a positive work ethic believes work is intrinsically good and rewarding and believes in such concepts as pride of workmanship. Like teamwork ability, this can be a difficult characteristic to pin down without actually seeing a person work. One way supervisors and human resources personnel attempt to gain an understanding of an applicant's work ethic is through the use of open-ended questions during the interview (as described earlier). For example, a supervisor might ask an applicant questions such as the following:

"Tell me how you feel about your work."

"Tell me what makes you proudest about your work."

Physical Examination

A physical examination should be part of the process. It offers two important benefits. First, it will answer questions about the applicant's health that might be out of order in an interview. Second, it will ensure that any preexisting health conditions are identified before an applicant becomes an employee. Such prior knowledge could be important if the employee ever files a worker's compensation claim. It can also ensure that an applicant is not employed in a position that is likely to exacerbate a preexisting condition.

Check References

This strategy has already been discussed but is included again to emphasize its importance. Candid comments from people who have worked with an applicant will sometimes tell you more than all the paperwork, test results, and interview questions combined.

It is said that in personal relationships you don't really know people until you have lived with them. Similarly, you don't really know about people's skills and work habits until you have worked with them. For this reason, talking not just with past employers, but also with past fellow workers can be helpful.

In talking with past employers and supervisors, make sure you determine the applicant's reemployment status. Would they hire this person again? If not, why not? In talking with past fellow employees, two questions can help you gain valuable insight into the likelihood of an applicant being a positive, contributing employee. Those two questions are:

"How would you feel about having this person for a supervisor?"

"How would you feel about supervising this person?"

The first question will provide insight into how well the applicant works with other people. Is he or she bossy? Cooperative? Pushy? Helpful? The second question will provide insight into the applicant's work habits. Fellow workers would want to supervise a person who has good job skills, is a good team player, and who is willing to put forth extra effort when it is called for.

Staffing is not an exact science, but the procedures set forth in this chapter can help make the process more effective and help make supervisors a more effective part of the process.

ORIENTATION

Too often the staffing process stops after the selection step. The successful applicant is put to work and expected to begin producing results. This approach is not sufficient in the modern workplace. There are too many potential inhibitors new employees will face to expect them to get a good start without some special attention and organized assistance. The final step in the selection process is orientation and it is critical.

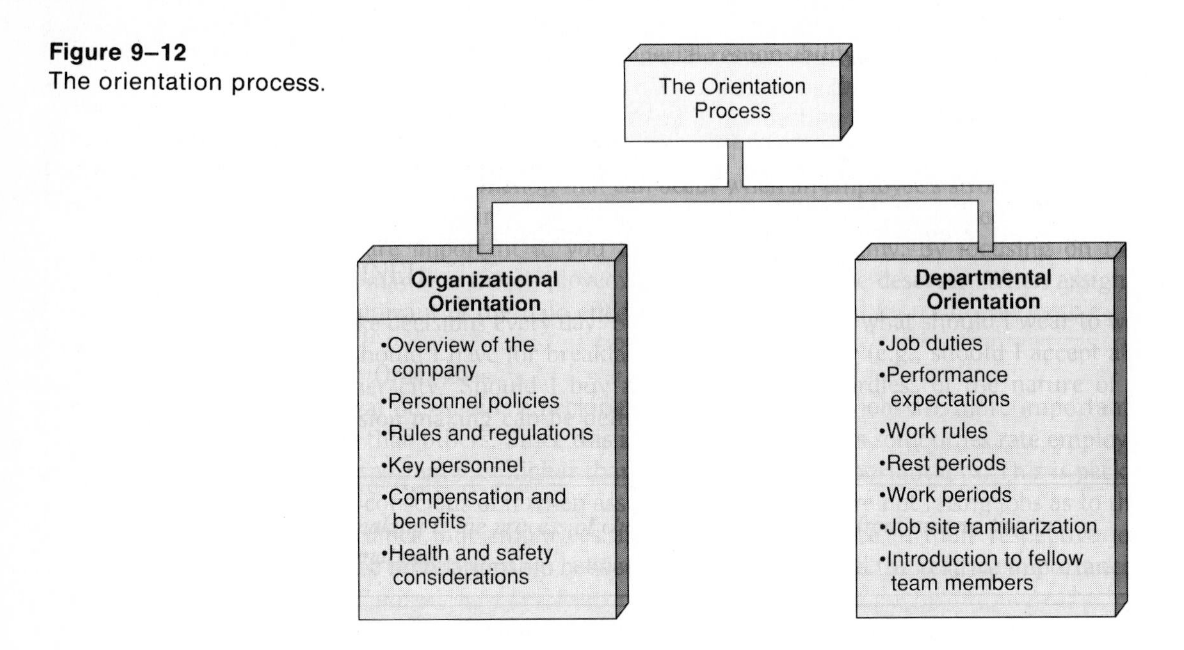

Figure 9–12
The orientation process.

Orientation involves helping new employees gain acceptance as a member of the team, gain a full understanding of where they fit in and what is expected of them, and learn the practical information all employees need to know. According to W. D. St. John, the orientation process can be divided into two parts: organizational orientation and departmental orientation.[15] They are summarized in Figure 9–12.

Organizational orientation makes new employees familiar with the company, its products, and its history. It also familiarizes them with policies, rules, and regulations they need to know as well as key personnel, health and safety, and compensation/benefits information. **Departmental orientation** helps them gain a foothold in becoming a fully accepted member of the team, an understanding of what is expected of them, and practical knowledge about day-to-day functioning. The human resources department typically provides the organizational orientation. Supervisors provide the departmental orientation.

Information that is old hat to the supervisor is brand new to the employee. Therefore, it is important to pace the orientation process so that new employees have time to take it all in. Overloading new team members can be just as bad as failing to orient them. Once the orientation process has been completed, supervisors should follow up periodically to ensure that the information provided has been digested and understood. Ask for feedback, encourage questions, and clarify where necessary.

OTHER STAFFING CONCERNS

In addition to the topics covered so far in this chapter, several other staffing issues are of concern to supervisors. Prominent among these are how to find employees during labor shortages, the benefits of checking references, how to handle layoffs, and termination by discharge. These staffing concerns are dealt with here.

Finding Employees During Labor Shortages[16]

With the 1990s came labor shortages that were the result of the baby bust of the 1970s, which followed the baby boom of the 1950s. In 1990 there were 7 million fewer people in the workforce between ages 18 and 24 than there were in 1980. Five strategies for overcoming labor shortages are as follows:

1. Hire immigrants.
2. Hire ex-convicts.
3. Hire older workers.
4. Hire the handicapped.
5. Hire temporary or part-time employees.

It is against the law to hire illegal aliens. However, immigration restrictions have been liberalized somewhat in this country, particularly for immigrants who are qualified for minimum wage service occupations. The U.S. Immigration and Naturalization Service is the agency to contact for information on hiring immigrants.

There are over a million ex-convicts who are repentant first-time offenders in the United States. As the prison system continues to overflow even more will be returned to society. Evidence suggests that such people can make excellent employees. When you hire ex-convicts, it is important to do a thorough background search so that the company does not become involved in a negligent hiring suit.

Older people can be excellent employees. More than other groups, they tend to have good work habits and a positive work ethic. Participation in the workforce drops off sharply in the 55–67 age group. Many people who retire during this period soon become restless and want to return to work.

The handicapped represent another source of good employees. At the same time there is a great deal of positive pressure on employees to hire the handicapped, particularly employers doing government work. Consequently, employers can benefit doubly by hiring handicapped workers. Information on and assistance with hiring handicapped workers is available from the following agency:

President's Committee on Employment of People with Disabilities
1111–20th Street, NW, Suite 636
Washington, D.C. 20036

Part-time and temporary employees represent another source of good workers. They can help you meet your needs less expensively than full-time workers. Most cities have companies that specialize in placing temporary and part-time workers. Modern supervisors should be familiar with these companies, which are listed in the yellow pages of the telephone directory.

Reference Checking

In today's society, reference checking is more important than ever. It has become increasingly difficult to terminate even incompetent employees. Therefore it is critical to ensure that applicants can do the job before hiring them. One way to do this is through extensive reference checking. Unfortunately too often reference checks are not done or are done in a cursory manner.

Linda Segall blames the lack of reference checking on two misconceptions about the process: (1) you check only the references supplied by the applicant; and (2) applicants only list good references anyway.[17] The first misconception is not true. Supervisors can contact additional references. It is a good idea to talk with the worker's previous supervisors. What they tell you, as well as what they don't can be invaluable in determining an applicant's real abilities.

The second misconception can be overcome using two simple strategies. First, ask references open-ended questions. Listen to what they say as well as what they don't say. Such things as long pauses, unnatural caution, and evasiveness are all signs of potential problems. Second, always conclude the discussion by asking if the employee is eligible for rehire. These strategies can turn even hand-picked references into sources of valuable information. Reference checking can ensure that an applicant who looks good on paper or does well in an interview performs equally well on the job.

Handling Layoffs Effectively[18]

The other side of staffing involves reducing staff. When layoffs are necessary, higher management typically decides how many must go, but supervisors decide who. A number of different criteria can be applied in making layoff decisions. Prominent among these are length of service, current performance, and proximity to retirement age. Regardless of the method used, layoff decisions should be objective, consistent business decisions.

Michael Smith recommends developing a set of objective job-related criteria that can be consistently applied, including current performance, productivity, quality of work, ability to work independently, past performance, willingness to follow directions, current/future need for a given skill, ability to perform multiple jobs, customer relations ability, and knowledge of the product or system.[19] Weights should be applied to each criterion so that all employees can be assessed and given a score.

There should be actual documentation for each employee's rating for each criterion. This approach will bring at least three benefits: (1) it will identify the employees who can best serve the company; (2) the objectivity built into the process puts the company on a better legal footing; and (3) the objective approach is more likely to be received well by employees than would a subjective approach.

Termination by Discharge

A final staffing concern is **termination by discharge** or as it is more commonly known, firing an employee. This subject is covered in more detail in Chapter 15. It is dealt with only briefly here to make the point that termination by discharge is an important consideration relating to staffing.

Termination by discharge is one of the most potentially problematic situations a supervisor must deal with. Increasingly, discharged employees are seeking redress in the courts claiming *wrongful discharge.* In addition, how the discharging is handled can affect the morale of employees who remain on the job. Correspondingly, if a problem-causing or nonperforming employee is removed appropriately, it can actually improve morale. "Appropriately" in the eyes of other employees is likely to mean that they view the process as being objective, impersonal, and humane.[20]

In order to discharge an employee, supervisors must first establish **cause.** Generally acceptable causes include the following: absence without appropriate leave, nonperformance of job duties, disobedience, poor discipline, disruptive behavior, lying, falsification, theft, damaging company property, and conduct off the job that is detrimental to the company. Before a determination of cause can be made, supervisors must have valid documentation, must have warned the employee about unacceptable behavior, and must have given the employee an adequate opportunity to improve or correct the negative behavior. This is known as **progressive discipline.** Every step in the process should be fully documented. When this has been done, the final step is the discharge interview. This step is covered in Chapter 15.

SUMMARY

1. Staffing is the process through which people are recruited, interviewed, selected, and put to work. The staffing process includes analyzing and specifying, forecasting, recruiting, interviewing, and selecting.
2. Primary legal considerations of staffing include equal employment opportunity, compensation and budgeting, health and safety, and industrial relations.
3. Forecasting is the process through which supervisors determine their personnel needs for a specified period or for a specific project.
4. Interviewing is the part of the process during which the supervisor's experience and expertise come into play. The purpose of an interview is to determine the real qualifications of people who, from their applications, appear to be qualified.
5. General guidelines for conducting an interview include: be relaxed but businesslike, ask open-ended questions, listen intently, do not create distractions, prepare, do not hurry, and remain neutral.
6. During an interview, supervisors should ask open-ended questions. This will elicit more information from applicants.
7. Questions in the following areas should not be asked: race, color, culture, national origin, gender, personal preferences, age, religion, marital status, children, place of residence, criminal record, disability, military service, and physical ability.
8. It is important to contact references. Past employers and past fellow employees can supply invaluable information about the applicant. If fear of litigation makes references reticent, one question can reveal a great deal of information: "Is this person eligible for reemployment with your company?"

9. Employment testing is a controversial issue. Tests related to job knowledge and skills must not have an adverse impact on any protected group. Integrity tests that attempt to predict such behaviors as stealing and lying are very controversial.
10. Once applicants have been interviewed and tested, the best applicant must be selected. Pointers for making the selection process more effective are as follows: put aside personal bias; make a checklist of skills, qualities, and characteristics you are looking for; require a physical examination; and check references.
11. Six different approaches might be used in making personnel selections: seniority, ability, experience, personality, education, and desire.
12. There are two components in the orientation process: organizational orientation, which is typically handled by the human resources department, and departmental orientation, which is handled by the supervisor.
13. Termination of an employee by discharge requires substantiation of cause through appropriate documentation, progressive discipline, and a discharge interview.

KEY TERMS AND PHRASES

Absenteeism
Adverse impact
Age Discrimination in Employment Act
Analyzing
Bona fide occupational qualification
Cause
Civil Rights Act
Compensation and benefits
Departmental orientation
Direct labor
Disparate impact
Disparate treatment
Drug testing
Employee Retirement Income and Security Act
Equal Employment Opportunity Act
Equal effort
Equal responsibility
Equal Pay Act
Equal skill
Fair Labor Standards Act
Federal Insurance Contribution Act
Forecasting
Human resources
Incentive system
Indirect labor
Industrial relations
Integrity testing
Interviewing
Labor
Landrum-Griffin Act
Lie detector testing
Merit pay system
National Labor Relations Board
National Labor Relations Act
Occupational Safety and Health Act
Open-ended question
Orientation
Organizational orientation
Personal bias
Personnel
Preemployment screening tests
Pregnancy Discrimination Act
Progressive discipline
Recruiting
References
Reliable
Selecting
Seniority
Similar working conditions
Staffing
Taft-Hartley Amendment
Termination by discharge
Valid
Wagner Act

CASE STUDY: Subjective Employment Practices[21]

The following real-world case study contains a staffing problem. Read the case study carefully and answer the accompanying questions.

After being rejected for a promotion four times, Clara Watson, a black employee, sued her employer under the provisions of Title VII of the Civil Rights Act of 1964. This act prohibits discrimination in employment on the basis of race, religion, gender, or national origin. Ms. Watson had a compelling case.

In all four instances the promotions had gone to white employees. Her employer had relied primarily on subjective evaluations in ruling Ms. Watson unfit for promotion. During the trial, evidence was presented showing that the employer had no black directors, only one black supervisor, and paid blacks lower wages. The employer did not challenge the claim that it based promotion decisions on subjective judgments.

Ms. Watson invoked the theory of disparate impact on charging her employer with discrimination. However, the court required her to claim disparate treatment instead because the employer had used subjective evaluation procedures.

The court denied Ms. Watson's claim, saying she had failed to prove intentional discrimination, a criterion for disparate treatment. The appellate court reviewed the case and upheld the lower court's ruling. However, the U.S. Supreme Court vacated the appellate court's decision and returned the case to the lower court to be retried on the basis of disparate impact. The Supreme Court ruled that subjective evaluations can be submitted to the test of disparate impact.

1. Do you think the case will be ruled differently using disparate impact as the basis? Why or why not?
2. Could the employer have handled this situation differently and avoided litigation? How?
3. What do you see as the supervisor's role in this situation?

REVIEW QUESTIONS

1. Define the term *staffing.*
2. What are the five steps in the staffing process?
3. What role do supervisors play in recruiting?
4. What role do supervisors play in selecting the best applicant?
5. Why is it so important to have the assistance of human resources personnel in the staffing process?
6. What are the protected classes of people with regard to equal employment opportunity?
7. What legislation established today's equal employment opportunity mandates?
8. Explain the concept of disparate treatment.
9. Explain the concept of disparate impact.

10. Explain the three-step process the courts use in handling cases of alleged disparate impact.
11. What is the key in determining disparate treatment?
12. What conditions must exist in a disparate treatment case before the burden of proof shifts to the employer?
13. What is the handbook used by human resources personnel to ensure compliance with equal opportunity laws?
14. Explain the major tenets of the following laws: Vocational Rehabilitation Act; Vietnam Era Readjustment Act; Age Discrimination in Employment Act.
15. Explain the concepts of exempt and nonexempt as set forth in the Fair Labor Standards Act.
16. What was the purpose of the Equal Pay Act of 1963?
17. What are the criteria used for determining if two jobs constitute equal work?
18. What is the purpose of ERISA?
19. What are the basic tenets of OSHA?
20. Why was the Wagner Act of 1935 tilted in favor of organized labor?
21. What was the purpose of the Taft-Hartley Amendment of 1947?
22. What law was enacted in 1959 to rid labor unions of corruption?
23. List the four steps used in forecasting staffing needs.
24. List six general guidelines for conducting interviews.
25. Give an example of an open-ended question.
26. What is a bona fide occupational qualification?
27. List five areas in which questions should not be asked in an interview.
28. If references are reluctant to give you information, what is the best question you can ask?
29. How are professional testing companies handling the issue of disparate impact in literacy testing?
30. What are the main complaints about integrity testing?
31. List four pointers for improving your chances of selecting the best qualified among a group of apparently qualified applicants.

SIMULATION ACTIVITIES

The following activities are provided to give you opportunities to apply what you learned in this chapter and previous chapters in solving simulated supervisory problems. The activities may be completed by individuals or in groups.

1. You are the floor supervisor in a manufacturing plant. You have been asked to forecast the staffing needs for a job that will involve producing 5,000 metal backpack frames over a period of just 60 days. Assume it takes one hour and 15 minutes to make one backpack frame. Forecast your staffing needs.
2. You are about to begin the selection process to fill two positions: a word processing specialist and a supervisor to replace you so that you can accept a promotion. Develop two separate checklists, one for each position. Make sure each checklist contains all the characteristics and experience you want applicants to have.

ENDNOTES

1. Wickliff, J. "Beyond Hiring: Staffing," *Personnel,* 1989, Volume 65, Number 5, pp. 52–56.
2. Ibid.
3. Wendover, R. W. "Plotting Your Recruitment Needs," *Supervisory Management,* February 1991, p. 6.
4. Hollman, R. W. "Let's Not Forget About New Employee Orientation," *Personnel Journal,* 1976, Number 55, pp. 244–250.
5. Twomey, D. P. *A Concise Guide to Employment Law: EEO & OSHA* (Cincinnati: South-Western Publishing Company, 1986), pp. 44–46.
6. Feldman, D. "Sexual Harrassment: Policies and Prevention," *Personnel,* 1987, Volume 64, Number 5, pp. 46–51.
7. U.S.C. Section 651(a).
8. U.S.C. Section 654(a).
9. Albanese, R. *Management* (Cincinnati: South-Western Publishing Company, 1988), pp. 344–354.
10. Koen, C. M., Jr. "The Pre-Employment Inquiry Guide," *Personnel Journal,* 1980, Number 59, pp. 825–829.
11. Ibid.
12. American Bar Association, *You and the Law* (Lincolnwood, IL: Publications International, 1990), pp. 384–385.
13. Hatakeyama, Y. *Manager Revolution!* (Cambridge, MA: Productivity Press, 1985), p. 34.
14. Ibid., pp. 34–40.
15. St. John, W. D. "The Complete Employee Orientation Program," *Personnel Journal,* 1980, Number 59, pp. 373–378.
16. Odiorne, G. S. "Beating the 1990s Labor Shortage," *Training,* July 1990, pp. 32–35.
17. Segall, L. J. "The Benefits of Reference Checking," *Supervisory Management,* March 1990, p. 5.
18. Ibid.
19. Smith, M. "Help in Making Those Tough Layoff Decisions," *Supervisory Management,* January 1990, p. 3.
20. O'Reilly, C. and Weitz, B. "Managing Marginal Employees: The Use of Early Warnings and Dismissals," *Administrative Science Quarterly,* 1980, Number 25, pp. 467–484.
21. Bersoff, D. N. "Should Subjective Employment Devices by Scrutinized?" *American Psychologist,* December 1988, pp. 1016–1018.

CHAPTER TEN

Control and the Supervisor

CHAPTER OBJECTIVES

After studying this chapter you will be able to explain or define the following topics:

- The Term *Control*
- The Historical Development of Control
- The Rationale for Control
- The Control Process
- What Should be Controlled
- Control Strategies for Supervisors
- How to Control Costs

CHAPTER OUTLINE

- Control Defined
- Historical Development of Control
- Rationale for Control
- The Control Process
- What Should be Controlled?
- Control Strategies for Supervisors
- Cost Control
- Summary
- Key Terms and Phrases
- Case Study Application Problem
- Review Questions
- Simulation Activities
- Endnotes

Control is an important management function of supervisors. It is the function through which supervisors ensure they are going where they planned to go and arriving when they planned to arrive. This chapter discusses the information needed by prospective and practicing supervisors to become effective controllers of their areas of responsibility.

CONTROL DEFINED

Control is a process supervisors use to ensure that they are effectively accomplishing the goals of their organization.[1] Figure 10–1 illustrates how and where control fits into the management process. Notice from this figure that control can enter the process at any step. With this background, **control** can be defined as follows (Figure 10–2):

> ***Control is a management process for ensuring that organizational output and behavior match organizational expectations as closely as possible.***

Notice that both **output** and **behavior** are controlled. This is a critical point because it emphasizes that supervisors must monitor both the accomplishment of goals and the behavior of employees in accomplishing them.

The most effective control system ensures that actual output and actual behavior match what is expected to the *maximum extent possible.* A perfect match is not the goal. This is because even with the best planning, organizing, staffing, and facilitating,

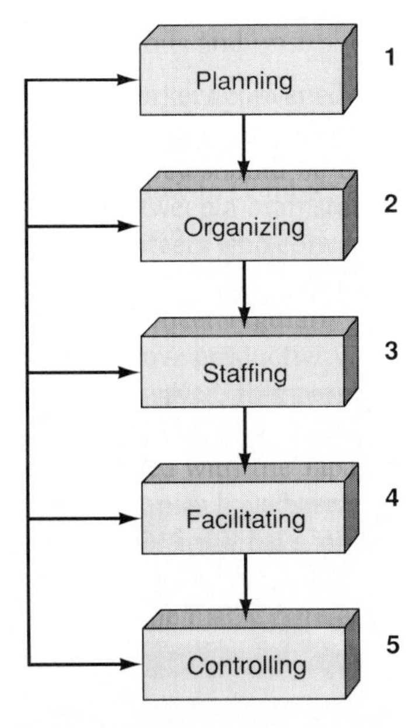

Figure 10–1
Major functions of management.

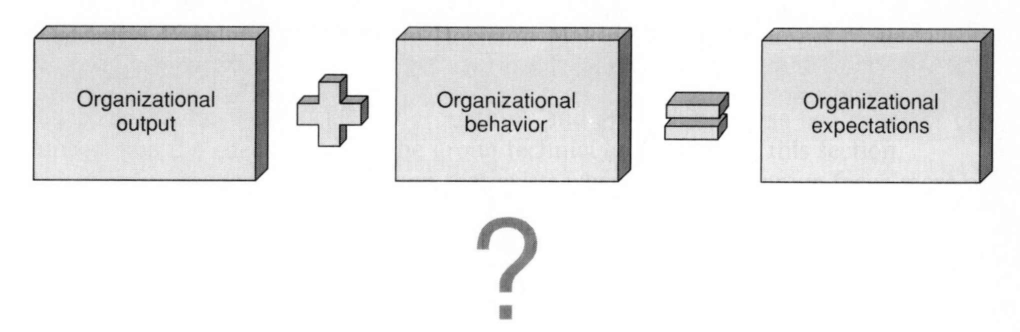

Figure 10–2
Do output and behavior match expectations?

external factors that mitigate against success will enter the picture. These are typically factors over which supervisors have little or no control.

No control system will be perfect so long as uncontrollable factors can affect it. Therefore, no control system is perfect.[2] A more realistic goal is to ensure that control systems seek to approach, to the maximum extent possible, perfect control.

HISTORICAL DEVELOPMENT OF CONTROL

Control was first introduced as a specific management function by Henri Fayol. The teachings of Fayol and Frederick Taylor established control as a legitimate management function as early as 1920.[3] Their separate but related work led to the development of the *traditional view* of control, which is that it should be applied from the top down. With such an approach executive managers control the output and behavior of managers, who control mid-managers, who control supervisors, who control employees.

External controls are the heart and soul of the traditional view (Figure 10–3). Such controls include rules, regulations, schedules, timetables, budgets, financial reports, absentee reports, and time cards. Supervisors use these mechanisms to monitor the output and behavior of their employees. Although the traditional view of control represents the dominant approach, a contemporary view of control is evolving in the modern industrial workplace. The contemporary view focuses on self-control. It is based on the theory that people respond to internally imposed controls but tend to resent externally imposed controls. This view is one of the basic tenets of the total quality management (TQM) philosophy and other innovative philosophies advocated by such management researchers as Peter Drucker and Tom Peters.

Advocates of the contemporary view argue that people respond better to internal controls because they appeal to their sense of self-respect, personal pride, and ego. Internal controls motivate employees by giving them the autonomy that people naturally desire.

In a modern industrial setting, the most effective control system is likely to be one that strikes an appropriate balance between externally and internally imposed controls

Figure 10–3
Traditional view of external control.

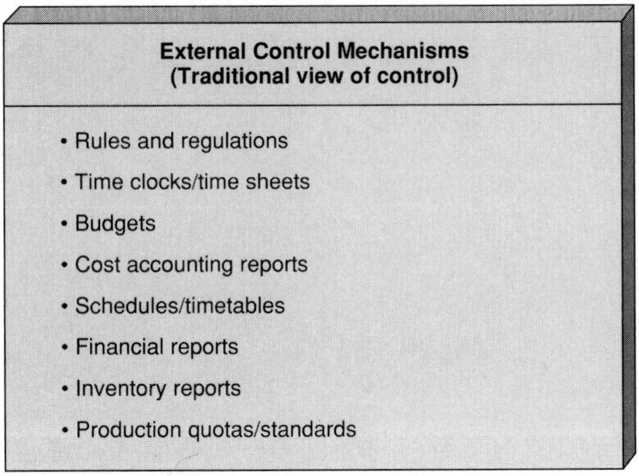

External Control Mechanisms
(Traditional view of control)
• Rules and regulations
• Time clocks/time sheets
• Budgets
• Cost accounting reports
• Schedules/timetables
• Financial reports
• Inventory reports
• Production quotas/standards

External control
Balance the two
Internal control (self)

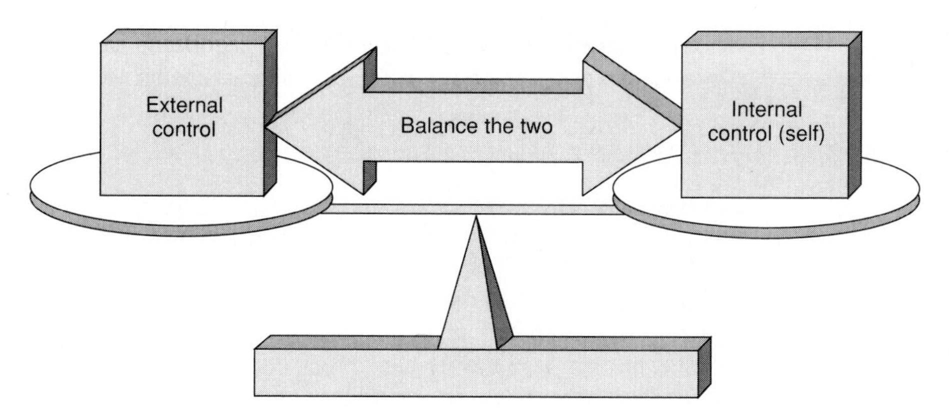

Figure 10–4
The proper balance between external and internal control is the optimum approach.

(Figure 10–4). This is because not all employees respond to internal control and because even the most self-controlled employees may not have sufficient information to control effectively in all cases. A good rule of thumb is that behavior lends itself more readily to self-control while output lends itself more readily to external control.

RATIONALE FOR CONTROL

The term *control* has come to be perceived as negative by some people. Those who feel this way see control as coercive and manipulative. For this reason, the term *monitor* is sometimes substituted. It is important for supervisors to understand the need for control regardless of what it is called. It is also important to develop control mechanisms that are not coercive or **manipulative**. For several reasons control is necessary and desirable in any organization:[4]

1. Employees have varied capabilities. In other words, not all workers are equally knowledgeable or equally skilled.
2. Employees have various levels of motivation and these levels increase and decrease in an unpredictable manner.
3. Employees' personal goals do not always match those of the organization.
4. Decision makers do not always agree on the best way to achieve organizational goals.
5. Organizations are almost always dependent on outside organizations over which they can exert only limited control.
6. Organizations are typically affected by external factors over which they have little or no control.

An important point for supervisors to understand and communicate to employees is that controls are necessary in order to keep resources effectively directed toward the accomplishment of organizational goals. They are not the result of a lack of trust in employees. If controls are stifling or manipulative, they are not being properly applied.

Capabilities of Employees

The purpose of the staffing process is to employ people who have the knowledge, skills, and attitudes to do the jobs needed. Every job requires specific capabilities and most people have certain capabilities. However, even under the best circumstances a perfect match between the two is rarely achieved. This fact has an effect on the accomplishment of organizational goals.

One of the major responsibilities of modern supervisors is to close the gaps between capabilities needed in employees and capabilities available. This is done through such strategies as on-the-job training, externally provided training, communication, incentive programs, and so on.

Because even with these gap-closing measures the capabilities of employees will still vary, it is necessary to have control mechanisms that ensure output and behavior match expectations. Using appropriate control mechanisms, supervisors can monitor continually and determine if their facilitation efforts are having the desired impact on output and behavior. If not, they can adjust accordingly.

Motivation of Employees

No two employees have equal motivation levels. Some are self-starters who can be depended on to go the extra mile when necessary. Others are less motivated and must be subjected to external motivation; some occasionally, some continually.

In addition, even employees who are usually motivated will have days when their motivation wavers. By applying appropriate control mechanisms and by monitoring continually supervisors can determine when motivation levels are having a negative impact and respond accordingly.

Personal Goals of Employees

During the staffing process, supervisors try to find qualified employees whose personal goals are compatible with those of the organization. Seldom is there a perfect match. Consequently, supervisors need to have control mechanisms in place that point out when the personal aspirations of employees are having a negative impact on the organization. At these times supervisors may be able to apply facilitation strategies to correct the situation.

Disagreement Over Strategies

Disagreements over how best to achieve organizational goals are common among decision makers. At some point in the process a strategy must be selected and implemented. If appropriate control mechanisms are in place, progress can be monitored to determine if the strategies selected are producing the desired results. If not, adjustments must be made.

Dependence on External Organizations

Most organizations depend on external organizations over which they have only limited control. For example, most companies depend on several different suppliers to provide the raw materials, parts, or piece goods they need. These materials must be delivered in the correct amounts, in accordance with specifications, and on time. The performance of suppliers and any other external organizations must be monitored so that appropriate adjustments can be made when they are not performing as expected.

Dependence on External Factors

Organizations depend on a variety of **external factors** over which they can exert little or no control, ranging from the weather to interest rates. For example, companies located in certain regions of the country see absenteeism and tardiness increase when the snow is particularly heavy.

With appropriate control mechanisms in place, supervisors are able to measure the negative impact external factors such as the weather are having. Having the information that is produced by control mechanisms allows supervisors to adjust, redirect resources, and coordinate efforts toward accomplishing organizational goals.

THE CONTROL PROCESS

Control is a process. Regardless of what is being controlled, there are four components in the process (Figure 10–5):[5]

1. Standards
2. Measurement mechanism

3. Comparison mechanism
4. Action mechanism

If the output being controlled is the number of subassemblies produced each week, the components of the control process might be as follows:

1. *Standard.* Produce 50 subassemblies that meet all quality standards each week for 10 weeks in each of 10 different manufacturing cells.
2. *Measurement mechanism.* A computer program has been developed that allows the supervisor to monitor production continually. Each time a completed subassembly passes the final acceptance check, its serial number is entered into the computer. The supervisor can call up or print out a running total at any time for any or all cells.
3. *Comparison mechanism.* The computer program contains a set of expected output levels for comparison against actual output at a given time. For example, after four hours on day one, 5 subassemblies should have been completed and accepted by each cell. By the end of day three, 30 subassemblies should have been completed and accepted for each cell or 300 overall. These expected output levels are compared against the actual levels.
4. *Action mechanism.* The supervisor monitors production levels by accessing the control program. To ensure that the supervisor knows when output is not in line with expectations, the control program has an action mechanism. If the supervisor displays the comparison report on a computer terminal, the actual number of subassemblies completed and accepted flashes on and off continually when output is not equal to or greater than the standard. If the report is printed, the actual number is preceded and followed by stars. In either case, the supervisor is given sufficient cues that corrective action may be necessary.

The Control Process

Step 1
Set the *standard* against which output and behavior will be measured.

Step 2
Develop a *measurement mechanism* for measuring output and behavior.

Step 3
Compare actual output and behavior with the standard.

Step 4
Take appropriate *action* when necessary to bring actual output and behavior in line with the standard.

Figure 10–5
The control process is a step-by-step process.

Types of Control Processes

Control systems can be structured in a number of different ways. Controls can be automatic or they can be manual and require human interaction. They can control after the fact or they can be predictive. Control systems can even combine characteristics of all of these (Figure 10–6).

Automatic controls are sometimes referred to as **closed-loop controls.** Closed-loop systems can often be developed for mechanical systems, but are less readily applied to humans. A closed-loop system might be a fully automated inventory control system programmed to transmit refill orders automatically to suppliers when inventory levels reach a specified point.

A manual or *open-loop* version of the same control system might have a supervisor conducting periodic visual inspections of inventory levels and placing orders to restock as needed. The primary difference between open- and closed-loop systems is that open-loop systems require human manipulation or human interaction in some form.

After-the-fact controls rely on **feedback** that is made available to supervisors only after the factor being controlled has not met the standard. Traditional quality control systems rely on after-the-fact feedback, as do most traditional control systems. With

Figure 10–6
Control systems available to supervisors.

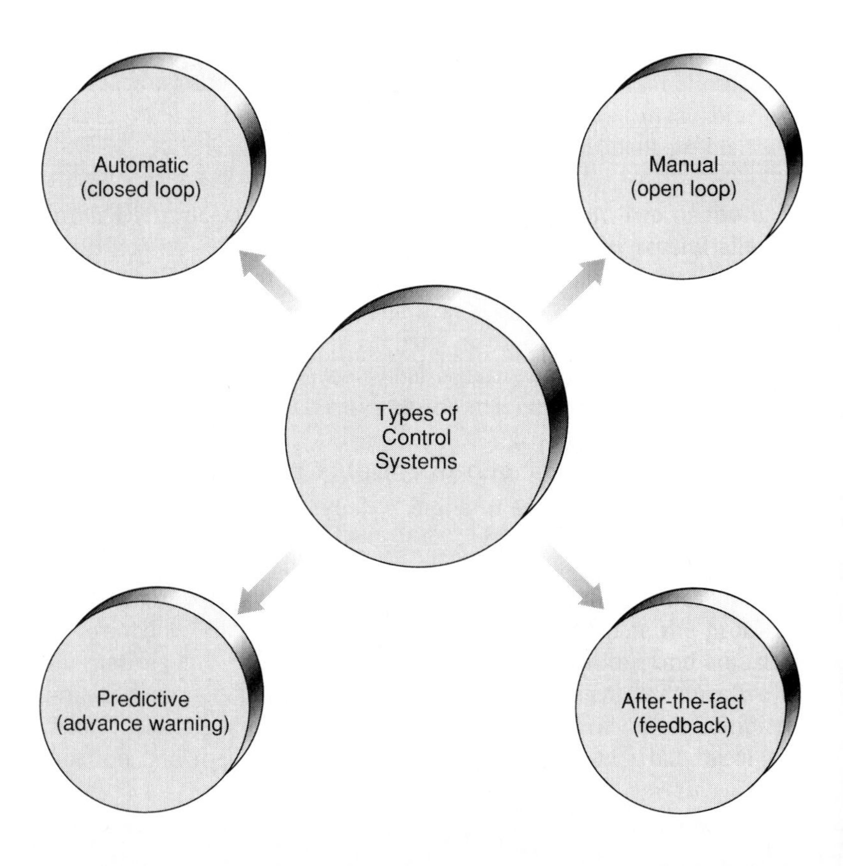

traditional approaches to quality control, completed parts are inspected against the standards set forth in the specifications.

Parts that do not pass all inspections are either returned for corrective work or rejected altogether. The problem with controls that rely on *after-the-fact feedback* is that by the time supervisors learn of problems, it is often too late to correct them. To extend the example of traditional quality control systems, rejected parts become waste, thereby reducing an organization's efficiency.

Problems with after-the-fact feedback systems have led to the development of *predictive* control systems. Such systems are structured to give decision makers information before problems occur so that they can be prevented or, at least, corrected as part of the process.

Two notable examples that build prediction and prevention into the process are *statistical process control* (SPC) and *total quality management* (TQM). SPC is a system for controlling any number of manufacturing processes. It combines the principles of statistics and quality control to establish performance limits for a given machine or process. Actual performance is plotted on a graph on which the upper and lower limits of acceptable performance are prominently shown (Figure 10–7). As long as the actual performance falls within the range of acceptable limits the quality of the performance is predictable and acceptable.

Notice in Figure 10–7 that at one point an actual performance plot exceeded the upper limit. The supervisor monitoring the process obviously saw a problem, took corrective action, and brought the process back in line since the next plots are clustered around the mean.

TQM is more than a predictive control system. It is a management philosophy that makes control every employee's responsibility. Any employee who sees or predicts a problem is encouraged to take appropriate action immediately. With TQM, the philosophy expressed in the phrase "It's not my job" is not allowed. All employees are encouraged to be "quality detectives" and to continually identify and help eliminate factors

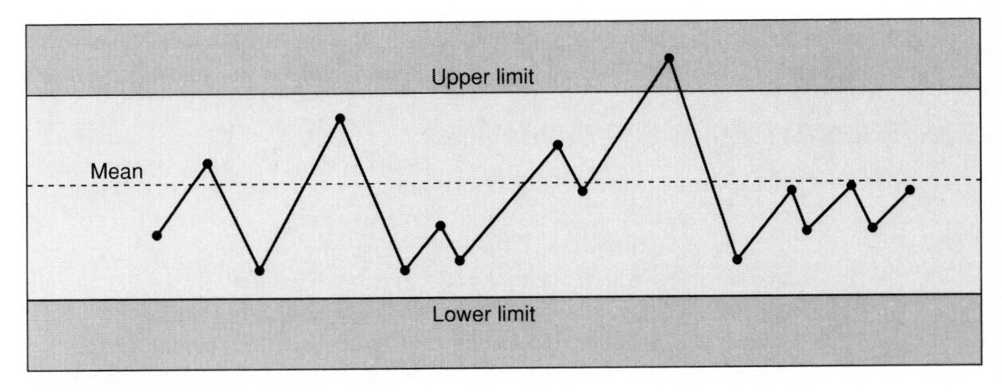

Figure 10–7
Statistical process control chart.

that hinder quality or that do not add to the value of the company's product or service. TQM also incorporates such control systems as SPC.

WHAT SHOULD BE CONTROLLED?

It has already been established that behavior and output are the factors supervisors seek to control. But what is output and how do we define behavior? In order to have effective controls it is necessary to break behavior and output down into more readily measurable components. There are four measurable characteristics of both behavior and output: **time, quality, quantity,** and **cost** (Figure 10–8).

Supervisors are concerned with the quality of an employee's work, the quantity of work, the amount of time spent working, and the quantity of work produced per unit of time; the quality of the output, the quantity of output, and whether or not output standards are met within the prescribed budget.

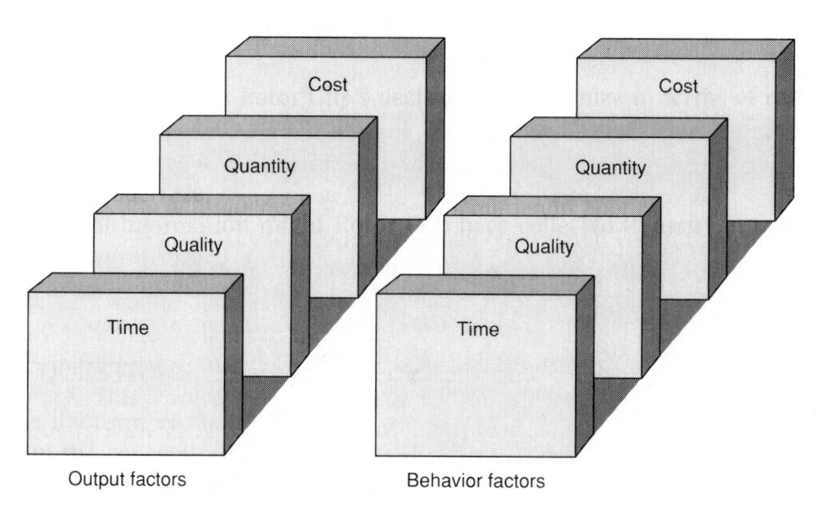

Figure 10–8
Output and behavior factors to consider.

CONTROL STRATEGIES FOR SUPERVISORS

Control strategies available to supervisors fall into three categories (Figure 10–9): personal controls, organizational controls, and peer/self-controls.

Personal controls are those imposed by the supervisor through leadership, by example, and through ongoing facilitation. Personal controls typically work best in small organizations that strongly identify with the supervisor.[7]

With the personal approach there is the potential for autocratic control, although this does not have to be the case. The effectiveness of personal controls depends on the characteristics of the supervisor. A supervisor who has the following characteristics can apply personal controls effectively and positively:

- Good communication skills
- Respect of employees
- High level of technical skills
- Cooperation of employees

Personal control does not work as well in larger organizations because the job becomes more than one person can handle effectively. In addition, with more people it becomes difficult for the supervisor to have enough personal contact to effectively apply the characteristics listed above.

Organizational controls are formal controls and standards used by organizations for controlling output and behavior. Typically, the larger the organization, the more formal and standardized the controls. Organizational controls consist of policies, rules and regulations, standard operating procedures, established day-to-day practices, standardized reporting mechanisms, budget monitoring, and so on.[8]

Organizational controls are sometimes referred to as **bureaucratic red tape.** In spite of this, organizational controls are necessary and desirable in all organizations that are too large for personal control—which means most organizations.

One of the ways organizations can have effective controls without bogging down the system is through effective **delegation.** Well-developed organizational controls lend themselves to delegation so that decision making within well-defined areas of responsibility can be handled routinely. Where some organizations get into trouble in applying organizational controls is in failing to delgate. As a result, all decisions are shuffled to the top along with a stack of paperwork that grows as it moves from point to point in the process.

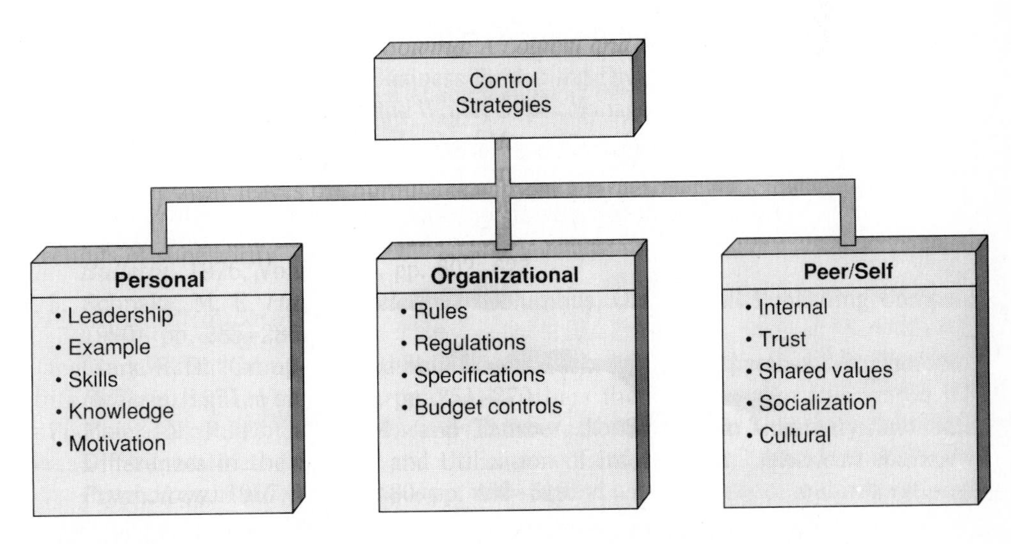

Figure 10–9
Types of control strategies.

Peer/self-controls should not be thought of as the only types of control applied in the smallest organizations. However, in organizations of all sizes they can be an effective way to supplement and complement other control strategies. The underlying methodology involves creating an environment or culture that lends itself to peer/self-control.

It is a two-step process. The first step involves building trust, mutual support, and shared commitment among team members. The second step involves instilling a set of expectations that define the organization's culture. For example, part of the organization's culture might be to produce the highest quality products or to have the best on-time delivery rate. If there is trust, mutual support, and a shared commitment among team members, both individual and peer pressure will serve as controls.

Peer/self-controls are the most difficult to implement and maintain. They require teamwork as opposed to individual achievement and just one uncooperative team member can upset the entire control system. However, this type of approach to control is gaining acceptance. Much of the success experienced by Japanese firms over the past two decades is attributed to these types of controls. Several well-known American companies have begun to use peer/self-controls with success. Prominent among these are Hewlett-Packard, IBM, Procter & Gamble, and Boeing.[9]

COST CONTROL

Supervisors are the first level of management. This means they are the first line in controlling costs. The modern supervisor should be proficient in keeping records, managing a budget, initiating cost improvement strategies, and motivating employees to share the responsibility for cost control.

Cost Control and Record Keeping

What supervisors need most when trying to control costs is information. How much did we budget for this project? How much have we spent? What costs remain? Is the balance sufficient? Through accurate record keeping supervisors are able to answer questions such as these.

Regardless of the type of company, supervisors deal with three basic types of costs: budgeted costs, allowable costs, and actual costs. *Budgeted costs* are those arrived at during the cost estimation process for a project or for the operation of a unit for a period of time. Whether costs are being budgeted for a specific project or for ongoing operation of an organizational unit, there will be specified *line items* or budget categories to which all estimated costs will be assigned.

Typical budget categories include personnel, expendable materials and supplies, equipment maintenance and repair, hardware purchase/lease, software purchase/lease, and so on. Costs in these and other categories are predicted and a budget is established. Budgeted costs identify the limits within which supervisors must operate. One of the main goals of cost control is to complete projects within budget.

Allowable costs are standard costs that have been determined to be acceptable for specified expenses. For example, a company may establish a chart of allowable costs for materials, direct labor, indirect labor, travel, and so on. This is similar to what many insurance companies now use as the basis for paying insurance claims filed by hospitals. Allowable costs let supervisors know what they should expect to pay for goods and services. This provides valuable standards against which actual costs can be compared. If a particular lot of material costs more than its allowable cost per unit, supervisors will know they need to identify a new supplier.

Actual costs are those actually incurred. For example, if a machining supervisor budgets $15,000 for stainless steel for a particular project but spends $20,000, the actual cost was $20,000. Actual costs are typically expressed in unit form (i.e., cost per ton, cost per batch, cost per roll, cost per hour, etc.). In the example of the machining supervisor, the $15,000 budgeted for stainless steel would have been arrived at by applying an expected unit price.

Did the supervisor in this case underestimate the amount of stainless steel that would be needed or the cost of the steel? Unit costs allow supervisors to answer such questions quickly.

Cost Control and Budgets

Cost control as a process might be described as follows: *Develop an accurate budget and stick to it.* Once a budget has been established, accurate records of expenditures should be kept so that the budget can be monitored. Expenditures are recorded and reported by the accounting department. Supervisors monitor the budget using these reports.

Figure 10–10 is an example of a six-month budget monitoring report on a project in which 6,000 parts are to be produced. Using this report as a monitoring tool, the

Budget Monitoring Report

Organizational Unit: Machine Shop — Job number: 1763 — Period: January--June
Total parts needed: 6,000 — Parts produced to date: 2,700 — Remaining work: 3,300
Parts per month projection: 1,000 — Current production per month: 900 — Difference: (-100)

		Actual expenditures					
Line item	Budgeted amount	January	February	March	April	May	June
Direct labor	60,000	10,000	10,000	10,000			
Indirect labor	5.400	900	900	900			
Material	13,200	2,195	3,156	1,032			
Operating supplies	3.000	1,200	-0-	296			
Equipment repair	5,400	-0-	-0-	3,600			
Total	87,000	14,295	14,056	15,878			

Figure 10–10
Sample budget monitoring report.

supervisor can quickly extract several items of useful information including the following:

- The machine shop is producing 100 fewer parts per month than projected. Consequently, this job is currently 300 parts behind schedule. If this situation is not corrected there will be cost overruns.
- Material costs are running slightly below projections (projected: $6,600; actual: $6,383). No problem here.
- Operating costs are running slightly below projections (projected: $1,500; actual: $1,496). No problem here.
- Equipment repair costs are on target (projected: $3,600; actual: $3,600).
- Direct and indirect labor are right on target. However, with the machine shop producing 100 fewer parts per month than projected, there may be a problem. If the production rate does not increase, the contract will have to be extended or overtime will have to be scheduled. In either case, there will be cost overruns in the direct and indirect labor items.

One way companies can simplify the budget monitoring process for supervisors is through the use of *variance reporting.* A variance report compares projected costs against actual costs and points out differences. Computer technology has made variance reporting a simple process.

In order to have the information for monitoring, the supervisor using the report in Figure 10–10 will have to make several calculations each month. A simple computer program can produce a variance report that will give supervisors the required information without the need for calculations.

An example of a monthly variance report based on the month of January from Figure 10–10 is shown in Figure 10–11. This report points out variances in a difference column. In this way supervisors are not burdened with the necessity of making calculations.

A potential weakness with monthly reports such as the example in Figure 10–11 is that they can provide too narrow a view. This can be overcome by making each successive report commutative over the entire budgeting period.

With variance reports, differences shown are not necessarily bad. A variance might represent a plus or minus difference. For example, in Figure 10–11 there is a minus $100 difference in production performance. A supervisor would hope to see either no variance or a plus variance in this column. Consequently a minus difference in this case represents a problem.

Now look at the equipment repair line item in Figure 10–11. It shows a minus $900 variance. In this case a negative variance is desirable because it means this line item is under budget.

Cost Control and Cost Reduction

One of the supervisors' major responsibilities with regard to cost control is cost reduction. By monitoring costs closely and continually supervisors can gain information that

Monthly
Budget Variance Report
January

Organizational unit Machine Shop **Month** January

Production performance

Projected	Actual	Difference
1,000	900	-100

Cost comparisons

Line item	Projected	Actual	Difference
Direct labor.	10,000	10,000	-0-
Indirect labor.	900	900	-0-
Material.	2,200	2,195	-5
Operating supplies.	500	1,200	+700
Equipment repair.	900	-0-	-900

Figure 10–11
Sample variance report.

might help them reduce costs. However, the areas where supervisors can have an impact in reducing costs are limited.

The areas that have the most potential for supervisors attempting to reduce costs are the following:

- *Waste.* Waste happens at the operating level. By monitoring processes closely where they take place, supervisors can identify factors that contribute to waste. These factors might be a lack of training, faulty equipment, or even faulty processes. Supervisors can have an impact in correcting these factors.
- *Labor costs.* Supervisors are in the best position to know where labor costs can be reduced. Is there a full-time employee when a part-timer would suffice? Are there employees who are frequently idle? Are there jobs that could be combined? Supervisors are in the best position to make such observations and correspondingly make recommendations for reducing personnel.
- *Process costs.* Because they work at the operational level, supervisors are in a position to observe the efficiency or inefficiency of processes. Does work rush through

one station only to back up at the next? Are processes sufficient to the task or have they become outdated? Are processes structured to minimize human movement? Careful observation will help supervisors answer such questions and react accordingly.

Other costs such as utilities, telephone, mailing, material, copying, tools, and equipment also represent areas that should be studied continually for potential cost reductions. Finally, the fastest way to reduce costs is to increase output. When more units are produced in the same amount of time or in less time, the cost per unit goes down.

For example, say a company's loaded labor rate (wages, benefits, and overhead) in its assembly shop is $19 per hour. In one month the employees in this unit worked a total of 1,730 hours and produced 150 assemblies. The unit cost of each assembly at this rate of output is $219.13 calculated as follows:

$$\$19 \times 1{,}730 = \$32{,}870 \div 150 = \$219.13$$

If the supervisor of the assembly shop is able to find ways to increase the output for the same amount of time, the unit cost of the assembly will drop correspondingly. For example, say next month's output is 175 assemblies. The unit cost will be reduced to $187.83 calculated as follows:

$$\$19 \times 1{,}730 = \$32{,}870 \div 175 = \$187.83$$

Another effective way to reduce costs is to cut back on inventory. This is the philosophy behind *just-in-time delivery* or JIT. Companies identify reliable suppliers who can deliver high-quality materials quickly and just when they are needed. Supplies are purchased from these companies over an extended period rather than going through the low-bid process and stockpiling large quantities that must be carried on the books as inventory. When reliable sources can be found, JIT can reduce costs substantially.

Even without implementing JIT, inventory costs can be controlled and reduced. Eugene F. Finkin, president of the Material Handling Group of Babrook Industries, Inc., recommends several strategies for controlling inventory expenses:[10]

- *Dispose of obsolete inventory.* If an item has been in inventory for six months or more and there are no immediate plans for it, dispose of it. To continue carrying the item in inventory is to tie up cash needlessly. Return such items to the supplier whenever possible. When this is not an option it is better to get whatever cash return you can, take the loss, and move on.
- *Stop overstocking.* Rather than building up a buffer by overstocking, companies should work with suppliers for reliable and fast delivery. Shortening internal lead times will also reduce the need for buffers.
- *Accurately track inventory.* One of the reasons inventories build up is that they are not accurately tracked. A company should know what items it has in inventory, how many, and where they are located. An accurate tracking system provides this information instantly, particularly if the system is automated.

- *Reduce inventory locations.* Typically, as a company grows so does the number of locations at which inventory is stored. This increases the amount of money tied up in inventory by increasing material handling costs and facility costs. Reducing the number of locations can save on material handling and facility costs and increase efficiency.

Cost Control and Employee Morale

It is not always fun living within a budget. Think of your own monthly budget. Do you sometimes find it frustratingly restrictive? Controlling costs requires a continuous, concerted effort on the part of employees. They must focus not just on doing their jobs effectively, but also efficiently.

This means continual efforts to reduce waste, increase output, improve processes, make wiser use of time, and reduce inventory (both incoming and outgoing). If cost control efforts are not handled carefully they can cause morale problems. Employees might view cost control measures as threatening.

Equity in cost cutting is particularly important. The case study in Chapter 6 told the story of the controversy created at General Motors when top management executives asked for substantial increases in their retirement pensions in the midst of a major companywide cost reduction program. If cost cutting measures are not perceived as equitable, they will cause morale problems and, more than likely, will fail.

What follows are several strategies supervisors can use to maintain morale while reducing costs:

- *Communicate* with employees. Let them know why cost reductions are necessary and how they will benefit in the long run.
- *Be specific.* Talk in terms the employees can relate to and give accurate information. Credibility is important.
- *Solicit input* on cost cutting strategies from employees. Let them participate in the process.
- *Keep employees up-to-date* on progress. Give periodic feedback so employees know where they stand and what results have been attained.

SUMMARY

1. Control is a management process for ensuring that organizational output and behavior match organizational expectations as closely as possible.
2. Control was recognized as a separate management function in the early 1900s. Traditional controls have a top-down orientation and are based on external controls. Contemporary approaches to control build in internal or self-controls.
3. The reasons for control are as follows: (a) varied capabilities of employees; (b) various motivation levels of employees; (c) personal goals of employees; (d) disagreement on how to accomplish goals; (e) dependence on external organizations; and (f) dependence on other external factors.

4. There are four components in the control process: standards, a measurement mechanism, a comparison mechanism, and an action mechanism.
5. Closed-loop control systems have automatic controls. They work best on mechanical systems. A fully automated inventory system is an example of a closed-loop control system.
6. Open-loop controls systems require human interaction and are used in the control of behavior and output.
7. After-the-fact control systems rely on feedback. Predictive control systems give decision makers advance information so that problems can be prevented or, at least, corrected as part of the process.
8. Supervisors are involved in controlling time, quality, quantity, and cost of both output and behavior.
9. Control strategies fall into three categories: personal controls, organizational controls, and peer/self-controls.
10. A supervisor with the following characteristics can apply personal controls effectively and positively: (a) good communication skills; (b) the respect of employees; (c) a high level of technical skills; and (d) the cooperation of employees.
11. Organizational controls can be made less cumbersome through effective delegation.

KEY TERMS AND PHRASES

Action mechanism
Automatic controls
Behavior
Bureaucratic red tape
Closed-loop controls
Comparison mechanism
Control
Cost
Delegation
External factors
Feedback
Manipulative controls
Measurement mechanism
Organizational controls
Output
Peer/self-controls
Personal controls
Quality
Quantity
Standards
Time

CASE STUDY: No Surprises Control[11]

The following real-world case study contains a control problem. Read the case study carefully and answer the accompanying questions.

T. J. Rodgers, president and CEO of Cypress Semiconductor Corporation of San Jose, California, uses a control system he calls "no surprises" control. This system gives managers the ability to continually monitor activity at all levels in the company, anticipate problems, and respond appropriately. Cypress' control system emphasizes discipline, accountability, and attention to detail.

At the heart of Cypress' control system is an automated goal setting/performance monitoring system. All 1,400 Cypress employees periodically set goals, establish dates for completing them, enter their goals into a common database, and update the databases on the status of previously set goals. All managers have access to the database and use it to monitor performance in their units. In this way managers know both past and current performance levels at all times. With this information they can make more accurate plans and predictions for the future.

The database and goals are updated weekly. Every Monday night new proposed goals are entered into the computer system. On Tuesday morning managers receive a printout of new goals for their areas of responsibility. Tuesday afternoon managers meet to adjust where necessary and finalize. Have the employees set realistic goals? Do one unit's goals detract from those of another unit? Are priorities set appropriately? These issues are worked out and the goals are finalized. Once finalized they become part of the company's goals and are monitored continually.

1. As a supervisor, how would you feel about all of your employees setting performance goals?
2. What do you see as the strengths and weaknesses of the Cypress system?
3. Could you implement such a system without a computer system?

REVIEW QUESTIONS

1. Define the term *control.*
2. Should supervisors expect perfect control of output and behavior?
3. Explain the traditional view of control.
4. Explain the contemporary view of control.
5. List three reasons why control is desirable and necessary.
6. Explain how external organizations can affect the behavior and output of an organization.
7. What are the four components of the control process?
8. Explain the term *after-the-fact controls.*
9. Explain the term *predictive controls.*
10. Differentiate between closed-loop and open-loop control systems.
11. What are the four components of output and behavior that can be controlled?
12. Explain briefly the strong and weak points of the following types of control strategies: personal control; organizational control; peer/self-control.
13. How can organizations keep formal organizational controls from bogging the system down?

SIMULATION ACTIVITIES

The following activities are provided to give you opportunities to use what you learned in this chapter and previous chapters in solving simulated supervisory problems.

1. As the supervisor of the shipping department for a personal computer manufacturer, you are interested in establishing peer/self control in your unit. Develop a brief plan explaining step-by-step how you intend to accomplish this.
2. Develop a budget monitoring report that you could use for monitoring your own monthly budget.
3. You are a supervisor of a ten-person line crew for an electrical utility company. Develop a cost reduction plan for your unit that will cut costs by 5 percent. Begin by visiting your local electrical utility company and doing the research necessary to develop a realistic but hypothetical budget for your unit. Apply your own cost reduction plan to this budget.

ENDNOTES

1. Giglioni, G. B. and Bedian, A. G. "A Conspectus Management Control Theory: 1900–1972," *Academy of Management Journal,* 1974, Volume 17, pp. 292–305.
2. Merchant, K. A., "The Control Function of Management," *Sloan Management Review,* Summer 1982, Volume 23, Number 4, pp. 43–55.
3. Child, J. *Organizations: A Guide to Problems and Practice* (London: Harper & Row, 1977), p. 119.
4. Albanese, R. *Management* (Cincinnati: South-Western Publishing Company, 1988), p. 554.
5. Ibid., p. 555.
6. Ibid., p. 559.
7. Child, J. *Organizations,* pp. 158–159.
8. Ouchi, W. G. "Markets, Bureaucracies, and Clans," *Administrative Science Quarterly,* 1980, Volume 25, pp. 129–141.
9. Peters, T. and Waterman, R. H., Jr. *In Search of Excellence: Lessons from America's Best Run Companies* (New York: Harper & Row, 1982), p. 348.
10. Finkin, Eugene F. "How to Limit Inventory Expenses," *Journal of Business Strategy,* January/February 1990, pp. 50–53.
11. Rodgers, T. J. "No Excuses Management," *Harvard Business Review,* July–August 1990, pp. 84–98.

CHAPTER ELEVEN

Decision Making, Creativity, and the Supervisor

CHAPTER OBJECTIVES

After studying this chapter, you will be able to define or explain the following topics:

- Decision Making
- Problems Associated with Decision Making
- The Decision-Making Process
- How to Use Decision-Making Models
- How to Use Employees in Decision Making
- The Role of Information in Decision Making
- The Role of Creativity in Decision Making

CHAPTER OUTLINE

- Decision Making Defined
- Problems and Decision Making
- The Decision-Making Process
- Decision-Making Models
- Involving Employees in Decision Making
- Information and Decision Making
- Management Information Systems
- Creativity in Decision Making
- Summary
- Key Terms and Phrases
- Case Study Application Problem
- Review Questions
- Simulation Activities
- Endnotes

Decision making is an important managerial responsibility of supervisors. Supervisors make decisions within specified limits that are defined by their range of authority. These limits should be clearly understood so there is no question as to what decisions supervisors are allowed to make.

DECISION MAKING DEFINED

All people make decisions every day. Some are minor (e.g., what should I wear to work today? What should I have for breakfast?). Some are major (e.g., should I accept a job offer in another city? Should I buy a new house?). Regardless of the nature of the decision, **decision making** can be defined as follows:

> ***Decision making is the process of choosing one alternative from among two or more alternatives.***

Decision making is one of the most critical tasks supervisors perform. Decisions can be compared to fuel in an engine. Decision making keeps the engine (organization) running. In a typical case, work cannot progress until a decision is made.

Consider the following example. Because a machine is down, the machine shop at Brown Machine Works has fallen behind schedule. The shop cannot complete an important contract on time without scheduling at least 75 hours of overtime.

The machine shop supervisor faces a dilemma. On the one hand, no overtime was budgeted for the project. On the other hand, there is substantial pressure to complete this contract on time since future contracts with this client could depend on an on-time delivery. The supervisor must make a decision.

In this case, as in all such situations, it is important to make the best decision. How do supervisors know when they have made the right decision? In most cases there is no *one* right choice. If there were, decision making would be easy. Typically there are several alternatives, each with its own advantages and disadvantages.

For example, in the case of Brown Machine Works, the supervisor had two alternatives: authorize 75 hours of overtime or risk losing future contracts. If the supervisor authorizes the overtime, his company's profit margin for this project will suffer, but its relationship with a client will be enhanced. If the supervisor does not authorize the overtime, his company's planned profit will be protected but its relationship with this client may be damaged.

Since it is not always clear at the outset what the best decision is, supervisors should be prepared to have their decisions criticized after the fact. It may not seem fair to critique, in the calm aftermath, decisions made during the heat of battle. However, having one's decisions evaluated is part of accountability and it can be an effective way to improve a supervisor's decision-making skills.

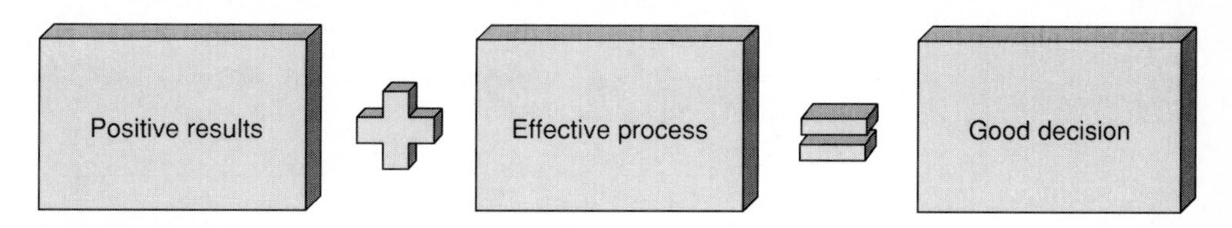

Figure 11–1
Components of a good decision.

Evaluating Decisions

There are two ways to evaluate decisions. The first is to examine the results. In every case the result of a decision should advance an organization toward accomplishing its goals. To the extent it does, the decision is usually considered good. Supervisors' decisions typically will be evaluated based on results.

However, this is not always the best way. Regardless of results, it is also wise to evaluate the process used in making a decision. This is because a positive result can cause you to overlook the fact that you used a faulty process. And in the long run, a faulty process will lead to more negative than positive results.

For example, say a supervisor must choose from among five alternatives. Rather than collect as much information as possible about each, weigh the advantages and disadvantages of each, and solicit informed input, the supervisor simply chooses randomly. There is one chance in five that she might choose the best alternative. Such odds will occasionally produce a positive result, but typically they won't.

This is why it is important to examine the process as well as the result, not just when the result is negative, but also when it is positive (Figure 11–1).

PROBLEMS AND DECISION MAKING

Everyone has problems. We have problems at home, at work, in relationships, and in every other form of human endeavor. But what is a problem? Ask any number of people to describe their biggest problem, and you will get an infinite number of responses. Here is a list of responses that might be given to this question:

"I don't make enough money."

"I am overweight."

"I need more education."

"I don't like my job."

What these responses have in common is that they point out a difference between what is desired and what actually exists. Such a condition is a problem. Therefore, a **problem** can be defined as follows:

> *A problem is the condition that exists when what is desired and what actually exists do not match.*

Obviously the greater the disparity, the greater the problem, with one exception. A key ingredient in determining the magnitude of a problem is the ability of the person with the problem to correct it. Even a pronounced disparity between what is desired and what exists does not represent a major problem if the person can eliminate the disparity. Correspondingly, even seemingly small disparities can represent big problems for people who do not have the ability to eliminate them.

To illustrate this point, consider the following example. A new machine has been installed on the shop floor that can turn out five times as many parts per hour as the one it replaced. But there is a problem. Nobody can remember the correct start-up sequence. Using the wrong sequence can damage the machine. For most operators this is a small problem. They will simply consult the operator's manual, reread the proper sequence, and follow it. However, for the operator who cannot read, this is a major problem. The difference is in the ability of the operator to solve the problem.

Characteristics of Problems

Problems can be classified according to their characteristics. H. H. Brightman identifies three such characteristics (Figure 11–2): structure, organizational level, and urgency.[1]

The **structure** of problems can vary from highly structured to no structure. A highly structured problem exists when the decision maker understands both the problem and how to solve it. An unstructured problem exists when the decision maker is unsure about alternative and solutions. Highly structured and unstructured problems represent opposite extremes. There are also problems with varying degrees of structure which fall at different points along a continuum connecting these extremes.

Highly structured problems are so predictable that decisions regarding them can be automatic. For example, when a tool on a milling machine wears down to the point that accuracy specifications cannot be met, the response is automatic: change the tool. In fact, tool wear represents such a highly structured and predictable problem that tools are usually changed automatically after a specified number of operating hours.

Unstructured problems are not so predictable, nor are responses to them automatic. For example, say the documentation supervisor for a small design firm temporarily loses her three best CAD technicians when their Army Reserve unit is suddenly called up to participate in a classified operation. She does not know where they are going or when they will return. Consequently, she does not know whether to defer action and hope they will return soon, put her remaining staff on overtime, advertise for temporary employees, or request an extension on the contract her team is trying to complete. As is always the case with unstructured problems, this supervisor will need to consider all alternatives carefully and seek informed input before making a decision.

Problems also vary according to **organizational level.** Executive-level decision

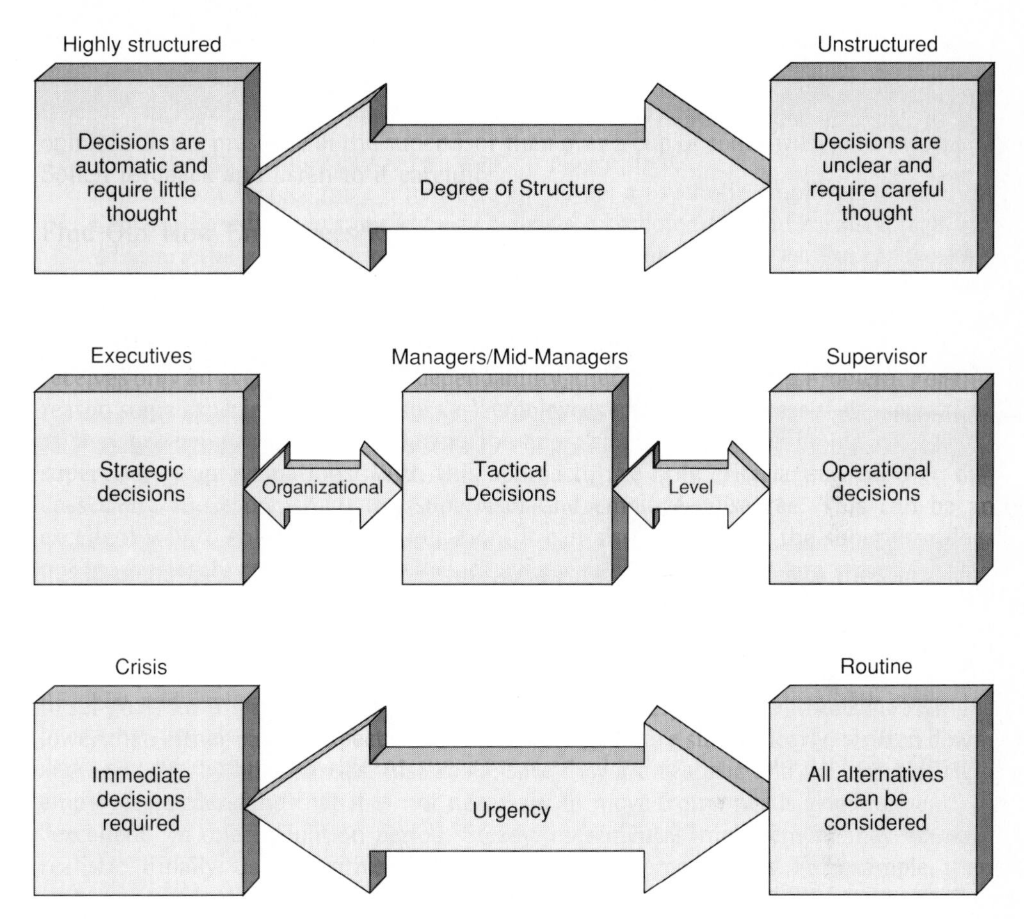

Figure 11–2
Decisions can be classified according to degree of structure, level, and urgency.

makers or top managers deal with strategic problems. Managers and mid-managers deal with tactical problems. Supervisors deal with operational-level problems, those that affect the day-to-day work of the organization. Supervisors are able to deal with problems within their span of authority without consulting higher management. Problems above the operational level are outside the scope of the supervisor's authority.

Problems can also be classified according to their degree of **urgency**. From this perspective problems range from routine to those representing a crisis. One of the underlying purposes of such management functions as planning and organization is to minimize the number of crisis problems decision makers must face.

Crisis problems require immediate attention and force decision makers to react. **Routine problems,** on the other hand, allow decision makers to study the situation, consider alternatives, brainstorm ideas, and make well-reasoned decisions. In decision making, it is always better to act than to react.

For this reason, supervisors should apply the three-step approach to minimizing crisis problems. The three steps are: (1) plan; (2) organize; and (3) learn.

Careful planning and thorough organization will minimize the number of crisis problems supervisors must deal with. However, even the best planning and organization will not completely eliminate crisis problems. This is why the third step—*learn*—is so important. Supervisors should learn from every crisis.

Could it be prevented? If so, how? Did I or my organization contribute to causing the crisis? Can changes be made to prevent a similar crisis from occurring in the future? These are questions supervisors should ask themselves after a crisis.

Careful after-the-fact analysis of a crisis can yield two important benefits. First, it can improve the planning and organization processes, thereby minimizing even further the number of future crises. Second, it can make you better prepared to handle a similar crisis should one occur.

THE DECISION-MAKING PROCESS

Decision making is a process. For the purpose of this textbook, the **decision-making process** is defined as follows:

> *The decision-making process is a logically sequenced series of activities through which decisions are made.*

Numerous decision-making models are available to supervisors. The one described here is adapted from one developed by J. R. Land, J. E. Dittrich, and S. E. White.[2] This model divides the decision-making process into three distinct steps (Figure 11–3):

1. Identify/anticipate the problem.
2. Consider the alternatives.
3. Choose the best alternative, implement, monitor, and adjust.

Identify/Anticipate the Problem

If supervisors can anticipate problems, they may be able to prevent them. Anticipating problems is like driving defensively. You never assume anything. You look, listen, ask, and sense. For example, if you hear through the grapevine that a key employee was injured in this weekend's company softball match, you can anticipate the related problems that may occur. She is likely to be absent or, if she can still work, her pace may be slow. The better supervisors know their employees, technological systems, products, and processes, the better able they will be to anticipate problems.

Even the most perceptive supervisors will not be able to anticipate all problems. A supervisor notices a "who cares" attitude among his team members. This supervisor might identify the problem as employee morale and begin trying to improve it. How-

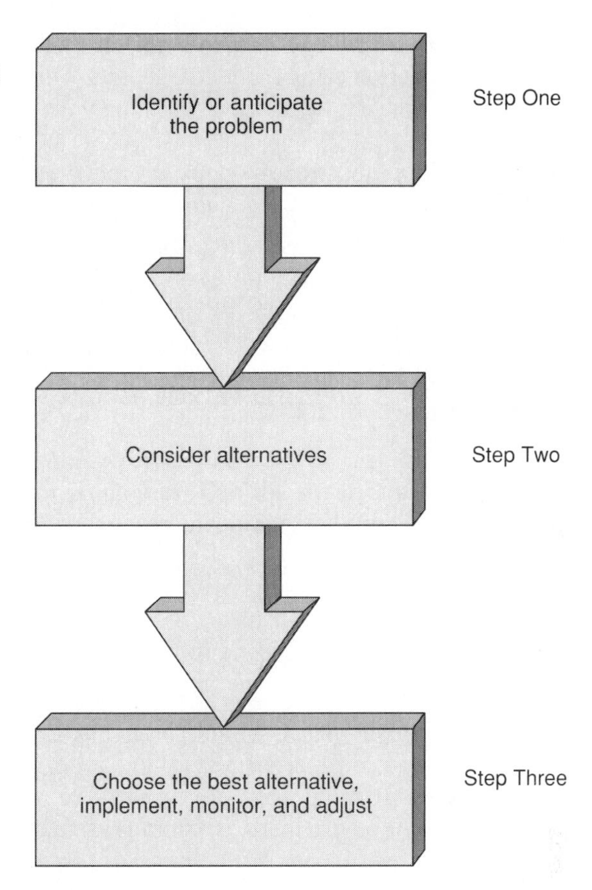

Figure 11–3
The problem-solving process.

ever, he would do well to identify what is behind the negative attitudes. It could be that employees are responding to an unpopular management policy.

The processes that might be at the heart of a problem include those for which a supervisor is responsible, such as scheduling and work processes, as well as a variety of others he or she does not control. These include purchasing, inventory, delivery of material from suppliers, in-house delivery, quality control, and work methods. For example, say a supervisor's organization is having problems meeting production deadlines. He or she might suspect sandbagging on the part of employees and take steps to solve that problem. However, the real problem could be the need to redesign production processes.

Resources that might be at the heart of a supervisor's problem include time, money, supplies, materials, personnel, and equipment. Is the problem caused by a lack of time? Insufficient funding? Poorly trained personnel? Outdated equipment? New, highly technical equipment that employees have not yet learned to operate proficiently? Poor quality materials? All of these are possible causes of the types of problems supervisors commonly deal with.

The purpose of the checklist in Figure 11–4 is to help supervisors look beneath the surface when attempting to identify the cause of a problem. This approach can save time and energy that might be wasted dealing with symptoms rather than actual causes.

Consider Alternatives

This is a two-step process. The first step is to list all the **alternatives** available. The second step is to evaluate each alternative. The number of alternatives identified in the first step will be limited by several factors, including practical considerations, the supervisor's range of authority, and the cause of the problem. Once the list has been developed, each entry on it is evaluated. The main criterion against which alternatives are evaluated is the desired outcome. If the problem is that a client's order is not going to be completed on time, will the alternative being considered solve the problem? If so, at what cost?

Cost is another criterion used in evaluating alternatives. The costs might be expressed in financial terms, in terms of employee morale, in terms of the organization's image, or in terms of a client's goodwill. In addition to applying objective criteria, supervisors will also need to apply their experience, judgment, and intuition when considering alternatives.

Choose the Best Alternative

Once all alternatives have been considered, one must be selected and implemented. Then supervisors monitor progress and adjust appropriately. Is the alternative having the desired effect? If not, what adjustments should be made?

Selecting the best alternative is an inexact process requiring logic, reason, intuition, guesswork, and luck. Occasionally the alternative chosen for implementation will not produce the desired results. When this happens and adjustments are not sufficient, it is important for supervisors to cut their losses and move on to another alternative.

Do not fall into the **ownership trap**. This happens when supervisors invest so much ownership in a given alternative that they refuse to change even when it becomes clear

Figure 11–4
Cause-and-effect checklist for supervisors.

People	Processes	Resources
• Higher management	• Purchasing	• Time
• Resource personnel	• Inventory	• Money
• Suppliers	• In-delivery	• Supplies
• Clients	• Out-shipping	• Equipment/technology
• Employees	• In-work processes	• Personnel
	• Quality control	• Material
	• Methods	
	• Scheduling	

the idea is not working. This can happen any time but is more likely to happen when a supervisor selects an alternative that runs counter to the advice he or she has received, is unconventional, or is unpopular. Remember, the supervisor's job is to solve the problem. Showing too much ownership in a given alternative can impede one's ability to do so.

DECISION-MAKING MODELS

There are many decision-making models available to supervisors. Most of them fall into one of two categories: objective or subjective. In actual practice, the model used by supervisors may have characteristics of both.

Objective Approach to Decision Making

The **objective approach** is logical and orderly. It proceeds in a step-by-step manner and assumes that supervisors have the time to systematically pursue all steps in the decision-making process (Figure 11–5). It also assumes that complete and accurate information is available and that supervisors are free to select what they feel is the best alternative.

Because of these assumptions, a completely objective approach to decision making is rarely used. Supervisors will infrequently, if ever, have the luxury of time and complete information. This does not mean supervisors should rule out objectivity in decision making. Quite the contrary—supervisors should be as objective as possible. However, it is important to understand that the day-to-day realities of the workplace

Figure 11–5
Components in objective decision making.

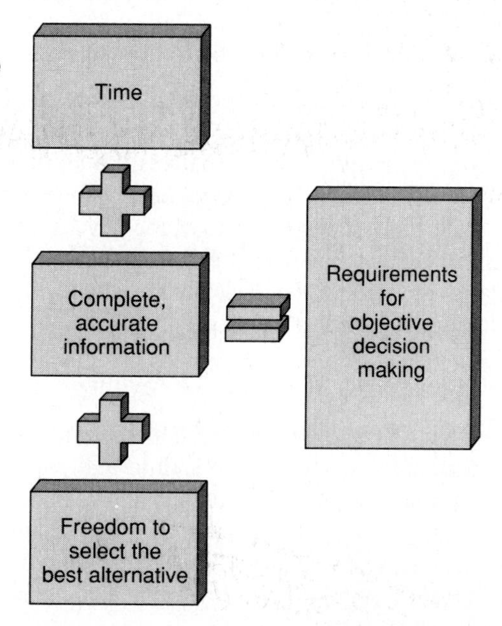

may limit the amount of time and information available. When this is the case, the degree of objectivity is decreased correspondingly.

Subjective Approach to Decision Making

Where the objective approach to decision making is based on logic and complete, accurate information, the **subjective approach** is based on intuition, experience, and incomplete information.

This approach assumes decision makers will be under pressure, short on time, and operating with limited information. The goal of subjective decision making is to make the best decision possible under the circumstances.

In using this approach, there is always the danger that supervisors might make quick, knee-jerk decisions based on no information and no input from other sources. The subjective approach does not give supervisors license to make sloppy decisions. If time is short, use the little time available to list and evaluate alternatives. If information is incomplete, use as much information as you have. Then call on your experience and intuition to fill in the rest of the picture.

INVOLVING EMPLOYEES IN DECISION MAKING

A strategy that can improve decision making is to involve workers who will have to carry out the decision or who are affected by it. Employees are more likely to show ownership in a decision they had a part in making and they are more likely to support a decision for which they feel ownership. There are advantages and disadvantages associated with involving employees in decision making.

Advantages of Employee Involvement

Involving employees in decision making can have a number of advantages for supervisors. It can result in a more accurate picture of what the problem really is and a more comprehensive list of alternatives. It can help supervisors do a better job of evaluating alternatives and selecting the best one to implement.

Perhaps the most important advantages are gained after the decision is made. Employees who participate in the decision-making process are more likely to understand and accept the decision and they will have a personal stake in making sure the alternative selected succeeds.

Disadvantages of Employee Involvement

Involving employees in decision making can also have its disadvantages. The major disadvantage is that it takes time and supervisors do not always have time. It takes employees away from their jobs and can result in conflict among team members.

Next to time, the most significant disadvantage is that employee involvement can lead to democratic compromises that do not necessarily represent the best decision. In addition, disharmony can result if the supervisor rejects the advice of the group.

There are several techniques available to help supervisors increase the effectiveness of group involvement. Prominent among these are brainstorming, the nominal group technique, and quality circles.

Brainstorming

In the **brainstorming** technique the supervisor serves as a catalyst in drawing out group members. Participants are encouraged to share any idea that comes to mind. All ideas are considered valid. Group members are not allowed to make judgmental comments or to evaluate the suggestions that are made.

Typically one member of the group is asked to serve as a recorder. *All* ideas suggested are recorded, preferably on a marker board, flip chart, or another medium that allows group members to review them continuously.

Once all ideas have been recorded, the evaluation process begins. Participants are asked to go through the list one item at a time weighing the relative merits of each. This process is repeated until the group narrows the choices down to a specified number. For example, supervisors may ask the group to narrow the number of alternatives down to three, reserving the selection of the best of the three to themselves.

Nominal Group Technique[3]

The **nominal group technique** (NGT) is a sophisticated form of brainstorming. It has five steps (Figure 11–6). In the first step the supervisor states the problem and clarifies if necessary to make sure all group members understand. In the second step, each group

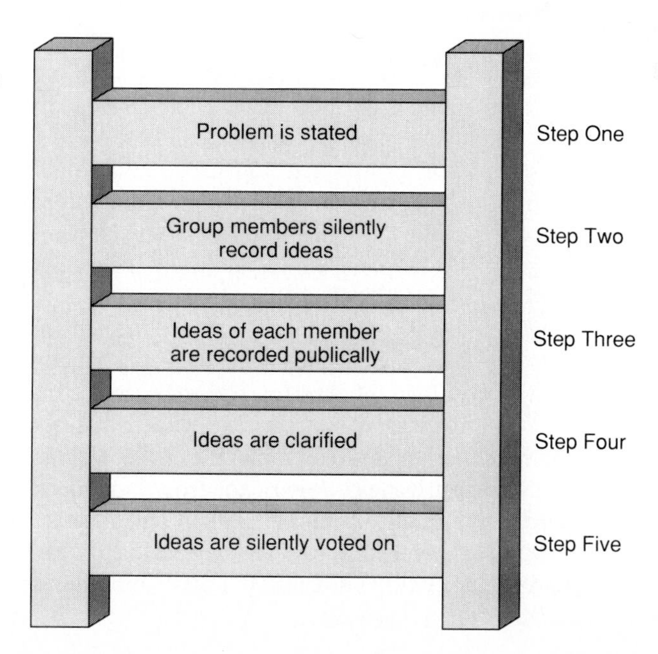

Figure 11–6
The nominal group technique.

member silently records his or her ideas. At this point there is no discussion among group members. This strategy promotes free and open thinking unimpeded by judgmental comments on peer pressure.

In the third step, the ideas of individual members are made public by asking each member to share one idea with the group. The ideas are recorded on a marker board or flip chart. The process is repeated until all ideas have been recorded. Each idea is numbered. There is no discussion among group members during this step. Taking the ideas one at a time from group members ensures a mix of recorded ideas, making it more difficult for members to remember what ideas belong to which member.

In the fourth step recorded ideas are clarified to ensure that group members understand what is meant by each. A group member may be asked to explain an idea, but no comments or judgmental gestures are allowed from other members. The member clarifying the idea is not allowed to make justifications. The goal in this step is simply to ensure that all ideas are clearly understood.

In the final step, the ideas are voted on silently. There are a number of ways to accomplish this. One simple technique is to ask all group members to record the numbers of their five favorite ideas on five separate 3 × 5 cards. Each member then prioritizes his or her five cards by assigning them a number ranging from one to five. The card that receives a 5 contains the best idea and so on down to the card that receives a 1.

The cards are collected and the points assigned to ideas are recorded on the marker board or flip chart. Once this process has been accomplished for all five cards of all group members, the points are tallied. The idea receiving the most points is selected as the best idea.

Quality Circles

A **quality circle** is a group of workers convened to solve problems relating to their jobs. The underlying principle of the quality circle is that people who do the work know the most about it. Consequently, they should be involved in solving problems relating to their work. A key difference between a brainstorming group and a quality circle is that members of the latter are volunteers who convene themselves without being directed to do so by the supervisor. In addition, they don't wait for a problem to occur before getting together. Rather, they meet regularly to discuss their work, anticipate problems, and identify ways to improve productivity. A quality circle does have a team leader who acts as the facilitator. However, this person is not necessarily the supervisor. In fact, it may be a different person each time a quality circle convenes.

Quality circles are associated with the Japanese. In fact, they are often cited as an example of why Japanese companies have been able to excel in the areas of productivity and competitiveness. American firms have occasionally been guilty of expecting too much from quality circles. They can be an excellent way to involve workers in problem solving; however, they are not the magic cure-all for the woes of American manufacturers, as is sometimes suggested. Rather, they can be one part of a much broader productivity improvement effect.

Potential Problems with Group Decision Making

Supervisors interested in improving decision making through group techniques should be familiar with the concepts of groupthink and groupshift. These two concepts can undermine the effectiveness of the group techniques set forth in this section.

Groupthink is the phenomenon that exists when people in a group focus more on reaching a decision than on making a good decision.[4] A number of factors can contribute to groupthink, including the following: overly prescriptive group leadership, peer pressure for conformity, group isolation, and unskilled application of group decision-making techniques. Mel Schnake recommends the following strategies for overcoming groupthink:[5]

- Encourage criticism.
- Encourage the development of several alternatives. Do not allow the group to rush to a hasty decision.
- Assign a member or members to play the role of devil's advocate.
- Include people who are not familiar with this issue.
- Hold "last-chance meetings." Once a decision is reached, arrange a meeting for a few days later. After group members have had time to think things over, call a last-chance meeting in case group members are having second thoughts.

Groupshift is the phenomenon that exists when group members exaggerate their initial position hoping that the eventual decision will be what they really want.[6] If group members get together prior to a meeting and decide to take an overly risky or overly conservative initial view, it can be difficult to overcome. Leaders can help minimize the effects of groupshift by discouraging reinforcement of initial points of view and by assigning group members to serve as devil's advocates.

INFORMATION AND DECISION MAKING

Information is a critical element in decision making. Although having accurate, up-to-date, comprehensive information does not guarantee a good decision, lacking such information can guarantee a bad one. The old saying that "knowledge is power" applies in decision making, particularly in a competitive situation. In order to make decisions that will keep their organizations competitive, supervisors need timely, accurate information.

Information can be defined as "data that have been converted into a usable format that is relevant to the decision-making process." Data that are relevant to decision making have an impact on the decision.

Chapter 4 explained that communication requires a sender, a vehicle, and a receiver. In this process, information is what is sent by the sender, transmitted by the vehicle, and received by the receiver. For the purpose of this chapter, supervisors are receivers of information who base decisions at least in part on what they receive.

Advances in technology have ensured that the modern supervisor can have instant access to information. Computers and telecommunications technology give supervisors a mechanism for collecting, storing, processing, and communicating information quickly and easily. Of course, the quality of the information depends on people receiving accurate data, entering it into these technological systems, and updating it continually. This condition gave rise to the saying "garbage in—garbage out" that is now associated with computer-based information systems. It means that information provided by a computer-based system can be no better than the data put into it.

Data Versus Information

Data for one person may be information for another. The difference is in the needs of the individual. A supervisor's needs will be dictated by the types of decisions he or she makes. For example, a computer printout listing speed and feed rates for a company's machine tools would contain valuable information for the machine shop supervisor, but the same printout would be just data to the warehouse supervisor.

In deciding on the type of information they need, supervisors should ask themselves these questions:

- What are my responsibilities?
- What are my organizational goals?
- What types of decisions do I have to make relative to these responsibilities and goals?

Value of Information

Information is a useful commodity. As such it has value. Its value is determined by the needs of the people who will use it and the extent to which it will help them meet their needs. Information also has a cost. Because it must be collected, stored, processed, continually updated, and presented in a usable format when needed, information can be expensive. Therefore supervisors must weigh the value of information against its cost when deciding what information they need to make decisions. It makes no sense to spend $100 on information needed to make a $10 decision.

Amount of Information

There used to be a saying that went like this: "A manager can't have too much information." This is no longer true. With advances in the information technologies, not only can managers have too much information, they frequently do. This phenomenon has come to be known as *information overload* or the condition that exists when people receive more information than they can process in a timely manner. *In a timely manner* means in time to be useful in decision making (Figure 11–7).

To avoid information overload, supervisors can apply a few simple strategies (Figure 11–8). First, examine all regular reports received. Are they really necessary? Do you receive daily or weekly reports that would meet your needs just as well if provided on a

Figure 11–7
Information overload can lead to a variety of problems.

Information Overload Can Cause...
• Confusion
• Frustration
• Too much attention given to unimportant matters
• Too little attention given to important matters
• Unproductive delays in decision making

monthly basis? Do you receive regular reports that would meet your needs better as **exception reports**? In other words, would you rather receive reports every day that say "Everything is all right" or occasional reports when there is a problem? The latter approach is called *reporting by exception.* It can cut down significantly on the amount of information supervisors must deal with.

Another strategy for avoiding information overload is *formatting for efficiency.* This involves working with personnel who provide information, such as management information systems (MIS) personnel, to ensure that reports are formatted for your convenience, not theirs. Supervisors should not have to wade through reams of computer printouts to locate the information they need. Nor should they have to become bleary-eyed reading rows and columns of tiny figures. Talk with MIS personnel and recommend an efficient report that meets your needs. Also, ask that information be presented graphically whenever possible.

Finally, make use of *on-line, on-demand information retrieval.* In the modern work setting most reports are computer-generated. Rather than relying on periodic hard copy reports, learn to retrieve information from the MIS database when you need it (on-demand), using a computer terminal or a networked personal computer (on-line).

Figure 11–8
Strategies for avoiding information overload.

Avoiding Information Overload
• Change daily reports to weekly reports wherever possible.
• Make weekly reports monthly reports wherever possible.
• Examine all on-going reports periodically. Eliminate unnecessary reports.
• Apply the reporting by exception technique.
• Format reports for efficiency.
• Use on-line/on-demand information retrieval.

MANAGEMENT INFORMATION SYSTEMS

The previous section referred to management information systems and MIS personnel. A **management information system** is a system used to collect, store, process, and present information used by managers in decision making. In a modern work setting a management information system is a computer-based system.

A management information system has three major components: hardware, software, and people (Figure 11–9). Hardware consists of the computer—be it a mainframe, mini, or microcomputer—all the peripheral devices for interaction with the computer, and output devices such as printers and plotters.

Software is the enabling component that allows the computer to perform specific operations and process data. It is primarily computer programs, but also consists of the database, files, and manuals that explain operating procedures. Systems software controls the basic operation of the system. Applications software controls the processing of data for specific computer applications (e.g., word processing, CAD/CAM, computer-assisted process planning [CAPP], spreadsheets, etc.).

A database is a broad collection of data from which specific information can be drawn. For example, a company might have a personnel database in which many different items of information about all employees are stored. From this database can be drawn a variety of different reports such as a printout of all employees in order of age, a breakdown of the workforce by race, or a printout of employees by zip code. Files are kept on computer disks or tape on which data are stored under specific groupings or file names.

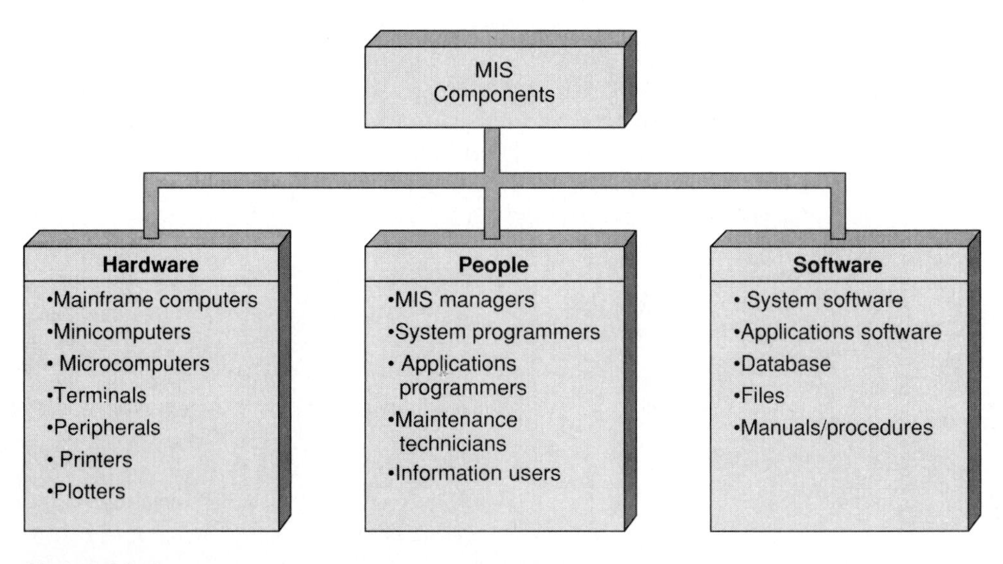

Figure 11–9
Management information system components.

The most important component is the people component. It consists of the people who manage, operate, maintain, and use the system, or users. Supervisors who depend on a management information system are users.

Supervisors should not view a management information system as the final word in information. It can do an outstanding job of providing information about predictable matters. However, many of the decisions supervisors have to make concern problems that are not routine. For this reason it is important to have sources other than the management information system from which to draw information.

CREATIVITY IN DECISION MAKING

The increasing pressures of a competitive marketplace are making it more and more important for industrial organizations to be flexible, innovative, and creative in decision making. In order to survive in an unsure, rapidly changing marketplace, industrial companies must be able to adjust rapidly and change directions quickly. To do so requires creativity at all levels of the organization.

Creativity Defined

Like leadership, **creativity** has many definitions and there are varying viewpoints concerning whether creative people are born or made. For the purposes of modern industrial organizations, a definition that works well is: "The process of developing original, imaginative, and innovative perspectives on situations."[7] Developing such perspectives requires that decision makers have knowledge and experience regarding the issue in question.

Creative Process

According to H. von Oech, the creative process proceeds in four stages: preparation, incubation, insight, and verification:[8]

- **Preparation** involves learning, gaining experience, and collecting/storing information in a given area. Creative decision making requires that the people involved be prepared.
- **Incubation** means giving ideas time to develop, change, grow, and solidify. Ideas incubate while decision makers drive, relax, sleep, and ponder. Incubation requires that decision makers get away from the issue in question and give the mind time to sort things out. Incubation is often accomplished by the subconscious mind.
- **Insight** follows incubation. It is the point when a potential solution becomes clear. This point is sometimes seen as a moment of inspiration. However, inspiration rarely occurs without having been preceded by the perspiration of preparation and incubation.
- **Verification** refers to reviewing the decision to determine if it will actually work. At this point, traditional processes such as feasibility studies and cost-benefit analyses are used.

Factors that Inhibit Creativity

Von Oech listed the following "mental locks" that can inhibit the creative process:[9]

- Looking for the right answer. There is seldom just one right solution to a problem.
- Focusing too intently on being logical. Creative solutions sometimes defy perceived logic and conventional wisdom.
- Adhering too closely to the rules. Sometimes the best solutions come from stepping outside the box and looking beyond the limits established by prevailing rules.
- Focusing too intently on practicality. Impractical ideas can sometimes trigger practical solutions.
- Avoiding ambiguity. Ambiguity is a normal part of the creative process. This is why the incubation step is so important.
- Avoiding risk. When organizations don't seem to be able to find a solution to a problem, it often means decision makers are not willing to take a chance on a risky decision.
- Forgetting how to play. Adults sometimes become so serious they forget how to play. Playful activity can stimulate creative ideas.
- Fear of rejection or looking foolish. This fear can cause people to hold back what might be creative solutions.
- Saying "I'm not creative." People who decide they are not creative won't be. Any person can think creatively and can learn to be even more creative.

SUMMARY

1. Decision making is the process of choosing one alternative from two or more. In most cases there is no one right choice. Rather, there are typically several alternatives, each with its own advantages and disadvantages.
2. There are two ways to evaluate decisions. The first is to examine the results. The second is to examine the process used in arriving at the decision.
3. A problem is the condition that exists when what is desired and what actually exists do not match. A key ingredient in determining the magnitude of a problem is the person's ability to solve it.
4. Problems can be classified according to their structure, organizational level, and urgency.
5. The three-step approach to minimizing the number of crisis problems a supervisor must deal with is plan, organize, and learn.
6. The decision-making process is a logically sequenced series of activities through which decisions are made. These activities are identify or anticipate the problem; consider the alternatives and choose the best; implement, monitor, and adjust.
7. Most decision-making models fall into one of two broad categories: objective or subjective decision making. Objective decision making is dependent on complete, accurate information and the decision makers's ability to select what he or she

thinks is the best alternative. Subjective decision making assumes the decision maker will be under pressure, short of time, and operating with only limited information.

8. There are advantages and disadvantages to involving employees in decision making. It can result in a more accurate picture of what the problem really is, a more comprehensive list of alternatives, and a more objective choice of the best alternative. It can also help employees understand and accept the decision better. On the negative side, it can be time consuming and it can lead to decisions that are popular but not good.
9. Techniques for involving employees in decisions are brainstorming, the nominal group technique, and quality circles.
10. Information can be defined as data that have been converted into a usable format that is relevant to the decision-making process. Data for one person are information for another and vice-versa. The difference is determined by the needs of the individuals.
11. Because information is a useful commodity, it has value. Because it must be collected, stored, processed, updated continually, and presented in a usable format, information has a cost. Consequently, decision makers have to weigh the value of information against its cost before deciding to pursue it.
12. Advances in information technology have created the potential for information overload. This is the condition that exists when people receive more information than they can process in a timely manner.
13. A management information system is a system used to collect, store, process, and present information. Such a system has three components: hardware, software, and people.
14. The creative process consists of four steps: preparation, incubation, insight, and verification.

KEY TERMS AND PHRASES

Alternative
Brainstorming
Creativity
Crisis problem
Data
Decision-making process
Decision making
Exception reports
Incubation
Information
Insight
Management information system
Nominal group technique
Objective approach
Organizational level
Ownership trap
Preparation
Problem
Quality circle
Routine problem
Structure
Subjective approach
Urgency
Verification

CASE STUDY: Modernization or Stagnation[10]

The following real-world case study contains a decision-making problem. Read the case study carefully and answer the accompanying questions.

Robert Slass, president and CEO of Rotor Clip Company, is an entrepreneur from the old school. Playing it safe while the competition moves ahead is not his style.

Rotor Clip Company manufactures low-cost retaining rings for the automobile industry. In the mid 1980s, Slass began to notice increased competition, particularly from Japanese manufacturers that had adapted such modern concepts as CAD/CAM and statistical process control (SPC). He faced a tough decision. Should Rotor Clip continue its traditional approach to production and risk stagnation or should it invest heavily in modernization and risk being overloaded with debt?

This is the type of situation that managers and supervisors face every day. Slass chose to modernize, and as a result Rotor Clip Company has become an award-winning manufacturer that is more than equal to its fierce international competition.

1. How would you evaluate Rotor Clip's decision to modernize in terms of results? In terms of process?
2. Describe how Rotor Clip might have used the brainstorming approach in making the decision to modernize.
3. What types of information might Rotor Clip have collected to assist in making the decision to modernize?

REVIEW QUESTIONS

1. Define the term *decision making.*
2. Explain the concept of right choice as it relates to decision making.
3. Briefly explain two ways to evaluate decisions.
4. What is a problem?
5. Explain the three characteristics of problems.
6. What is a crisis problem?
7. Explain how supervisors can minimize the number of crisis problems they have to deal with.
8. Define the term *decision-making process.*
9. Briefly explain the three steps in the decision-making process.
10. Compare and contrast the objective and subjective approaches to decision making.
11. What are the advantages of involving employees in decision making?
12. What are the disadvantages of involving employees in decision making?
13. Briefly explain the following strategies for involving employees in decision making: brainstorming; NGT; quality circles.
14. Define the term *information.*

15. Explain the difference between data and information. What determines the difference?
16. Explain the concept of value versus cost relative to information.
17. Explain the concept of information overload and how supervisors can avoid it.
18. What is a management information system?
19. Explain the three major components of an MIS.
20. List and explain the four stages of the creative process.

SIMULATION ACTIVITIES

The following activities give you opportunities to apply what you learned in this chapter and previous chapters in solving simulated supervisory problems.

1. Form a group of eight to ten fellow students. Use NGT to deal with the following problem: How to solve the federal deficit problem facing the country.
2. Form a group of eight to ten fellow students. Select a controversial subject about which a decision should be made (e.g., reinstituting the draft). Conduct a brainstorming session. Ask one nongroup member to be an observer and to record examples of groupthink, groupshift, creativity, and other group-related phenomena. Discuss his or her observations.
3. Form a group of five to eight students. Initiate a creative process of identifying as many uses as possible for discarded plastic milk bottles (gallon size). Spread the ideas over a week to allow for incubation.

ENDNOTES

1. Brightman, H. H. *Problem Solving: A Logical and Creative Approach* (Atlanta, GA: Georgia State University, Business Publishing Division, 1980), p. 89.
2. Land, J. R., Dittrich, J. E., and White, S. E. "Managerial Problem-Solving Models: A Review and a Proposal," *Academy of Management Review,* 1978, Volume 3, p. 103.
3. Delbecq, A. L., Van DeVen, A. H., and Gustafson, D. H. *Group Techniques for Program Planning* (Glenview, IL: Scott, Foresman and Company, 1975), p. 126.
4. Myers, D. G. and Lamm, H. "The Group Polarization Phenomenon," *Psychological Bulletin,* 1976, Volume 85, pp. 602–627.
5. Schnake, M. E. *Human Relations* (Columbus, OH: Merrill Publishing Company, 1990), pp. 285–286.
6. Clark, R. D. "Group-Induced Shift Toward Risk: A Critical Appraisal," *Psychological Bulletin,* 1971, Volume 80, pp. 251–270.
7. Maier, N. R. F., Julius, M., and Thurber, J. "Studies in Creativity: Individual Differences in the Storing and Utilization of Information," *American Journal of Psychology,* 1967, Volume 80, pp. 492–519.
8. Von Oech, H. *A Whack on the Side of the Head,* (New York: Warner, 1983), p. 77.
9. Ibid.
10. Slass, R. "The Hard Road Is the Best Road," *Nation's Business,* April 1990, p. 6.

PART THREE

The Supervisor as a Counselor

CHAPTER TWELVE

Performance Appraisal and the Supervisor

CHAPTER OBJECTIVES

After studying this chapter, you will be able to define or explain the following topics:

- The Rationale for Performance Appraisal
- What Constitutes an Effective Performance Appraisal
- The Supervisor's Role in Performance Appraisal
- How to Complete a Performance Appraisal Instrument
- How to Conduct a Performance Appraisal Interview
- How to Give Corrective Feedback to Employees
- How to Conduct Upward Evaluations
- The Legal Aspects of Performance Appraisal
- The Supervisor's Coaching Role

CHAPTER OUTLINE

"How are we doing?" "Where can we do better?" "What are our strengths?" "What are our weaknesses?" "Who should be rewarded for performance?" "Who needs additional development?" These are questions supervisors must answer if they are going to achieve the continual improvement their companies need to compete in the modern marketplace.

Why then is the performance appraisal process typically viewed in negative terms by both supervisors and employees? One reason is that performance appraisal systems are often poorly designed, improperly administered, and inappropriately used. This chapter will help supervisors learn to develop and implement effective performance appraisal systems that motivate employees to continually improve their performance.

RATIONALE FOR PERFORMANCE APPRAISALS[1]

The rationale for **performance appraisal** is simple: to **improve performance.** Consequently, every aspect of an organization's appraisal system should serve this purpose either directly or indirectly. An appraisal system should identify weak areas so that they can be corrected. It should identify areas of strength so that they can be capitalized on.

If the results are used as the basis for rewards or promotions, this should be done in such a way as to be an incentive for improved performance. When designing a new appraisal system, keep this purpose in mind at every step. When working with an existing system, examine all aspects of it critically. In both cases, ask the following question every time you make a decision about some element of an appraisal system: How will this element serve the purpose of improved performance? Any element that does not serve this purpose should be eliminated.

EFFECTIVE PERFORMANCE APPRAISAL[2]

Effective performance appraisals are those that improve the performance of the employees being evaluated. This is easy to say, but can be difficult to accomplish. Before evaluating an employee's performance it is important to ensure that both the supervisor and the employee clearly understand what the job entails.

The best way to clearly delineate what a job entails is through the development of a well-written **job description.** Before attempting to evaluate an employee's performance, supervisors should make sure the employee has a clearly stated job description and that he or she understands it. If there isn't one, write one. If there is, make sure it meets the following criteria:

- State work descriptions in an action-oriented format (e.g., writes CNC programs; loads application software; operates five-axis mill, etc.).
- State all information clearly in as few words as possible.
- Give comprehensive descriptions (list all work required of the employee).
- If machines or tools are to be used, give specific types and models.
- Indicate which work tasks are more important than others. List them in order of importance or with notation.

Once a job description is complete, validate it by asking other supervisors and senior employees to review it. Then make sure all employees performing that job have a copy and that they understand it.

The next step involves using the information in the job description as well as other sources such as company rules and regulations, production standards, and widely accepted practices to develop an appraisal instrument. This step is covered later in this chapter. At this point it is necessary to understand only that there should be no surprises in performance appraisals. Employees should know what is expected of them and how they will be evaluated. This is fundamental to making performance appraisal systems effective.

SUPERVISOR'S ROLE IN PERFORMANCE APPRAISAL[3]

Supervisors play a key role in a company's performance appraisal system. The system cannot succeed unless supervisors play their role effectively. Figure 12–1 summarizes the supervisor's responsibilities with regard to performance appraisal. These responsibilities fall into three broad categories: (1) preparing for the appraisal; (2) doing the appraisal; and (3) following up the appraisal.

Supervisor's Responsibilities in Performance Appraisals

Preparing for the Appraisal

- Writing or updating job descriptions
- Reviewing and revising appraisal instruments
- Making sure employees know what is expected of them and how they will be evaluated

Doing the Appraisal

- Completing the appraisal instrument
- Ensuring objectivity
- Preparing for the appraisal interview
- Listening and soliciting employee input
- Conducting the appraisal interview
- Ensuring against interruptions
- Ensuring privacy

Following Up the Appraisal

- Encouraging
- Assisting
- Monitoring

Figure 12–1
Checklist of the supervisor's responsibilities in performance appraisals.

Preparing for the Appraisal

This step, as we saw in the previous section, involves writing new job descriptions or updating existing ones, reviewing and revising appraisal instruments, and making sure that employees know what is expected of them and how they will be evaluated. This step must be accomplished fully before an appraisal is actually conducted.

Doing the Appraisal

In this step the **appraisal instrument** is completed and the appraisal interview is conducted. In completing the instrument, it is important to be objective. In conducting the appraisal interview, it is important to prepare, give the employee opportunities to speak, listen carefully, meet in private, guard against interruptions, and set specific goals for improvement. Both components of this step are covered in more depth in the next two sections of this chapter.

Following Up the Appraisal

Following up means doing what is necessary to ensure that improved performance results from the performance appraisal. In this step supervisors monitor the performance of employees with special attention to the goals for improvement set during the appraisal interview. Supervisors should assist and encourage employees.

DEVELOPING/COMPLETING THE APPRAISAL FORM[4]

Supervisors should be able to both develop a new appraisal form and complete an existing one. It is more likely that supervisors will have opportunities to revise existing forms than to actually develop a new form from scratch. However, if supervisors can develop a form, they can revise one. Therefore this section describes how to develop a performance appraisal form. All performance appraisal forms should have at least the following components: (1) a list of performance criteria; (2) a way to rate performance according to each criteria; (3) space for written comments; (4) an employee response section; and (5) a supervisor's report section.

Performance Criteria

Performance criteria are statements of expectation relative to the employee's performance on the job. Typical performance criteria include quality of work, quantity of work, consistency, job skills, job knowledge, adaptability, attitude, punctuality, dependability, judgment, ability to innovate, initiative, safety practices, personability, team skills, and growth potential. These are just some of the many performance criteria that might be included on an appraisal form. Others can be added based on the individual situation.

In deciding what performance criteria to include on an appraisal form, supervisors should be careful to select those that are truly important relative to the job in question.

An appraisal form might be challenged in court or before a grievance board. If this happens, supervisors might be called upon to justify their performance criteria. Keep this in mind when developing a new form and when revising existing forms. A good rule of thumb is to include only those criteria that an objective panel of your peers would agree have a definite relationship to the job.

Rating Methodology

An appraisal form must contain a **rating methodology** for judging an employee's performance. There are a number of different ways of doing this. Numerical rating scales ranging from 1 to 5 or 1 to 10 are common. Continuums ranging from *poor* to *outstanding* are widely used. Some take the simple approach and use just two ratings for each criterion: *satisfactory* and *unsatisfactory.* Another approach is to follow each criterion with a set of several descriptive statements. The supervisor circles the statement that comes closest to describing the employee's performance for that category.

For example, if the criteria statement concerns quality of work, it might be accompanied by statements such as the following:

Always meets or exceeds standards

Usually meets or exceeds standards

Occasionally below standards

Does not meet standards

Regardless of the rating system used, make sure that the approach is simple, easy to understand, and lends itself to factual documentation. Subjectivity should be avoided, even in the comments section.

Comments Section

The **comments section** gives supervisors an opportunity to expound on why a certain performance criterion shows a low rating, make suggestions for improvement, comment on strengths, or clarify when necessary. Caution is in order when adding written comments to an appraisal form. As with the performance criteria, supervisors should have factual documentation to back up their comments.

Employee's Response Section

Although the **employee's response section** is not always included on appraisal forms, it should be. This section gives employees an opportunity to respond to an appraisal and make their response a matter of record. Supervisors should work with employees to reach agreement on an evaluation. However, there will be times when the supervisor and the employee cannot agree. Having a response section for the employee can keep such situations from developing into a problem. Giving employees the opportunity to state their disagreement in writing tells them their point of view is important.

Supervisor's Report Section

In the **supervisor's report section** the supervisor has the final word. In this section supervisors can comment on the employee's response to the evaluation and his or her attitude. Does the employee agree or disagree with selected ratings? What was his or her attitude toward the evaluation process? Did the employee take a positive approach? Will the employee make a sincere effort to improve? Supervisors answer questions such as these in this section. Figure 12–2 is an example of a performance appraisal form used in a modern industrial setting.

Keeping Performance Appraisals Objective

Objectivity is important when conducting performance appraisals. This is easier to say than it is to accomplish, since people are not objective by nature. However, objectivity is the best approach both from a legal perspective and from the perspective of improved performance. Strategies that will help enhance objectivity are summarized in the paragraphs that follow and in Figure 12–3.

Review Performance Standards

Appraisal instruments typically use rating scales to describe **performance standards** (i.e., satisfactory/unsatisfactory, above average, average, below average; 0–5; etc.). Before assigning a rating, establish clearly in your mind what each option on the rating scale means to you. Share your interpretation with every employee you plan to evaluate. Make sure all employees understand how you interpret the rating scale. Then stick to this interpretation.

Base Ratings on Facts

Do not assign a rating for a given criterion unless you have actual evidence on which to base the rating. **Documentation** is just as important with high ratings as it is with low ratings. One of the surest ways to ensure objectivity in performance evaluations is to base ratings on facts. If the facts are not available, respond with the phrase "insufficient information." But do not do this to avoid assigning a rating. If you do not have the facts for this rating period, make sure you get them before the next evaluation.

Avoid Personality Bias

It is difficult to be impartial when evaluating people who are different from us. Correspondingly, there is a natural human tendency to rate those who are like us more easily. Keep **personality bias** in mind when conducting performance appraisals. Before assigning a rating, ask yourself, "Is this rating influenced in either direction by personality factors?"

Avoid Extremes in Assigning Ratings

Some supervisors develop reputations as tough evaluators; others are known as soft touches. Either extreme should be avoided. If supervisors develop a reputation regarding performance appraisals, it should be one of fairness and objectivity. If you find your

Overall Evaluation

Total Performance Rating ________

( ) Does not meet standards. (0-50)	( ) Partially meets standards. (51-70)	( ) Meets standards. (71-84)	( ) Exceeds standards. (85-94)	( ) Always exceeds standards (95-100)

Performance Improvement

What are employee's specific strengths, including training received?

____

____

____

____

What are specific recommendations that can improve the employee's performance, including training recommendations?

____

____

____

____

Use This Space for Additional Comments if Needed.

Certification of Rater: I hereby certify that this evaluation constitutes my best judgment of the performance of this employee and is based on personal observations and knowledge of the employee's work.

Signature: ________________ Date: ________

Certification of Employee: I hereby certify that I have personally reviewed this evaluation.

Signature: ________________ Date: ________

Figure 12–2
Sample performance evaluation form.

Performance Evaluation

Employee: ______________ Division: ______________ Date: ______________

Job title: ______________ Dept.: ______________

Rating Factors (Use any rating number from one to the maximum shown in each category.)

Quality of Work/TQM Activity

3	6	9	12	15	Rating
Almost always makes errors. Quality is someone else's responsiblity.	Quite often makes errors. Quality not important; just show.	Make errors, but equals job standards. Employee is quality oriented.	Makes few errors, has high accuracy. Makes quality and productivity suggestions.	Almost never makes errors. Actively participates in performance enhancement activities.	_____

Comments:

Quantity of Work

3	6	9	12	15	
Almost never meets standards.	Quite often doesn't meet standards.	Volume of work is satisfactory.	Quite often produces more than required.	Always exceeds standards; exceptionally productive.	_____

Comments:

Job knowledge

3	6	9	12	15	
Inadequate	Requires considerable assistance/ training.	Adequate grasp of essentials; some assistance required.	Knowledge thorough enough to perform without assistance.	Expert in all phases of work expected.	_____

Comments:

Attendance/Punctuality

3	6	9	12	15	
Excessively absent or tardy.	Frequently absent or tardy.	Occasionally absent or tardy. No uncompensated time.	Infrequently absent or tardy. No uncompensated time. Has accumulated sick leave and vacation.	Almost never absent or tardy. Never abuses sick leave, vacation or LOA.	_____

Comments:

Cooperation & Relationships

3	6	9	12	15	
Does not get along with others; does not follow instructions.	Has difficulty getting along with others. Shows reluctance to cooperate.	Gets along adequately with others. Average skill in human relations.	Gets along well with others. Above average skill in human relations.	Goes out of way to cooperate with others. Excellent attitude.	_____

Comments:

Initiative

2	4	6	8	10	
Requires constant supervision. No self-improvement.	Too frequently requires supervision. Little self-improvement.	Requires average supervision. Some self-education.	Works independently with limited supervision. Takes courses/reads books pertaining to job improvement.	Consistent self-starter. Needs minimal supervision. Participates in college classes.	_____

Comments:

Figure 12–2
(*Continued*)

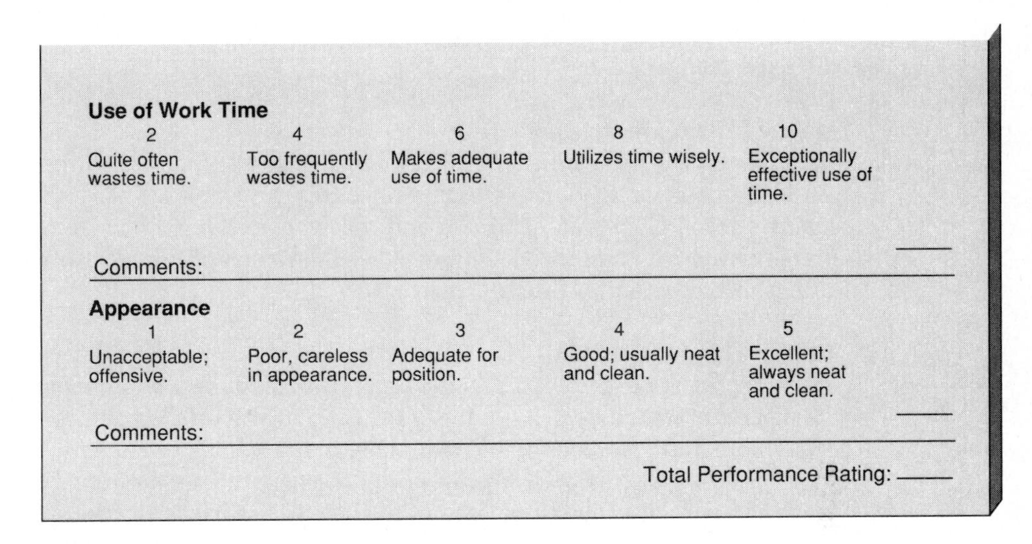

Use of Work Time

2	4	6	8	10
Quite often wastes time.	Too frequently wastes time.	Makes adequate use of time.	Utilizes time wisely.	Exceptionally effective use of time.

Comments: ______

Appearance

1	2	3	4	5
Unacceptable; offensive.	Poor, careless in appearance.	Adequate for position.	Good; usually neat and clean.	Excellent; always neat and clean.

Comments: ______

Total Performance Rating: ______

Figure 12–2
(*Continued*)

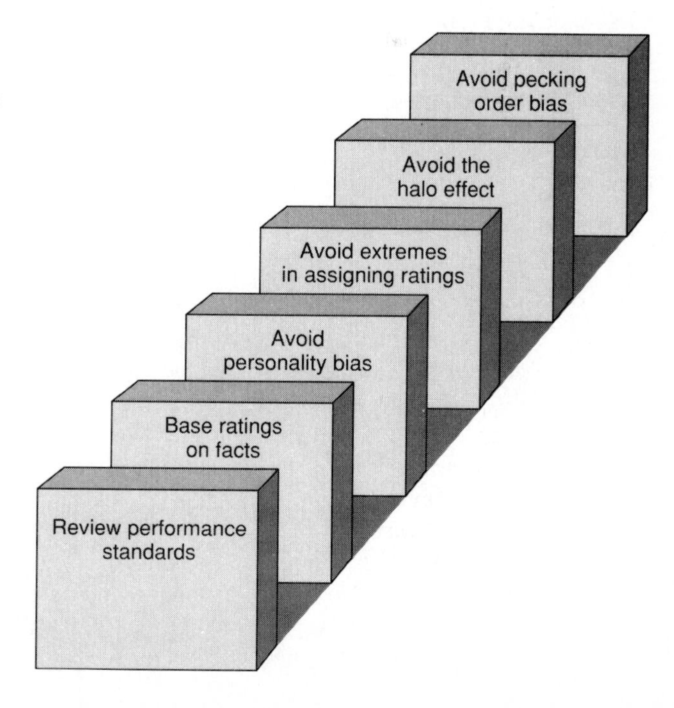

Figure 12–3
Strategies for keeping performance appraisals objective.

natural tendency is toward either extreme, be conscious of the fact and work to overcome the tendency.

Avoid the Halo Effect

The **halo effect** is a phenomenon that can occur when an employee's strong points cause you to overlook weak points. It happens like this. An employee has a couple of strong qualities that are important to you and that you rate highly. By focusing on these strengths you may rate the employee higher than he or she deserves. When assigning ratings, be cognizant of the halo effect.

Avoid Pecking Order Bias

In any organization there is a pecking order; that is, some jobs are more important to the supervisor than others. Since this is the case, supervisors sometimes rate employees in the more important jobs higher than those in the less important jobs. This is **pecking order bias.** Be conscious of it when assessing ratings. You are not rating jobs as to their relative importance, but employees as to their performance in their respective jobs. There should be no relationship between ratings in a job and the relative importance of the job.

Once the appraisal instrument has been completed, the next step is to prepare for the appraisal interview. This step is covered in the next section.

CONDUCTING THE APPRAISAL INTERVIEW[5]

The **appraisal interview** is an important part of the appraisal process, perhaps the most important. In this step supervisors share the results shown on the appraisal form with the employee. This serves several vitally important purposes: communication, feedback, counseling, and planning for improvement (Figure 12–4).

Communication

If not handled properly, performance appraisal interviews can become contentious. Employees or supervisors can become angry. Employees might question ratings and make it difficult for supervisors to justify them. This potential for conflict can lead supervisors to give the completed appraisal form to the employee and leave without any discussion, or worse yet, send it through interoffice mail. Employees are expected to read, sign, and return the form. This approach is more likely to lead to conflict than avoid it. The reason for this is lack of communication.

Communication is an important part of the appraisal interview. Employees need to know not just how they have been rated, but why. On a higher level, they need to fully understand the purpose of the performance appraisal. In addition, employees should be given an opportunity to question, comment, explain, and generally share their views. All of these things are the communication aspect of performance appraisals and they happen during the appraisal interview.

Figure 12–4
The purposes of performance appraisals.

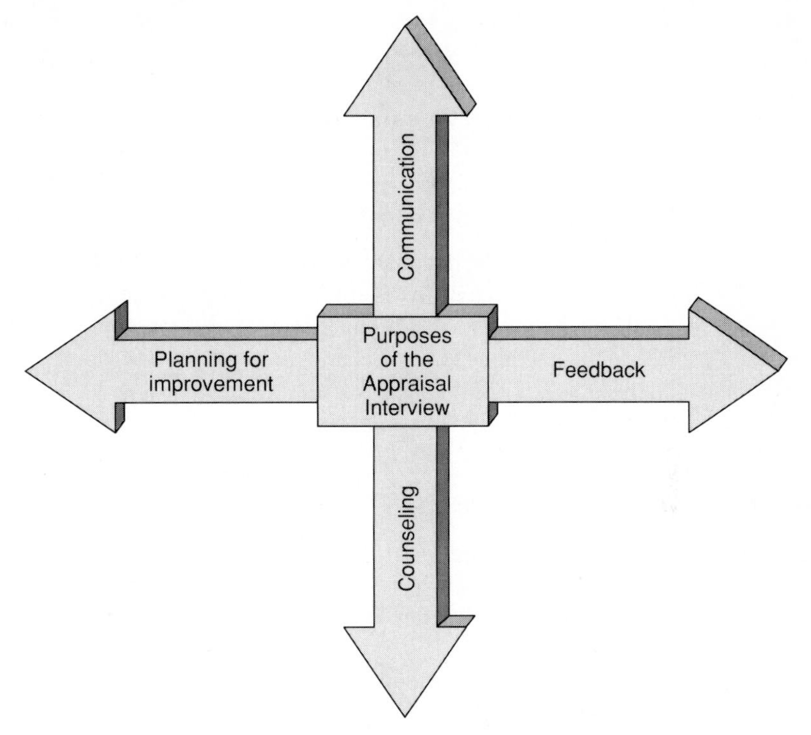

Feedback

The appraisal interview allows supervisors to give employees specific **feedback** about goals for improvement set during the last performance appraisal. It also gives employees opportunities to give feedback concerning how the supervisor has helped or hindered them in their improvement efforts. Two-way feedback is an important aspect of the appraisal interview. It is fundamental to improved performance.

Counseling

Counseling employees is an important responsibility of the modern supervisor. One of the most appropriate times for counseling is during the appraisal interview. In this context, **counseling** should be viewed as helping employees take a positive, mature approach in doing their job and toward continually improving their performance. Supervisors should keep in mind that the most important counseling skill is listening.

Be a good listener during the appraisal interview. Practice until the following strategies become habitual:

- Concentrate on what the employee says (verbally and nonverbally). Do not let your mind wander.
- Hold eye contact with the employee.

- Continue to listen even if you do not agree with what the employee is saying.
- Let the employee say what is on his or her mind without interrupting, giving negative nonverbal cues, or glancing at your watch.
- Follow the employee's train of thought. Do not jump ahead to where you think he or she is heading.

Planning for Improvement

Supervisors should never set goals for improvement for employees, nor should employees set them unilaterally. **Planning for improvement** should be a participatory process. Goals for improvement are more likely to be accomplished when they are mutually set. In addition, when supervisors and employees work together, the goals they set are more likely to be the right goals. The appraisal interview is the most appropriate occasion for mutually setting goals for improvement.

FACILITATING THE APPRAISAL INTERVIEW

This section presents several strategies supervisors can use to improve the effectiveness of the appraisal interview. These strategies will help supervisors and employees work together so that the interview is a positive experience for both.

Explain the Purpose of the Performance Appraisal

Do not assume that even the most experienced employees understand the purpose of performance appraisals. Begin the interview with a brief review of its purpose. Give employees an opportunity to share their thoughts and concerns. Take a positive attitude toward the process and stress its positive outcomes.

Discuss the Ratings

The purpose of the interview is to share the results of the appraisal. Do not prolong the employee's suspense; get right to it. Share the results openly and in a straightforward manner. A question often asked by supervisors at this point is, "Should I discuss strengths first or weaknesses?" There are proponents of both approaches. One side recommends building employees up with the positive before discussing the negative. The other side recommends getting the negative out of the way first so that the interview concludes on a positive note. Another point of view recommends intermixing so that a negative is always followed by a positive. Perhaps the best advice for supervisors is to select the approach that seems to work best with the individual employee. This may take some experimenting, but over time supervisors will learn what works best with their employees.

Solicit Feedback

Although some employees will react angrily to ratings that do not meet their expectations, others will not react at all. It is particularly important to solicit feedback from the latter. Do not assume that no response means acceptance. Some employees find it difficult to question the boss face to face. It is particularly important to draw out this type of employee during the interview. It is better to have employees express their opinions in the presence of the supervisor than over a cup of coffee with fellow workers. Solicit feedback and listen to it carefully.

Find Out How Employees Rate Themselves

Trouble is most likely to occur when the supervisor and employee disagree on a given rating. This is particularly true in the case of a criterion that is especially important to an employee. For example, if an employee is particularly proud of being dependable but receives only an average rating on dependability, there is going to be a problem. For this reason some experienced supervisors ask employees to rate themselves at the same time as they are being rated. Then, during the appraisal interview, the employee and the supervisor swap evaluations. With this approach, the only criteria that need be discussed are those on which the supervisor and employee disagree. This can be an excellent way to enhance communication. It can also ensure that the supervisor does not inadvertently rate employees low in areas where they feel they are strong.

Set Goals for Improvement

Since improved performance is the overall goal of performance appraisal, it is important to set goals for improvement. This should be done in every category where the rating is lower than either party's expectations. Make sure goals are stated clearly, written down, and agreed to by both parties. Also make sure they are realistic and measurable. Help employees understand that it is not necessary to move from "needs improvement" to "excellent" in one evaluation period. Steady, incremental improvement may be more realistic. Finally, use a positive tone in setting improvement goals. For example, if an employee needs to become more independent, do not say, "I want you to stop coming to me for instructions so often." Instead say, "I would like you to work on being more of a self-starter." Then convert this into a measurable goal or goals.

Follow-Up and Feedback

The appraisal interview should be viewed as a beginning rather than an ending. Once improvement goals have been set, supervisors should **follow up** and give employees continual feedback on their progress. This will help both parties stay focused on the goals they have set. It also shows that supervisors are interested in improved performance and want to help. Finally, follow-up allows supervisors to identify problems employees are having so that they can be eliminated right then.

GIVING CORRECTIVE FEEDBACK TO EMPLOYEES

If appraisals are to improve performance, supervisors must be skilled at giving corrective feedback to employees. Robert A. Luke, Jr. recommends six strategies for making the process more positive and more effective (Figure 12–5):[6]

Figure 12–5
Strategies for giving corrective feedback.

Summary of Strategies for Giving Corrective Feedback to Employees
• Be positive
• Be prepared
• Be realistic
• Do not focus only on the negative
• Make feedback a two-way process
• Listen first

- *Be positive.* Do not take an apologetic, "I hate to tell you this" approach. Be positive and say, "I have some feedback that might make you even more productive."
- *Be prepared.* Have your documentation readily available and be familiar with it. Also, focus only on the behavior to be corrected. Do not bring up old or undocumented problems.
- *Be realistic.* Make sure you are complaining about problems the employee is able to change. Do not hold an employee accountable for factors over which he or she has no control. Also, do not recommend corrective action that is beyond the employee's ability to carry out.
- *Do not focus only on the negative.* Soften the blow for the employee by also complimenting some aspect of his or her behavior. Kenneth Blanchard recommends giving the bad news first and then closing with a compliment to let the employee know he or she is important to you.[7]
- *Make feedback a two-way process.* Let the employee have opportunities to give feedback. There may be factors with which the supervisor is not familiar. Encourage employees to give their side of the story.
- *Listen first.* Consider letting the employee give his or her side of the story first. If there are factors of which the supervisor is not aware, it is better to hear them early. This strategy may keep the employee from becoming defensive at the outset. If so, the interview will work better for both the supervisor and the employee.

CONDUCTING UPWARD EVALUATIONS

Performance evaluation has traditionally been a top-down process. In recent years some companies have begun to add bottom-up appraisals to the process. Employee evaluations of supervisors can enhance both the credibility and effectiveness of the overall performance appraisal process. Because of this, modern supervisors should be prepared not just to evaluate their employees, but to be evaluated by them. In fact, the modern supervisor is well advised to get out front and take the lead on this issue. If your company does not conduct both top-down and bottom-up evaluations, recommend this approach.

Designing Bottom-Up Evaluation Forms

When employees evaluate their supervisors, what should the criteria be? J. E. Osborne recommends the following supervisory performance criteria:[8]

- *Decision-making ability.* Does the supervisor make timely decisions that are in the best interest of the unit and the company?
- *Supervising tasks.* Is the supervisor effective at assigning, monitoring, facilitating, and managing work?
- *Organizing and controlling.* Is the supervisor effective at organizing resources to maximize them? Is the supervisor effective at controlling the use of resources allotted to his or her unit?
- *Delegating.* Is the supervisor an effective delegator? Is work delegated appropriately and fairly?
- *Technical ability.* Does the supervisor keep his or her technical skills up to date? Can the supervisor fill in for employees? Can the supervisor train a new employee?
- *Development.* Does the supervisor encourage and support the ongoing development of employees?
- *Motivation.* Is the supervisor an effective motivator? Does he or she set a positive example?
- *Communication.* Is the supervisor an effective communicator? Is he or she a good listener? Is communication under this supervisor truly a two-way process?

The evaluation instrument used should clearly identify the supervisor's level of performance with regard to each of these criteria. Choices such as *always, sometimes, seldom,* and *never* are an effective way to accomplish this. Completing forms anonymously and having a third party summarize them can be an effective way to alleviate the fears subordinates may have about evaluating their supervisors.

When employee evaluations have been summarized they should be shared with the supervisor. At this point in the process, goals for improvement are mutually developed. These goals should also be shared with employees. Do not make a secret of evaluation results. Use them as the basis for communication.

LEGAL ASPECTS OF PERFORMANCE APPRAISALS[9]

In today's litigious society it is important for supervisors to be familiar with the legal aspects of performance appraisals. The **Freedom of Information Act** now forces companies to approach performance appraisals carefully, objectively, and according to the book. Do supervisors need to be attorneys? No. But supervisors should be familiar with several important rules of thumb that will ensure that their performance appraisals meet the test of law. These rules are discussed in the following paragraphs.

Keep Comprehensive Records

It is important to document every rating on every performance appraisal, both good and bad. Keep accurate notes, mark dates on a calendar, and use these records when discussing performance appraisals with employees. Was an employee late several times? Document it in writing. Is an employee's attitude consistently negative? Counsel the employee, mark the date on a calendar, and follow up with a memorandum that goes to the employee with a copy to his or her personnel file. Documentation takes time, but the resulting **record keeping** can protect supervisors and their companies if subjected to litigation (Figure 12–6).

Focus on Performance, Not Personality

Keep all comments, both verbal and written, objective and on a professional level. Criticize an employee's performance in a constructive way, but do not critcize the individual. Critiques that focus on performance are likely to pass the test of law. Critiques that appear to focus on personality or to be biased in any way will not. Good people can have a bad performance rating. Focus on the performance, not on the person.

Be Positive, Constructive, and Specific

Make sure that all comments, both verbal and written, are positive, constructive, and specific. Remember, the purpose of the performance appraisal is to improve performance. In order to do so, appraisals must be positive. They should build employees up, not tear them down. Be specific in your critiques and in goals for improvement. Do not say, "This employee needs to learn how to get to work on time like everybody else." Instead say, "This employee should work on increasing his on-time record to 98 percent." This is positive, constructive, and specific.

Be Honest and Treat All Employees the Same

Do not hold back out of fear of a lawsuit. Your job is to conduct a performance appraisal that will help improve the performance of individual employees and, as a result, the

Figure 12–6
Keep comprehensive records.

Record-Keeping Checklist for Supervisors

- Have you kept notes documenting your interactions with employees?
- Have you marked dates of meetings, conferences, counseling sessions, etc on a calendar?
- Have you kept copies of correspondence sent to employees?
- Have you retained evidence that documents good and bad behavior/performance?
- Do you have copies of previous performance appraisals showing goals for improvement?
- Have you periodically reviewed the documentation on file in the Human Resource Department?

organization. Give an honest and straightforward appraisal of performance and make sure that you treat all employees the same. Playing favorites is an easy trap to fall into and a difficult one to get out of. If an appraisal is ever challenged by an employee, past appraisals as well as those of other employees may be examined. Supervisors who are honest, straightforward, and equitable are on a solid legal footing.

Apply Objective Standards

The supervisor does not necessarily control the standards used in performance appraisals. However, by knowing that the courts expect **objective standards**, supervisors can at least let higher management know that potential legal problems exist if subjective standards are used. If an objective standard cannot be found for a given aspect of an employee's job, use a subjective standard (i.e., leadership, assertiveness, etc.) but give factual examples that illustrate the employee's shortcomings. For example, if a supervisor has an employee who lacks initiative, it may be necessary to comment on it as part of a performance appraisal. In such a case, do not say, "This employee needs to be a better self-starter." Instead say, "On the following occasions this employee failed to begin work until told to do so." Then list the actual dates this behavior was exhibited.

THE SUPERVISOR AS A COACH

The most fundamental goal of the performance appraisal process is improved performance. When employees improve their performance they win, the supervisor wins, and the organization wins. This is the rationale behind the philosophy that supervisors should be "career coaches" for their employees.[10] According to G. M. Sturman, "The fundamental rule of career coaching is that each person is unique and has individual needs. . . . The job of the career coach is primarily to offer direction so that the employer doesn't get stuck with some aspect of the career management process."[11]

Career management is the process through which employees take responsibility for continually developing their job-related knowledge and skills. Career coaching encourages this process, thereby helping employees do a more effective job of helping themselves.

Sturman recommends the following five-point plan supervisors can use to promote career management among their employees:[12]

- *Assess.* Help employees get a clear picture of their strengths and weaknesses.
- *Investigate.* Help employees investigate all potential avenues for growth available to them within the organization.
- *Match.* Help employees match their strengths and weaknesses with the appropriate opportunities available to them in the organization.
- *Choose.* Work with employees to determine if they have made appropriate choices.
- *Manage.* Help employees prepare a development plan that will result in the accomplishment of their goals.

The payoff for the time invested in career coaching can be substantial. According to Sturman, "The investment in time and effort is quite small in comparison to the potential long-term benefits in employee satisfaction, effectiveness, and productivity."[13]

SUMMARY

1. The rationale for performance appraisal is simple: to improve performance. All other uses (promotions, salary increases, terminations, etc.) should directly or indirectly tie to improved performance of the organization.
2. Effective performance appraisal is that which leads to improved performance. In evaluating any aspect of the performance appraisal process, ask the question, "How does this component contribute to improved performance?"
3. The supervisor plays a key role in performance appraisal. Major responsibilities include: preparing for the appraisal, completing the appraisal instrument, conducting the appraisal interview, and following up. Supervisors should also work with higher management to continually improve the performance appraisal process and ensure that it can pass the test of law.
4. Performance appraisal instruments should have the following components: performance criteria, rating methodology, comments section, employee's response section, and supervisor's report section.
5. Performance appraisals should be objective. Strategies for enhancing objectivity include the following: review performance standards, base ratings on facts, avoid personality bias, avoid extremes in assigning ratings, avoid the halo effect, and avoid pecking order status.
6. The appraisal interview should serve the following purposes: communication, feedback, counseling, and planning for improvement.
7. Strategies for ensuring a successful appraisal interview include the following: explain the purpose of the performance appraisal, discuss the ratings, solicit feedback, find out how employees rate themselves, set goals for improvement, follow up and give feedback.
8. Strategies for ensuring that performance appraisals pass the test of law include the following: keep comprehensive records; focus on performance, not personality; be positive, constructive, and specific; be honest and treat all employees the same; and apply objective standards.

KEY TERMS AND PHRASES

Appraisal interview
Appraisal instrument
Comments section
Communication
Counseling
Documentation
Employee's response section
Feedback
Follow-up
Freedom of Information Act
Halo effect
Improved performance
Job description
Objective standards

Objectivity
Pecking order bias
Performance appraisal
Performance criteria
Performance standards
Personality bias
Planning for improvement
Rating methodology
Record keeping
Supervisor's report section

CASE STUDY: Simplifying Performance Appraisals[14]

The following real-world case study illustrates a performance appraisal problem. Read the case study carefully and answer the accompanying questions.

L. W. Looney & Son Manufacturing is a Department of Defense contractor located in Crestview, Florida. Its product line consists of airtight aluminum munitions containers of various sizes and configurations. The workforce consists of 150 employees. Key production employees are machinists and welders.

CEO L. W. Looney likes to keep things simple. This is the reason he tried and discarded several performance appraisal systems he judged to be too complicated. In Looney's opinion, a performance appraisal form should be easy to complete and easy for both the supervisor and the employee to understand.

Out of frustration with the performance appraisal forms he had seen, Looney decided to design his own. The result is a form that contains five simple criteria:

- Attendance record
- On-time record (also includes breaks and lunches)
- Dependability of work
- Accuracy of work
- Effort to improve performance

1. In your opinion, are there sufficient criteria to constitute a comprehensive evaluation? Why or why not?
2. Would you add criteria? If so, what would you add?
3. After completing a form with these criteria, what would you do to get the employee to improve?

REVIEW QUESTIONS

1. Explain the rationale for performance appraisal.
2. What is meant by effective performance appraisal?
3. List five criteria for a good job description.
4. Explain the supervisor's role in performance appraisal.

5. List the components that should be included in a performance appraisal instrument.
6. Explain four strategies for keeping performance appraisals objective.
7. Explain the halo effect.
8. Explain pecking order bias.
9. What are the four basic purposes of the appraisal interview?
10. List five strategies for improved listening.
11. List and explain four strategies for facilitating the appraisal interview.
12. List and explain five strategies for ensuring that performance appraisals can pass the test of law.

SIMULATION ACTIVITIES

The following activities simulate problems supervisor may face on the job. The activities may be completed by groups or by individual students. Apply what you have learned in this and previous chapters to solve these problems.

1. You have rated an employee lower than expected and, in spite of her protestations, you do not intend to change the rating. She is clearly upset. What should you do?
2. You are having troubie being objective in evaluating an employee. You really don't like this guy. His work is mostly satisfactory, but he rubs you the wrong way. How can you perform an objective appraisal?
3. During an appraisal interview you find your mind drifting. The employee has talked earnestly for five minutes and you don't even know what she said. Before the next interview, you want to get focused. What should you do?
4. A new supervisor has never conducted an appraisal interview. In an hour he will conduct his first. He is nervous and unsure of what to do. What advice or directions would you give him?
5. Your company has been sued by a disgruntled former employee over performance appraisals in her personnel file. As a result, you have been asked to give a short talk to all supervisors entitled "How to Conduct a Performance Appraisal That Will Stand Up in Court." What pointers will you give your colleagues?

ENDNOTES

1. Latham, G. P. and Wexley, K. N. *Increasing Productivity Through Performance Appraisal* (Reading, MA: Addison-Wesley Publishing Company, 1981), p. 67.
2. Ibid.
3. Bernardin, H. J. and Beatty, R. W. *Performance Appraisal: Assessing Human Behavior at Work* (Reading, MA: Kent Publishing Company, 1984), p. 94.
4. Bernardin, H. J. "Behavioral Expectation Scales versus Summated Scales: A Fairer Comparison," *Journal of Applied Psychology,* 1963, Volume 47, pp. 89–94.
5. Rice, B. "Performance Review: The Job Nobody Likes," *Psychology Today,* September 1985, pp. 30–36.

6. Luke, R. A., Jr. "How to Give Corrective Feedback to Employees," *Supervisory Management,* March 1990, p. 7.
7. Blanchard, K. *The One Minute Manager* (New York: William Morrow, 1982), p. 36.
8. Osborne, J. E. "Upward Evaluations: What Happens When Staffers Evaluate Supervisors," *Supervisory Management,* March 1990, pp. 1, 2.
9. Cascio, W. F. and Bernardin, H. J. "Implications of Performance Appraisal Litigation for Personnel Decisions," *Personnel Psychology,* 1981, Volume 34, pp. 211, 226.
10. Sturman, G. M. "The Supervisor as a Career Coach," *Supervisory Management,* November 1990, p. 6.
11. Ibid.
12. Ibid.
13. Ibid.
14. Looney, L. W. From an interview with the author, March 11, 1990.

CHAPTER THIRTEEN

Problem Employees and the Supervisor

CHAPTER OBJECTIVES

After studying this chapter you will be able to explain the following topics:

- How to Handle Employees with Personal Problems
- How to Deal with Employee Theft
- How to Deal with Emotionally Disturbed Employees
- How to Handle Employee Sabotage
- How to Handle Humiliators and Intimidators
- How to Handle Procrastinators
- How to Handle Angry Employees
- How to Handle Absenteeism
- How to Handle Substance-Abusing Employees
- How to Handle AIDS on the Job
- How to Supervise Low Conformers

CHAPTER OUTLINE

- Handling Employees with Personal Problems
- Dealing with Employee Theft
- Dealing with Emotionally Disturbed Employees
- Dealing with Employee Sabotage
- Handling Humiliators and Intimidators
- Handling Procrastinators
- Handling Angry Employees
- Handling Absenteeism
- Substance Abuse by Employees
- Dealing with AIDS on the Job
- Supervising Low Conformers
- Summary
- Key Terms and Phrases
- Case Study Application Problem
- Review Questions
- Simulation Activities
- Endnotes

As the first level of management, supervisors are the managers most likely to know when employees have problems, ranging from minor personal problems, to severe emotional disturbances, to substance abuse problems. Supervisors need to know how to recognize and handle problem employees, what they can do themselves, and when they should refer problem employees to professionals. This chapter will help prospective and practicing supervisors learn how to deal appropriately with problem employees.

HANDLING EMPLOYEES WITH PERSONAL PROBLEMS[1]

Are the **personal problems** of employees any of your business? No, you have no right to know the details of an employee's personal life. These things are none of your business unless the employee chooses to share them with you. However, how an employee's personal problems affect his or her work *is* your business.

Every year employees with personal problems cost American employers as much as $150 billion. These costs are the result of absenteeism, tardiness, accidents, mistakes, turnover, disruptive conduct, and other related factors. For this reason, it is important for supervisors to be able to help employees cope with personal problems.

Mario Alonso recommends several strategies for dealing with employees who have personal problems.[2] These strategies are summarized as follows:

- *Be alert to the signs of the troubled worker.* Various signs suggest an employee is experiencing personal problems. One whose temper is uncharacteristically short or whose attendance suddenly falls off may be suffering personal problems.
- *Document problems.* Employees may deny that they have personal problems, even to themselves. Until a person can admit to having a problem, the problem cannot be solved. Therefore supervisors should note dates of problematic behavior on a calendar and keep a log containing examples of the behavior. **Documentation** may help the employee see that a problem exists.
- *Focus on documented evidence.* Before meeting with the employee organize your documentation and refamiliarize yourself with it. Focus on **job performance**, not the personal problem itself. Your concern as a supervisor is the employee's performance on the job.
- *Keep the discussion on a professional level.* Personal problems are potentially touchy. Therefore, it is important for supervisors to take a positive, constructive, professional approach. Do not let your personal feelings color your judgment or demeanor. Performance, not the personal problem, should be the focal point of the discussion.
- *Solicit input and listen.* The employee should do most of the talking and the supervisor most of the listening. Listen to what is said and what is not said. Observe nonverbal cues. Intervene when necessary to keep the discussion on track, but as long as the discussion is focused on the issue, let it continue.
- *Ask for solutions.* It is tempting when counseling troubled employees to propose a solution and, in fact, this may be necessary at times. However, it is best to let them

find their own solutions. Supervisors can suggest outside sources of help, though. This allows you to be helpful without being overly directive.

- *Explain potential consequences.* It is important to let troubled employees know that their behavior can have negative **consequences** and what those consequences are. Supervisors walk a fine line here. The purpose is to inform, not threaten. Don't say, "If you can't get your act together we're going to fire you." Instead say, "I want to help you understand the consequences of poor performance." They can include a low performance rating, suspension without pay, and even termination.
- *Plan for improvement.* Supervisors should work with troubled employees to develop definite, written plans for improvement. The plans should specify timetables and be readily measurable. Do not dictate plans to the employee. Instead, solicit the employee's input and develop these plans jointly.
- *Be a supervisor, not a therapist.* Supervisors should avoid the temptation to play the role of the amateur therapist. Your job is to improve performance, not to try to solve marital, alcohol, or drug problems. Focus on performance and refer employees to qualified professionals for help with their problems.

DEALING WITH EMPLOYEE THEFT[3]

Employee theft of cash, merchandise, and time costs employers in the United States as much as $15 billion annually. Cash theft is more prevalent in businesses that conduct daily cash transactions. But in all businesses where cash changes hands, a certain amount is lost through theft.

Merchandise theft is very common. In an industrial firm it typically involves theft of company supplies, materials, and small equipment, perhaps nothing more than "borrowing" a legal pad or a computer disk for personal use. Perhaps the most common form of employee theft is theft of time: cheating on sick leave, altering time cards, or stretching out the recovery period for an illness or injury.

There are several strategies companies can use to prevent employee theft. Read Hayes recommends four:[4] (1) screen employees carefully before hiring them; (2) establish control procedures that make theft difficult; (3) make sure employees are aware of the costs and potential consequences of theft; and (4) establish an atmosphere of honesty in the workplace and make it part of the corporate philosophy. Supervisors can have the most impact in carrying out the last strategy by setting a positive example, discussing the issue with team members, and persuading all team members to model honest behavior.

In addition to the strategies recommended by Hayes, the following practical strategies can be used by supervisors to prevent employee theft:

- *Rotate security guards.* Guards who work too long in one area can become so familiar with employees that fraternization causes them to become less vigilant.
- *Double-checking incoming shipments* to prevent collusion between drivers and receiving employees.

- *Control locks and keys* so that material in controlled storage stays put.
- *Supervise trash collection and pickup* to ensure that the trash disposal system does not become a conduit for moving material out of the organization.

DEALING WITH EMOTIONALLY DISTURBED EMPLOYEES

The term **emotionally disturbed** sounds rather ominous. However, anyone can become emotionally disturbed, from those who are temporarily reacting to stress in their lives to those with serious and chronic emotional problems. Supervisors may be able to help the former but should not even attempt to help the latter, at least not beyond referring them to qualified professional care. The key here is in recognizing the difference between normal people who are having trouble adjusting to stress in their lives and those with a serious mental illness.

Recognizing Normal Employees

Emotional disturbances in normal people typically result when they are not able to deal with stress or some type of emotional trauma in their lives. Symptoms may take many forms. Their performance level might suddenly change, they might become unusually irritable, or their attendance might suddenly drop off (Figure 13–1). These symptoms can be summarized as **uncharacteristic behavior.**

When such behavior affects an employee's performance, the supervisor should take the steps explained in the first section of this chapter. If progress is not made after two attempts, the employee may be having more than temporary adjustment problems. If you suspect this is the case, it is time to seek professional help.

Recognizing Psychotic and Neurotic Employees

Supervisors are not qualified to deal with **psychotic** employees. The most common psychotic condition is schizophrenia or having multiple personalities. When a schizophrenic person has trouble adjusting to emotional stress he or she withdraws and one of the other personalities may be revealed. Regardless of the nature of the schizophrenic

Figure 13–1
Supervisors should be aware of these indicators.

Trouble Signs in Employees

- Sudden behavioral changes
- Drop in attendance
- Sudden change in performance
- Rapid mood swings
- Uncharacteristic fatigue
- Constant daydreaming and preoccupation
- Frequent accidents
- Isolation
- Dropping out of regular activities

person's behavior, supervisors should refer him or her to professional help via the company's human resources department.

Neurotic employees exhibit exaggerated levels of otherwise normal behaviors, such as fear, aggressiveness, timidity, irritability, hostility, and so on. When do such behaviors become exaggerated to the point that they represent neurosis? This is a question only mental health professionals can answer. However, from the supervisor's perspective they are a problem when the adversely affect performance.

Again, if two sessions with the employee fail to make a difference, refer him or her to professional help. In most companies this will be handled by the human resources department. However, a supervisor is not out of line in encouraging an employee to seek professional help.

Who Are Mental Health Professionals?

Mental health professionals to whom emotionally disturbed employees might be referred are psychiatrists, psychologists, and counselors. Supervisors need to know the similarities and differences among these categories of mental health professions.

Psychiatrists are medical doctors (physicians) who have chosen psychiatry as their specialization. Psychiatrists are able to diagnose, prescribe for, and treat patients with mental illness. **Psychologists** hold doctorate degrees, but are not physicians. Consequently, they do not prescribe medicine as part of their treatment. **Counselors** typically hold master's degrees and work under the supervision of a psychologist or a psychiatrist.

Psychiatrists use many different forms of treatment and there are different schools of thought within the profession. The same is true of psychologists. A key difference between their apprcaches is that psychiatrists may try to help patients understand why they behave as they do. Psychologists are more likely to focus on helping clients adjust appropriately without delving into reasons for their behavior.

DEALING WITH EMPLOYEE SABOTAGE[5]

Employee sabotage has long been an issue with employers, although not one that is discussed widely or openly. There are several reasons for this: (1) fear of copycat activities; (2) unwanted publicity; and (3) fear of a negative impact on investors. In spite of this reticence to confront employee sabotage openly, supervisors should be aware of the problem and know what they can do to help prevent sabotage.

Reasons for Employee Sabotage

Robert Giacalone lists six causes of employee sabotage:[6] vengeance, turf protection, laziness, competition, boredom, and frustration (Figure 13–2).

Angry employees will sometimes resort to sabotage as an act of vengeance. The employee might be angry at the boss, a fellow employee, a company policy, or any number of perceived injustices. Sabotage as a form of turf protection can result from a variety of circumstances. An example might be an employee destroying an expensive

Figure 13–2
Sabotage can result from these factors.

Typical Causes of Employee Sabotage
- Vengeance
- Turf protection
- Laziness
- Competition
- Boredom
- Frustration

piece of equipment to keep it from being moved to another unit. Laziness is another common cause of employee sabotage. One of the most effective ways to slow the pace or get a paid break from work is to damage key equipment. In order to keep a fellow employee from meeting a deadline or breaking a productivity record, another employee might sabotage his or her work. Such self-promotion efforts are most likely to occur when there is intense competition between employees for promotions or other incentives.

Boredom is dangerous in a work setting. Bored employees might choose to direct their pent-up energy toward negative activities, particularly if they are dissatisfied with other aspects of their job. Criminal justice professionals list boredom as a key contributing factor in juvenile delinquency. The same phenomenon applies on the job.

Frustration is another cause of employee sabotage. Rapid technological change is often the source, since one way to relieve the frustration of continually having to learn how to use new technologics is to sabotage them. As technology continues to change, this form of sabotage is likely to increase.

A form of sabotage closely associated with advanced technology is computer software sabotage. This involves striking back at a company by damaging critical software. One particularly vindictive form of computer software sabotage is the *computer virus.* Computer viruses are developed by software experts who have a grudge against a company. Once introduced into a computer system, a virus can eliminate critical data in the company's database, thereby rendering the computer system useless, at least temporarily.

Preventing Sabotage

There is no way to eliminate employee sabotage completely. However, supervisors can help prevent much of it. Giacalone recommends three strategies for reducing the risk and the incidence of employee sabotage:[7]

- *Screen applicants carefully.* Careful background and reference checking can reduce the probability of risk by screening out applicants who have participated in sabotage against other employers. Supervisors should work closely with human resources personnel to look for this type of behavior in an applicant's past.

- *Keep accident logs.* Supervisors should keep up-to-date, accurate accident logs. By noting what happened, where, and who was involved, supervisors might be able to identify patterns. Do incidents of damage to machines or equipment happen more in one unit than in others? Are one or more employees consistently involved? Who had access and opportunity each time damage occurs?
- *Implement security precautions.* Make sure appropriate security measures are in place, then change them occasionally and at irregular intervals. Security precautions might include a widely circulated company policy making sabotage grounds for immediate dismissal, controlled access to sensitive equipment, and surveillance.

HANDLING HUMILIATORS AND INTIMIDATORS[8]

Supervisors occasionally will have to deal with people who like to humiliate others or who try to intimidate others. **Humiliators** typically resort to threats, either verbal or physical. If they are not handled appropriately these people can cause serious morale problems. The challenge confronting supervisors is to stop the behavior, maintain a professional demeanor, and ensure that the negative behavior does not interfere with other employees who are trying to get the job done. The strategies for handling humiliators and intimidators are not the same.

Handling Humiliators

Anita Jacobs offers several strategies for handling humiliators.[9] A key point is not to react angrily or in a negative way. When this happens the humiliator has won and you have lost. Instead, try these strategies:

- Walk away. Never attempt to justify, defend, or rationalize your position with a humiliator.
- Defuse the situation using self-effacing humor, even sarcasm.
- Maintain your poise and confidence.

Walking away leaves the humiliator without a target. Self-effacing humor turns the tables on the humiliator. Finally, poise and confidence take away the humiliator's feelings of superiority.

Handling Intimidators

Jacobs makes the point that intimidators should be confronted rather than ignored.[10] **Intimidators** hope to instill fear and make others their victims. For this reason it is important to confront them. However, never respond in kind. This will only lead to escalating hostilities.

Instead, maintain your composure and respond with confidence and in a rationale manner. This is particularly important if there are onlookers. For example, how would you handle the following situation: You are trying to speed up an employee whose work is lagging behind and holding everyone else up. Suddenly, the employee blurts out, "Get off my back! If you want it done faster, do it yourself!" Work comes to a halt and all eyes are on you. Don't become angry. Instead, look the employee in the eye and say, "I do want the work done faster and I'll bet that you and I working together can get it done faster."

HANDLING PROCRASTINATORS

One of the most frustrating employees supervisors have to deal with is the **procrastinator.** Employees who are not self-starters or who continually put off work can cause dissension within the team. Berating procrastinators rarely works. They usually feel bad enough about their problem as it is. Supervisors with procrastinators on their team should begin with the reasons people give most frequently for why they procrastinate.

Theodore Kurtz lists several of these reasons:[11] (1) procrastinators simply do not want to do what they are supposed to do; (2) procrastinators do not see sufficient reward in their work; (3) procrastinators let their behavior become habitual; (4) procrastinators feel they need to save their energy for more important tasks; (5) procrastinators sometimes deny to themselves the potential negative consequences of their behavior; and (6) procrastinators convince themselves that if enough work stacks up, someone will come along and rescue them.

Helping Procrastinators Change

Supervisors who understand why people procrastinate have taken the first step necessary to help them change. Confront procrastinators openly and objectively, but not judgmentally. Have a one-on-one conference and work with the employee to determine his or her reasons for procrastinating. Deal with each reason in a straightforward manner and set definite goals for improvement.

If employees do not see the reward in doing their work, help them see it. If they are denying to themselves the potential negative consequences of their procrastination, let them know their pay, promotions, and even their jobs could be at risk. If employees procrastinate because they simply do not like doing the work, they may need to consider a job change. In such cases, ask them if they have thought about changing jobs.

A straightforward, businesslike, nonjudgmental approach is best when dealing with procrastinators. Judgmental reactions are more likely to exacerbate the problem than solve it.

HANDLING ANGRY EMPLOYEES[12]

One of the most potentially explosive situations supervisors face is trying to handle an angry employee. A factor that complicates this situation is that the natural human reaction to anger is to return it in kind. Supervisors should suppress this urge when confronted by an angry employee.

Instead, listen. Then, when a natural opening occurs, acknowledge the anger. For example, say an employee stomps up to you, places his hands on his hips, and yells,

"This is the tenth change order I've gotten this week! Don't change orders go to anyone but me? When are those clowns in engineering going to get their act together?" Your first reaction might be to yell back, "Don't you think I know it's the tenth change order this week? Who do you think gives them to you?"

All this will do is shift the employee's anger from the engineering department to you and increase it. Instead, maintain a calm exterior and don't react at all. Let the employee vent his or her anger. When it appears to be about ready to run its course, acknowledge the anger by saying, "You sound pretty angry about this." Typically, when an employee has expressed his or her anger to a nonjudgmental listener and the anger has been acknowledged, it will begin to subside. If it doesn't, continue to listen and acknowledge. Eventually, the steam will blow off and you can begin dealing with the facts.

Once the employee begins stating factual information, let him know you hear and understand by paraphrasing back what he has said. In this example, one of the things the employee said was, "Don't change orders go to anyone but me?" There is a clue to the employee's anger in this question.

On the surface, the employee appeared to be complaining about the engineering department. But is that really the problem or is it that this employee feels he is given an inordinate percentage of the change orders that come to his team? By paraphrasing, you can find out. You might say, "What I hear you saying is that I am giving you too many of the change orders that come to our unit." The employee will either acknowledge your perception or correct you. In either case you have taken the next step toward solving the problem.

To continue narrowing down the problem, ask open-ended questions that draw the employee out, letting him do most of the talking and you most of the listening. Once you think you have identified the real problem, state it and let the employee confirm it. It is important to ensure that the supervisor and employee agree on what the problem is before attempting to solve it.

Once the problem has been clearly delineated, supervisors should avoid the temptation to rush into a solution or even suggest one. Instead, ask the employee what he feels the solution should be. You might be surprised to find no further action is needed. In the example of the change orders, the employee who felt he was given more than his share of the work might have felt better just knowing the supervisor understood he was doing more work than his colleagues. The supervisor might have been giving him more work because he was the employee most able to get it done and, as a result, most likely to receive the next promotion or a raise. However, even if this is not the case, a jointly developed solution is more likely to work than one prescribed by the supervisor.

HANDLING ABSENTEEISM

Absenteeism is a critical problem in the modern workplace. All of a supervisor's efforts to make employees more productive are negated if they don't come to work. When employees are absent, their team members must work even harder to get the job done. This can lead to serious morale problems.

Consequently, supervisors should know how to hold absenteeism in their units to an absolute minimum. The following strategies will help:

- Have an attendance policy and make sure employees understand it.
- Keep accurate, comprehensive attendance records.
- Counsel employees on attendance in groups and one-on-one.
- Create an attendance-conscious attitude.
- Cross-train employees.

Attendance Policy

Most companies have an **attendance policy**, but are employees familiar with it? Publish your company's policy and display it prominently in the workplace. Have periodic team meetings to review the policy. Make attendance part of the employee performance evaluation and discuss attendance during the performance appraisal interview.

Make sure all employees know the rules, then stick to them and enforce them objectively. Be fair and impartial and apply the rules evenly. If you ever bend the rules for an employee, make sure you have a well-documented justification so that charges of favoritism are not warranted. For example, if a particularly dependable and productive employee has worked overtime several times without pay, you might overlook her coming in late one day. However, make sure that the overtime days are documented and that she understands the reason you are bending the rule in this case.

Record Keeping

Most companies have a system for verifying attendance. It might be a time clock, time sheet, or daily attendance roster. Supervisors should monitor attendance records carefully and continually. It is a good idea for supervisors to have monthly and yearly attendance summaries on each employee and for their overall unit (see Figures 13–3 and 13–4).

In monitoring attendance records, it is a good idea to consider not just the number of absences, but also **absence trends**. Is an employee in the habit of being absent on Fridays and Mondays? Does an employee stay out the day before or after holidays? It is not uncommon for absentee rates to be higher on these days.

Supervisors who can document these trends with thorough record keeping have the information they need to deal with them. Team meetings in the middle of the week in which information on Friday and Monday trends is shared can be an effective strategy. The same strategy can be used prior to holidays.

Counsel Employees

If you want to reduce absenteeism it is important to talk about it. Team brainstorming sessions can be an effective strategy. During these sessions fellow employees should be encouraged to talk about how absenteeism affects them personally. Supervisors should disclose the team's absentee numbers without giving names, so that everyone understands the problem.

Monthly Attendance Record Month______													
Employee ______													**Total**
Vacation													
Excused absence													
Unexcused absence													
Late													
Left early													
Total													

Figure 13–3
Sample record-keeping tool for attendance monitoring (monthly).

Work with the team to set an attendance goal. Efforts such as this can create a "don't be absent" attitude among team members. Peer pressure can be a strong motivator. If individual employees feel they are letting the team down by staying out, they will be more likely to come to work.

In addition to group sessions, hold one-on-one meetings with employees. The periodic performance appraisal interview is a good time, but don't limit yourself to this occasion. It is a good idea to counsel all employees regularly about attendance, not just those with poor attendance records. It is just as important to recognize good attendance as it is to question poor attendance.

Annual Attendance Record Year ______													
Employee ______	**Jan**	**Feb**	**Mar**	**Apr**	**May**	**Jun**	**Jul**	**Aug**	**Sept**	**Oct**	**Nov**	**Dec**	**Total**
Vacation													
Excused absence													
Unexcused absence													
Late													
Left early													
Total													

Figure 13–4
Annual attendance record.

It is particularly important to spend time with employees whose attendance records suddenly take a turn in the wrong direction. This usually means there is a problem in some aspect of the employee's job or personal life. Whatever the problem is, supervisors will want to work with the employee to solve it.

Attendance-Conscious Attitude

One of the reasons attendance in Japanese companies is so high is the Japanese attitude about absenteeism. Japanese workers feel they are letting their team members down if they don't come to work. In Japan, absenteeism is *personalized.* This is the key to creating an **attendance-conscious attitude** in any organization. Effective strategies for personalizing absenteeism include the following:

- Set unit and individual attendance goals.
- Recognize good attendance through a formal recognition program.
- Talk about attendance in groups and one-on-one.
- Maintain frequent personal contact with employees on a continual basis.
- Solicit input from employees in groups and one-on-one about improving the work environment.
- Set a good example of perfect attendance yourself.

Cross-Train Employees

Even if you could achieve perfect attendance, which is highly unlikely, there are still times when employees will be off the job (i.e., on vacation, away at meetings, involved in training, etc.). Regardless of how justified they may be, absences cost the organization. The consequences of absenteeism are predictable: quality goes down, morale goes down, and costs go up (Figure 13–5).

Because of this, supervisors should not be satisfied just to reduce avoidable absenteeism, it is also important to minimize the cost of unavoidable absenteeism. An effective strategy for accomplishing this goal is to **cross-train** employees.

The more jobs each employee in your unit can do the less costly absenteeism will be for the unit. Multiskilled employees give supervisors the flexibility to cover for absent workers. To illustrate this point, consider the example of the documentation and drafting department at a company that manufactures prefabricated metal buildings. On a normal day when everyone is present, the work is distributed as shown in Figure 13–6.

Figure 13–5
Some of the potential consequences of absenteeism.

Consequences of Absenteeism

- Productivity goes down
- Quality goes down
- Morale goes down
- Costs go up
- Competitiveness goes down

Employee Capability Summary				
Job: ______				
Employee	Framing plans	Details	Bill of materials	Specifications
Jones, A.	X			
Mathews, B.	X			
Garner, H.		X		
Myling, O.		X		
Ramierez, J.			X	
Peters, R.			X	
Johnson, B.				X
Abduhl, S.				X
Totals	2	2	2	2

Figure 13–6
Supervisors can use tools such as this to identify employee capabilities.

For each job there are two employees capable of doing it. This system works well until the two people who do a given job are absent at the same time or when one is absent, but there is more work than one person can do. By cross-training so that each employee can do two or three jobs the negative impact of absenteeism can be reduced.

SUBSTANCE ABUSE BY EMPLOYEES[13]

Substance abuse by employees may be one of the most difficult problems supervisors will have to deal with. Substance abuse is widespread in today's society. Its effect on productivity can be devastating. The negative impact of substance abuse includes such factors as:

- Increased absenteeism and tardiness
- Diminished performance on the job
- Decreased quality
- Increased number of accidents
- Increased insurance and health care costs

The signs of the substance abuse problem in the workplace cannot be pinned down and neatly presented in a tabular format. The truth is, we don't know the extent of the problem. However, the following estimates will help supervisors put the substance abuse problem in perspective:

- The Alcohol, Drug Abuse, and Mental Health Administration estimates that substance abuse costs business and industry almost $100 billion annually in lost productivity.

- In 1982, 65 percent of Americans between the ages of 18 and 25 who responded to a survey indicated they use illegal drugs.
- The National Institute of Drug Abuse estimates that there are 5,000 new cocaine users in the United States every day.
- The American Council for Drug Education estimates that over 5 million people in this country use cocaine.
- The Research Triangle Institute completed a study in 1981 that revealed the following: (1) approximately 60 percent of high school seniors use marijuana; (2) cocaine use in this country had tripled since 1976; and (3) daily amphetamine use in this country has tripled since 1976.

These are only estimates. The actual numbers are difficult to quantify. However, there is clearly a substantial substance abuse problem in the modern workplace and supervisors will have to deal with it.

What Are the Most Widely Used Drugs?

The modern supervisor should be familiar with the most widely used drugs, how they are used, and the recognizable physical symptoms they produce. The chart in Figure 13–7 summarizes this information for **cannabis, cocaine, amphetamines, barbiturates, hallucinogens,** and **opiates.** In addition to the drugs listed in this chart, there is the most widely used drug of all, **alcohol.** Symptoms of alcohol abuse are widely recognized. They include slurred speech, impaired physical movement, and impaired thinking and judgment.

How Do You Spot Substance Abuse?

Supervisors can apply the following rule of thumb (Figure 13–8) to spot a substance abuser:

> *Watch for impaired work performance or inappropriate personal behavior.*

Remember, neither inappropriate behavior nor impaired work performance automatically means that an employee is a substance abuser. These are just indicators that there may be a problem. If you suspect substance abuse there are very specific actions to take and equally specific actions to avoid.

What the Supervisor Should and Should Not Do

A substance abuse situation is different from most supervisory problems. It *must* be handled correctly and there is little room for error. Actions supervisors should take include the following:

- *Keep accurate, up-to-date, written records* that document the suspicious behavior and your interaction with the employee concerning the behavior.
- *Get to know your employees* well enough so that you will immediately recognize inappropriate behavior and impaired performance.
- *Evaluate job performance regularly* so that performance and behavior are well documented.

Classification of Drugs	Common Forms	Common Methods of Use	Some Short-Term Symptoms
Cannabis	■ Marijuana ■ Hashish ■ Hash oil	■ Smoked ■ Swallowed ■ Mixed with food	■ Increased pulse ■ Reddened eyes ■ Impaired logical thinking ■ Impaired physical manipulation
Cocaine (stimulant)	■ Powder ■ Crack ■ Free-base	■ Inhaled ■ Injected ■ Smoked ■ Sniffed	■ Bizarre behavior ■ Talkativeness ■ Faster breathing ■ Restlessness ■ Paranoia
Amphetamines (stimulants)	■ Benzedrine ■ Dexedrine ■ Neodrine ■ Preludin	■ Pill swallowed ■ Capsule swallowed ■ Injected	■ Reduced appetite ■ Dilated pupils ■ Sleeplessness ■ Restlessness ■ Sweating
Barbiturates (sedatives)	■ Seconal ■ Nembutal ■ Amytal ■ Quaalude	■ Pill swallowed ■ Capsule swallowed	■ Slurred speech ■ Uncommonly relaxed ■ Staggering
Hallucinogens	■ LSD ■ PCP (angel dust)	■ Pill swallowed ■ Capsule swallowed ■ Injected	■ Poor muscle control ■ Dilated pupils ■ Exhilaration ■ Panic ■ Prolonged anxiety
Opiates	■ Heroin ■ Morphine ■ Codeine ■ Opium	■ Sniffed ■ Injected ■ Eaten ■ Smoked	■ Detachment ■ Nausea ■ Contentment

Figure 13–7
Summary of forms, methods of use, and symptoms for commonly used drugs.

Figure 13–8
Supervisors concerned about a potential substance-abusing employee should . . .

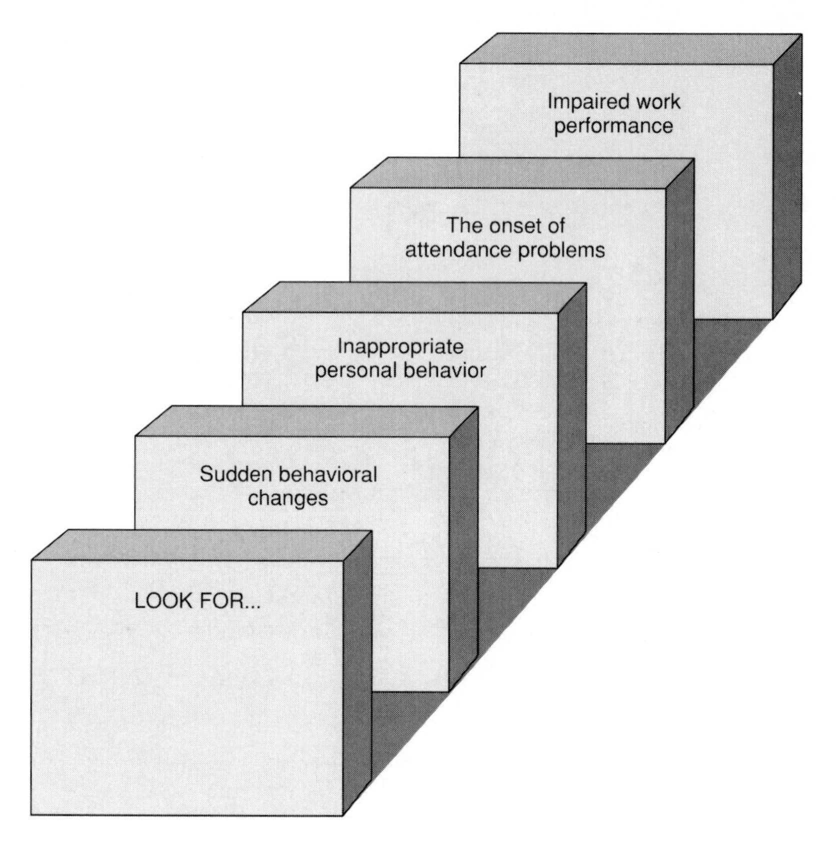

- *Take immediate action* whenever you notice inappropriate behavior or impaired performance, regardless of the suspected cause.
- *Know your company's substance abuse policy* and respond in strict accordance with the policy.
- *Know when to get help for substance-abusing employees* so that you are prepared should an employee ask for help.
- *Notify your supervisor or the appropriate person according to the company policy* the minute a problem is suspected. Keep this person up-to-date throughout the entire life of the problem.

In addition to taking these actions it is important to avoid a number of others. Actions that could lead to legal problems or exacerbate the abuse problem can be summarized as follows:

- *Do not play amateur psychologist* and attempt to diagnose the substance abuser's problem.

- *Do not confront the employee* and accuse him or her of having a substance abuse problem.
- *Do not discuss a suspected substance-abusing employee with other employees* or anyone else not specified in company policy.
- *Do not state in writing that an employee has a substance abuse problem.* Instead, refer to the abnormal behavior or impaired job performance.
- *Do not try to pressure suspected employees* into making a confession or making admissions of any kind.
- *Do not allow employees who are behaving erratically or who appear to be impaired in any way to operate dangerous equipment* or any other machine that could cause harm to them or to fellow workers.
- *Do not try to handle a suspected substance abuse problem by yourself.*
- *Do not allow yourself to be pulled too deeply into an employee's personal problems.*
- *Do not go to fellow supervisors for help* in handling a substance abuse situation unless this is an approach sanctioned by company policy.
- *Do not fire an employee yourself* even if you catch him or her in the act of substance abuse.

Supervisors should commit the do's and don'ts in this section to memory. Substance abuse problems should be handled in a straightfoward, timely manner. However, they must be handled properly and within the limits set by company policy and the law.

DEALING WITH AIDS ON THE JOB

One of the modern supervisor's most difficult challenges is how to deal with the situation when an employee has AIDS. Supervisors need to know the facts about AIDS, how to counsel infected employees, how to ease the fears and even paranoia of other employees, and how to protect other employees from infection.

Facts about AIDS[14]

The first cases of acquired immunodeficiency syndrome or **AIDS** were reported in 1981. As of this writing over 170,000 people have died of a disease that has rapidly become an epidemic. Because there is no known medical cure for AIDS, the disease has caused a strong public reaction. AIDS is feared, misunderstood, and very controversial. Supervisors need to know the facts about AIDS.

AIDS and various related conditions are caused when humans become infected with the human immunodeficiency virus or HIV. This virus attacks the human immunity systems, rendering the body incapable of repelling disease-causing microorganisms. Symptoms of the onset of AIDS are (Figure 13–9):

- Enlarged lymph nodes that persist
- Persistent fevers
- Involuntary weight loss

Figure 13–9
Summary of common symptoms of AIDS.

When should Employees Suspect They Have AIDS?

Common symptoms include:

- Enlarged lymph nodes that persist
- Persistent fever
- Involuntary weight loss
- Pronounced fatigue
- Diarrhea that does not respond to standard medications
- Purplish spots or blotches on the skin or in the mouth
- White, cheesy coating on the tongue
- Night sweats
- Forgetfulness

- Pronounced fatigue
- Diarrhea that does not respond to standard medications
- Purplish spots or blotches on the skin or in the mouth
- White, cheesy coating on the tongue
- Night sweats
- Forgetfulness

HIV is transmitted in any of the following three ways: (1) sexual contact, (2) blood contact, and (3) mother to child during pregnancy or childbirth. Any act in which bodily fluids are exchanged can result in infection if either partner is infected. The following groups of people are at the highest level of risk with regard to AIDS: (1) homosexuals who do not take appropriate precautions; (2) IV drug users; (3) people with a history of multiple blood transfusions; (4) sexually promiscuous people who do not take appropriate precautions; and (5) hemophiliacs (Figure 13–10).

Counseling Infected Employees

The employee who learns he or she has AIDS will be angry, frightened, and confused. Supervisors who are confronted by such an employee should proceed as follows:

- Listen.
- Maintain a nonjudgmental attitude.
- Make the employee aware of the company's policy on AIDS.
- Respond in accordance with company policy.

Listen carefully as you would to an employee who comes to you with a problem. If it is necessary to question for clarification, do so but be objective, professional, and nonjudgmental. Make the employee aware of the company's policy on AIDS and respond in strict accordance with the policy. The company policy aspects of AIDS are covered in Chapter 20 of this book.

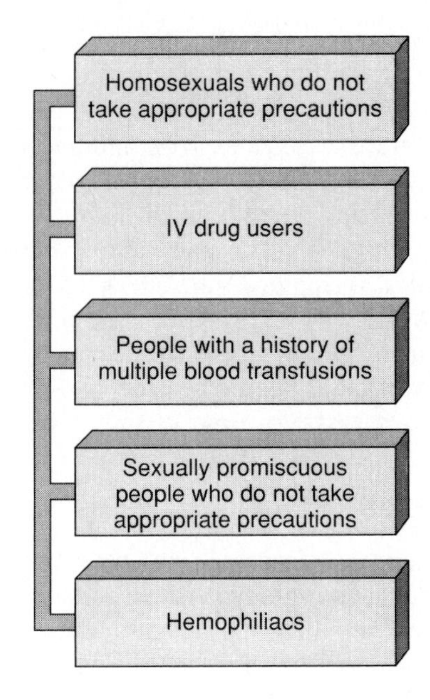

Figure 13–10
Groups at high risk for contracting AIDS.

Easing Employee Fears about AIDS[15]

Fear, panic, and hysteria are all common reactions to AIDS. If a team member discovers he or she has AIDS, fellow employees are likely to respond in this manner. Consequently, supervisors need to know how to ease the fears and misconceptions associated with AIDS. The following strategies will help:

- Work with higher management to establish an AIDS education/awareness program that covers the following topics at a minimum: (1) how HIV is transmitted, (2) precautions that workers can take; and (3) concerns about AIDS testing.
- Conduct a group roundtable discussions that allow employees to express their concerns.
- Correct inaccuracies, rumors, and misinformation about AIDS as soon as they occur.

Protecting Employees from AIDS

Supervisors should be familiar with the precautions that will protect employees from HIV infection on and off the job. OSHA's guidelines for preventing exposure to HIV infection identify three categories of work-related tasks: Categories I, II, and III. Jobs that fall into Category I involve routine exposure to blood, body fluids, or tissues that might be HIV infected. Category II jobs do not involve routine exposure to blood, body fluids, or tissues, but some aspects of the job may involve occasionally performing Category I tasks. Category III jobs do not normally involve exposure to blood, body fluids, or tissues.

Most industrial workers fall into Category III. However, employees who go to the aid of another who is cut, electrocuted, or has had a heart attack might expose themselves to HIV infection. Because of this, it is important for all employees to understand how AIDS is transmitted and to be familiar with appropriate precautions.

To guard against contact with blood, employees should not assist a bleeding person without first putting on latex or vinyl gloves. Additional protection can be provided by plastic aprons or gowns. The appropriate reaction in such cases should be spelled out in company policy. If employees are allowed to assist injured co-workers who are bleeding, gloves, gowns, and/or aprons should be made available and be readily accessible.

For mouth-to-mouth resuscitation of heart attack or electrocution victims, one-way valved face masks should be used. If company policy allows employees to assist co-workers in such cases, face masks should be made available and be readily accessible.

SUPERVISING LOW CONFORMERS

Low-conforming employees do not have to be low-performing employees. In fact, if properly handled low conformers can actually improve the performance of an organization by finding new and innovative ways to get work done and by continually challenging the status quo. However, for the harried supervisor, the additional pressures of dealing with continual nonconforming behavior can be stressful. Therefore, it is important for modern supervisors to know how to handle low-conforming employees in a positive manner. Edward Glassman recommends the following strategies:[16]

- Make sure low conformers are aware of the unit's goals.
- Learn to tolerate their honesty, even when it hurts.
- Let low conformers know they are valued members of the team.
- Do not let low conforming be interpreted as a lack of loyalty.
- Promote the flow of ideas so that low conformers are not pressured into going along with their peers.
- Interact with low conformers on an informal basis.
- Intervene when low conformers begin to wear thin with other team members.
- Intervene when low conformers are being overly pressured or criticized by their peers.

SUMMARY

1. Strategies for dealing with employees with personal problems include: be alert to signs of trouble, document problems, discuss the problem, focus on documented evidence, keep the discussion on a professional level, solicit input and listen, ask for solutions, explain potential consequences, and plan for improvement.
2. Strategies for dealing with employee theft include: screen, establish control procedures, make employees aware of negative consequences, and establish an atmosphere of honesty.
3. Emotionally disturbed employees range from normal to neurotic to psychotic peo-

ple. Supervisors should deal with normal people who are having trouble adjusting to problems in their lives. However, supervisors are not qualified to deal with neurotic or psychotic employees. They should be referred to qualified mental health professionals including counselors, psychologists, and psychiatrists.

4. Employee sabotage is typically an act of vengeance, turf protection, laziness, competition among employees, boredom or frustration. To reduce sabotage, supervisors should screen applicants carefully, keep accident logs, implement security precautions, and change the precautions at irregular intervals.
5. When dealing with a humiliator, supervisors should walk away, defuse the situation with self-effacing humor, and maintain poise and confidence.
6. When dealing with intimidators, maintain composure and respond in a confident, rational manner.
7. When dealing with procrastinators, avoid judgmental responses. Instead, use a straightforward, businesslike approach.
8. When dealing with angry employees, listen, acknowledge the anger, let the employee vent his or her anger, paraphrase the employee's complaints, and solicit the employee's input as to solutions.
9. To reduce absenteeism and the negative impact it can have, apply the following strategies: make sure employees understand the company's attendance policy; keep accurate comprehensive attendance records; counsel employees in groups and one-on-one; create an attendance-conscious attitude; and cross-train employees.
10. The negative impact of substance abuse includes increased absenteeism and tardiness, diminished performance, decreased quality, increased accidents, and increased costs.
11. The most widely used drugs are cannabis, cocaine, amphetamines, barbiturates, hallucinogens, and opiates. Each has its own forms, methods of use, and symptoms.
12. One way to spot substance abuse is to watch for impaired work performance or inappropriate behavior.
13. Actions the supervisor should take for dealing with substance abuse include: keep accurate records, know employees, evaluate job performance regularly, take immediate action, know your company's substance abuse policy, know when to get help, and notify appropriate company personnel.
14. Actions the supervisor should *not* take include: play amateur psychologist, accuse employees, discuss the problem with other employees or peers, state in writing that an employee is a substance abuser, seek confessions, apply discipline unevenly, let suspicious employees operate dangerous machines, act alone, get pulled too deeply into an employee's personal problems, go to fellow supervisors for help, or fire an employee themselves.

KEY TERMS AND PHRASES

Absence trend
Absenteeism
AIDS
Alcohol
Amphetamines
Attendance policy

Attendance-conscious attitude
Barbiturates
Cannabis
Cocaine
Consequences
Counselor
Cross-training
Documentation
Emotionally disturbed
Employee theft
Employee sabotage
Hallucinogens
Humiliators
Intimidators
Job performance
Neurotic
Opiates
Personal problems
Procrastinators
Psychiatrist
Psychologist
Psychotic
Substance abuse
Uncharacteristic behavior

CASE STUDY: Conflict over AIDS[17]

The following real-world case study illustrates a problem employee situation. Note that because of the sensitivity of the subject matter the names of the company and its employees have been changed. Read the case study carefully and answer the accompanying questions.

John Lewis was a well-liked employee of XYZ Corporation who requested sick leave on short-term disability at full pay. The company tried to keep his illness confidential, but it became known in the company that Lewis had AIDS.

After two months of treatment Lewis was pronounced well enough to return to work. The company approved his return but had to create a new position for him since his old job had been filled. Employee reaction to Lewis's return was very negative.

In an attempt to alleviate the fear, the company brought in a well-respected expert on AIDS to hold discussions with managers and employees. Discussion was intense, but eventually the employees, with the exception of one, were convinced there was no danger that Lewis would infect co-workers. This employee, Jean Long, asked to be transferred away from Lewis. The request was denied.

Long responded by taking an immediate vacation and never returned. Claiming that conditions at work forced her to leave, Long filed for unemployment compensation, which was granted. The company appealed.

1. If you were a supervisor with this company, what questions would you have asked the AIDS consultant?
2. If Jean Long was your employee, what counsel would you give her to prevent this situation.
3. As a supervisor, how would you alleviate the fears of your employees concerning AIDS?

REVIEW QUESTIONS

1. List and explain ten strategies for dealing with employees who have personal problems.
2. What are four strategies for reducing employee theft?
3. Explain how to recognize a neurotic employee; a psychotic employee.
4. Explain the differences among psychiatrists, psychologists, and counselors.
5. What are the most common causes of employee sabotage?
6. List and explain three strategies for reducing employee sabotage.
7. Explain the difference between handling a humiliator and an intimidator.
8. Give a brief description of how to handle procrastinators.
9. Mary Anderson is an assembly supervisor who has trouble handling angry employees. Her immediate reaction is to return the anger in kind. If Mary came to you for advice, what would you tell her?
10. List and briefly explain the strategies you would use to reduce absenteeism if you were a supervisor.
11. Explain how to personalize absenteeism.
12. Summarize the negative impact of substance abuse as it relates to the workplace.
13. List the most common forms and short-term symptoms of the following drugs: cocaine; cannabis; barbiturates.
14. What is the simplest way to spot a substance abuser?
15. List and explain five actions supervisors should take when dealing with suspected substance abuse.
16. List and explain five actions supervisors should not take when dealing with suspected substance abusers.

SIMULATION ACTIVITIES

The following activities simulate problems supervisors may face on the job. They may be completed as group activities or by individual students. Your task is to apply what you have learned in this and previous chapters in formulating an appropriate response to each problem.

1. How would you handle the following situation? Your best employee is suddenly irritable and frequently late. As a result, productivity in your unit has fallen off.
2. At a social occasion you see one of your employees who is obviously drunk. This same employee has been uncharacteristically late, absent, and unproductive lately. As his supervisor, what should you do?
3. One of your employees always seems to put off work assignments until the last moment. In her defense, she usually gets the work done on time, but not without a hectic and disruptive mad rush. This is upsetting other employees and getting their work off schedule. How should you handle this situation?
4. You walk into the restroom next to the plastics processing unit and catch one of your employees in the act of snorting cocaine. How should you handle this situation?

5. Your boss chews you out because attendance in your shop on Mondays and Fridays is the lowest in the company. How can you improve it?
6. It is widely suspected that one of your employees has AIDS. As a result, attendance in your unit has fallen off significantly. Your employees are angry and frightened. How should you handle this situation?

ENDNOTES

1. Alonso, M. "When an Employee Has Personal Problems," *Supervisory Management,* April 1990, p. 3.
2. Ibid.
3. Hayes, R. "Employee Abuse of the Business," *Supervisory Management,* February 1990, p. 7.
4. Ibid.
5. Giacalone, R. A. "Employee Sabotage: The Enemy Within," *Supervisory Management,* July 1990, pp. 6–7.
6. Ibid.
7. Ibid.
8. Jacobs, A. I. "Handling the Humiliators and the Intimidators," *Supervisory Management,* April 1990, p. 6.
9. Ibid.
10. Ibid.
11. Kurtz, T. "10 Reasons Why People Procrastinate," *Supervisory Management,* February 1990, p. 1.
12. Conley, C. "What to Do When Someone's Yelling," *Supervisory Management,* February 1990, pp. 6–7.
13. Bureau of Business Practice. *Drugs in the Workplace: Solutions for Business and Industry* (Waterford, CT: Simon & Schuster, 1989), p. 17.
14. Brown, K. C. and Turner, J. G. *AIDS: Policies and Programs for the Workplace* (New York: Van Nostrand Reinhold, 1989), pp. 1–8, 15, 16.
15. Ibid., pp. 106, 107, 116, 117.
16. Glassman, E. "Understanding and Supervising Low Conformers," *Supervisory Management,* November 1990, p. 10.
17. Meer, J. "Anatomy of an AIDS Dispute," *Across the Board,* September 1986, pp. 62–63.

CHAPTER FOURTEEN

Handling Employee Complaints

CHAPTER OBJECTIVES

After studying this chapter, you will be able to explain the following topics:

- Why It Is Important to Handle Complaints Properly
- The Role of Listening in Handling Complaints
- How to Handle Employee Complaints
- How to Handle Habitual Complainers
- How to Involve Employees in Resolving Complaints
- How to Turn Complaints into Improvements
- How to Handle Wage and Salary Complaints

CHAPTER OUTLINE

- Why Complaints Must Be Handled Properly
- Role of Listening in Handling Complaints
- Handling Employee Complaints
- Handling Habitual Complainers
- Involving Employees in Resolving Complaints
- Handling Complaints about Wages
- Turning Complaints into Improvements
- Summary
- Key Terms and Phrases
- Case Study Application Problem
- Review Questions
- Simulation Activities
- Endnotes

The process recommended in this chapter for handling employee complaints will serve three purposes. First, it will keep complaints from escalating into problems. Second, it will help transfer complaints into improvements. Finally, it will establish a mechanism that can keep chronic complainers from damaging morale in the organization.

WHY COMPLAINTS MUST BE HANDLED PROPERLY

There is a saying, "Some employees aren't happy unless they are complaining." Of course there are people who seem to complain habitually and supervisors have to deal with them no matter how frustrating the process. However, most employees are not constant complainers. **Complaints** from most employees are evidence of a problem or a potential problem. To ignore them is to say, "I don't care about your problems." Correspondingly, complaints can represent opportunities for improvement.

The difference between a complaint being a problem or an opportunity can lie in how the supervisor handles it. Consequently, it is important for supervisors to take employee complaints seriously and handle them properly, even when they are from constant complainers.

ROLE OF LISTENING IN HANDLING COMPLAINTS[1]

The first and most important step in handling employee complaints is to *listen.* No matter how hurried supervisors are, no matter how bogged down in the everyday details of getting the job done, when an employee has a complaint supervisors should take the time to listen.

There are several reasons, all related to improving employees' performance, why supervisors should listen to complaints:

- *The employee may just be frustrated and in need of an opportunity to ventilate.* In this case, a few minutes spent listening can get the employee back to normal and his or her job performance will not suffer. On the other hand, an angry or frustrated employee will not do his or her best work. Therefore, just listening can help maintain and even improve job performance.
- *The complaint might be evidence of a bigger problem or the potential for one.* For example, a complaint by one employee about a certain working condition might be evidence of dissatisfaction among a much larger number of employees. Listening in this case might head off labor/management problems.
- *The complaint might be the employee's way of pressing a hidden agenda.* For example, one employee's complaints about another employee might just be an attempt to build himself up in the competition for promotion. By listening carefully, supervisors can identify **hidden agendas** and respond accordingly.
- *The complaint might reveal a legitimate need for improvement.* This is particularly true when the complaint comes from a high-performing employee. When such employees complain about factors that are having a negative effect on the job, supervisors have an opportunity to make **productivity improvements.**

Listening Tips

- Maintain eye contact and assume a posture that says, "I am listening."
- Give the employee your undivided attention. Eliminate distractions.
- Do not interrupt except for clarification.
- Listen to what is said and what is not said.
- Read the employee's body language and tone of voice.
- Maintain a professional bearing. Do not become angry or hostile.
- Try to remember without taking notes.
- Paraphrase and repeat back to show you have heard and understand.

Figure 14–1
Strategies for better listening.

Figure 14–1 contains tips on how to be an effective listener when an employee has a complaint.

In addition to listening, it is important to ask questions properly to clarify and enhance understanding, and to avoid wasting time solving the wrong problem. Here are some tips for asking questions of a complaining employee:[2]

- *Ask questions for clarification.* If you don't understand a point the employee is trying to make, find an opening and ask for clarification. For example, if an employee complains about the company's overtime policy, you might need to clarify whether she means there is too much or too little overtime.
- *Ask questions to clear up inconsistencies.* Inconsistencies can mean that the employee is frustrated and confused, or they can also reveal a hidden agenda or another unspoken problem. Therefore, it is important to clear them up.
- *Ask questions to gain more complete information.* Is the employee skirting the real issue? Does he seem to be holding back? Probing questions may be necessary to gain complete information. For example, an employee who complains in general terms rather than giving specifics may have to be questioned in order to identify the real problem.
- *Ask questions to determine what the employee would like you to do.* Does the employee want you to do something, or is she just blowing off steam?

Once the session seems to be winding down, ask, "What would you like me to do?" Occasionally you will find you have already done it.

HANDLING EMPLOYEE COMPLAINTS[3]

It is important to have a structured procedure for handling complaints and to make sure that all employees are familiar with it. This will serve two purposes. First, it will reassure well-intentioned employees that their complaints will be given the attention

Figure 14–2
Step-by-step approach for handling employee complaints.

Five-Step Procedure for Handling Employee Complaints

- Listen
- Investigate
- Act
- Report
- Follow up

they deserve. Second, it will discourage habitual complainers and whiners who simply use complaining as a way to get attention. The following **five-step procedure** is an effective way to handle employee complaints (Figure 14–2):

1. Listen
2. Investigate
3. Act
4. Report
5. Follow up

It is important to proceed through the steps in the order specified and to ensure that no step is left out. Each step in the process is explained in the following paragraphs.

- *Listen.* This step was explained in the previous section; its importance is reiterated here. Remember, if you do this step well, the others may not be necessary.
- *Investigate.* During the listening session you should have been able to form a thorough picture of the problem from the perspective of the complaining employee. However, every story has at least two sides. **Investigating** will turn off employees who complain simply to gain attention. Knowing that all sides will be heard is a deterrent to such employees. Talk with other employees to fill out the story. Talk with fellow supervisors to see if they have dealt with similar complaints and, if so, how they handled them. Talk with your supervisor to get his or her input and to clearly define your authority relative to the complaint.
- *Act.* Once all of the pertinent information relating to the complaint has been collected, you can take the appropriate action: (1) correct the problem; (2) work with higher management to have the problem corrected; (3) hold the situation in abeyance while collecting additional information; or (4) take no action.
- *Report.* It is important to keep the complaining employee fully informed as to what has been done, will be done, or will not be done. **Reporting** is easy when you can say the problem has been solved or will be solved. However, it can be difficult when you are forced to say that you do not intend to take any action. When this is the case, it is important to deliver the bad news in a way that does not damage the employee's morale or the supervisor-employee relationship. One way to do this is to put the employee's complaint in the context of the goals and the overall good of the organiza-

tion. It is also important to thank the employee for bringing the complaint to your attention and to encourage him to do so in the future.

- *Follow up.* The supervisor's job is not over once the reporting step has been accomplished. Regardless of the course of action taken and reported to the employee, supervisors should **follow up** periodically. By following up you can ensure that the good news you reported is having the desired effect and that the bad news is not having a detrimental effect.

HANDLING HABITUAL COMPLAINERS

Even a well-structured process for handling complaints won't always do the job in discouraging **habitual complainers.** This type of employee can present a difficult challenge for the supervisor. Here are some tips for handling habitual complainers (Figure 14–3):

Figure 14–3
Strategies for handling habitual complainers.

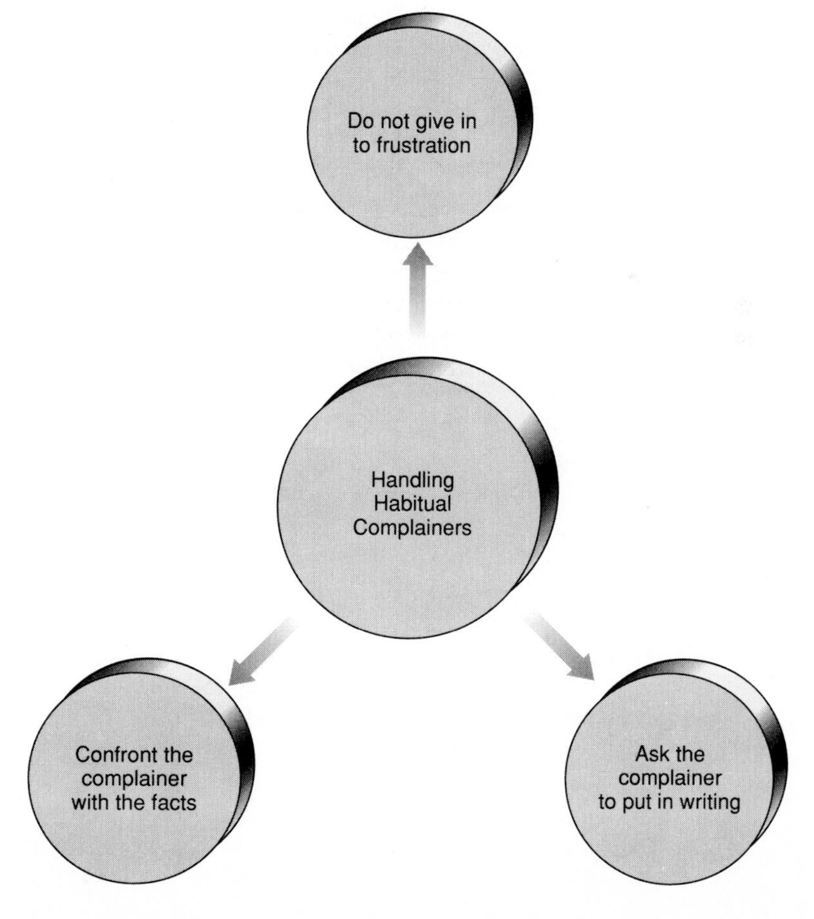

- *Do not give in to frustration.* This may be precisely what the habitual complainer wants you to do. Constant complaining is sometimes a sign of a deeper problem. The employee may be seeking attention, a special relationship with the supervisor, or any one of a number of items on a hidden agenda. Maintain your composure and stay with the process.
- *Ask the complainer to put it in writing.* When the habitual complainer is doing all the complaining and you are doing all the work, try turning the tables. Ask the complainer to give you, in writing, a detailed explanation of the problem. This approach will usually discourage the employee and cut the complaints down to those that are legitimate.
- *Confront the complainer with the facts.* At some point the supervisor simply needs to confront habitual complainers with the fact that they complain too much. This should be done in an objective, professional manner. Avoid anger and judgmental statements. Point out that you appreciate the employee's input, but that he or she complains much more than other employees. Each complaint takes a significant amount of your time, and like the employee, your time is limited. You might even convey to the habitual complainer that your best employees are those who solve problems first and then inform you of the solution, rather than those who just complain.

INVOLVING EMPLOYEES IN RESOLVING COMPLAINTS

If the entire team, not just one employee, has a complaint, the entire team should participate in resolving it. An advantage to involving employees in **complaint resolution** is that it multiplies the chances of finding an acceptable solution. However, in order to get employees to participate as a group, supervisors must first lay some groundwork.

Preparing employees to be meaningful participants in complaint resolution is a matter of accomplishing the following tasks (Figure 14–4):

- Ensuring that employees know you value their participation and want their input.
- Ensuring that employees understand the types of problems that are and are not appropriate for group participation.

Preparing employees to participate in group complaint resolution is half of the groundwork; the other half involves preparing for and conducting group sessions. Before scheduling a session:

- Do your research. Come to the meeting fully prepared and armed with information. If there is information of benefit to the entire group, copy it and provide copies to each participant at the beginning of the meeting.
- Distribute an outline or put one on a flipchart or marker board.

During a complaint resolution session supervisors should accomplish the following tasks:

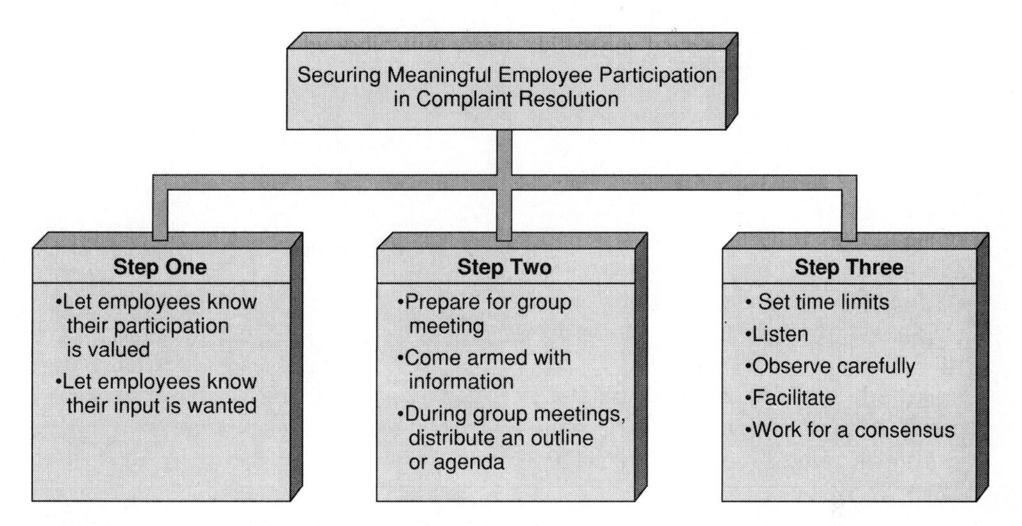

Figure 14–4
Step-by-step approach for ensuring meaningful employee participation in resolving complaints.

- *Set time limits* and hold all participants to them.
- *Listen* and encourage the other participants to listen.
- *Observe carefully.* Look for signs that participants understand or do not understand each other. Question as necessary to bring out additional information that will clarify.
- *Facilitate.* Draw everyone out to ensure maximum participation. Do not allow one or two employees to dominate the session.
- *Work for a consensus.* Then close the meeting by describing the action you will take to implement the group's recommendations. In subsequent meetings give the group feedback about progress you have made.

HANDLING COMPLAINTS ABOUT WAGES[5]

Wages are likely to be a common source of complaints in any company. You will not be a supervisor long before an employee will come to you to complain about money. The same strategies used for handling any complaint apply in the case of wage problems. In addition, there are special strategies with which supervisors should be familiar (Figure 14–5):

- *Do not allow employees to draw you into comparison debates.* Employees invariably discuss wages and make comparisons among themselves. As soon as employee X finds he is paid less than employee Y, there is likely to be a complaint. The supervisor's approach to such situations should be that individual wage rates are: (1) confidential;

Figure 14–5
Handling complaints about wages.

Rules of Thumb to Remember When Handling Complaints About Wages
• Wages are confidential and not to be discussed. What individuals are paid is the business of the supervisor and the employees in question.
• Wages are based on performance as rated by the supervisor. Do not debate who makes how much and why with employees.
• Listen to complaints, but do not debate the issue.
• Know company policy concerning wages to be able to correct invalid assumptions and/or complaints.

and (2) based on performance. Listen and let the employee state his case, but do not make or debate comparisons.

- *Do not discuss raises given to individual employees.* This is another case in which complaints are sure to arise. It usually happens like this. José and Marie, who are both printed circuit board assemblers, are talking during a break. José asks, "Did you get a raise this year?" Marie enthusiastically responds, "Yes! A good one; 12 percent." José, whose raise was 6 percent, quickly finishes his coffee and marches straight to the supervisor's office. The supervisor should listen and let José blow off some steam, but she should not discuss anyone's raise with José except his.
- *Have a detailed knowledge of the wage levels and wage histories of all employees as well as the company's wage policies.* Be well enough informed to refute false claims or misconceptions. For example, if an employee demands to know "why Ethel makes twice what I make," the supervisor should know the salary schedule well enough to refute this claim.

TURNING COMPLAINTS INTO IMPROVEMENTS

There is a natural human tendency to respond defensively to complaints. Modern supervisors must work to overcome this tendency for two reasons. First, complaints left unresolved may develop into bigger problems. Second, if handled properly complaints can be turned into **improvements.** All of the material in this chapter is intended to help supervisors turn complaints into improvements.

The most important step is deciding to take this approach. Once you have decided to view complaints as opportunities, follow the five-step procedure described earlier in this chapter:

- Listen
- Investigate
- Act

- Report
- Follow up

There are no shortcuts. Undertake each step in the order specified and do so every time. Every complaint is not legitimate, but many that do not appear to be on the surface will turn out to be after you have listened and investigated. If a complaint is legitimate, act on it promptly, report results back to the employee, and follow up periodically.

An additional strategy that will help turn complaints into improvements is to maintain a complaint log. Log in every complaint that is acted on along with the final resolution. Complaints that result in positive changes such as improved productivity, quality, competitiveness, or morale should be given special recognition as should the employee who originally brought the complaint. Supervisors who make up their minds to view complaints as opportunities; who learn to listen, investigate, act, report, and follow up; and who give appropriate recognition when complaints result in needed change will be successful in turning complaints into improvements.

SUMMARY

1. The most important step in handling employee complaints is to listen. The employee may just need to blow off steam, the complaint might be evidence of a bigger problem, the complaint might be the employee's way of pressing a hidden agenda, or the complaint might reveal a legitimate need for improvement.
2. In addition to listening, it is important for supervisors to ask questions. Questions are asked to clarify, clear up inconsistencies, gain more complete information, and determine what the employee would like you to do.
3. The five-step procedure to handling employee complaints is listen, investigate, act, report, and follow up.
4. When dealing with a habitual complainer remember the following rules of thumb: do not give in to frustration, ask the complainer to put it in writing and confront the complainer with the facts.
5. When involving employees in the resolution of complaints on a group basis, begin by preparing them. This involves ensuring they know you really want their involvement, ensuring that employees understand the types of problems that will be dealt with in a group format, and ensuring that employees understand the procedure. Before the meeting, do your research. During the meeting, distribute an outline, set time limits and stick to them, listen and encourage other participants to listen, observe carefully, ensure maximum participation by drawing everyone out, and work for a consensus resolution.
6. When handling complaints about wages, do not allow employees to draw you into comparison debates, do not discuss raises given to other employees, and have your facts in order.
7. To turn complaints into improvements, view them as opportunities, apply the five-step approach, and give recognition when a complaint results in an improvement.

CASE STUDY: Complaints about Break Periods[6]

The following real-world case study illustrates a supervisory problem relating to employee complaints. Read the case study carefully and answer the accompanying questions.

Everset, Inc.'s lunch period policy had been in effect for many years but employees often ignored it. The policy allowed thirty-minute lunch periods but employees often took more time and returned to work late. In an attempt to solve this problem, the company began requiring employees to punch in and out on a time clock. Angered by this decision, employees complained to their union and the matter was referred to arbitration. Union representatives and the company attorney argued their respective cases.

1. If you were a supervisor at Everset, how would you have dealt with the employee complaints about the time clock?
2. What are some alternative solutions the company might have used instead of requiring employees to punch in and out on the time clock?

REVIEW QUESTIONS

1. Explain briefly why it is important to handle complaints properly.
2. Give four reasons why it is important for a supervisor to listen when an employee brings a complaint.
3. What are the reasons for questioning employees who have a complaint?
4. Explain the five steps for handling employee complaints.
5. How would you handle a habitual complainer?
6. What are the steps for preparing employees for involvement in group conflict resolution?
7. Explain how to prepare for and conduct a group complaint resolution session.
8. How is handling a complaint about wages different from handling other complaints?
9. Briefly explain how you would turn complaints into improvements.

KEY TERMS AND PHRASES

Clarification
Comparison debates
Complaint
Complaint resolution
Five-step procedure
Follow up
Habitual complainers
Hidden agenda
Improvements
Inconsistencies
Investigate
Productivity improvements
Report

SIMULATION ACTIVITIES

The following activities simulate problems supervisors may face on the job. They may be completed as group activities or by individual students. In either case, apply what you have learned in this and previous chapters in formulating an appropriate response to each problem.

1. How would you handle the following situation? An employee bursts into your office, slams the door, and says, "I am tired of this place! It's so disorganized. It's a miracle we ever get anything done!"
2. An employee asks for an appointment to discuss a problem with you. She is calm and in control, but clearly something is wrong. Once in your office she says, "I just learned that John McCoy makes more than I do. We both know I am a better technician than he is." How would you handle this situation?
3. Van Tram Ly is a constant complainer. He has just walked into your office with his tenth complaint of the week. How should you handle the situation?
4. Joan Gomez is one of your best employees. She rarely complains and when she does it is usually legitimate. She has just informed you that the new four-day work week is causing morale problems among team members. How would you handle this situation?

ENDNOTES

1. Bureau of Business Practice, Inc. *Handling Complaints: Turning Employee Gripes into Innovations and Improvements* (Waterford, CT: 1985), p. 3.
2. Ibid., p. 5.
3. Ibid., p. 7.
4. Ibid., pp. 12, 13.
5. Fuller, G. *Supervisor's Portable Answer Book.* (Englewood Cliffs, NJ: Prentice Hall Business and Professional Publishing Division, 1990), pp. 15, 16.
6. Baderschneider, E. "Elastic Break Period," *Supervisory Management,* May 1990, p. 3.

CHAPTER FIFTEEN
Discipline and the Supervisor

CHAPTER OBJECTIVES

After studying this chapter, you will be able to explain the following topics:

- The Rationale for Disciplining Employees
- The Fundamentals of Disciplining Employees
- The Guidelines for Disciplining Employees
- The Discipline Process
- The Concept of Employment at Will
- The Concept of Wrongful Discharge

CHAPTER OUTLINE

- Rationale for Disciplining Employees
- Fundamentals of Disciplining Employees
- Guidelines for Disciplining Employees
- The Discipline Process
- Summary
- Key Terms and Phrases
- Case Study Application Problem
- Review Questions
- Simulation Activities
- Endnotes

Even the best supervisor will have to discipline employees occasionally. This chapter provides prospective and practicing supervisors with the information they need to discipline effectively.

RATIONALE FOR DISCIPLINING EMPLOYEES

Behavior that has a negative impact on the accomplishment of organizational goals—that is, behavior that is **disruptive** to productivity, quality, and/or competitiveness—must be corrected. Dealing with such behavior is known as **disciplining.** Since some

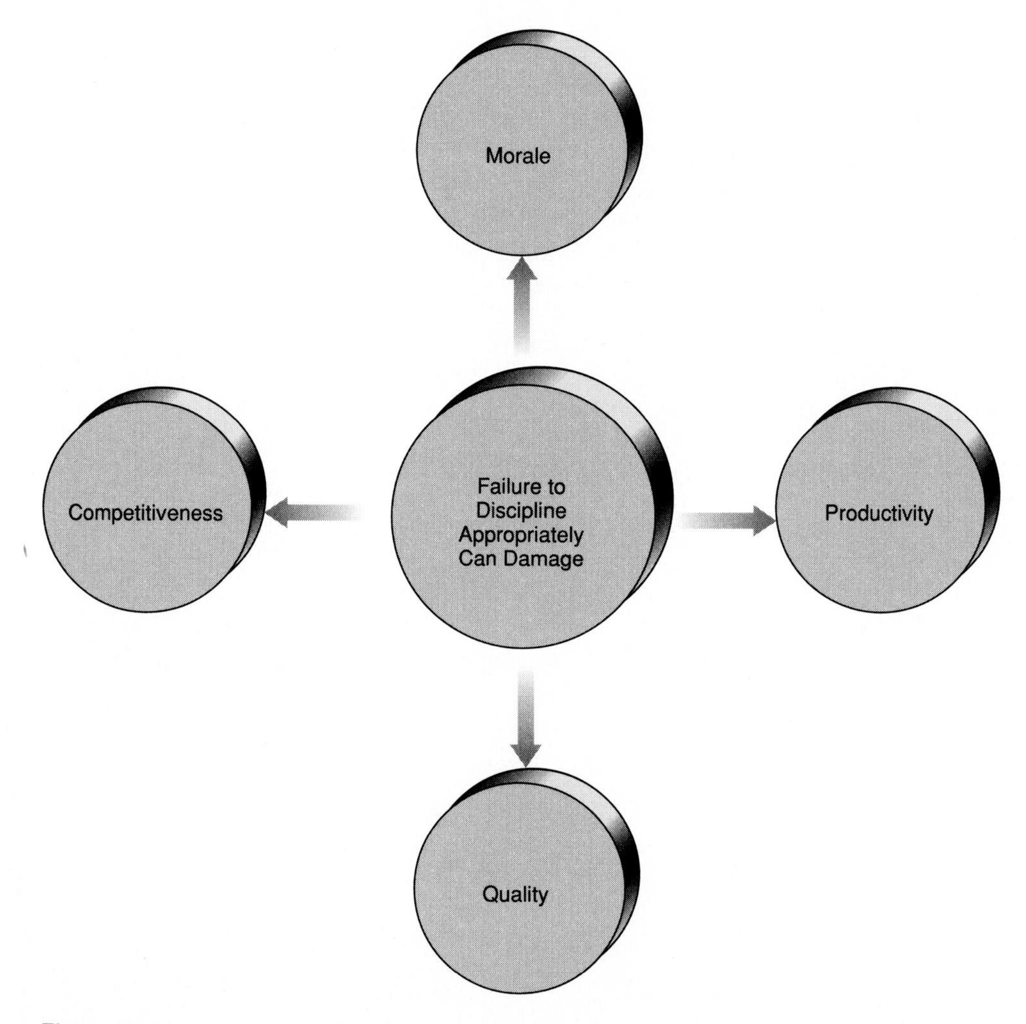

Figure 15–1
Disciplining employees is never easy, but failure to discipline can be much worse.

disruptive, nonproductive behavior is almost inevitable, supervisors should know how to discipline appropriately and properly. To do otherwise is to risk damaging morale, productivity, quality, and competitiveness (Figure 15–1).

This is the basic rationale for disciplining disruptive employees. In addition, external discipline properly applied can lead to **self-discipline**. When this happens, the discipline system can be judged successful.

Stated simply, a properly applied discipline system will lead to a self-disciplined team in which individual members know the organization's rules and regulations, abide by them, and expect co-workers to abide by them. The key to achieving self-discipline among team members is to do the following (Figure 15–2):

- *Personalize disruptive behavior* by helping team members see how the disruptive behavior of one is detrimental to all. Relate losses in productivity and quality to potential losses of customers and contracts. The obvious result is the potential for loss of jobs, raises, and the benefits associated with continued growth.
- *Apply discipline in a step-by-step manner.* People can be very forgiving in their attitudes toward offenders. Therefore, to dismiss an employee on the first offense is more likely to damage morale than to enhance it. Everyone is likely to err occasionally. Consequently, employees will sometimes respond to disciplinary measures applied to co-workers by mentally putting themselves in their place. They will want to see discipline applied to a co-worker in the way they would want it applied to them. A step-by-step process that is reasonable and has built-in opportunities for improvement will be well received by most employees.

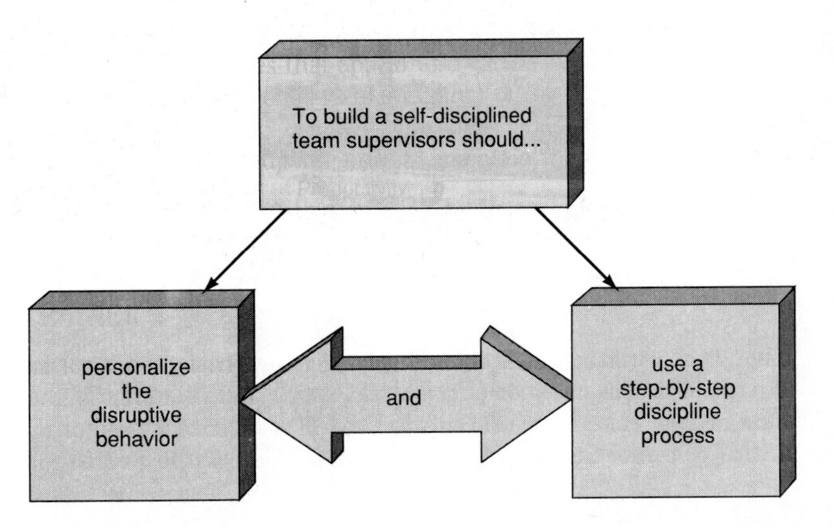

Figure 15–2
The goal of disciplining is to help team members become self-disciplined.

FUNDAMENTALS OF DISCIPLINING EMPLOYEES[1]

Regardless of the actual process set forth in company policy, there are some universal rules of thumb about discipline with which supervisors should be familiar. Applying the following rules will enhance the supervisor's effectiveness regardless of the type of discipline process used (Figure 15–3).

- *Learn your company's rules and regulations and commit them to memory.* The company's rules and regulations translate company policy into expectations for everyday behavior. Supervisors cannot help to enforce rules and regulations if they don't know them.
- *Get to know your employees and their personalities.* Prevention can be an important part of a discipline program. Supervisors who know the personalities of their employees can notice even subtle changes in behavior that might be evidence of the fact that a problem is brewing. One-on-one conferences with such employees might prevent problems that would require disciplinary measures.
- *Prevent discipline problems through education.* Make sure employees know the rules, understand the reasons for them, and know the consequences of breaking them. Rules should be posted, distributed, and periodically discussed, and the reasons for them should be emphasized. In addition, the consequences of breaking the rules should be discussed. There should be no surprises in a company's discipline program.
- *Set a positive example.* The "do as I say, not as I do" approach will not work with employee discipline. Supervisors must set a positive example of knowing the rules and abiding by them. If supervisors lose credibility, the discipline program loses credibility.

Before Undertaking the Discipline Process, Supervisors Should . . .
- Familiarize themselves with the organization's rules and regulations.
- Get to know their employees.
- Educate team members about the organization's rules and regulations.
- Set a positive example.

Figure 15–3
Rules of thumb for disciplining employees.

GUIDELINES FOR DISCIPLINING EMPLOYEES

Robert N. Lussier has developed some guidelines for ensuring that discipline is effective, summarized here as follows:[2]

- *Understand your authority.* Never undertake a disciplinary measure that is not within your span of authority.

- *Understand the rules and the reasons for them.* Workers want to know the parameters within which they must operate and the reasons for them. "Because I said so" is not a sufficient reason in the modern workplace.
- *Communicate the rules to all employees.* Do not assume they know. Post the rules, distribute them, and periodically discuss them.
- *Avoid negative comments about the rules.* If you don't believe in the rules, your employees won't either. Disagree with higher management in private if you must, but among employees, support the rules.
- *Follow the rules yourself.* Set a positive example and never veer from it. To your employees you should be the embodiment of the rules.
- *Never act on hearsay.* Before beginning the discipline process, make sure you have hard facts. Never act on hearsay, gossip, or second-hand information.
- *When rules are broken, act.* If discipline is to have the desired effect, it must be acted on. But this should be done in private. Publicly administered discipline will belittle the recipient and may damage the morale of fellow workers.
- *Discipline at the end of the day.* Employees who are disciplined probably will not be able to keep their mind on their work. Disciplining at the end of the day will give emotions time to settle down and the worker time to get his or her mind back on the job.
- *Document discipline problems and corresponding actions.* In today's litigious society, a good rule of thumb is to document, document, and document. In addition to alleviating legal concerns, documentation can help employees see where and when they have broken the rules.
- *Do not hold grudges.* Once disciplinary measures have been taken, the supervisor/employee relationship should return to normal. Do not hold grudges and do not allow employees to do so.
- *Give advance warning.* This is a matter of communicating the rules and the consequences of breaking them to all employees. There should be no surprises.
- *Discipline immediately.* In order for a discipline measure to work it must be administered in a timely manner.
- *Be consistent.* Favoritism or even forgetfulness will rob the discipline program of its credibility. Consistency is fundamental to the success of the discipline process.
- *Be objective and professional.* Do not allow any step in the discipline process to be affected by personalities, yours or the employee's. Focus on the behavior, not the person.

THE DISCIPLINE PROCESS[3]

The best way to ensure that discipline measures adhere to the fundamentals and guidelines listed in this chapter is to have a step-by-step process that takes all of these considerations into account. The following five-step approach to discipline accommodates all applicable guidelines (Figure 15–4):

Figure 15–4
The discipline process should proceed in a series of logical steps.

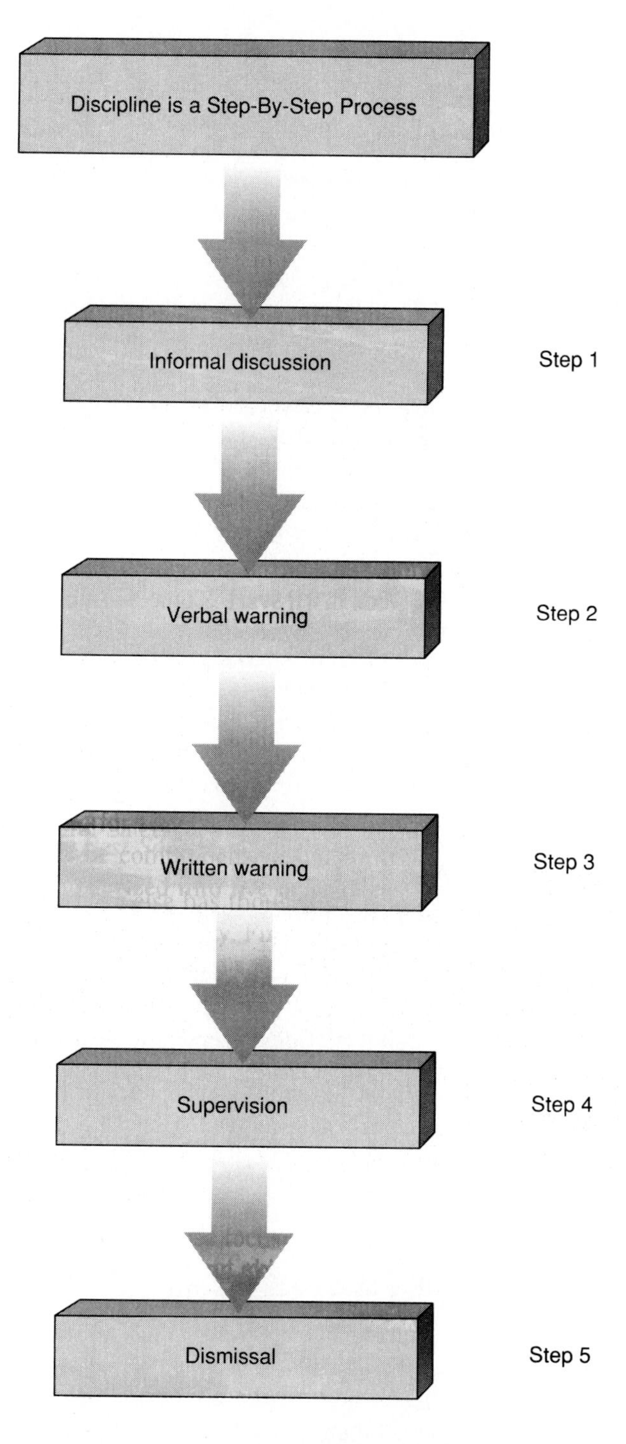

1. Informal discussion
2. Verbal warning
3. Written warning
4. Suspension
5. Dismissal

Informal Discussion

Informal discussion should always be the first step in the process for first-time offenses or minor transgressions. It is intended to head off problems before additional measures are required. During an informal discussion the supervisor reminds the employee that a rule has been broken. No action will be taken. This should be explained to the employee. However, it is a good idea to mark the date and time of the informal discussion on a calendar or in a log book for future reference.

Verbal Warning

If the formal discussion does not stop the disruptive behavior, the next step is a **verbal warning.** Before issuing a verbal warning get the facts. What rule has been broken, when, and how? Who knows about it? Were there witnesses? Who is to blame? What action does company policy specify?

When you have all the facts, make a decision as to whether or not a verbal warning is in order. If so, bring the employee in for a private meeting, preferably at the end of the day. The following guidelines may enhance the effectiveness of the meeting (Figure 15–5):

- *Have a definite objective* and share it with the employee. Let the employee know what rule has been broken and what the consequences of future infractions will be.
- *Have supportive material on hand and readily available.* Whatever evidence you have of an infraction plus any evidence of past problems (i.e., old warnings, material from a log book) should be readily available to share with the employee.
- *Explain what rule has been broken and the reason for the rule.* Be specific. Keep the conversation **objective**, nonjudgmental, and professional. Give the employee an opportunity to offer input or to rebut any evidence you may have. If at the end of the meeting you are still convinced a verbal warning is in order, issue it. A verbal warning should simply inform the employee that further rule breaking may result in specific consequences (as set forth in company policy). Document the verbal warning and let the employee know it has been documented.

Figure 15–6 is an example of an appropriate way to document a verbal warning. Put this type of document in a file and give the employee a copy. Before concluding the meeting, ask the employee the following questions:

"Do you understand why you are receiving this warning?"

"What can we do to prevent further infraction?"

"Do you understand the consequences of further infraction?"

Figure 15–5
These steps will make verbal warnings more effective.

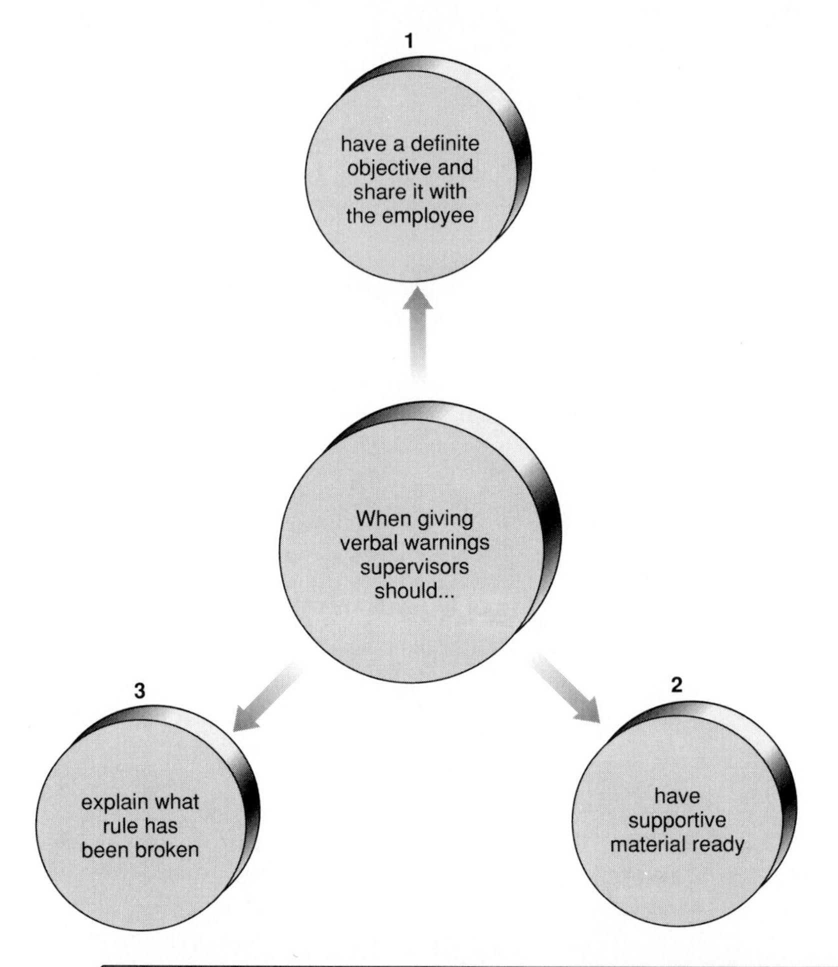

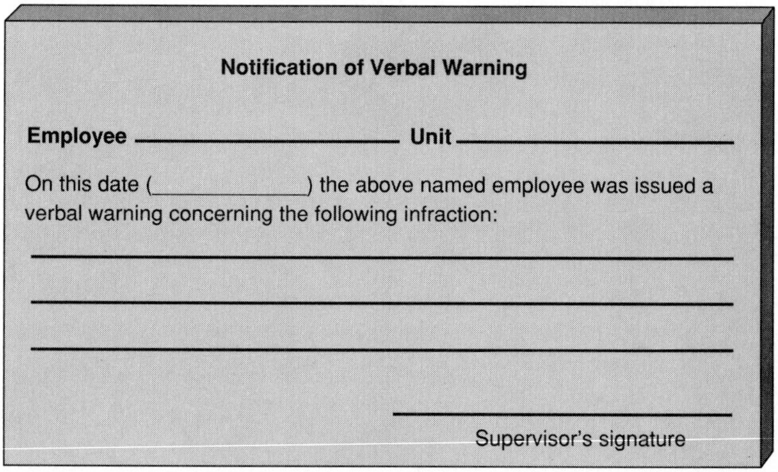

Figure 15–6
Forms such as this can be used for recording that an employee was verbally warned about unacceptable behavior.

Notification of Verbal Warning

Employee ____________________ **Unit** ____________________

On this date (____________) the above named employee was issued a verbal warning concerning the following infraction:

____

____

____

____
Supervisor's signature

Written Warning

For most employees the informal discussion is enough to correct the negative behavior. When this does not work the documented verbal warning usually solves the problem. However, if it doesn't, a **written warning** is the next step.

Handle the meeting for giving the written warning in the same manner as the meeting for the verbal warning. But at the end of this meeting, give the written warning. Sign the warning and have the employee sign it. Both of you should keep a copy. Figure 15–7 is an example of a form that can be used for issuing a written warning.

Notice in Figure 15–7 that the employee is required to acknowledge the written warning. The employee may or may not agree that the warning is justified. By signing the form, he or she is acknowledging receipt of the warning, not admitting guilt. As with the verbal warning, the written warning should be specific.

Figure 15–7
Forms such as this can be used for issuing written warnings.

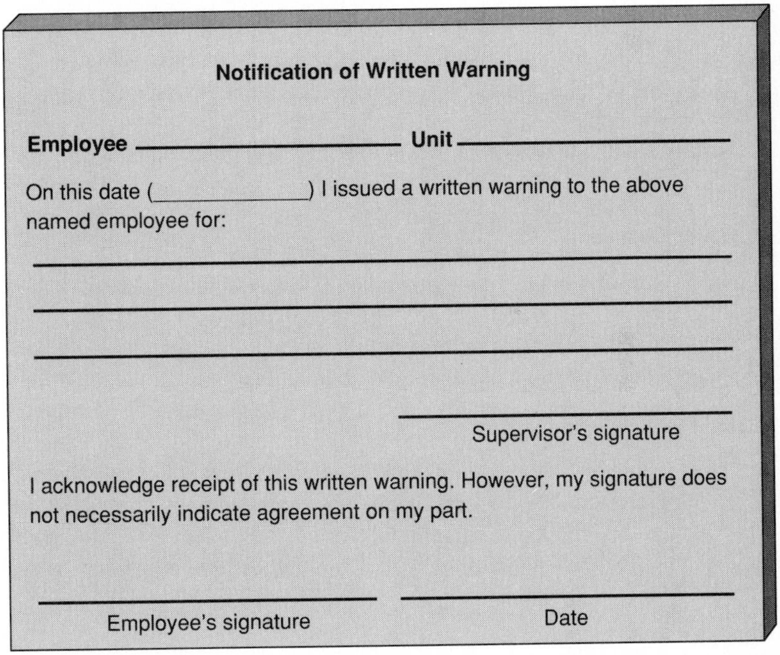

Notification of Written Warning

Employee ______________________ **Unit** ______________________

On this date (______________) I issued a written warning to the above named employee for:

____

____

____

____
Supervisor's signature

I acknowledge receipt of this written warning. However, my signature does not necessarily indicate agreement on my part.

______________________ ______________________
Employee's signature Date

Suspension

If verbal and written warnings do not work, the next step is **suspension**.[4] This is a serious step. It involves requiring the employee to take a specified number of days off from work, usually without pay. Typically the supervisor will need the approval of higher management before suspending an employee. Before recommending a suspension, ask yourself the following questions:

- Am I overreacting?
- Am I acting in accordance with company policy?
- Has suspension ever been used for a similar infraction?
- Do I have sufficient documentation to stand up to a formal challenge?
- How will be suspension be received by the employee's co-workers?

If you are not overreacting and your recommendation is clearly supported by company policy, you have a foundation for suspension. If suspensions have been used for similar infractions, a **precedent** has been set. If you have sufficient documentation to withstand a formal challenge, your recommendation is more likely to be well received by higher management. Finally, if the employee's co-workers will not react negatively to the suspension, you have a strong case for moving forward.

Dismissal

The final step in the discipline process is **dismissal.**[5] When all other steps have failed to change the negative behavior, **termination** is an appropriate action. However, it should not be rushed into. Before dealing with the specifics of how to handle a dismissal, two related concepts must be explained: employment-at-will and wrongful discharge (Figure 15–8).

In the past employers could fire workers without cause or notice. In the modern world this is no longer the case. Civil rights legislation provides a measure of protection for employees in protected classes and the Equal Employment Opportunity Commission is available to assist employees who feel they have been unjustly fired. Ethical reasons for the fair treatment of employees aside, susceptibility to charges of wrongful discharge is causing employers to look carefully at their dismissal practices.

The term **employment-at-will** means the employee works at the will of the employer and has no say in the matter, no rights or protections. The concept of **wrongful discharge**, on the other hand, establishes that employees have rights relative to continued employment and should be protected from arbitrary and unjust termination.

This concept is supported by federal civil rights legislation and a steadily growing body of state legislation. The result is that employers throughout the country are being successfully sued by ex-employees who feel they were unjustly treated. The key to properly dismissing an employee is knowing why you can and cannot terminate. An employee can be terminated for **cause**.

Figure 15–8
These two concepts can clash when it comes to dismissing an employee.

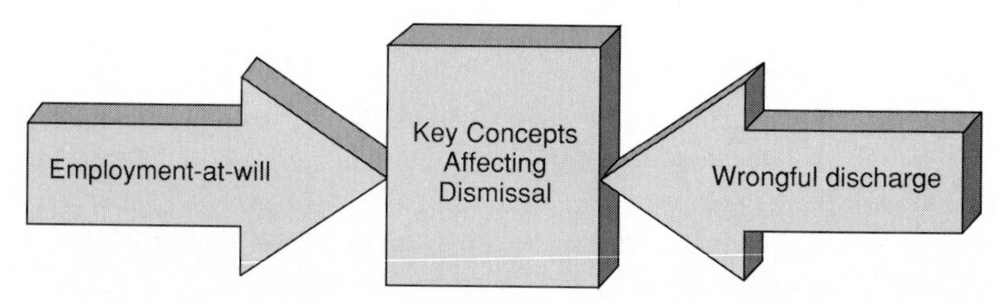

Causes should be listed in the company's policy manual and supervisors should be familiar with them. Typical causes include the following (Figure 15–9):

- Substance abuse on the job
- Theft of company property
- Fighting on the job
- Insubordination
- Sabotage
- Sleeping on the job
- Excessive absences and/or tardiness
- Falsification of time cards or other records
- Commission of a felony on or off the job

These examples and others may not necessarily be included in a company's policy manual, but supervisors should know what causes of termination are part of their company's approved and published policy.

Figure 15–9
Notification to employees of potential causes of termination.

Shalimar Aircraft Re-Work, Inc.
900 Eglin Parkway
Shalimar, Florida 32548

Notice to all Employees of SAR, Inc.

Behavior of the types set forth herein can result in termination *for cause:*

- ▶ Substance abuse
- ▶ Theft of company property
- ▶ Physical violence on the job
- ▶ Insubordination
- ▶ Sabotage of company property
- ▶ Sleeping on the job
- ▶ Excessive absenteeism
- ▶ Excessive tardiness
- ▶ Altering or otherwise falsifying timecards, records, or other paperwork
- ▶ Commission of a felony

Note:
For additional details, consult the SAR, Inc. policy manual or see your supervisor.

It is also important to know reasons that *cannot* be used as cause for dismissing an employee. These reasons include the following (Figure 15–10):

- Attempting to unionize fellow employees
- Race, religion, gender, national origin, culture, or sexual preference
- Pregnancy, childbirth, or related medical problems
- Age (if the employee is between 40 and 70 years of age)
- Refusal to perform unsafe tasks
- Refusal to work in a hazardous environment
- "Blowing the whistle" about health and safety issues
- Physical or mental handicap

These specific areas are protected by the same laws that prohibit discrimination in hiring practices (i.e., the Civil Rights Act of 1964 and subsequent amendments, the Pregnancy Discrimination Act of 1978, the Age Discrimination in Employment Act of 1967, the Occupational Safety and Health Act of 1970, and the Vocational Rehabilitation Act of 1973). As with hiring practices, dismissal practices should focus strictly on the job performance issues.

With this background, supervisors should be prepared to approach the termination step from a well-informed perspective. Rarely would a supervisor actually dismiss an employee. Typically, he or she will recommend termination to higher management. Once approval to proceed has been secured, the employee must be properly informed. This may be done by the supervisor, by human resources personnel, or in an interview where both are present. Such procedures should be set forth in company policy and strictly adhered to.

If company policy requires the supervisor to conduct the dismissal interview, the supervisor should know the proper steps to follow. These steps are summarized in the following list:

- Set up a private interview with the employee, but have a representative of the human resources department present. This person can serve as a witness for both the supervisor and the employee.
- Ask the employee to explain his or her side of the problem in question and listen carefully. This gives you one last opportunity to ensure that there are no relevant facts of which you are not aware.
- Explain that based on the facts, you have decided to terminate the employee for cause. Explain the cause and present your documentation of the facts. Take full responsibility for the termination decision. *Do not pass the buck.*
- Inform the employee in writing of the effective date of dismissal. Be prepared for the employee to react out of anger and quit on the spot. One effective way to handle such situations is to have a resignation letter prepared and dated. If the employee says, "You can't fire me; I quit!" or words to that effect, ask him or her to sign the letter.
- Explain or ask the human resources representative to explain the details of severance pay, vacation pay, sick leave, insurance, pensions, and other related concerns.

Pheadix Plastic Recycling Company
100 John Sims Parkway
Niceville, Florida 32578
904/555-3726

MEMORANDUM

To: All supervisors

From: Amanda Barksdale, President

Subject: **Wrongful discharge**

Date: April 2,1991

By now you have heard of the settlement PPR made in the recent wrongful discharge suit brought against us by Camillo Vargas. The lawsuit was a costly, time-consuming, and frustrating process for the PPR management team. For future reference, please be advised that the following factors may NOT be used for cause in dismissing an employee:

- Union organizing
- Refusing to perform an unsafe task
- Race, religion, gender, national origin, culture or sexual preference
- Refusing to work in a hazardous environment
- Pregnancy, childbirth, or related medical problems
- Age
- "Whistle-blowing"
- Physical handicap
- Mental handicap

Figure 15–10
Factors that cannot be used for cause in terminating an employee.

- Have the employee sign any exit forms required by company policy.
- Explain the company's appeal process and refer the employee to the appropriate office should he or she wish to appeal.

Once the termination has been accomplished, the supervisor must inform other members of the team. This is an important step. Do not try to make a secret of a dismissal. Co-workers will be curious and morale may suffer if they are not informed. This puts the supervisor in a touchy position. On the one hand, you must protect the right of privacy of the terminated employee. On the other hand, you must give remaining team members enough information to protect morale.

An effective way to do this is to call a team meeting and announce the dismissal. However, rather than getting into details that might violate privacy, explain that the employee was terminated for continual violations and then only after an informal discussion, verbal warning, written warning, and suspension failed to correct the disruptive behavior.

Finally, check your documentation once again and make sure it is in order and readily available should the employee file a wrongful discharge suit. While you are at it, check the documentation (good and bad) maintained on all of your employees. To maintain documentation on just one employee might make termination of that employee appear to be a personal vendetta in the eyes of the court.

SUMMARY

1. Dealing with disruptive behavior to correct it is known as disciplining.
2. Failure to correct disruptive behavior can be damaging to morale, productivity, quality, and competitiveness. This is the basic rationale for disciplining.
3. The key to achieving self-discipline among team members is to personalize disruptive behavior, be consistent, and apply discipline in a step-by-step manner.
4. Some fundamental rules of thumb for applying discipline are: know your company's rules, regulations, and policy; know your employees; prevent discipline problems through education; and set a positive example.
5. More specific guidelines for disciplining include the following: understand your authority; communicate rules and regulations to employees; support the rules; follow the rules yourself; never act on hearsay; when rules are broken, act; discipline in private; discipline at the end of the day; document; apply the appropriate degree of punishment; do not hold grudges; give advance warnings; discipline immediately; be consistent; be professional; be objective.
6. There are five steps in the discipline process; informal discussion, verbal warning, written warning, suspension, and dismissal.
7. In the past employers could dismiss workers without cause or notice. This concept was known as employment-at-will.
8. The concept of wrongful discharge established that employees have rights relative to continued employment and can be protected from arbitrary and unjust termination.

9. If you believe a dismissal is in order, the termination interview should proceed as follows: set up a private interview; let the employee state his or her case; tell the employee he or she is being dismissed for cause and explain the cause; inform the employee in writing of the termination date; explain severance pay, insurance, vacation pay, and related issues; have the employee sign any required forms; and explain the company's appeal process.
10. When an employee is dismissed, call a team meeting and inform remaining team members of the decision without revealing specifics that might violate the employee's right to privacy.

KEY TERMS AND PHRASES

Authority
Cause
Company policy
Consistency
Credibility
Disciplining
Discipline process
Dismissal
Disruptive behavior
Documentation
Employment-at-will
Favoritism
Hearsay
Informal discussion
Objectivity
Positive example
Precedent
Prevention
Rules and regulations
Self-discipline
Suspension
Termination
Verbal warning
Written warning
Wrongful discharge

CASE STUDY: To Discipline or Discriminate?[7]

The following real-world case study contains a disciplinary problem of the type supervisors face on the job. Read the case study carefully and answer the accompanying questions.

Johnson Controls, Inc. is the largest manufacturer of automobile batteries in the United States with thirteen plants located in thirteen different states—a bona fide business success story. But in 1990 Johnson found itself embroiled in a far-reaching sex discrimination case. At issue was Johnson's policy of barring women from potentially hazardous jobs. The company did not allow women to hold positions that required working with lead, a principal ingredient in the manufacture of automobile batteries. Therefore women were denied promotional opportunities.

Johnson said its "fetal protection" policy was designed to protect unborn babies from lead poisoning. The policy was supported by the U.S. Department of Commerce and several business and religious groups. Opponents, who included a number of Johnson employees, the United Auto Workers Union, and the American Civil Liberties Union, claimed the policy was a flagrant case of sex discrimination. They claimed that since lead is dangerous to both men and women, women should not be singled out.

Johnson countered that it had spent over $15 million on improvements to make the work environment safe for adult men and women, but that working with lead was still dangerous to unborn children. To get around the policy and keep her job, one woman actually had herself sterilized, an act she later regretted. In March 1991 the U.S. Supreme Court found Johnson's policy to be discriminatory.

In the midst of the turmoil, supervisors at Johnson had to get the job done on a day-to-day basis.

1. Assume you are a supervisor at Johnson. An employee has missed three days of work (unauthorized) to march in a picket line protesting the company's fetal protection policy. How would you handle this situation?
2. Assume you are a supervisor at Johnson. You agree wholeheartedly with the case against the company. Would this change how you would handle the above situation? Explain your reasoning.
3. Assume you are a supervisor at Johnson. One of your team members is married to another Johnson employee who feels she has been denied a promotional opportunity because of the fetal protection policy. As a result his work has suffered measurably. An informal conference, a verbal warning, a written warning, and a three-day suspension without pay have not corrected his behavior. What would you do now?

REVIEW QUESTIONS

1. Define the term *discipline.*
2. What is the rationale for disciplining employees?
3. Briefly explain three strategies for promoting self-discipline.
4. Explain how to ensure there will be no surprises when applying discipline.
5. Why is it important for the supervisor to set a good example?
6. List and briefly explain what you feel are the ten most important guidelines for disciplining employees.
7. What are the five steps in the discipline process?
8. Explain the proper way to issue a verbal warning.
9. Explain the proper way to give a written warning.
10. What questions should you ask yourself before suspending an employee?
11. Explain the concepts of employment-at-will and wrongful discharge.
12. List five causes that might be given for terminating an employee.
13. List five reasons that *cannot* be used to justify dismissing an employee.
14. List and briefly explain the steps to follow in conducting a dismissal interview.
15. Explain how to properly inform remaining team members that one of them has been terminated.

SIMULATION ACTIVITIES

The following activities simulate the types of discipline problems supervisors often face. Apply what you have learned in this and previous chapters in solving these problems. This may be done individually or in groups.

1. In the middle of a team meeting you have called to announce the need for overtime work, an employee interrupts and says, "I don't care what you say, I'm not working any overtime!" Then he stomps out of the meeting. How would you handle this situation?
2. Your predecessor was lax in following and enforcing the company's rules and regulations. Some employees don't even know what they are. As a result, you have inherited a shoddy situation. How do you plan to instill self-discipline in the team?
3. You hear through the grapevine that an employee occasionally sleeps on the job. How do you handle this situation?
4. You have gone through every step in the discipline process with an employee and nothing has worked. She is a valuable employee, but her disruptive behavior is damaging team morale. How do you handle this situation?
5. Your predecessor left after subjecting the company to three separate wrongful discharge suits, all won by the terminated employees. You have been hired and given the charge to turn things around while ensuring that any future wrongful discharge suits are won by the company. Explain your plan for accomplishing this goal.

ENDNOTES

1. Bureau of Business Practice. *Discipline: A Step-by-Step Guide for the Supervisor* (Waterford, CT: Simon & Schuster, 1984), pp. 4–6.
2. Lussier, R. N. "16 Guidelines for Effective Discipline," *Supervisory Management,* March 1990, p. 10.
3. Bureau of Business Practice. *Discipline,* pp. 4, 5.
4. Ibid., pp. 13, 14.
5. Bureau of Business Practice. *Employment at Will: When You Can—And Cannot—Fire Employees* (Waterford, CT: Simon & Schuster, 1986), pp. 4, 5.
6. Bureau of Business Practice. *Discipline,* pp. 15, 16.
7. Cohen, S. "Women Irked Over Job or Child Choice." Associated Press, October 7, 1990.

PART FOUR

The Supervisor as an Innovator

CHAPTER SIXTEEN
Training and the Supervisor

CHAPTER OBJECTIVES

After studying this chapter you will be able to define or explain the following topics:

- Training
- The Need for Training
- How to Assess Training Needs
- How to Provide Training
- How to Evaluate Training
- The Use of Supervisors as Trainers
- Training the Supervisor
- Workforce Literacy Training
- How Training in the United States Compares with Training in Other Countries
- The Changing Role of Training in the United States

CHAPTER OUTLINE

- Training Defined
- Need for Training
- Assessing Training Needs
- Providing Training
- Evaluating Training
- The Supervisor as a Trainer
- Training the Supervisor
- Workforce Literacy and the Supervisor
- Training in the United States Compared with Training in Other Countries
- Changing Role of Corporate Training
- Summary
- Key Terms and Phrases
- Case Study Application Problem
- Review Questions
- Simulation Activities
- Endnotes

It has always been important to have well-trained, highly skilled employees. In the competitive age of advanced technology it is more important than ever. Evidence of the importance of training can be seen in the fact that American companies spend almost $50 billion each year providing training for employees. Two factors drive the unprecedented need for training: (1) international competition is becoming continually more intense; and (2) the quality of the workforce is not keeping pace. The modern supervisor must be proficient in assessing the need for training and for planning, arranging, providing, and evaluating training.

Tom Peters summarizes his philosophy about training in six brief statements, paraphrased as follows:[1]

- Companies must invest in human capital as much as in hardware.
- Companies must train entry-level people and retrain them as necessary.
- Companies must train all employees in problem-solving techniques so they can contribute to quality improvement.
- Companies must train extensively following promotions to the first managerial job and then train managers every time they advance.
- Companies must use training as a vehicle for instilling commitment to a strategic thrust.
- Companies must insist that training be need-driven from the bottom up.

TRAINING DEFINED

It is common to hear such terms as *training, education,* and *development* used interchangeably in discussions about enhancing the productivity of a company's workforce. All three concepts can contribute to productivity improvement but distinctions can be drawn among them.

In this book **training** is defined as follows:

> ***Training is an organized, systematic series of activities designed to enhance work-related knowledge and skills.***

Training has the characteristics of **specificity** and **immediacy.** This means that teaching and learning are geared toward the development of specific job-related knowledge and skills that have immediate application on the job.

Development typically describes activities aimed at enhancing the potential of management-level personnel. Developmental activities typically have less specificity and immediacy than training activities. For example, sending a manager back to school to complete an MBA program is a developmental activity.

Education typically describes teaching and learning that takes place in a more formal setting where students must meet some type of entrance requirements and teachers are required to have specific credentials, certification, or accreditation. Education is typically thought of as being more theoretical than training and having less specificity and immediacy.

The distinctions among these concepts are not black and white, however. For example, going back to school to complete a college degree in management and supervision (education) would be a developmental activity for a supervisor. Additionally, college- and university-based technical and business programs are always under pressure from employers to make their teaching more relevant, by which is meant more up-to-date, specific, and immediate (more like training). This chapter focuses on training for employees that is specific and immediate.

NEED FOR TRAINING

The need for training is the result of the necessity of industry to be productive and competitive. This is why U.S. industry spends almost $50 billion per year on the training, development, and education of employees at all levels. Several factors combine to intensify the need for training. Prominent among these are the following:

- Intensely competitive nature of the modern workplace
- Rapid and continual change
- Technology transfer problems
- Changing demographics

Unfortunately and in spite of the huge sums spent by industry on training each year, many employers still do not understand the role of training in productivity improvement. John Hoerr, writing in *Business Week,* paints a grim picture of the attitudes of U.S. employers toward training.[2] Hoerr cites a study conducted by the National Center on Education and the Economy which compares the education and training of workers in the United States with that of workers in competing countries.

This study concludes that less than 10 percent of U.S. companies plan to use a flexible approach requiring better trained workers as a way to improve productivity. This approach is standard practice in Japan, Germany, Denmark, and Sweden. In addition, less than 30 percent of U.S. firms have special training programs for women, minorities, and immigrants in spite of the fact that 85 percent of all new workers come from these groups. It is critical to the survival of U.S. industry that employers understand the need for training that results from intense international competition, rapid and continual change, technology transfer problems, and changing demographics.

In his book *Thriving on Chaos,* Tom Peters says, "Our investment in training is a national disgrace."[3] According to Peters, 69 percent of U.S. companies with 50 or more employees provide training for middle managers and 70 percent train executive-level personnel. However, only 30 percent train production personnel. Peters contrasts this with the Japanese and Germans who outspend U.S. firms markedly in providing training for skilled personnel.

According to Peters, in order for a company to succeed in a competitive global marketplace every employee must be prepared to contribute ideas for improving quality and productivity and to work under less supervision. The only way to accomplish this, says Peters, is through constant training.

Competition in the Marketplace

American industrial firms, even some of the smallest, find themselves competing in a global marketplace and the **competition** is intense. A small manufacturer of automotive parts in Michigan might find itself competing with companies in the United States as well as in Korea, Japan, Taiwan, and Europe. In order to win the competition, this small Michigan firm must make its products both better and less expensively than the competitors.

According to Philip Harkins,

> The Japanese, the West Germans, and others are battling fiercely for the lead technological and manufacturing positions. The international business world is being led by managers who are committed to increasing productivity and quality through advanced educational processes and methods. The result: the best-trained workforce will gain the long-term competitive advantage. No longer are we locked in a cold war. Today, the battle is for economic survival and dominance—the most powerful weapons on this new global business battlefield are the education and training of our workforce.[4]

Every employee of the automotive parts company mentioned above must be well trained, highly skilled, and more productive than similar employees in competing companies here and abroad. One of the key components in achieving maximum productivity levels is high-quality, ongoing training.

Rapid and Continual Change

Rapid and continual change is a fact of life in the modern workplace. Skills that were current yesterday may be obsolete today. Knowledge that is valid today may not be tomorrow. The only way working people can stay current is to undergo periodic training on a lifelong basis.

Technology Transfer Problems

Technology transfer is the movement of technology from one arena to another. There are two steps in the process. The first step is the *commercialization* of new technologies developed in research laboratories or by individual inventors. This is a business development issue and does not involve training.

However, the second step does. This step, known as technology diffusion, is training dependent. **Technology diffusion** is the process of moving newly commercialized technologies into the workplace where they can be used to enhance productivity and competitiveness.

This step breaks down unless the workers who must use the technology have been trained to use it efficiently and effectively. This is critical because new technologies by themselves do not enhance productivity. A word processing system given to an untrained traditional secretary will be nothing more than an expensive typewriter with a screen. In order to take maximum advantage of the capabilities of new technologies, workers must know how to use them effectively. Knowing how comes from training. Two of the major inhibitors of effective technology transfer are fear of change and lack of know-how. Both of these inhibitors can be overcome by training.

Changing Demographics

The demographics of the workforce in the United States are changing. The days when the workforce consisted primarily of Caucasian males are past. The typical worker of tomorrow will be a female, a minority, or an immigrant. During the 1990s approximately 85 percent of all new workers entering the labor force will come from these groups, which have their own special training needs. In order to compete, United States firms will have to provide for these needs.

ASSESSING TRAINING NEEDS

What knowledge, skills, and attitudes do our employees need to have in order to be more productive than the employees of our competitors? What knowledge, skills, and attitudes do our employees currently have? The difference between the answers to these questions identifies training needs (Figure 16–1).

Supervisors may become involved in **needs assessment** at both the unit level and the individual level. Assessing training needs at these two levels is not difficult. Supervisors work closely enough with their team members to see their capabilities firsthand. *Observation* is one method supervisors can use for assessing training needs.

A more structured way to assess training needs is to ask employees to state their needs in terms of their job knowledge and skills. Employees know the tasks they must perform every day. They also know which tasks they do well, which they do not perform well, and which they cannot do at all. A *brainstorming* session focusing on training needs is another method supervisors can use.

The most structured approach supervisors can use to assess training needs is the **job task analysis** survey. With this approach, a job is analyzed thoroughly and the knowledge, skills, and attitudes needed to perform it are recorded. Using this information, a survey instrument such as the one in Figure 16–2 is developed.

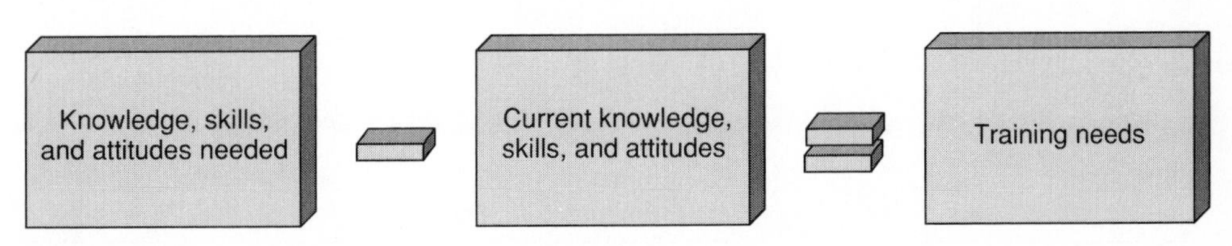

Figure 16–1
Identifying training needs.

Job Knowledge and Skills Survey

The following is a list of tasks that are performed by some people who work in the same job as yours, or who work in related manufacturing positions. In some cases, the task will be part of your job, while in other cases it will not. For each item on the list, *Please circle the number that indicates how frequently, if at all, you perform this task yourself.* If you do not understand the question indicate this by *circling # 5.*

	I perform this task				
	Never	Rarely, but only as a backup or in an emergency	Sometimes, as a part of my regular job	Frequently, as a part of my regular job	I don't understand this question
1. Use operating manuals, repair manuals or other written instructions.	1	2	3	4	5
2. Read blueprints.	1	2	3	4	5
3. Interpret geometric dimensions and tolerances (e.g., true positions, datums, flatness, circularity, or perpendicularity).	1	2	3	4	5
4. Read electrical schematics.	1	2	3	4	5
5. Use charts or graphs to convert from one measure to another (e.g., from inches to metrics or calculate speed and feed rates).	1	2	3	4	5
6. Convert measures manually.	1	2	3	4	5
7. Convert measures with a calculator.	1	2	3	4	5
8. Use manually operated processing machines (e.g., lathes, drill presses, sewing or fabric cutters).	1	2	3	4	5
9. Use computer numerically controlled processing equipment (e.g., CNC milling machine, brake presses, or fabric cutters).	1	2	3	4	5
If yes, how many types of equipment?	1	2	3	4	5 or more

Figure 16–2
Sample training needs assessment instrument.

Employees respond by indicating which skills they have and which they need. The survey results can be used in two ways: (1) to identify unitwide training needs; and (2) to identify individual training needs. Both types of needs must be converted into training objectives.

The following example demonstrates how supervisors might go about assessing an organization's training needs. Keltran, Inc. is a manufacturer of low voltage power supplies for military and civilian aircraft. Production workers at Keltran install components on printed circuit boards, wrap wires on transformers, and build up wire harnesses for cables. John Harris, production supervisor, is convinced that Keltran's production teams would benefit from specialized training aimed at improving quality and productivity.

After mulling over the issue for a few days, Harris decides to assess the training needs of Keltran's three production teams: Team A (printed circuit boards), Team B (transformers), and Team C (cables). Using what he learned from reading about needs assessment, Harris proceeds as follows:

1. Harris asks himself, "What are my most persistent production problems or bottlenecks and what could be causing them?" He decides that incorrect components installed on printed circuit boards and improperly matched wires in cables are high on his list of frequently occurring problems. In his opinion, these problems are caused by a lack of blueprint reading skills and an inability to adequately read technical manuals and specifications. He lists blueprint reading and general reading improvement as potential training needs.
2. Harris asks the lead employee in each team to undertake the same process he has completed and to do so individually without discussing the issue among themselves. The lead employees list their perceptions of the training needs of their team members and submit them to Harris.
3. Harris then calls a meeting of all production workers and explains that he is trying to assess their training needs and wants their input. After promising confidentiality, he distributes a form to each employee that has this statement typed across the top: "I could do my job better if training were provided for me in the following areas:". All employees are asked to complete the statement and drop their forms into a box.
4. Harris analyzes the input he has received from the lead employees in each team and from the employees themselves. Adding his own, Harris compiles a master list of training needs. Those needs listed most often are given the highest priority.

As a result of this process, Harris determines that the highest priority training needs of Keltran's production employees are:

- General reading improvement
- Interpretation of technical standards
- Proper use of specifications
- Blueprint reading

Sample Training Objectives

Upon completion of the training employees should be able to:

1. Add, subtract, multiply and divide whole numbers.
2. Add, subtract, multiply and divide common fractions.
3. Add, subtract, multiply, and divide decimal fractions.
4. Convert everyday shop problems into algebraic expressions.
5. Solve right triangles.
6. Solve nonright triangles.

Figure 16–3
Sample training objectives.

Writing Training Objectives

The first step in providing training for employees is to write **training objectives**. This responsibility will fall in whole or in part to the supervisor. Some companies will have training personnel who can assist, others will not. In either case, modern supervisors should be proficient in writing training objectives. The key to success lies in learning to be specific and to state objectives in behavioral terms. For example, say a supervisor has identified a need for training in the area of shop math. She might write the following training objective: *Employees will learn shop mathematics.*

This training objective, as stated, lacks specificity, nor is it stated in behavioral terms. Shop math is a broad concept. What does the training objective encompass? Arithmetic? Algebra? Geometry? Trigonometry? All of these? In order to gain specificity, this objective must be broken down into several objectives.

In order to be stated in behavioral terms, these more specific objectives must explain what the employee should be able to do after completing the training. **Behavioral objectives** contain action verbs. The sample training objectives in Figure 16–3 are stated in behavioral terms, they are specific, and they are measurable. The more clearly training objectives are written, the easier it is to plan training to meet them.

PROVIDING TRAINING

Nell P. Eurich, a trustee of the Carnegie Foundation, describes the following dimensions of corporate education and training:[5]

- In-house education programs are being used more frequently to meet the general education needs of industry and business.
- Corporations such as Motorola, Xerox, and RCA have established their own education and training facilities.
- Corporations are owning and operating their own degree-granting institutions, which are accredited by the same associations that accredit public colleges and

universities. For example, Wang offers a master's degree through its corporate university and Rand offers a Ph.D. degree. Many community colleges offer degree-granting partnership programs with industrial firms.

- Use of the Satellite University through which instruction is provided by satellite to local downlink sites is becoming common. The National Technological University (NTU) of Fort Collins, Colorado transmits instruction at levels through the master's degree to corporate classrooms nationwide.

According to Tom Peters, several U.S. firms learned the value of education and training and, as a result, are reaping the benefits.[6] Peters cites IBM, Nissan (in Tennessee), and Motorola as examples. At IBM, training immediately follows each promotion. All IBM employees must complete at least 40 hours of training each year. Today, as IBM faces intense pressure from foreign competitors, training is at the heart of its strategy for confronting the challenge.

Before Nissan opened its plant in Smyrna, Tennessee, it spent $63 million training about 2,000 employees. This amounts to approximately $30,000 per employee, which is about average for a Japanese company. Motorola stays competitive in the semiconductor business by investing 2.5 percent of its annual payroll in education and training. In addition to its employees, Motorola also trains its suppliers.

Nell Eurich of the Carnegie Foundation cites IBM and AT&T as leaders among U.S. firms in using education and training to enhance competitiveness.[7] According to Eurich, before divestiture in 1984 AT&T spent $1.7 million for education and training. This investment produced 12,000 courses in 1,300 different locations and required 13,000 trainers and training support workers. In 1984, IBM invested $700 million in education and training.

There are several different ways for supervisors to provide training (Figure 16–4). All fall into one of the following three broad categories: *internal approaches, external approaches,* and *partnership approaches.* Regardless of the approach used, supervisors will want to maximize their training resources. Carolyn Wilson recommends five strategies for doing this:[8]

1. *Build in quality from the start.* Take the time to do it right from the outset.
2. *Design small.* Do not try to develop courses that are all things to all people. Develop specific activities around specific objectives.
3. *Think creatively.* Do not assume that the traditional classroom approach is automatically best. Videotapes, interactive video, or one-on-one peer training may be more effective.
4. *Shop around.* Before purchasing training services, conduct a thorough analysis of specific job training objectives. Decide exactly what you want and make sure the company you plan to deal with can provide it.
5. *Preview and customize.* Never buy a training product (e.g., videotape, self-paced manual) without previewing it. If you can save by customizing a generic product, do so.

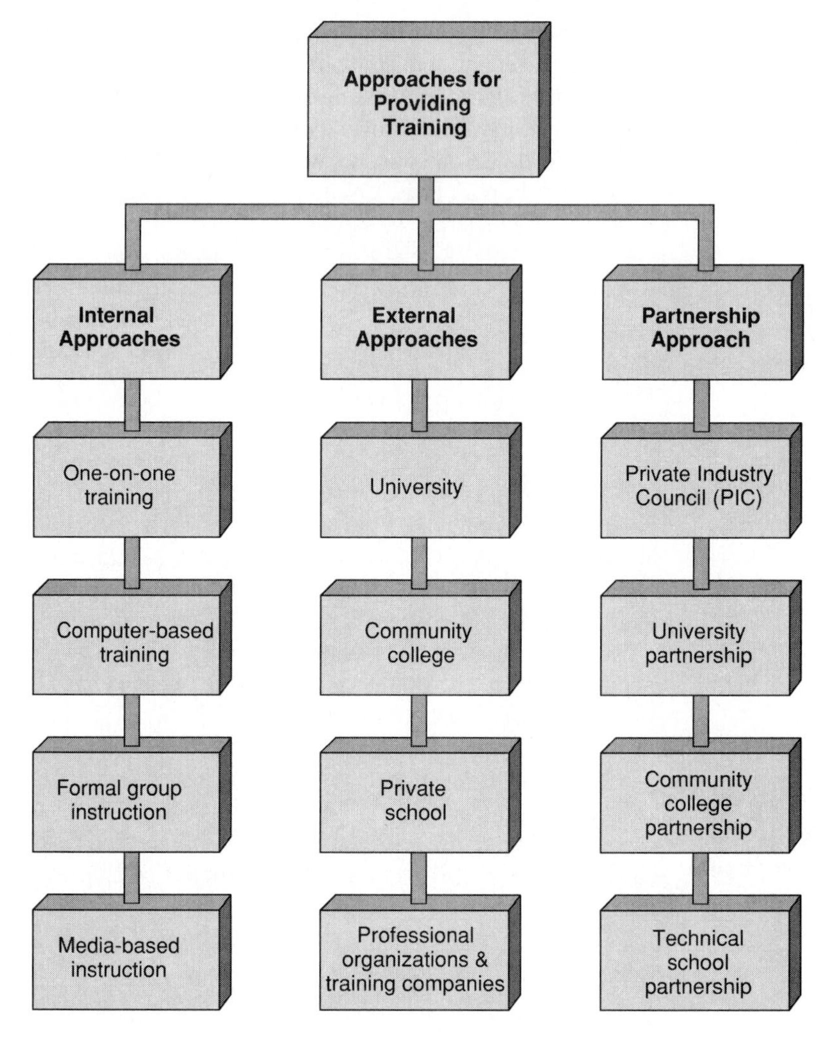

Figure 16–4
There are a variety of approaches for providing training.

Internal Approaches

Internal approaches provide training on-site in the organization's facilities. These approaches include one-on-one training, computer-based training, and media-based instruction. **One-on-one training** involves placing a less skilled employee under the instruction of a more skilled employee. This approach is often used when a new employee is hired. It is also an effective way to prepare a replacement for a high-value employee who plans to leave or retire.

Computer-based training (CBT) has proven to be an effective internal approach. Over the years it has been continually improved and is now a widely used training method. It offers the advantages of being self-paced, individualized, and able to provide immediate and continual **feedback** to learners. Its best application is in developing general knowledge rather than in developing company-specific job skills.

Formal group instruction in which a number of people who share a common training need are taught together is a widely used method. This approach might involve lectures, demonstrations, multimedia use, hands-on learning, question/answer sessions, role playing, and simulation.

Media-based instruction has become a widely used internal approach. Private training companies and major publishing houses produce an almost endless list of turnkey media-based training programs. The simplest might consist of a set of audio tapes. A more comprehensive package might include videotapes and workbooks. Interactive laser disk training packages that combine computer, video, and laser disk technology are also a popular internal approach.

An example of an extensive internal training program is Motorola University, the in-house educational institution operated by Motorola, Inc. Motorola University consists of institutes for manufacturing and engineering workers, middle managers, and senior managers and an instructional design center. Between 1981 and 1989 more than 54,000 Motorola employees completed training in Motorola University. The company's goal is to have employees spend a minimum of 2 percent per year of their time in training.

External Approaches

External approaches are those that involve enrolling employees in programs or activities provided by public institutions, private institutions, professional organizations, and private training companies. The two most widely used approaches are (1) enrolling employees in short-term training (a few hours to a few weeks) during work hours; and (2) enrolling employees in short- or long-term training and paying all or part of the costs (i.e., tuition, books, fees). External approaches encompass training methods ranging from seminars to college courses.

Partnership Approaches

In recent years, public community colleges, universities, and technical schools have begun to actively pursue partnerships with employers through which they provide customized training. Those **training partnerships** combine some of the characteristics of the previous two approaches.

Customized on-site training provided cooperatively by community colleges and private companies or associations have become very common. According to Nell Eurich of the Carnegie Foundation, the 1,274 community colleges have built extensive networks of alliances with business and industry.[9] Eurich cites the example of General Motors Corporation, which has contracts with 45 community colleges throughout the country to train service technicians. General Motors contributes to the partnership by training the instructors.

According to Eurich, between 1980 and 1982 the American Association of Community and Junior Colleges and the Small Business Training Network brought together 86 colleges and local offices of the U.S. Small Business Administration. These partnerships produced 2 million hours of short-term training in 47 states.

Tom Peters cites the example of a partnership between Amdahl, a California computer manufacturer, and DeAnza Junior College.[10] Amdahl and DeAnza jointly offer courses in quality assurance, materials in process, production, inventory control, management principles, and accounting. Employees who complete these courses receive up to 41 credits toward an associate degree.

Many universities, community colleges, and technical schools have continuing education or corporate training divisions that specialize in providing training for business and industry. Supervisors should know the adminstrator responsible for training partnerships at colleges, universities, or technical schools in their communities.

Partnerships with institutions of higher education offer several advantages to supervisors who need to arrange training for their employees. Representatives of these institutions are education and training professionals. They know how to transform training objectives into customized curricula, courses, and lessons; how to deliver instruction; how to design application activities that simulate real-world conditions; and how to develop a valid and reliable system of evaluation and use the results to chart progress and prescribe remedial activities when necessary. They have access to a wide range of instructional support systems (i.e., libraries, media centers, and instructional design centers).

In addition to professional know-how, institutions of higher education have resources that can markedly reduce the cost of training for an organization. Tuition costs for continuing education activities are typically much less than those associated with traditional college degree coursework. If these institutions do not have faculty members on staff who are qualified to provide instruction in a given area, they can usually hire a temporary or part-time instructor who is qualified.

Other advantages institutions of higher education can offer in the training arena are credibility, formalization, standardization, and flexibility in training locations. Employers sometimes find their attempts at customized training hampered because their employees have been conditioned to expect formal grade reports, transcripts, and certificates of completion. These things tend to formalize training in the minds of employees and can make it more real for them. Associating with a community college, university, or technical school can formalize a company's training program and give it credibility.

Another problem employers sometimes experience when providing their own customized training is a lack of standardization. The same training provided in three different divisions might produce markedly different results. Professional educators can help standardize the curriculum and evaluation systems. They can also help standardize instruction by providing **train-the-trainer workshops** for employees who are serving as in-house instructors.

Another potential partner for supervisors who need to provide training for employees is the local **Private Industry Council** or PIC. The PIC is the local implementing agency for the federal Job Training Partnership Act (JTPA). The purpose of the JTPA is to make training available to economically disadvantaged people through formal educa-

tion or on-the-job training. It can be an excellent resource for companies that hire women, minorities, and immigrants.

The PIC can help supervisors decrease the cost of training new employees; it cannot be used for updating employees who are already on the payroll. The PIC targets economically disadvantaged, unemployed people and matches them with employers who have job openings. The PIC will pay a percentage of the employee's salary for a contracted period of time if the employer provides **on-the-job training** (the employee learns on the job from a fellow worker). Since many new employees require on-the-job training anyway, supervisors can decrease their training costs by filling vacant positions with PIC clients. Your local PIC can be found in the telephone directory under Private Industry Council.

Regardless of the approach used in providing training, there is a widely accepted rule of thumb with which supervisors should be familiar: *People learn best when the learning approach involves them in seeing, hearing, speaking, and doing.*

Educational practitioners hold that the following percentages apply regarding what learners remember and retain:

10 percent of what is read

20 percent of what is heard

30 percent of what is seen

50 percent of what is seen and heard

70 percent of what is seen and spoken

90 percent of what is said while doing what is being talked about

Clearly, in order for learning to be effective it must involve activity on the part of learners, must be interactive in nature, and must involve reading, hearing, seeing, talking, and doing.

EVALUATING TRAINING

Did the training provided satisfy the training objectives? This can be a difficult question to answer.

Evaluating training requires that supervisors begin with a clear statement of purpose. What is the overall purpose of the training? This broad purpose should not be confused with training objectives. The objectives translate this purpose into more specific, measurable terms.

The purpose of training is to improve the individual productivity of employees and the overall productivity of the organization so that the organization becomes more competitive. In other words, *the purpose of training is to improve performance.*

In order to know if training has improved performance, supervisors need to know: (1) Was the training provided valid? (2) Did the employees learn? (3) Has the learning

Organization of Course

Teacher__________________
Course# __________ Sec# ______
Course title ________________
Date: Month______ Day______Year______

Instructions: On a scale from 1 to 5 (1 = highest rating to 5 = lowest rating),rank your teacher on each item. Leave blank any item which does not apply.

1. Objectives-- Clear to Unclear
2. Requirements-- Challenging to Unchallenging
3. Assignments-- Useful to Not Useful
4. Materials-- Excellent to Poor
5. Testing Procedures-- Effective to Ineffective
6. Grading Practice--Explained to Unexplained
7. Student Work Returned-- Promptly to Delayed
8. Overall Organization-- Outstanding to Poor

Comments:

Teaching Skills

9. Class Meetings-- Productive to Nonproductive
10. Lectures-- Effective to Ineffective
11. Discussions-- Balanced to Unbalanced
12. Class Proceedings-- To-the-Point/Wandering
13. Provides Feedback-- Beneficial/Not Beneficial
14. Responds to Students--Positively/Negatively
15. Provides Assistance-- Always to Never
16. Overall Rating of Instructor's Teaching Skills-- Outstanding to Poor

Comments:

Substantive Value of Course

17. The course was-- Intellectually Challenging/Too Elementary
18. The instructor's command of the subject was-- Broad and Accurate/Plainly Defective
19. Overall substantive value of the course-- Outstanding to Poor
20. What grade do you expect from this class?

Comments:

Figure 16–5
Form for evaluating instruction.

made a difference? Valid training is training that is consistent with the training objectives.

Evaluating training for **validity** is a two-step process. The first step involves comparing the written documentation for the training (i.e., course outline, lesson plans, curriculum framework, etc.) with the training objectives. If the training is valid in design and content, the written documentation will match the training objectives. The second step involves determining if the actual training provided is consistent with the documentation. Training that strays from the approved plan will not be valid. Student evaluations of instruction conducted immediately after completion can provide information on consistency and the quality of instruction. Figure 16–5 is an example of an instrument for evaluating instruction.

Determining if employees have learned is a matter of building evaluation into the training. If the training is valid and employees have learned, the training should make a difference in their performance, which should improve. This means quality and productivity should be enhanced. Supervisors can make determinations about performance using the same indicators that told them training was needed in the first place.

"Can employees perform tasks they could not perform before the training? Is waste reduced? Has quality improved? Is setup time down? Is in-process time down? Is the on-time rate up? Is the production rate up? Is the throughput time down?" These are the types of questions supervisors can ask to determine if training has improved performance. Gilda Dangot-Simpkin of Dynamic Development suggests a checklist of questions for evaluating purchased training programs:[11]

- Does the program have specific behavioral objectives?
- Is there a logical sequence for the program?
- Is the training relevant for the trainees?
- Does the program allow trainees to apply the training?
- Does the program accommodate different levels of expertise?
- Does the training include activities that appeal to a variety of learning styles?
- Is the philosophy of the program consistent with that of the organization?
- Is the trainer credible?
- Does the program provide follow-up activities to maintain the training on the job?

THE SUPERVISOR AS A TRAINER

It is not unusual for the modern supervisor to serve as an instructor in a company-sponsored training program. In fact, this is becoming so common that it is important for the supervisor to have at least basic instructional skills. Supervisors need to understand the basic principles of learning (Figure 16–6) and the basic four-step teaching approach (Figure 16–7).

Basic Principles of Learning

1. People learn best when they are ready to learn.
2. People learn more easily when what they are learning can be related to something they already know.
3. People learn best in a step-by-step manner.
4. People learn by doing.
5. The more often people use what they are learning, the better they will remember and understand it.
6. Success in learning tends to stimulate additional learning.
7. People need immediate and continual feedback to know if they have learned.

Figure 16–6
Basic principles of learning.

Principles of Learning

The **principles of learning** summarize what is known and widely accepted about how people learn. Trainers can do a better job of facilitating learning if they understand the following principles:

- *People learn best when they are ready to learn.* You cannot *make* employees learn anything; you can only make them *want to learn.* Therefore, time spent motivating employees to want to learn will be time well spent. Explain why they need to learn and how they will benefit personally from learning.
- *People learn more easily when what they are learning can be related to something they already know.* Build today's learning on what was learned yesterday and tomorrow's learning on what was learned today. Begin each new learning activity with a brief review of the activity that preceded it.
- *People learn best in a step-by-step manner.* An extension of the preceding principle, this means that learning should be organized into logically sequenced steps that proceed from the concrete to the abstract.
- *People learn by doing.* This is probably the most important principle for trainers to understand. Inexperienced trainers tend to confuse talking (i.e., lecturing or demonstrating) with teaching. These things can be part of the teaching process, but they do

Figure 16–7
The four-step teaching approach.

Four-Step Teaching Approach

1. Preparation
2. Presentation
3. Application
4. Evaluation

little good unless they are followed up with application activities that require the learner to *do* something. To illustrate this point, consider the example of teaching employees how to roller-skate. You might present a thorough lecture on the principles of roller-skating and give a comprehensive demonstration on how to do it. However, until the employees put on the skates and begin taking those first tentative steps they will not learn how to roller-skate. They learn by doing.

- *The more often people use what they are learning, the better they will remember and understand it.* How many things have you learned in your life that you can no longer remember? People forget what they do not use. Trainers should keep this principle in mind. It means that repetition and application should be built into the learning process.
- *Success in learning tends to stimulate additional learning.* This principle is a restatement of a principle of management that success breeds success. Organize learning into short enough segments to allow learners to see progress, but not so short that they become bored.
- *People need immediate and continual feedback to know if they have learned.* Did you ever take a test and get the results back a week later? That was probably a week later than you wanted them. People in a learning setting want to know immediately how they are doing. Feedback can be as simple as a nod, a pat on the back, or a comment such as "Good job!" It can also be more formal, such as a progress report or a graded paper. Regardless of the form it takes, trainers should concentrate on giving continual feedback.

Four-Step Teaching Approach

Regardless of the setting, teaching is a matter of helping people learn. One of the most effective approaches for facilitating learning is not new, innovative, gimmicky, or high tech in nature. It is known as the **four-step teaching approach** and it is an effective approach to use in a corporate training setting. The four steps are explained in the following paragraphs.

- **Preparation** encompasses all tasks necessary to get students prepared to learn, trainers prepared to teach, and facilities prepared to accommodate the process. Preparing students means motivating them to want to learn. Personal preparation involves planning lessons on preparing all the necessary instructional materials. Preparing the facility involves arranging the room for both function and comfort, checking all equipment to ensure it works properly, and making sure that all tools and other training aids are in place.
- **Presentation** is a matter of presenting the material students are to learn. It might involve giving a demonstration, presenting a lecture, conducting a question/answer session, helping students work with a computer or interactive video disk system, or assisting students who are proceeding through self-paced materials. Regardless of the format, certain rules of thumb apply. Victor Parachin recommends ten strategies for giving an effective presentation:[12]

Begin dramatically.

Be brief.

Be organized.

Use humor.

Keep it simple.

Take charge.

Be sincere.

Consider conditions.

Tell stories.

- **Application** refers to arranging for learners to use what they are learning. Application might range from simulation activities in which learners role play to actual hands-on activities in which learners use their new skills in a live format.
- **Evaluation** means determining the extent to which learning has taken place. In a training setting, evaluation does not need to be a complicated process. If the training objectives are written in measurable, observable terms evaluation is simple. Employees are supposed to learn how to do *X, Y,* and *Z*. Have them do *X, Y,* and *Z* and observe the results. In other words, have employees **demonstrate proficiency** and observe the results.

TRAINING THE SUPERVISOR

In addition to ensuring that their employees are trained, supervisors should be attentive to their own training needs. Training for supervisors is more important in the current technology age than ever. It should be comprehensive and continual.

Supervisory training can be categorized into five broad areas:

- Leadership skills
- Management skills
- Counseling skills
- Innovation skills
- Legal concerns

Supervisors must be leaders. This means they must develop communication and motivation skills. They must also understand the role ethics plays in leadership. These things can be learned.

Management training for supervisors should focus on developing, planning, organizing, staffing, controlling, and decision-making skills. The counseling aspects of a supervisor's training should focus on how to conduct effective performance appraisals, how to properly handle problem employees and employee complaints, and how to discipline properly.

Innovation training for supervisors should focus on helping them learn how to arrange, develop, present, and evaluate training; improve productivity; and enhance

quality. Finally, supervisory training should focus on the legal concerns of the job including equal opportunity, health, safety, and labor relations.

WORKFORCE LITERACY AND THE SUPERVISOR

In recent years industry has been forced to face a tragic and potentially devastating problem: **adult illiteracy**, which is having a major impact on the productivity of industry in the United States. It is estimated that over 60 million people or approximately one-third of the adult population in this country are marginally to functionally illiterate.[13]

People are sometimes shocked to learn that the number of illiterate adults is so high. Ernest Fields and colleagues list several reasons why this number has been obscured in the past and why there is now a growing awareness of the adult illiteracy problem:[14]

1. Traditionally the number of **low-skill jobs** available has been sufficient to accommodate the number of illiterate adults.
2. Faulty research methods for collecting data on illiterate adults have obscured the reality of the situation.
3. Reticence on the part of illiterate adults to admit they have a problem and to seek help has further obscured the facts.

The reasons we are now becoming more aware of the adult illiteracy problem, according to Fields, are as follows:

1. Basic skill requirements are being increased by technological advances and the need to compete in the international marketplace.
2. Broader definitions of literacy that go beyond just reading and writing also include speaking, listening, and mathematics.
3. Old views of what constitutes literacy no longer apply.

Impact of Illiteracy on Industry

The basic skills necessary to be productive in a modern industrial setting are increasing steadily. At the same time the national high school dropout rate continues to increase as does the number of high school graduates who are functionally illiterate in spite of their diploma. This means that while the number of **high-skilled jobs** in modern industry is increasing, the number of people able to fill them is on the decline. The impact this will have on industry in the United States can be summarized as follows:

1. Difficulty in filling high-skill jobs.
2. Lower levels of productivity and, as a result, a lower level of competitiveness.
3. Higher levels of waste.
4. Higher potential for damage to sophisticated technological systems.
5. Greater number of dissatisfied employees in the workplace.

What Industry Can Do

Industry in the United States has found it necessary to confront the illiteracy problem head on. This is being done by providing remedial education for employees in the workplace. Some companies contract with private training firms, others provide the education themselves, and others form partnerships with public community colleges or vocational schools.

In 1987 the National Center for Research in Vocational Education conducted a study of industry-based adult literacy training programs in several industrial firms.[15] Figure 16–8 summarizes several key points from this study. Examining the three examples in this figure can be instructive for supervisors.

Texas Instruments requires math, verbal and written communication, and basic physics of its employees. Physics skills have not traditionally been viewed as required for functional literacy. However, in order to function in this modern high tech company, employees must have these skills. The approach used by Texas Instruments to provide literacy training can be one of the least expensive. By working in conjunction with public community colleges or vocational schools, companies can provide literacy training at little and, in some cases, no cost. Modern industrial supervisors should establish a close working relationship with representatives of local community colleges and vocational schools.

Company	Skills Taught	How Training is Provided
Texas Instruments	■ Math ■ Verbal communication ■ Written communication ■ Basic physics	State-certified teachers provided by a community college
Rockwell International	■ Math ■ Algebra ■ General science ■ Physics and chemistry	Company-employed state-certified teachers
Polaroid	■ Physics ■ Algebra ■ Geometry ■ Trigonometry ■ Statistics ■ Computer literacy ■ Communication ■ Problem solving	Company-hired private consultants who were qualified in adult education but not necessarily certified

Figure 16–8
Training provided by major industrial firms.

Rockwell International also defines literacy more stringently than has been typical in the past. In order to function effectively at Rockwell International, employees must be skilled in chemistry and physics. This is another example that is indicative of the need for higher skill levels in order to be functionally literate. Notice in Figure 16–8 that Rockwell hired its own certified teachers.

Of the three companies identified in Figure 16–8, Polaroid takes the most aggressive approach in defining functional literacy. The skills taught in Polaroid's program are also indicative of the trend, especially in the areas of statistics, problem solving, and computer literacy.

The need for these skills is technology driven. Knowledge of statistics is needed in order to use statistical process control (SPC), which is becoming a widely used quality control methodology in automated manufacturing settings. Few employees in a modern industrial firm get by without using a computer on the job, hence the need for computer literacy training.

Problem-solving skills are becoming critical as companies implement total quality management (TQM) programs. Such programs involve all employees in identifying and correcting problems that have a negative impact on quality or do not add value to the company's products.

What Supervisors Should Know About Literacy Training

In their study of industry-based literacy training programs, Fields and colleagues made recommendations to assist in the planning and implementation of such programs.[16] Several of these recommendations have relevance for supervisors:

1. The definition of literacy should be driven by the needs of the company.
2. Companies should establish an environment in which employees feel comfortable having their literacy skills assessed and in seeking help to raise those skills.
3. Whenever possible companies should establish programs to raise the skills of existing employees rather than laying them off and hiring new employees.
4. Whenever possible companies should collaborate with educational institutions or education professionals in providing literacy training.

As industrial companies continue to enhance their technological capabilities, the skill levels of the workforce will have to increase correspondingly. Because the number of people in the labor force who have high-level skills is not increasing, the need for workforce literacy training will be a fact of life supervisors will have to deal with for some time to come.

Figure 16–9 is a copy of an education and training agreement between a manufacturing firm and a community college. This agreement was developed on the basis of input provided by the company's various departmental supervisors working in conjunction with continuing education personnel of the college. Notice the assessment and literacy components of the partnership agreement.

Ver-Val Enterprises, Inc./Okaloosa-Walton Community College
Education and Training Agreement
Partnership

The following agreement formalizes the partnership between Ver-Val Enterprises (VVE) and Okaloosa-Walton Community College (OWCC) for the provision of customized training to be jointly provided in the facilities of VVE's main office on Hill Avenue in Fort Walton Beach Industrial Park, Florida. The partnership has the following three main components: *assessment, instruction,* and *evaluation.*

Assessment
All employees will be assessed using the Test of Adult Basic Education (TABE) prior to beginning instruction. Based on the TABE scores and the needs of the employee's work assignment a Personal Development Plan will be established which sets forth all training to be undertaken by the employee.

Instruction
Depending on the needs of the individual employee and VVE, instruction will be provided by OWCC in the following areas: adult basic education (literacy), shop math, blueprint reading, computer numerical control (CNC) programming, computer literacy, desktop publishing, and industrial supervision.

Evaluation
Employees will be evaluated in each course attempted and the results will be placed in their personnel files. Performance evaluations by supervisors will be conducted periodically to determine if an employee's training is having the desired effect.

______________________	______________________
President—OWCC	Date
President—VVE	Date

Figure 16–9
Sample training agreement.

TRAINING IN THE UNITED STATES COMPARED WITH TRAINING IN OTHER COUNTRIES[17]

The Commission on the Skills of the American Workforce, cochaired by Ray Marshall and William E. Brock—both of whom have served as U.S. Secretary of Labor—issued a report in June 1990 entitled "America's Choice: High Skills or Low Wages." This report concluded that U.S. businesses place an alarmingly low priority on the skills of their employees. A survey of over 400 companies yielded the following results:

- Less than 10 percent of the companies planned to increase productivity by reorganizing work in ways that call for employees with broader skills.
- Only 15 percent expressed concern over the potential for shortages of skilled workers.
- Less than 30 percent intended to offer special training programs for women and minorities although these groups account for 85 percent of all new workers.
- Over 80 percent were more concerned about workers' attitudes than their skills.

Based on their findings, the commission concluded that business was not taking a proactive role in demanding the improvements that are so badly needed in American education and training. Contrasting what is found in the United States with other industrialized countries, the commission also concluded that education and training for non–college-bound students is much better in such countries as Germany, Japan, Sweden, and Denmark. Consequently, American students rank near the bottom in indicators of school performance when compared with students from these countries. The commission described the U.S. system for transitioning students from school to work as "the worst of any industrialized country." Many feel the commission's findings are exaggerated. However, clearly there is a problem.

Other industrialized countries are creating what the commission described as a "high-performance workplace" by reorganizing work in ways that call for multiple skills and high levels of reading, math, science, and problem-solving skills. As a result, workers are better able to continually develop new skills as technology changes. The ability of such workers to adapt quickly and continually also allows their employers to introduce new products on shorter cycle times and make more frequent changes in production runs. By contrast, American industry is still attempting to use automation to accomplish what has come to be known as "dumbing down the workplace."

Clearly, the attitudes of U.S. business and industry toward training must change if America is going to have a world-class workforce. Supervisors can play a key role in bringing this about by communicating this message to higher management and by helping carry out the decision to build a world-class workforce.

CHANGING ROLE OF CORPORATE TRAINING

Corporate training in the United States is over 100 years old. However, it got its modern-day start near the end of World War II with the establishment of the American Society for Training and Development (ASTD). According to Dick Schaaf, the ASTD grew out of "wartime manufacturing businesses that placed a premium on rapidly bringing new industrial workers up to speed on large-scale processes and machines that made things."[18]

Since the mid-1940s, interest in corporate training has grown steadily so that today,

> those that employ more than 100 workers spend some $45 billion a year on developing their human resources, and training directors command comfortably middle-management-level salaries. The field has professionalized, become specialized, developed its own alphabet soup of acronyms and special interest groups, not to mention a handful of formulas to calculate training's organizational role and clout.[19]

Schaaf created a matrix that summarizes the past, present, and current roles of corporate training in terms of such factors as mission, focus, and imperative. For example, Schaaf claims that the mission of corporate training has evolved from a focus on developing skills and discipline to a new focus on improving values and motivation. The future focuses, according to Schaaf, will be on enhancing service and quality.[20]

The author's view is that the mission of corporate training for the present and in the future is to continually improve productivity, quality, and competitiveness. Such a mission encompasses the past, present, and future missions set forth by Schaaf. In other words, the evolving role of corporate training is to develop skills, knowledge, discipline, values, motivation, and anything else necessary to continually improve productivity, quality, and competitiveness. A shorter way of stating the same mission is that the role

Figure 16–10
Purpose of modern corporate training.

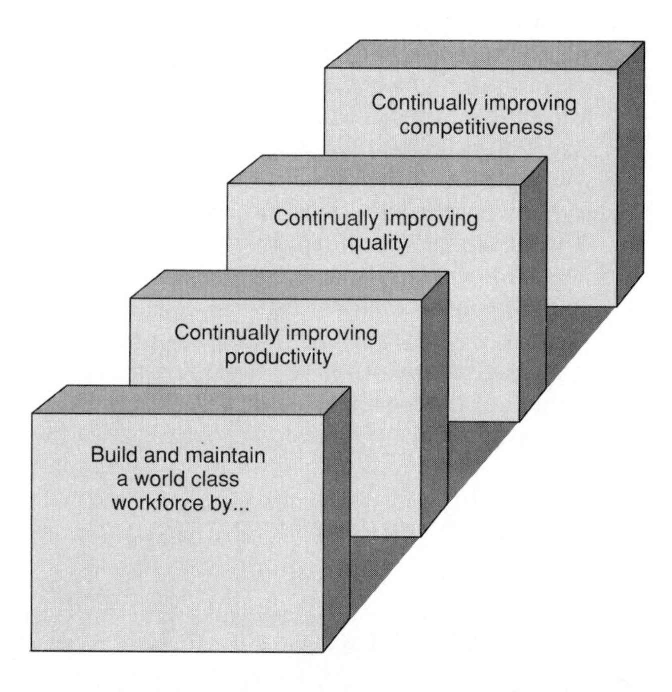

of corporate development is to build and maintain a world-class workforce (Figure 16–10).

SUMMARY

1. Two factors drive the demand for industry-based training programs: (1) the need to compete in an intensely competitive international arena; and (2) the skill levels of the general labor force are not keeping pace with technological advances.
2. Training can be defined as an organized, systematic series of activities designed to enhance work-related knowledge and skills. Training has the characteristics of specificity and immediacy.
3. The demand for training grows out of a company's need to be productive and, as a result, competitive. Factors that continue to intensify the need for training are: (1) increasing level of competition in the modern marketplace; (2) rapid and continual change; and (3) technology transfer problems.
4. Once training needs have been identified, training objectives are written. A training objective should be specific and stated in behavioral terms that can be readily measured.
5. Training can be provided internally using company personnel as instructors; externally by sending employees to workshops, seminars, or educational institutions; or through partnerships with public or private educational institutions or agencies.
6. Since the overall purpose in providing training is to improve performance, it is important to evaluate training. This is a two-step process: (1) evaluate to determine if employees learned as a result of the training; and (2) evaluate to determine if what employees learned actually makes a difference in performance.
7. The modern supervisor may be called on occasionally to serve as an instructor in a company-sponsored training program. In these cases supervisors should understand the basic principles of learning and the four-step teaching approach.
8. Approximately one-third of the adult labor force is marginally to functionally illiterate. As a result, workforce literacy training is becoming a necessary part of doing business. The reasons we are becoming more aware of the adult illiteracy problem are: (1) basic skill requirements are being increased by technological advances; (2) broader definitions of what constitutes literacy are coming into play; and (3) a realization that what used to constitute literacy no longer applies is beginning to sink in.

KEY TERMS AND PHRASES

Adult illiteracy
Application
Behavioral objectives
Competition
Computer-based training
Demonstrate proficiency
Development
Education
Evaluation
Feedback
Four-step teaching approach
High-skill jobs

Immediacy
Job task analysis
Low-skill jobs
Media-based instruction
Needs assessment
One-on-one training
On-the-job training
Preparation
Presentation
Principles of learning
Private Industry Council
Specificity
Technology transfer
Technology diffusion
Train-the-trainer workshop
Training
Training objectives
Training partnerships
Validity

CASE STUDY: A Commitment to Training[21]

The following real-world case study contains an example of workplace training. Read the case study carefully and answer the accompanying questions.

Motorola, Inc. has made ongoing training for employees one of its highest priorities and the foundation of its quality and competitiveness effort. The company spends more than $50 million annually on the education and training of employees at all levels. A corporate commitment at Motorola is that at least 2 percent of the time of all employees will be spent in training each year.

The cornerstone of Motorola's training program is Motorola University. This in-house educational institution consists of an instructional design center and three institutes: manufacturing and engineering, middle management, and senior management. Training focuses on achieving a 99.99 percent perfect quality rate in all of these areas. All training provided is evaluated at four levels: end of course and instructor evaluation; follow-up telephone survey to determine the extent to which newly learned skills are being used; assessment of the impact of training on the job; and assessment of the impact of training on the bottom line.

1. Assume you are a supervisor at Motorola, Inc. You have been asked to assess the training needs of your unit. How would you proceed?
2. As a supervisor you suspect that employees in your unit may not be fully literate. How would you proceed to (1) clearly determine if a problem exists; and (2) solve the problem?
3. As a supervisor you have identified specific training needs for your unit. However, higher management has rejected the projected training costs. How can you decrease the costs and still provide the training?

REVIEW QUESTIONS

1. What is driving the acute need for training in modern industry?
2. Define the term *training.*
3. Distinguish between education and training.
4. Why does industry in the United States spend about $50 billion each year on education, training, and development?
5. Explain the impact of international competition relative to the training needs of industry.
6. Explain the term *technology transfer* and the impact it has on training needs.
7. What is technology diffusion and how does it relate to training needs?
8. Explain the most structured approach for assessing training needs.
9. Write a sample training objective that can be readily measured.
10. Explain briefly the following approaches to providing training: internal; external; partnership.
11. What is the Private Industry Council and how can it help decrease a company's training costs?
12. Explain the two steps in determining the validity of training.
13. List four principles of learning.
14. Explain the four-step teaching approach.
15. Why has the adult literacy problem been obscured in the past?
16. Why are we now becoming more aware of the adult illiteracy problem?
17. Summarize the impact of adult illiteracy on industry in the United States.

SIMULATION ACTIVITIES

The following activities simulate the types of training-related problems supervisors often face. Apply what you have learned in this and previous chapters in solving these problems. This may be done individually or in groups.

1. Assume you are a supervisor for a manufacturer of consumer electronics products. You suspect your unit is in need of training. Explain how you plan to go about assessing the training needs.
2. Write a set of training objectives for a work area with which you are familiar.
3. Explain how you plan to maximize your limited training resources.
4. Your company has contracted with a private firm to provide training for all production personnel. Explain how you plan to evaluate the effectiveness of the training.
5. You have been asked to serve on a committee that will plan and implement literacy training for your company. What guidelines do you intend to apply and recommend that the committee apply?

ENDNOTES

1. Peters, T. *Thriving on Chaos* (New York: Harper & Row, 1987), p. 386.
2. Hoerr, J. "Business Shares the Blame for Workers' Low Skills," *Business Week,* June 15, 1990, p. 71.
3. Peters, T. *Thriving on Chaos,* pp. 388, 389.
4. Harkins, P. J. "The Changing Role of Corporate Training and Development," *Corporate Development in the 1990s* (A Supplement to *Training*), 1991, p. 27.
5. Eurich, N. P. *Corporate Classrooms: The Learning Business* (Lawrenceville, NJ: Princeton University Press, 1985), pp. x, xi.
6. Peters, T. *Thriving on Chaos,* pp. 389, 390.
7. Eurich, N. P. *Corporate Classrooms,* p. 8.
8. Wilson, C. *Training for Non-Trainers* (New York: AMACOM, 1990), pp. 18–19.
9. Eurich, N. P. *Corporate Classrooms,* p. 16.
10. Peters, T. *Thriving on Chaos,* pp. 390, 391.
11. Dangot-Simpkin, G. "How to Get What You Pay For," *Training,* July–August 1990, pp. 53, 54.
12. Parachin, V. "10 Tips for Powerful Presentations," *Training,* July–August 1990, pp. 71–83.
13. Fields, E. L., Hull, W. L., and Sechler, J. A. *Adult Literacy: Industry-Based Training Programs* (Columbus, Ohio: The National Center for Research in Vocational Education, 1987), p. vii.
14. Ibid.
15. Ibid., pp. 5–31.
16. Ibid.
17. Hoerr, J. "Business Shares the Blame," p. 71.
18. Schaaf, D. "The Changing Role of Training: From Rags to Riches?" *Corporate Development in the 1990s* (A Supplement to *Training*), 1991, p. 5.
19. Ibid.
20. Ibid., p. 7.
21. Wiggenhorn, W. "Motorola U: When Training Becomes an Education," *Harvard Business Review,* July–August 1990, pp. 71–83.

CHAPTER SEVENTEEN

Productivity and the Supervisor

CHAPTER OBJECTIVES

After studying this chapter, you will be able to define or explain the following topics:

- Productivity as a Concept
- How Productivity is Measured
- U.S. Productivity Compared with That of Other Nations
- The Supervisor's Role in Productivity Improvement
- How Technology Can be Used to Improve Productivity
- How People Can Improve Productivity
- How to Use Work Measurement Techniques
- Employee Participation Strategies
- How Training Can Affect Productivity
- How to Use Incentives
- How to Use Work Measurement Techniques
- How to Apply Method Improvement Techniques
- How to Encourage Innovation

CHAPTER OUTLINE

- Productivity Defined
- Measuring Productivity
- Productivity: The International View
- Supervisor's Role in Productivity Improvement
- Technology and Productivity Improvement
- People and Productivity Improvement
- Work Measurement and Productivity
- Employee Participation and Productivity
- Training and Productivity
- Incentives and Productivity
- Work Improvement and Productivity
- Innovation and Productivity
- Summary
- Key Terms and Phrases
- Case Study Application Problem
- Review Questions
- Simulation Activities
- Endnotes

One of the key measures of a company's competitiveness is its productivity. Better productivity means better competitiveness. This chapter provides the information supervisors need to help improve their company's productivity.

PRODUCTIVITY DEFINED

Productivity is the name given to the concept of comparing the output of goods and/or services to the input of resources needed to produce or deliver them. In simpler terms, *productivity is the ratio of output to input.* It can be a ratio of output to input of capital, labor, energy, or any combination of these. Modern industrial firms are interested in achieving maximum output from their input.

Productivity and quality are the key ingredients in competitiveness. In today's intensely competitive international marketplace, industrial firms must continually improve productivity in order to survive. If Company A is more productive than Company B and quality levels are the same, Company A will win the competition.

Historically industry has focused on improving the productivity of technicians, skilled workers, and unskilled workers, those traditionally known as blue-collar employees. In the modern workplace it is necessary to improve the productivity of all employees. For example, a Cambridge, Massachusetts, consulting firm, the Eddy-Rucker-Nickels Company, has developed an incentive program designed to improve the productivity of blue-collar and white-collar workers. Known as the Rucker Plan, it has proven capable of increasing productivity by as much as 20 percent in just 18 months.[1]

MEASURING PRODUCTIVITY

Productivity is expressed as a ratio of output to input as in the following equation:

$$\frac{\text{Output}}{\text{Input}} = \text{Productivity}$$

Input consists of the resources needed to produce the product or deliver the service. Input resources might be capital, labor, material, or energy. **Output** is the actual product or service. For example, the productivity of an assembly team that assembles low-voltage power supplies might be three power supplies per hour. This is a ratio of 3 : 1 or three power supplies (output) for every hour of labor (input).

To understand how important productivity is, supervisors need to understand the concept of **value added**, also referred to as **production value**. Value added is calculated by determining the difference between what it costs to produce a product and what the product will sell for. For example, if it costs a furniture manufacturer $25 to produce a rocking chair that it sells to distributors for $75, the value added by the production process is $50.

To further illustrate this concept, let's return to the earlier example of the power supplies. The assembly team produced three power supplies per hour or one power supply in 20 minutes. If all costs (material, overhead, labor, etc.) are converted into a **loaded labor rate** of $90 per hour for the assembly team, the cost to produce each power supply is $30 ($90 × .33 of an hour). If each power supply sells for $150, the value added by production is $120 per power supply.

If an assembly team in another company is more productive—say it can produce four power supplies in an hour at the same loaded labor rate—what is the effect on the value added? In this case the cost to produce one power supply is $22.50 ($90 × .25 of an hour). The power supplies still sell for $150 each. Therefore, this assembly team adds value in the amount of $127.50 per power supply.

The higher level of productivity gives the second company distinct advantages when competing against the first company. Because it is more productive, it can sell its power supplies for less and still make more money. Hence, it can outbid the first company in the marketplace. The higher level of productivity of the second company gives it a $7.50 per power supply advantage over the first company.

Supervisors should be able to recognize when productivity in their units is improving or declining. The following rules of thumb can be used for making such determinations.

Productivity is declining when:

- Output declines and input is constant.
- Output is constant, but input increases.

Productivity is improving when:

- Output is constant, but input decreases.
- Output increases and input is constant.

A word of caution is in order here regarding the measurement of productivity. Nothing has been gained if productivity is increased to the detriment of quality. As productivity is improved, quality must be held constant or also improved. Competitiveness requires both productivity and quality (Figure 17–1). Quality is the subject of the next chapter.

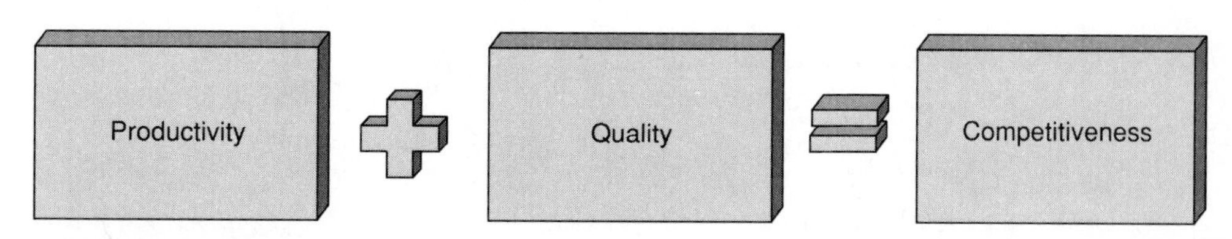

Figure 17–1
Formula for competitiveness.

PRODUCTIVITY: THE INTERNATIONAL VIEW[2]

Historically the United States has been the most productive industrial nation in the world. However, since the early 1970s this competitive advantage has steadily eroded until now such countries as Germany and Japan actually outperform the United States in the international marketplace.

To determine the extent to which U.S. productivity has slipped in recent years, the National Council for Occupational Education and Wharton Econometric Forecasting Associates compared manufacturing productivity in this country with that of Japan and several other industrialized nations. The comparison revealed a disturbing trend. In 1972, productivity in the U.S. was 56 percent higher than that of Japan. In 1987 this advantage had slipped to just 6 percent.

In the years following World War II, productivity in the U.S. improved annually by approximately 3 percent. This trend continued until 1973. Between 1973 and 1982 the rate slowed to less than 1 percent per year. In 1983 U.S. productivity began to improve again. However, taking the long view, other industrialized nations have shown productivity improvements well above those achieved in the United States. Figure 17–2 shows increases in the output per hour of workers in various industrialized countries between 1950 and 1983.[3]

The practical result of losing the edge in productivity is that the losing company (or country) prices its goods and services out of the market. More productive companies produce more for less and, as a result, can sell their products for less. Less productive companies produce less for more and, therefore, must sell their products for more. This

Figure 17–2
Productivity improvement in selected countries.

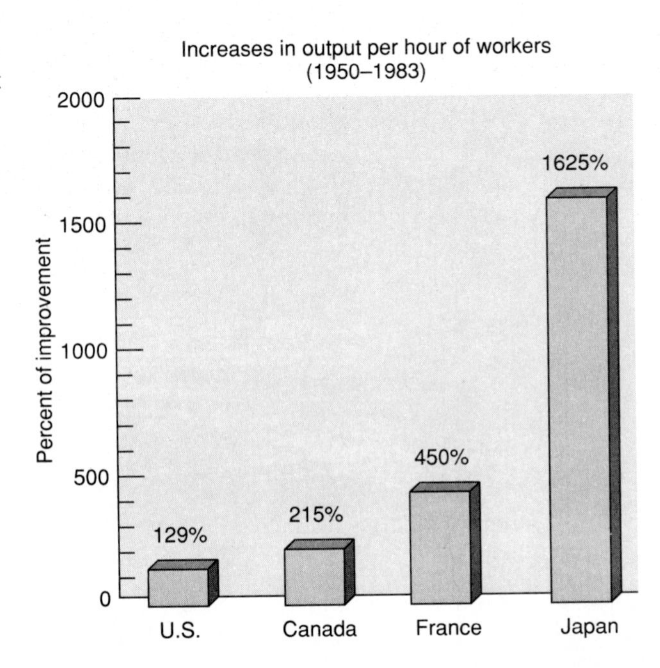

fact, plus quality, are why so many Japanese automobiles, computers, and consumer electronics goods are sold in the United States.

There is a direct correlation between productivity and the quality of life of a country. On a smaller scale, there is a direct correlation between productivity and a company's competitiveness and, in the long run, survival. Therefore, supervisors should be concerned about productivity and able to play a key role in continually approving it.

SUPERVISOR'S ROLE IN PRODUCTIVITY IMPROVEMENT

Competition between companies, domestic and international, is a lot like a foot race. A competitor will set a record this year and another will break it the next year. For this reason, it is important to improve productivity on a continual basis. What was good enough to win yesterday may not be good enough tomorrow. A supervisor can play an important role in continually improving the productivity of his or her employees and organization.

Historically, attempts at improving productivity have focused primarily on technological improvements. First there was the **mechanization** of work in the Industrial Revolution. Then there was **automation** of mechanized equipment and systems that could be controlled by computers. Now we are in the age of **integration** in which computer-controlled systems are being tied together electronically in companywide networks that encompass all subunits of the company (e.g., production, sales and marketing, management, accounting, etc.).

However, even in the age of high technology, people still hold the key to significant productivity improvement. The reason for this is simple. Theoretically, two competing companies can both purchase the same technological systems. Consequently, the only way one can gain a competitive edge over the other is by making more efficient and effective use of the technology. This is a people issue. What this says to the supervisor is that in order to make significant and lasting productivity improvements it is necessary to continually improve both people and technology (Figure 17–3).

To understand the importance of the **people/technology partnership** in improving productivity, consider the example of the word processing system. By now most people are aware of the significant improvements to office productivity the word processor can bring when compared with the typewriter. However, two offices using the same word

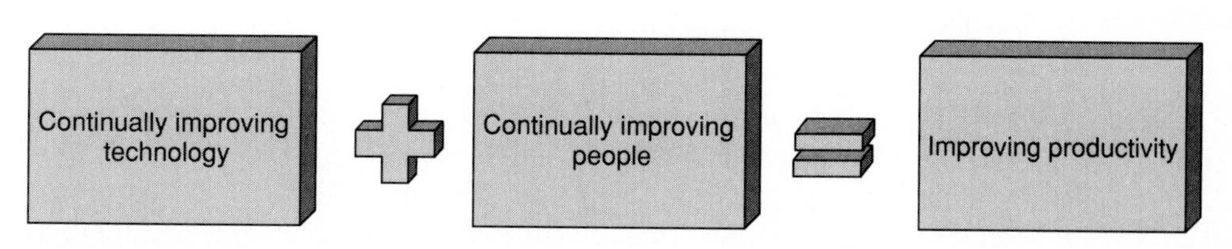

Figure 17–3
The most significant productivity improvements come from people and technology.

processing system to do the same work might achieve markedly different results. The difference is in the knowledge, skills, and attitudes of the users or, in other words, the ability of users to make efficient and effective use of the technology. The companies that will win the daily competition are those that continually improve both technology and people.

TECHNOLOGY AND PRODUCTIVITY IMPROVEMENT

The technology component consists of the tools, machines, equipment, materials, and processes used to produce the product or deliver the service. Supervisors can help improve the contribution technology makes to productivity improvement by applying the following rules of thumb:

- Be alert to design changes in the product that will make it simpler to produce. This concept is known as **design for manufacture.** Continual communication between the design and production units is the key here.
- Be alert to **modernization** concerns. Would facility or technology upgrades improve the productivity of your unit? If so, work with higher management to have the necessary upgrades made.
- Be aware of ways in which automation and integration might improve productivity. Can an assembly process be automated? Can a robot do a repetitive task more efficiently and effectively? Will computer control of a machine or process improve productivity?
- Be alert to potential improvements to **work methods.** Is there a better way to accomplish certain tasks? You and your employees are your best source of this type of information. Empower employees by encouraging them to recommend improvements to work methods.
- Be alert to potential improvements to **process layout.** Can wasted motion be eliminated or reduced by altering the layout of a process? Ensure that processes are laid out to promote efficiency and effectiveness in producing the product or delivering the service rather than for other reasons unrelated to productivity ("This is the way it has always been," or "It's done this way for the convenience of support personnel," etc.).
- Be alert to changes in materials that might improve productivity. Is there a different type of paint that would go on easier and dry faster? Is there a different grade of metal that will hold welds better?

In the age of high technology, several technological developments are having a significant impact on industrial productivity worldwide. Examples of these developments are manufacturing resources planning, industrial robots, and computer-aided design/computer-aided manufacturing (CAD/CAM). Modern industrial supervisors should be familiar with these developments and the impact they are having on productivity.

Manufacturing Resources Planning and Productivity

An early effort designed to cut down on raw material and work-in-process inventory was called *material requirements planning.* This concept turned out to be too limited in scope and was superseded by **manufacturing resources planning** or MRP. Oliver Wight, president of Oliver Wight Companies of Newbury, New Hampshire, describes MRP as a systematic way of asking four questions:[4]

- What are we going to make?
- What does it take to make it?
- What have we got?
- What do we need to get?

According to Wight, MRP used to be just an ordering system used to replenish stock. However, with the introduction of computers MRP began to evolve into a scheduling technique. Eventually MRP developed into a much broader concept encompassing a company's production plan, master schedule, capacity plan, vendor schedule, shop schedule, and financial plan. With these components integrated through the use of computers, the latest version of MRP evolved, known as **MRPII.**

MRPII proceeds in four phases, each encompassing three different functional areas, as illustrated in Figure 17–4. Supervisors are involved at the operational level or Phases

Figure 17–4
MRPII process.

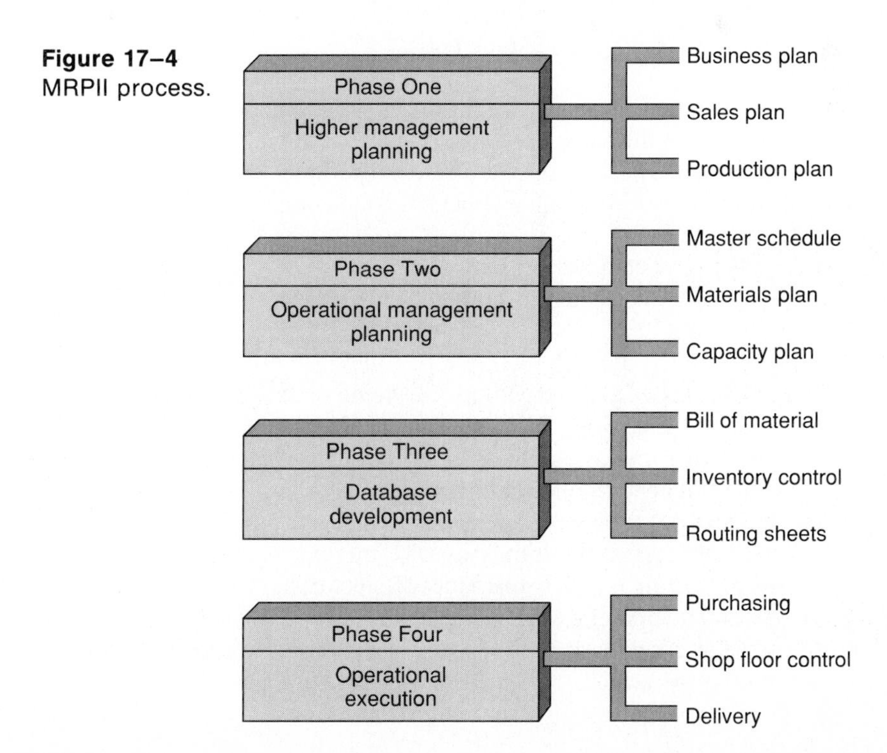

Two, Three, and Four of the process.[5] When properly implemented, MRPII can be an effective technological approach to improving productivity in manufacturing firms.

The Bently Nevada Corporation in Minden, Nevada, implemented MRPII after a fire in one of its fabrication facilities exacerbated an already critical on-time delivery problem.[6] After solving some initial problems, Bently Nevada Corporation began to turn the corner. As a result of the implementation of MRPII, the company increased productivity markedly. Prior to implementation the company employed 480 manufacturing personnel and shipped $29 million worth of products. After implementation, Bently Nevada Corporation was able to ship $50 million worth of products using 340 manufacturing employees.

Industrial Robots and Productivity

Repetition and danger are inhibitors of productivity. When people become bored with repetitive tasks, productivity and quality fall off. The safety precautions required when human workers perform dangerous tasks can also have a negative impact on productivity. In some cases where repitition and/or safety concerns inhibit productivity improvement, **industrial robots** can be a solution.

A robot is a programmable, multifunctional machine that can perform a variety of manufacturing tasks such as material handling, assembly of discrete parts, painting, welding, and others. Robots are designed to imitate the capabilities of human workers. Modern industrial robots move, sense, see, and think while performing their duties. They have the following characteristics:[7]

- A manipulator consisting of an arm, a wrist, and a hand
- Flexibility of movement
- Attachments that can grip and release workpieces
- Manual controls that allow a person to operate the manipulator
- A memory
- An electronic communications device
- Operating speeds equal to human performance
- Reliability

Thousands of industrial robots are now being used in such applications as spot welding, arc welding, palletizing, stocking, loading of conveyors, loading and unloading of bins, spray painting, and assembly.

One of many examples of companies that have improved productivity through the use of industrial robots is the TV Business Division of General Electric in Portsmouth, Virginia.[8] Concerned about the excessive amount of time injection molding press operators spent waiting for their machines to eject parts, GE's management team decided to try robots. The first one was installed in 1979. By 1980, a total of six robots were used to unload GE's injection modeling presses. The robots brought a 10 to 15 percent increase in productivity, which translated into a $370,000 per year savings. Figures 17–5 and 17–6 show examples of modern industrial robots.

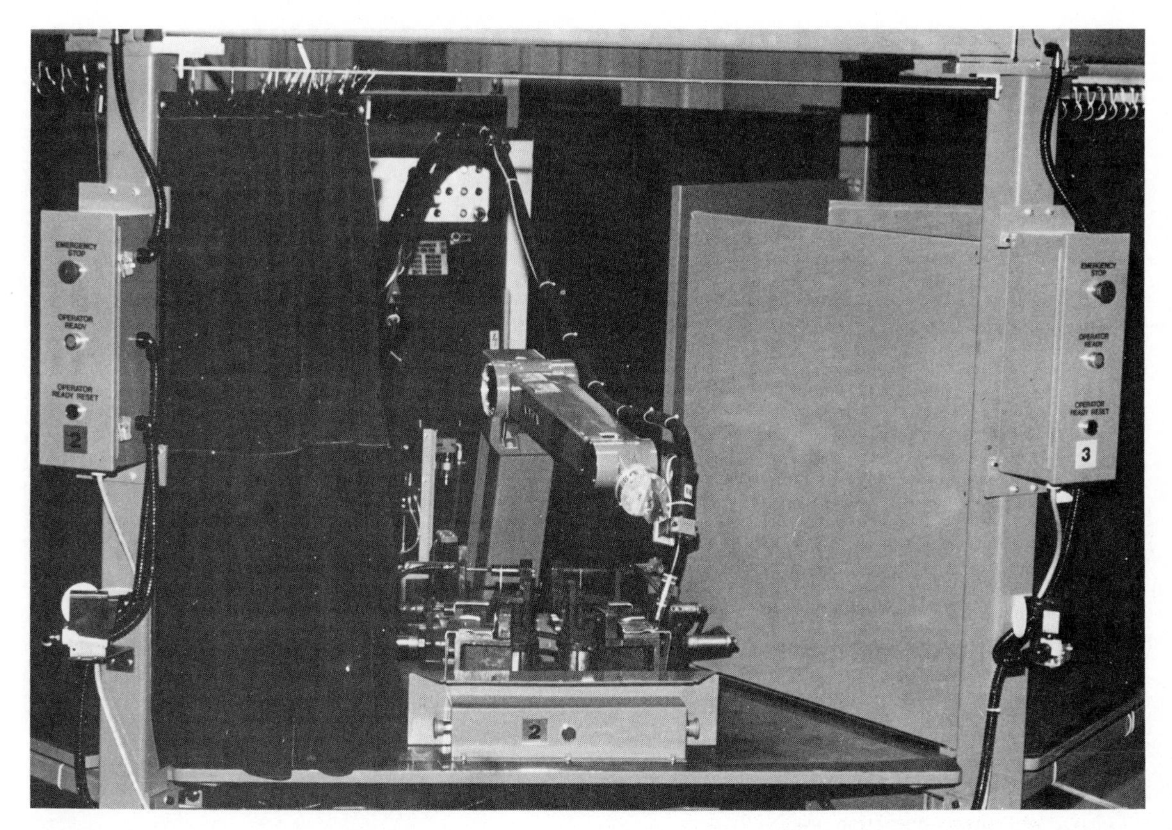

Figure 17–5
IRb-LG robot with four fixed tables.
Courtesy of ESAB Automation, Inc.

CAD/CAM and Productivity[9]

CAD/CAM is now a widely used approach to productivity improvement in design and manufacturing. It represents the automation of design and the communication interface between the design and manufacturing components. According to Edwin N. Nilson, manager of Technical Management Data Systems for United Technologies, 5 : 1 productivity improvement ratios are common with CAD/CAM. Other advantages include fewer design errors and faster product development.

The New England Aircraft Products Division of Homet Turbine Components Corporation achieved a 10 : 1 improvement in productivity by adopting CAD/CAM, enabling the company to survive in a competitive market. Prior to the adoption of CAD/CAM, the company used enlarged models as guides in manufacturing airfoils. All the production work was done manually. With CAD/CAM the handmade models have been replaced by three-dimensional computer models. These models are used to create the numerical

Figure 17–6
Two overhead IRb 2000 robots.
Courtesy of ESAB Automation, Inc.

control code needed to machine the parts. The use of three-dimensional computer models saves time in both design and manufacturing. This is a typical example of how CAD/CAM has been used to enhance productivity.

PEOPLE AND PRODUCTIVITY IMPROVEMENT

Obviously technology can have a positive effect on the continual improvement of productivity. However, the supervisor's impact with regard to technology-based improvements is limited. You can observe, identify problems, and recommend solutions. But in most cases, recommending is as far as you can go. For example, if you become convinced that industrial robots could enhance productivity in your unit, you can recommend the purchase, but higher management will make the decision whether or not to purchase the robots.

On the other hand, supervisors can have a direct impact on improving the people side of the productivity equation. Even in the age of high technology it is still the people component that can have the most far-reaching and most immediate impact on produc-

Figure 17–7
Strategies supervisors can use to improve the productivity of employees.

Strategies for Improving People Productivity

- Use work measurement techniques.
- Encourage employee participation.
- Implement quality circles.
- Arrange training for employees.
- Implement incentive programs.
- Use work measurement techniques.
- Use method improvement techniques.
- Encourage innovation.

tivity. There are many different ways in which supervisors can improve the productivity of their employees. Some of the more widely used are summarized in Figure 17–7. The remainder of this chapter is devoted to an examination of these strategies.

WORK MEASUREMENT AND PRODUCTIVITY

An important component in any productivity improvement program is **work measurement**, the process whereby determinations are made as to how long it takes to perform specific tasks. Work measurement provides a yardstick for quantifying productivity improvements. Unless it is known how long tasks take to complete, there is no basis for making improvements.

The most widely used work measurement technique is the **time and motion study**. Job tasks are broken down into individual motions and the various motions are timed. By adding the times required for individual motions, total times for specific tasks are determined. The times or averages taken from several such studies become standards.

To improve productivity, supervisors focus on improving performance as measured against the time standards. An added advantage of a time and motion study is that those conducting it are frequently able to identify motions that can be eliminated altogether or at least shortened substantially.

Another work measurement technique is **work sampling**. This is a quick and easy way for supervisors to get an idea of the productivity levels of their employees. Work sampling involves observing employees working and taking note of delays and interruptions. For example, say an observed task takes fifteen minutes to complete and there are nine interruptions and/or delays. One way to accomplish the task in less than fifteen minutes (increase productivity) is to eliminate the delays and interruptions.

An example of work sampling would be to observe a machine operator at irregular intervals ten times a day for one week (five days). This is a total of 50 observations. If during 17 of the observations the machine operator was idle, it can be assumed that he or she is idle 34 percent of the time. Obviously, this situation should be studied more carefully to determine the reasons for the high percentage of idle time and then strategies for decreasing idle time could be developed.

A word of caution is in order here. Work sampling gives supervisors enough information to indicate that a closer look is called for, but it is not a scientific approach. The results of work sampling should be viewed as approximations rather than hard facts.

A further word of caution relating to both time and motion studies and work sampling is to *let employees know what you are doing.* It can be disconcerting for employees to have a supervisor standing over them with a clipboard and a stopwatch. Not only might they resent being studied, but their actions are not likely to be normal if they are being watched or timed. The natural human reaction in these cases is to work harder and faster than normal. If the pace set while being timed is unrealistic, the time standards derived will also be unrealistic.

Before conducting either type of study, talk with employees and let them know what is going to happen and when. Let them know that you are interested in work processes, not their individual performance. Explain how the information will be collected and how it will be used. Show employees how such studies can be used to simplify their jobs. Figure 17–8 is an example of a work sampling summary sheet for a team of

Figure 17–8
Work sampling summary for a team of six welders.

Work Sampling Summary

Unit studied Container Assembly

Date samples taken February 15, 1991

Observer Andrew Carnathan

Times of observations

Employee	8:35	9:52	1:17	3:12	4:39
Jones.	1	3	8	3	6
Smith.	2	3	5	3	1
Menendez.	1	4	6	3	4
Guiliano.	3	2	1	4	3
DuVo.	4	7	1	4	3
Modewleski.	8	5	3	2	6

Activities key

1 = Retrieving precut aluminum pieces
2 = Placing precut pieces in fixtures
3 = Welding joints
4 = Unclamping fixtures
5 = Removing container from fixture
6 = Stacking container
7 = Away from work station
8 = At work station but idle

welders. Each welder assembles precut pieces of aluminum into shipping containers. The process involves fastening the individual pieces into fixtures, welding all joints, loosening the fixture clamps, and removing the assembled container.

In this sample, the supervisor made five random observations on February 15, 1991. By summarizing the number of times employees were observed performing each activity, the supervisor can get an idea of how much time is probably devoted to each. For example, activity number 3 was observed 9 times out of 30 observations (six employees × five random observations). This tells the supervisor that the container assembly workers, all highly skilled welders, spend only one-third or 33 percent of their time actually welding. Could a robot or less skilled workers be used for the other activities? This is the type of question that might result from a work sampling study.

EMPLOYEE PARTICIPATION AND PRODUCTIVITY

An effective way to increase the productivity of people is to **empower** them through **employee participation** programs. Empowering workers involves letting them participate in making decisions that affect them or that they will have to carry out. This approach offers two advantages.

First, any attempt at improving productivity depends for its success on the people who will implement it. If they do not support the effort it will fail, no matter how well conceived it is. Second, the employees closest to the work are the ones most likely to know the factors that inhibit productivity.

Empowerment and participation are not the same thing and it is important for supervisors to know the difference. Participation means involving people in making decisions that affect them. Empowerment means giving real weight to their input. Tom Peters says empowering means taking people seriously.[10]

A number of different approaches to participation can be used to empower employees (Figure 17–9). The most widely used of these are as follows:[11]

- Suggestion boxes
- Open-door policy
- Management by objectives (MBO)
- Communications councils
- Team building
- Job enrichment
- Quality circles
- Task forces
- Self-managed work teams

Suggestion Boxes

Suggestion boxes represent one of the oldest forms of employee participation. This is one of the easiest strategies to implement and it can be effective. However, it can also fail miserably. The difference between success and failure with suggestion boxes is empowerment. Are the suggestions taken seriously and acted on?

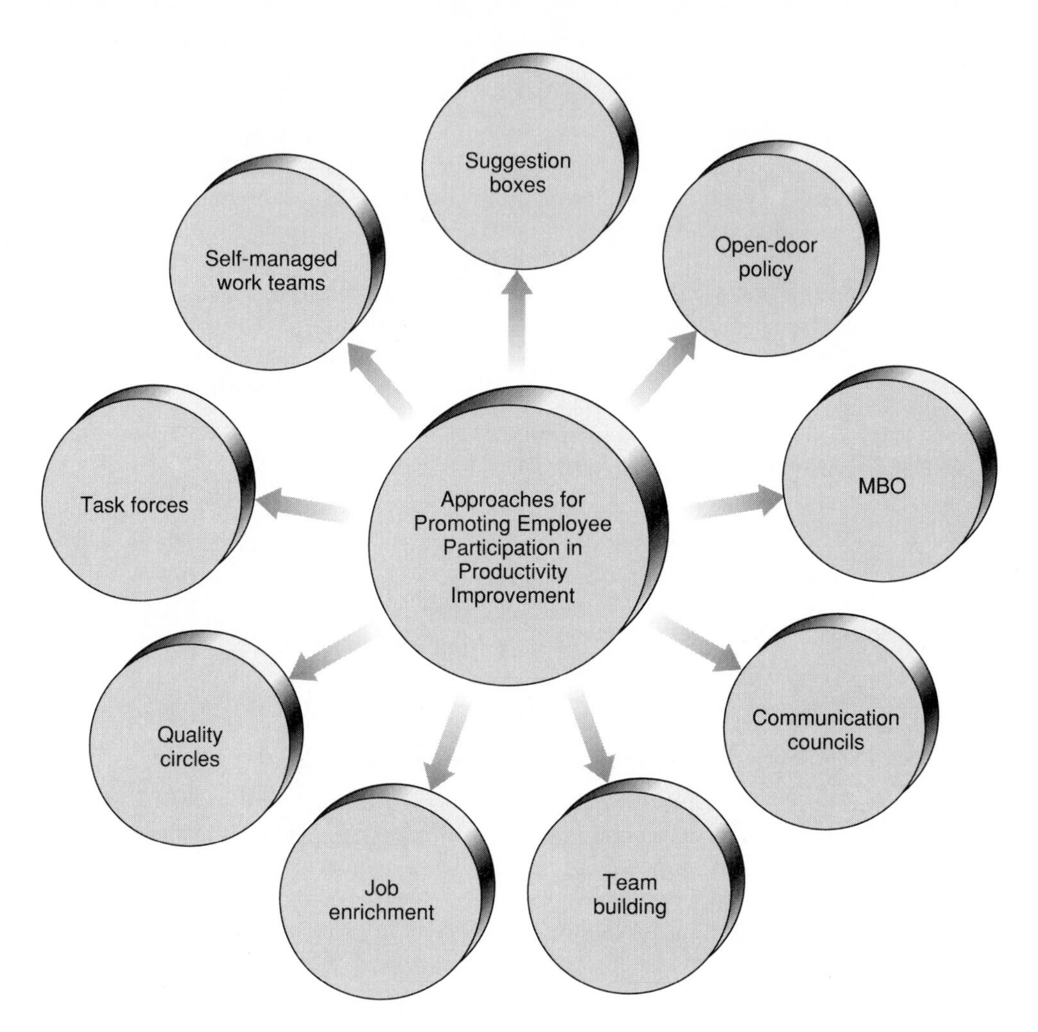

Figure 17–9
Different approaches for promoting employee participation in productivity improvements.

In order to enhance the effectiveness of suggestion boxes, some companies reward employees whose suggestions result in productivity improvements. This strategy can bring two benefits. First, it provides an added incentive for employees to make suggestions. Second, it shows employees that their suggestions are not just taken seriously, but carried out.

Open-Door Policy

An **open-door policy** encourages employees to approach any level of management to discuss potential productivity improvements. It can be an excellent way for higher management to stay in touch with the operational level. However, unless employees, supervisors, mid-managers, and executive management understand the purpose of an open-door policy and accept it in practice, it will backfire and do more harm than good. Before implementing an open-door policy, there should be open discussions involving all levels of personnel. Employees need to understand that they should take their suggestions to the most appropriate person, the one most likely to be able to implement the suggestion. Supervisors need to put aside feelings of defensiveness and all employees need to understand the importance of communicating openly and frequently.

Management by Objectives

The **management by objectives (MBO)** concept was developed by Peter Drucker in 1954 and described in his book, *The Practice of Management.* It involves supervisors and employees working together to set work goals. If used properly, it can be an effective way to improve productivity. However, there is a great deal of potential for abuse by both the employee and the supervisor.

MBO will fail if employees purposely set their goals low. Correspondingly, it will also fail if the supervisor turns it into a prescriptive exercise with no real employee involvement. According to Tom Peters, MBO is a good idea that is typically "neutered" by "bureaucrats."[12] Peters lists four criteria relating to the objectives of MBO that are necessary for success. The objectives must: (1) be simple; (2) focus on what is really important; (3) be the result of real participation by empowered employees; and (4) be a contract rather than a pointless exercise in paperwork.[13]

Communications Councils

A **communications council** is a group of employees representing all components of a company and all levels from operational through executive management. A council can be an excellent way to: (1) keep all levels and components of a company informed of policy matters and decisions that have been made; (2) solicit input concerning proposed policy recommendations; and (3) "test the water" before announcing a major decision.

Team Building

Team building at the operational level involves cross-training employees so that they can perform more than one job. This allows the supervisor to build a solid team of employees who can cover for each other during absences and can double up with each other when a given process gets backed up. According to Peters, one way to increase the effectiveness of teams is to base performance evaluations on team performance rather than on the performance of individuals.[14] This approach creates a sense of peer pressure among team members to achieve peak performance levels. The key to building strong teams is to build interdependence and camaraderie among members.

Job Enrichment

Boredom and resultant factors such as absenteeism and turnover can have a substantially negative impact on productivity. **Job enrichment** is the name given to any strategy that attempts to make work more challenging and varied. Typical strategies include giving employees more responsibility, cross-training and team building, and rotating work assignments.

Quality Circles

A **quality circle** is a group of employees from within the same unit or from different units who do the same type of work. Such groups meet regularly to discuss their work and solve problems relating to it. The goal of a quality circle is to identify the causes of problems and eliminate them.

Quality circles work only if the participants are truly empowered to identify causes and eliminate them. According to Peters, all participants in a quality circle need to be trained in cause-and-effect analysis and group problem-solving techniques.[15] This is an important point because there is a danger that participants might identify symptoms rather than causes. Effort wasted in this way will cause quality circles to fail.

Quality circles are closely associated with Japanese management. Traits exhibited by Japanese workers that contribute to the success of quality circles include a feeling of being part of a group, a belief that work has inherent value, a determination to achieve peak performance levels continually, a strong sense of loyalty to the company, and self-discipline. Traits traditionally associated with workers in the United States include a strong sense of individualism, a tendency to associate the amount of effort one should put forth with the potential reward, loyalty to the company that pays the most, and a tendency to dislike discipline.

On the surface these traits would appear to doom quality circles in the United States from the outset. However, this does not have to be the case. Independence can be a positive factor in a quality circle. It can ensure that the hard questions are asked and that all sides of an issue are examined. Participation in a quality circle can engender loyalty to the company and the peer interaction of a quality circle can build self-discipline in individual participants. The team building and team evaluation strategies spoken of earlier can also help enhance the potential for success with quality circles.

Task Forces

A **task force** is a group of employees established for a single purpose and disbanded once the purpose has been served. Such groups are sometimes referred to as ad hoc committees. As with all other participatory strategies, the use of task forces will not enhance productivity unless participants are empowered to do what they have been convened to do.

Self-Managed Work Teams

A **self-managed work team** is a team that has full responsibility for producing an entire product. Team members make all decisions concerning the production of the product. They select their own leaders and handle their own discipline problems. The key to success with self-managed work teams is giving the teams full authority to make the decisions that must be made.

TRAINING AND PRODUCTIVITY

Nothing is more important to the improvement of productivity than **training** (Figure 17–10). In his book *Thriving on Chaos,* Tom Peters makes the following statement about training:

> Work-force training and constant retraining . . . must climb to the top of the agenda of the individual firm and the nation. Value added will increasingly come through people, for the winners. Only highly skilled—that is, trained and continuously retrained—people will be able to add value.[16]

Supervisors interested in improving the performance of their employees should train, train, and train some more. The most successful companies in the United States are those that have grasped the significance of training and are using it as a key element in their productivity improvement efforts. IBM, Disney, Federal Express, Motorola, GE, and Millikin are all committed to training as the cornerstone of productivity.

Tom Peters lists the ten elements of a good training program:[17]

- Extensive entry-level training focuses on the skills in which a company wants to be distinctive.

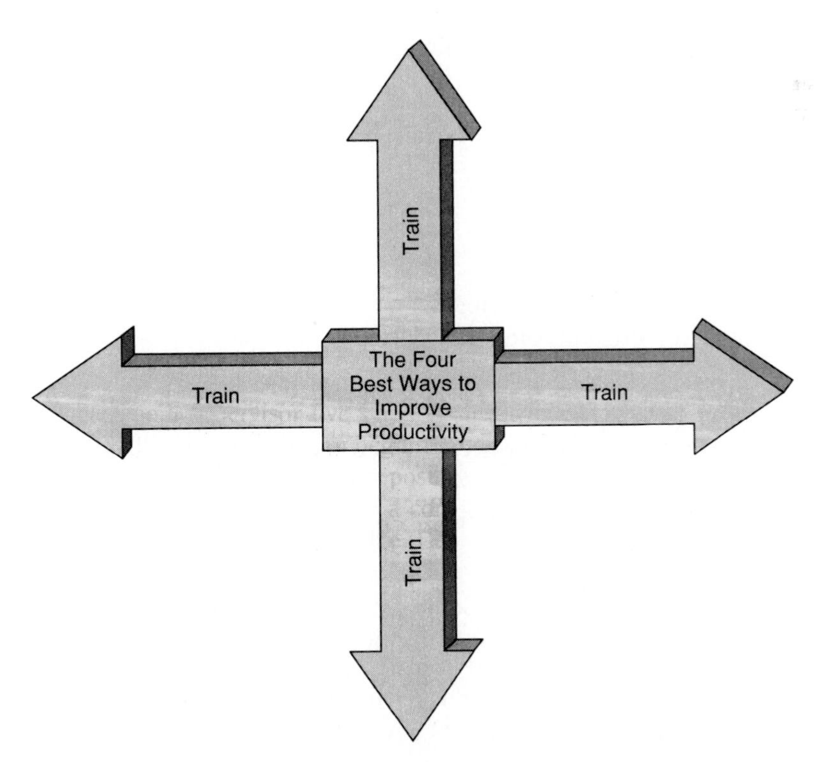

Figure 17–10
The key to improving productivity.

- All employees are treated as potential career employees.
- Regular retraining is built in.
- Both time and money are generously expended.
- On-the-job training counts.
- There are no limits on the skills that can be taught to all levels of employees.
- Training is used to promote companywide commitment to a new strategic thrust.
- Training is emphasized at a time of crisis.
- All training is employee-need driven.
- Training is used to communicate the organization's vision and values.

Not all companies are committed to training, but the most successful ones are. Consequently, supervisors must be prepared to play a leadership role in moving their companies in the right direction on this issue. All other productivity improvement efforts will fail if employees do not have the knowledge, skills, and attitudes necessary to be productive.

INCENTIVES AND PRODUCTIVITY

J. Lamar Roberts, a dynamic bank executive, is a firm believer in using **incentives** to improve productivity. His philosophy concerning incentives is simple and straightforward: "I cannot make my employees excel, but I can make them want to."

Making employees want to excel is the purpose of all incentives. Incentive programs are the responsibility of higher management, but supervisors should be familiar with them because they are used to improve productivity. Incentive pay is not a new concept. However, the emphasis is shifting from individual incentives to companywide incentives. The shift in emphasis is the result of a realization that one individual's outstanding performance is to no avail if the overall bottom line of the company is not improved. Companywide incentives that apply to industry fall into three broad categories:[18]

- Profit sharing
- Cost savings sharing
- Production value sharing

Profit Sharing Incentive Plans

Profit sharing involves distributing profits periodically based on merit ratings, job performance ratings, seniority, or other related criteria. Profit is typically defined as the net income after all operating and administrative expenses have been deducted.

Profit sharing says to employees, "When the company makes more money, you will make more money." The profit motive is supposed to encourage employees to work harder and smarter. Companywide incentives can be effective at enhancing productivity. However, there are weaknesses with which supervisors should be familiar:

- The system breaks down when factors over which employees have no control intervene and the company does not make a profit.

- The system can break down if the incentive pay period is too long. For example, the effect is diminished if incentives are paid only once each year.

Cost Savings Sharing

Cost savings sharing involves distributing a portion of the money saved from cost-cutting strategies or increased productivity among the employees who saved it. This is typically handled at the cost center or unit level. For example, say a new contract is based on producing a given part at a cost of $3,000 each. If the responsible unit finds a way to produce each part for $2,500, it has saved the company $500 per part. If the contract is to produce 1,000 parts, the total savings is $500,000. This amount is divided between the company and the employees on a predetermined basis. The part that goes to employees is typically set aside in a holding account and distributed at the next incentive pay period.

Production Value Sharing

Production value sharing involves distributing among employees the gains from increased production value or value added. Recall that the production value is the value added to a product by the production process. For example, if it costs a company $150 to produce a product and the company can sell that product for $200, the production value (value added) is $50.

The advantage of production value sharing is that it accounts for all the various factors that can increase the production value (i.e., increased output, materials savings, labor cost savings, equipment savings, and quality improvements).

To illustrate, go back to the example of the product that costs the company $150 to product. This product sells for $200 with a resultant production value of $50. This value-added figure can be increased by cutting the $150, increasing the $200, or both. The major advantage of production value sharing is that the employee's incentives are not affected by changes in selling, research, administrative expenses, or volume levels.

WORK IMPROVEMENT AND PRODUCTIVITY

Work improvement involves changing the way work is accomplished in ways that lower cost, decrease time, increase quality, or in other words, increase the production value of a product. There are many different names for the concept of work improvement. Regardless of the name, they all involve finding ways to increase the production value by improving the way work is done.

Supervisors can identify ways to improve work by paying close attention to *how* it is done, *why* it is done that way, *who* does it, *when* it is done, *where* it is done, and even *what* is done. An even more effective strategy for identifying ways to improve work is to motivate employees to ask these questions. Encourage them to adopt a "How can we do this better?" attitude toward their work.

Helping Employees Improve Work Methods

Supervisors must do more than just encourage employees to continually look for ways to improve work methods, although this is critical. Supervisors should also help employees become skilled work improvers. This can be done by providing work improvement training, through brainstorming sessions, and through one-on-one interaction. Regardless of the approach used, supervisors should help all employees learn how to: (1) divide work into stages, and (2) apply standard work improvement criteria to each stage.

Most work can be divided into three stages: gear up, work, and break down. *Gearing up* is preparing to do the work. The next step is actually accomplishing the work tasks. *Breaking down* involves all the follow-on tasks that must be accomplished before the worker can gear up again. Dividing work into these stages makes it easier to accomplish the next step, applying work improvement criteria.

Work is improved by **eliminating** unnecessary steps, **combining** steps, changing the **sequence** of steps, and/or **simplifying** steps. The key is to get employees into the habit of mentally asking themselves the following questions about all three stages of the work:

- What steps can I eliminate?
- What steps can I combine?
- Can the sequence be changed for the better?
- How can I simplify these tasks?

When supervisors and their employees come to apply these work improvement techniques as a matter of habit, productivity can be continually improved.

INNOVATION AND PRODUCTIVITY

Innovation involves finding ways no one else has thought of to solve problems. In the context of this chapter, the problem is productivity. Finding innovative ways to improve productivity is an area where supervisors can have a significant impact. It is a matter of "stepping out of the box" and thinking creatively. Sometimes the best solutions are found when a person steps back and looks at the problem from a different angle.

Every supervisor and every employee can think creatively and find innovative solutions to problems. The key is to encourage creative thinking rather than discourage it. Supervisors should be adept at thinking creatively themselves and at getting employees to think creatively.

Thinking Creatively

Although all people are able to think creatively, not everyone knows how. The following strategies will help supervisors improve their creative thinking abilities and help employees learn to do the same:

- Never let "we have always done it this way" be a reason for doing anything.
- Begin by focusing on the problem at hand. Write down the specific problem you are trying to solve and concentrate on it.

- Read the professional literature in your field. Has someone else faced this problem? If so, how did they handle it? Even if your specific problem is not dealt with, reading may trigger ideas.
- Find out when you think best—while driving, walking, sleeping? Keep a notepad near to jot down ideas.
- Give ideas time to germinate and grow.
- Don't rule out any idea at this time no matter how outrageous it may seem on the surface. In fact, one way to get started is by asking yourself, "What is the most outrageous solution I could propose?"
- Never say, "This solution is too simple. Surely someone has already tried it."
- Be patient. Not all solutions are found overnight. Persist in your efforts and they will eventually pay off.

Soliciting Creative Ideas from Employees

Your employees represent an excellent source of **creative input** and innovative solutions. Supervisors can tap this resource by applying the following strategies:

- Stress the need for their ideas. Encourage employees to make suggestions for improving productivity. Then follow up on their suggestions.
- Share the information from the preceding section with employees so they can become more proficient at thinking creatively.
- Hold periodic brainstorming sessions that are dedicated to the solution of a specific problem. During these sessions encourage all ideas and record every suggestion made. Do not criticize suggestions or allow others to criticize. Let the ideas flow freely, recording them as they come. Once all ideas have been recorded, begin the process of discussing, eliminating, combining, and adding to the ideas that have the most merit.

SUMMARY

1. Productivity is a comparison of the output of goods and/or services to the input of resources needed to produce or deliver them. It can be expressed as a ratio (output ÷ input = productivity).
2. The terms *value added* and *production value* are used interchangeably. They mean the difference between what it costs to produce a product and what the product will sell for.
3. Productivity is declining when output declines and input is constant or when output is constant but input increases.
4. Productivity is improving when output is constant but input decreases or when output increases and input is constant.
5. In 1972, productivity in the United States was 56 percent higher than in Japan. In 1987, this advantage had slipped to just 6 percent.
6. Supervisors can play a key role in improving the productivity of their units and individual employees because, even in the age of technology, the most significant improvements will be made by people.

7. To improve the technology side of the productivity equation supervisors should promote the concept of design for manufacture, be alert to modernization concerns, be aware of how automation and integration can help, continually improve work methods, be alert to process layout improvements, and be alert to material changes that might make a difference in productivity.

KEY TERMS AND PHRASES

Automation
CAD/CAM
Combining
Communications council
Competitiveness
Cost savings sharing
Creative input
Design for manufacture
Eliminating
Employee participation
Empowerment
Incentives
Industrial robot
Innovation
Input
Integration
Job enrichment
Loaded labor rate
Management by objectives (MBO)
Manufacturing resources planning
Mechanization
Modernization
MRPII
Open-door policy
Output
People/technology partnership
Process layout
Production value sharing
Production value
Productivity
Profit sharing
Quality circle
Self-managed work team
Sequence
Simplifying
Suggestion box
Task force
Team building
Time and motion study
Training
Value added
Work improvement
Work measurement
Work sampling
Work methods

CASE STUDY: Productivity Improvement in Action[19]

The following real-world case study contains a productivity-related problem of the kind supervisors may confront on the job. Read the case study carefully and answer the accompanying questions.

Ver-Val Enterprises is an award-winning minority-owned manufacturing firm located in Fort Walton Beach, Florida, near Eglin Air Force Base. Its product line consists of ground support equipment for aerospace and military applications, including carts, munitions, containers, armament transport vehicles, and mobile field laundries.

Ver-Val's management team had always been interested in improving productivity, but in 1990 this interest increased markedly. As were all Department of Defense contractors, Ver-Val was faced with the prospect of drastic cuts in the DOD budget. This meant significantly less work available. Clearly, the competition for the limited amount of work would be intense.

CEO Nate Smith knew that Ver-Val would have to achieve peak performance levels in order to survive and succeed. To accomplish this they adopted a two-pronged, integrated approach: technology-based improvements plus people-based improvements.

On the technology side, Ver-Val automated its design and drafting department through the adoption of CAD/CAM, hired consultants to help improve speed and feed rates on machine tools, and redesigned its shop floor layout. On the people side, Ver-Val worked with the Center for Manufacturing Competitiveness at Okaloosa-Walton Community College to establish the VVE Total Quality Institute (TQI).

TQI is a customized in-house education and training program operated as a partnership of Ver-Val and the college. Courses offered include basic literacy development, problem solving, computer numerical control programming, computer literacy, shop math, industrial supervision, total quality management, statistical process control, just-in-time delivery, and others as needs were identified.

1. Is the "technology and people" approach adopted by Ver-Val the approach you would recommend to a company in Ver-Val's position?
2. How can Ver-Val get its employees to suggest innovative ways to improve productivity?
3. What else might be included in the curriculum of Ver-Val's Total Quality Institute?

REVIEW QUESTIONS

1. Define the term *productivity.*
2. Explain how productivity is measured.
3. Explain the concept of value added.
4. Explain how you know when productivity is declining and when it is improving.
5. Explain the concept of competitiveness.
6. Briefly describe the state of productivity in the United States as compared with productivity in Japan.
7. What is the supervisor's role in productivity improvement?
8. What is the technology component of productivity improvement?
9. List five strategies for improving the technology component of productivity.
10. Briefly explain the following technological concepts: MRPII; industrial robot; CAD/CAM.
11. How does work measurement relate to productivity improvement?
12. Explain how work sampling is done.
13. Explain the following approaches to improving people productivity: open-door policy; MBO; team building; quality circle; self-managed work team.
14. Explain the relationship of training to productivity improvement.

15. List five elements of a good training program.
16. Briefly explain the following types of incentive programs: profit sharing; cost savings sharing; production value sharing.
17. Explain how supervisors can help employees become skilled at improving work methods.
18. Explain five strategies for improving your creative thinking ability.
19. Explain how supervisors can get their employees involved in creative problem solving.

SIMULATION ACTIVITIES

The following activities simulate the types of productivity problems supervisors often face. Apply what you have learned in this and previous chapters in solving these problems. This may be done individually or in groups.

1. You have just been promoted to supervisor of the design shop for a large printing company. Having worked in the shop for five years, you have an idea that productivity could be improved by making technology-oriented improvements. How do you plan to proceed?
2. You supervise 20 forklift drivers in a large manufacturing firm. In an attempt to get a handle on productivity, you want to conduct a work sampling study. Explain how you plan to conduct the survey.
3. Higher management has decided they want to get operative-level employees involved in productivity improvement. You have been asked to make recommendations. What do you plan to recommend?
4. Your company's training program is not popular with employees and management does not think it is yielding the desired results. You are chairperson of a special ad hoc committee to evaluate the training program. Explain how you plan to carry out your task.
5. Your company has decided to implement a production value sharing incentive program in an attempt to improve productivity. The employees in your unit are anxious about it. They do not understand how it will affect them. What do you plan to tell them?
6. You are stumped as to how to solve a problem that has been bothering you for weeks. You want to solicit ideas from employees in your unit. How should you proceed?

ENDNOTES

1. Bureau of Business Practice. *Increasing Productivity with Plantwide Incentives* (Waterford, CT: Simon & Schuster, 1984), pp. 69–71.
2. National Council for Occupational Education and American Association of Community and Junior Colleges. *Productive America: Two Year Colleges Unite to Improve Productivity in the Nation's Workforce* (Washington, DC: 1990), pp. 1–12.
3. Ibid., p. 7.

4. Bureau of Business Practice. "From MRP to MRPII—A Productivity Improvement System Evolves," *BBP Handbook of Productivity Improvement Strategies* (Waterford, CT: Simon & Schuster, 1984), p. 95.
5. Ibid., p. 98.
6. Bureau of Business Practice. "A New Commitment of Productivity of Bently Nevada Corporation," *BBP Handbook of Productivity Improvement Strategies* (Waterford, CT: Simon & Schuster, 1984), p. 108.
7. Bureau of Business Practice. "The Anatomy of a Productivity-Improving Robot," *BBP Handbook of Productivity Improvement Strategies* (Waterford, CT: Simon & Schuster, 1984), p. 115.
8. Bureau of Business Practice. "Five Case Studies of Productivity Improvement," *BBP Handbook of Productivity Improvement Strategies* (Waterford, CT: Simon & Schuster, 1984), p. 126.
9. Bureau of Business Practice. "Harness CAD/CAM Technology to Aid Productivity Growth," *BBP Handbook of Productivity Improvement Strategies* (Waterford, CT: Simon & Schuster, 1984), pp. 138, 153, 154.
10. Peters, T. *Thriving on Chaos: Handbook for a Management Revolution* (New York: Harper & Row, 1987), p. 525.
11. Bureau of Business Practice. "Employee Participation: What It Is, How It Works," *BBP Handbook of Productivity Improvement Strategies* (Waterford, CT: Simon & Schuster, 1984), p. 28.
12. Peters, T. *Thriving on Chaos,* p. 603.
13. Ibid.
14. Ibid., p. 262.
15. Ibid., p. 92.
16. Ibid., p. 386.
17. Ibid., pp. 391–394.
18. Bureau of Business Practice. "A Closer Look at Plantwide Incentives and the Rucker Plan," *BBP Handbook of Productivity Improvement Strategies* (Waterford, CT: Simon & Schuster, 1984), pp. 73–75.
19. Hansen, R. "Total Quality Management at Ver-Val." Unpublished paper, January 1990.

CHAPTER EIGHTEEN
Quality and the Supervisor

CHAPTER OBJECTIVES

After studying this chapter you will be able to define or explain the following topics:

- Quality
- Quality-Related Terms
- The Cost of Quality
- The Quality Philosophies of the United States and Japan
- The Supervisor's Role in Quality Improvement
- The Employee's Role in Quality Improvement
- The Attributes of a Good Quality Program
- Contemporary Quality-Related Concepts
- How to Build a Quality Workforce
- Inspection Systems
- Quality in Nonproduction Settings
- The Malcolm Baldrige National Quality Award

CHAPTER OUTLINE

- Quality Defined
- Quality-Related Terms
- Cost of Quality
- Quality Philosophies of the United States and Japan
- Quality Performance Comparisons
- Supervisor's Role in Quality Improvement
- Employee's Role in Quality Improvement
- Attributes of a Good Quality Program
- Contemporary Quality-Related Concepts
- Building a Quality Workforce
- Inspection Systems and Quality
- Quality in a Nonproduction Setting
- Malcolm Baldrige National Quality Award
- Summary
- Key Terms and Phrases
- Case Study Application Problem
- Review Questions
- Simulation Activities
- Endnotes

Chapter 17 discussed the productivity component of the competitiveness formula. This chapter deals with the quality component. In reality, these two concepts are inseparable. The modern marketplace demands both. So do you. Think about it. Do you go back to a restaurant where the quality of the food was bad and the service worse? Of course not. Would you continue to take your car to an unreliable mechanic? Of course not. If the car you buy turns out to be a lemon, would you buy that kind of car again? Of course not. You want quality.

In his book *Thriving on Chaos* Tom Peters summarizes his conclusions about quality as follows:[1]

1. Customers will pay for quality, especially best quality.
2. Firms that provide quality will thrive.
3. Employees at all levels will become energized by the opportunity to produce a top-quality product or provide a top-quality service.
4. Quality is relative. No product has a safe quality lead. Quality must be improved continually.

Clearly, quality is a critical concern of the modern supervisor. And supervisors play a key role in the continual improvement of quality.

QUALITY DEFINED

Quality can be defined as a measure of the extent to which a product or service conforms to expectations. Tom Peters makes the point that quality should be defined in terms of customer perceptions.[2] Customer expectations can be set forth in written specifications or they may be more subjective. In any case, a quality product or process is one that meets or exceeds customer expectations.

Certain products and/or companies have become synonymous with quality (for example, Mercedes-Benz automobiles, IBM computers, Federal Express package delivery, and Harvard University education). Customers have learned they can trust these companies and/or products to meet or exceed their expectations.

A key point to understand about quality is that it is both relative and ever changing. The concepts of *best* and *better* make quality relative. Is your product the best product on the market? Is it better than the product of Company X, but not better than the product of Company Y? These are relative comparisons. Quality is forever changing because it is tied to the perceptions of the customer. These perceptions can change as new producers find ways to improve on the best or redefine what is better.

QUALITY-RELATED TERMS

The language of quality continues to evolve. Key terms include *quality control, quality assurance, acceptable quality level,* and *inspection.* To this list we must now add more contemporary terms including *total quality management, total quality control, zero quality control systems, statistical quality control, statistical process control, judg-*

Figure 18–1
Important quality-related concepts.

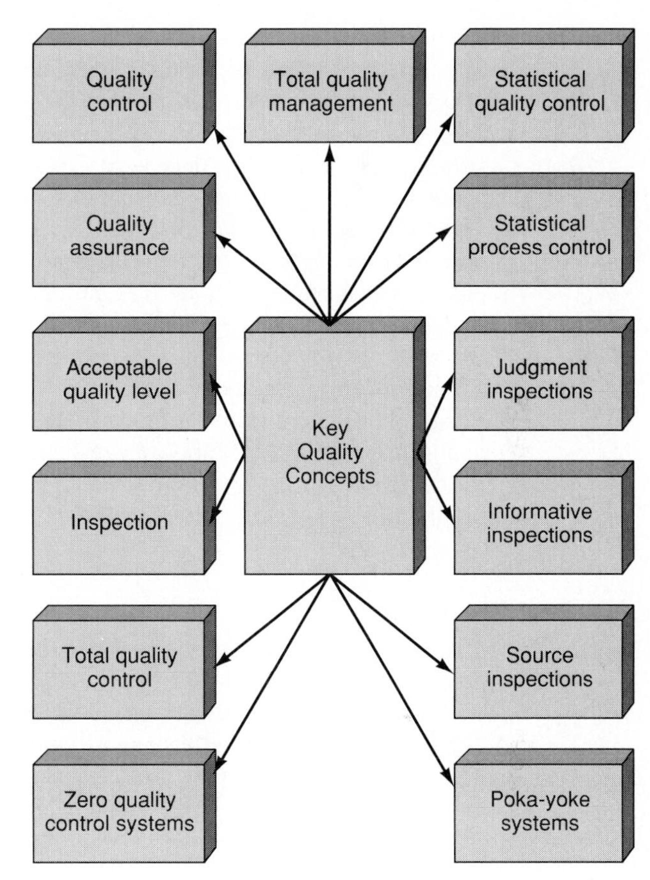

ment inspections, informative inspections, source inspections, and, from the Japanese, *Poka-Yoke systems* (Figure 18–1). These terms are briefly defined in the following paragraphs.

- **Quality control** is the traditional term used to encompass all of a company's efforts to ensure that products, processes, and/or services conform to expectations.
- **Quality assurance** is a term that was developed in an effort to eliminate the negatives associated with the word *control* in quality control. It is viewed by some as a more palatable term for the concept of quality control.
- *Inspection* encompasses a variety of methods used to examine a product or process to determine if it conforms to specifications. **Judgment inspections** discover discrepancies. **Informative inspections** reduce discrepancies. **Source inspections** eliminate discrepancies. Inspection is fundamental to quality control.
- **Total quality management** or **TQM** is an all-encompassing philosophy for maximizing quality by integrating all aspects of the quality program (planning, maintenance,

improvement) and all components of an organization. The goal of TQM is to achieve full customer satisfaction (quality) at the most economic level. **Total quality control (TQC)** is another term used to describe the same concept.

- **Zero quality control systems** encompass a variety of approaches all of which are aimed at the goal of zero defects. It involves both detection and prevention methods applied to both products and processes.
- **Statistical quality control** combines the use of informative inspections and statistical principles to control quality. Control limits are set to define normal and abnormal operations. A statistically derived number of samples is taken to detect abnormal situations that are corrected. **Statistical process control** or **SPC** is an example of a statistical quality control system that applies the principles of statistics to the measurement of process quality.
- **Poka-Yoke systems** (Japanese for mistake proofing) are those espoused by Japanese quality expert Shigeo Shingo. Poka-Yoke systems involve 100 percent inspections, immediate feedback, and immediate corrective action through the use of mistake-proofing devices at the source of the work.

COST OF QUALITY

In the past, the typical attitude toward quality was that it was a costly concept. Such strategies as sampling and statistical quality control are attempts at controlling quality economically. A more contemporary view of quality is that *poor quality costs.*

In a competitive marketplace, poor quality will quickly and surely lead to a loss of market share. Companies that produce defective products must add the cost of corrective measures to the cost of the products. More and more they must also add the cost of litigation as the number of defective product lawsuits continues to rise. As a result, many companies have come to realize that although it costs to build in quality from beginning to end, it costs more not to do so. Said another way, "it is less expensive to do it right the first time than to have to do it over."

The **cost of quality** can be divided into two categories: the cost of doing it right the first time (building in quality) and the cost of doing it over (correcting defects and paying for resultant litigation) (Figure 18–2).

- *Building in quality.* Costs associated with building in quality are those generated by such activities as planning, carrying out, and continually upgrading the quality program (i.e., development of quality procedures; inspection, feedback, and corrective action; and education/training).
- *Correcting defects and deficiencies.* Costs associated with corrective action include the cost of wasted material and time, rework, warranted work, customer service personnel, insurance premiums, and litigation.

The Japanese have proven conclusively that quality doesn't cost, it pays. In 1960, the term "made in Japan" meant shoddy and cheap. In less than 20 years the same term came to be synonymous with quality. As a result of its dedication to quality, this tiny

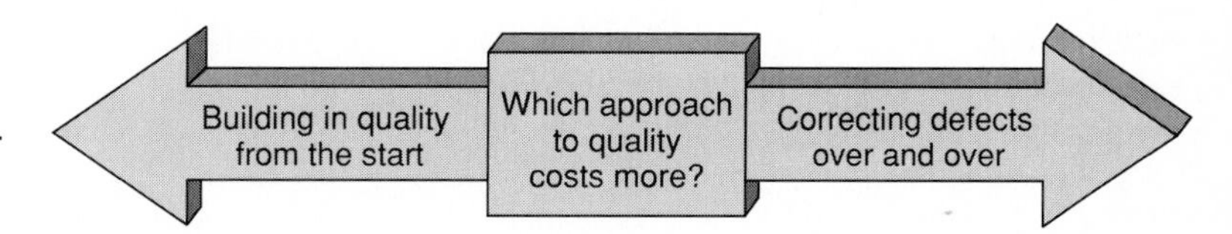

Figure 18–2
The cost of quality can be divided into two categories.

nation that must import most of its necessities transformed itself into an economic superpower.

This phenomenon is often referred to as the **Japanese miracle**, but, in fact, it is no miracle at all. It's just common sense. The Japanese have based much of their success on the belief that people are willing to pay more up front for quality, particularly when this approach is viewed as being less expensive in the long run.

This philosophy was borne out by a 1985 Gallup survey conducted for the American Society for Quality Control. The survey revealed that most consumers would be willing to pay substantially more for quality in such products as a new car, dishwasher, television, sofa, or shoes.[3]

QUALITY PHILOSOPHIES OF THE UNITED STATES AND JAPAN

It is often said that Japan's ability to produce world-class products is a matter of culture. Harmonious labor/management relations and the loyalty of the individual worker to the company are said to be indigenous in the Japanese culture. These are factors, of course, that contribute to quality. While it is true that culturally inspired and nurtured **work motivation** has been important in the success of Japanese companies, it is equally true that work motivation represents only half the story.

The other half is the **work methods** of the Japanese. These, too, differ markedly from those traditionally associated with the United States and they are not tied to the Japanese culture. Noted Japanese quality expert Shigeo Shingo makes this point in his book *Zero Quality Control, Source Inspection, and the Poka-Yoke System.*

Work Motivation Compared[4]

Shingo compares work motivation factors in the United States and Japan in five key areas: employment, salary systems, labor unions, group activities/individualism, and communication (Figure 18–3). His conclusions are summarized here:

- *Employment.* In Japan, companies subscribe to the philosophy of lifelong employment for workers. Even during down periods it is very difficult for Japanese companies to make personnel cuts. This engenders a sense of loyalty and positive labor/management relations. In the United States companies that experience economic downturns often respond immediately by laying off workers. This makes it difficult to maintain company loyalty and contributes to adversarial labor/management relations. Many American workers feel their job security is good only for the length of the company's current contract.

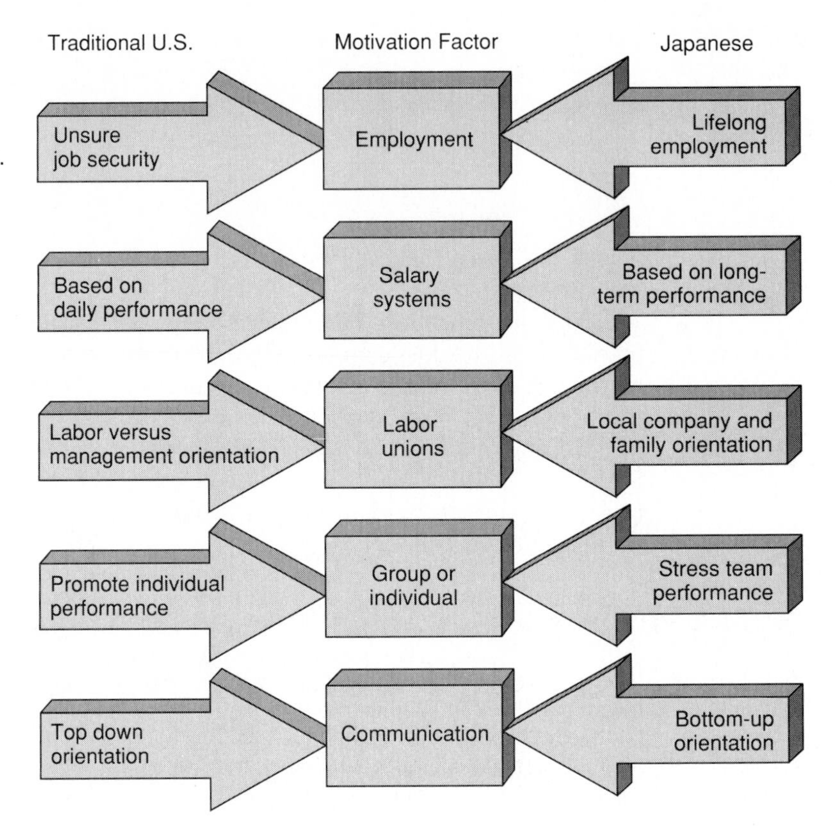

Figure 18–3
Japanese work motivation factors compared with the traditional U.S. approach (which is slowly changing).

- *Salary systems* in Japan are on a monthly basis, which allows performance to rise and fall without fear of an immediate impact on income. In the United States workers' wages are based on daily performance. Therefore, a bad day can have an immediate negative impact on an American worker's pay.
- *Labor unions* in Japan are company unions. These are strictly internal organizations. As a result, labor/management negotiations are more family-oriented and less likely to arouse hard feelings. Labor unions in the United States are typically part of a larger organization that can call strikes without regard to conditions in a given company, which might be harmonious. External forces such as this virtually ensure negative negotiations.
- *Group activities/individualism* represent a major area of difference. Japanese companies and Japanese society stress the group and team approach. Companies in the United States and American society have traditionally stressed individualism. This can make it difficult for U.S. companies to successfully undertake team-oriented quality enhancement strategies.
- *Communication* in Japan has a bottom-up orientation, whereas in the United States it typically is top-down. Japan's communication system results in a long, drawn-out

decision-making process based on consensus building. Once a decision is finally made, every person involved in carrying it out supports it. The American approach results in quick decisions that are implemented without the support of those who must carry them out.

These comparisons point out differences that relate, at least indirectly, to cultural factors. These factors are widely viewed as major contributors to the success Japanese companies have enjoyed in producing quality products. Changes in these areas may come slowly for U.S. companies because of cultural issues. However, differences in work methods do not have as many cultural entanglements and could be implemented more readily in the United States.

Work Methods Compared[5]

Shingo compares work methods in the United States and Japan in several key areas: inventory, lot sizes, tooling setups, quality control, equipment maintenance, productivity, and operations/process characteristics. His conclusions are summarized here:

- *Inventory ties up money.* Consequently, Japanese companies strive to develop production systems that do not require large inventories. Such production systems must have the following characteristics: short setup times, short lead times, and low defect rates. **Just-in-time (JIT)** is an inventory elimination strategy widely used in Japan. Materials and parts are delivered to the production site "just-in-time" to be used. In the United States the traditional approach to materials and parts has been to order large quantities at low bid prices and stockpile a large inventory. Some U.S. companies have begun to adopt JIT and other inventory reduction strategies, hoping to eliminate long setup times, long lead times, and defects. Since these are work methods rather than cultural factors, they can be adopted by U.S. companies as quality improvement strategies. The trend in the U.S. is toward inventory elimination/reduction.
- *Production lot sizes/tooling setups.* Japanese companies have adopted flexible production strategies such as **flexible manufacturing cells (FMCs)**. An FMC is a group of computer-controlled machines integrated with an automated material-handling system that produces a product or part. The machine tools in an FMC are designed for quick setup and rapid tool changes. This gives Japanese companies the flexibility to produce a small lot of one part and then retool and reprogram quickly to produce a small lot of another part. The importance of this flexibility cannot be overstated in a marketplace where the customer wants to be able to have small lot sizes and to customize within product lines.

  Many U.S. companies are still using traditional transfer line/assembly line technology that requires long setups and, in turn, large lot sizes and no customization. Since it is estimated that most lot sizes for manufacturers are now 3,000 or less (considered small), U.S. manufacturers that have not adopted flexible manufacturing strategies are operating at a disadvantage. As a result, the current trend in the United States is toward flexible manufacturing.

- *Quality control.* Japanese companies build quality in and make it everyone's responsibility. **Defects** are identified at every step in a process and corrected immediately. The use of the zero quality control approach and various defect prevention strategies hold the number of quality problems to a minimum. In contrast, the traditional approach to quality in the United States has been to establish an allowable level of defects and to make quality the responsibility of the quality control department with after-the-fact inspections. A worker might say, "Finding defects is not my job. That's the QC inspector's problem." Of course, rather than prevent defects, this approach guarantees them. The trend now in the United States is toward total quality control, statistical process control, and the other more contemporary approaches to quality explained in this chapter.
- *Equipment maintenance.* Japanese companies are committed to preventive maintenance. Although the concept of preventive maintenance was exported to Japan from the United States, the American attitude toward breakdowns has become relatively tolerant. Companies in the United States tend to view equipment breakdowns as a normal part of doing business. Japanese companies view them as intolerable and do everything possible to prevent them. This is an area in which U.S. firms still need improvement.
- *Productivity.* In Japan emphasis is placed on people-oriented **productivity** improvement. U.S. firms have traditionally emphasized technological improvements. This has led to the concept of **dumbing down**, in which machines are designed so that workers don't have to think. The approach has put U.S. companies at a disadvantage since the largest gains in productivity in the modern workplace will come from people. This is a critical concept for the modern supervisor to understand. Better technology can certainly improve productivity if efficiently and effectively used. However, once two competitors have both purchased the same technologies, the one with the best people will outperform the other.
- *Operations versus process.* In Japan the emphasis is on the **work flow** through the overall process. Machines and equipment are grouped to optimize this flow. This approach eliminates bottlenecks and encourages flow. In the United States companies have traditionally focused on improving the operation of individual machines (i.e., speed and feed rates) as opposed to improving the overall flow. It is not uncommon to find machines grouped by type (i.e., mills together, drills together, etc.) instead of for optimization of flow. Improving the operation of one machine in a process therefore creates a bottleneck at the next machine. Consequently, the trend in the United States is toward process optimization. Operations improvements, which are still vital, are undertaken as part of a larger and well-coordinated process optimization effort.

Whether the United States can or even should adopt Japanese approaches to work motivation can be debated. However, U.S. companies can adopt the work methods that Japanese companies have used so successfully, many of which actually originated in the United States in the first place. Supervisors should be familiar with the work methods described in this section and how they might be used to enhance quality in their organizations.

QUALITY PERFORMANCE COMPARISONS

Modern supervisors should be familiar with how the quality performance of U.S. companies compares with that of other industrialized nations. Perhaps the most instructive base for comparisons is the automobile industry. U.S. companies compete with those of Japan and Europe and, increasingly, Korea, Mexico, and Brazil.

The Office of Technology Assessment of Congress published comparisons of the quality performance of selected countries (for automobile manufacturing) in its publication, *Making Things Better: Competing in Manufacturing.*[6] These comparisons are summarized in Figure 18–4.

The figure compares the best and worst automobile manufacturers in terms of the number of assembly defects per 1,000 vehicles. Notice that Japanese plants, Japanese plants in the United States, and a composite of the plants in East Asia, Mexico, and Brazil outperform U.S. plants. Only European plants have a higher defect rate than U.S. plants. The pattern holds for both the best and the worst companies.

Tom Peters compares the quality of U.S. automobile manufacturers with that of Japan using two measures of performance: "**things gone wrong**" or **TGWs** in the first eight months of new car ownership per 100 cars produced; and customer opinions of vehicle reliability and dependability.[7] According to Peters, the worst Japanese automobile manufacturer has a quality performance record comparable to the best U.S. manu-

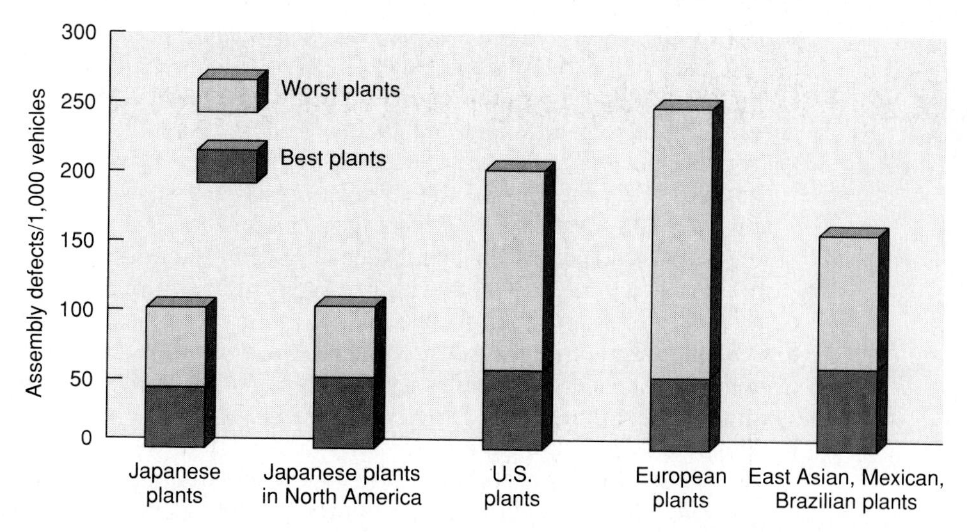

Figure 18–4
Quality performance of automobile manufacturers worldwide.

facturer when TGWs are the basis of comparison. Peters also points out that customer opinion surveys rank the United States far behind its Asian competitors in both dependability and reliability.

Just as important as the facts about quality in the United States are the perceptions about quality. Quality performance can be improved much faster than negative perceptions can be changed. This fact makes quality improvement imperative. Supervisors can play a key role, discussed in the next section.

SUPERVISOR'S ROLE IN QUALITY IMPROVEMENT

Supervisors can play a key role in **quality improvement** regardless of whether a company uses the traditional U.S. approach to quality or the more contemporary total quality approach being used successfully in Japan and in some U.S. firms. In *What Every Supervisor Should Know,* Lester Bittel and John Newstrom list seven points for the supervision of quality that were adapted from the works of noted quality expert W. Edwards Demming.* They summarize Demming's seven points:[8]

- *Do not accept commonly applied levels of delay, mistakes, or defects.* Work with employees to do better. Improve on **acceptable quality levels** by striving for zero quality control performance levels and encourage higher management to adopt the same approach companywide.
- *Continually improve the system.* Supervisors are in an excellent position to see an overall process as well as individual operations within it. Use this simultaneously broad and focused view to identify and correct systemic problems on a continual basis.
- *Focus supervision efforts on helping employees do a better job.* Because of pressure to produce both output and quality, supervisors can emphasize the wrong things. Supervisors who focus their efforts on helping employees do a better job will, as a result, improve both output and quality.
- *Provide tools and techniques that promote pride of workmanship.* Employees need the tools to do the job and the techniques that will allow them to use their tools efficiently and effectively. When the right tools and best techniques are used, supervisors can instill and nurture an attitude of pride in the work among employees. This means supervisors must work with management to ensure that they invest properly in modern tools and continually seek ways to improve techniques.
- *Eliminate fear through two-way communication.* Make sure that employees are empowered to make suggestions for improving performance. Communicate with employees continually and *listen, listen, listen.*
- *Eliminate barriers among departments.* Quality is everybody's responsibility. Establish a close working relationship with supervisors in other departments. Communicate with them frequently and encourage the establishment of cross-departmental teams. Quality cannot be achieved as long as it is someone else's problem.

---

* Demming, an American, is widely accepted as the father of the Japanese quality movement.

- *Institute a comprehensive program of education and training* aimed at simultaneously improving quality, productivity, and competitiveness. Noted American quality expert Phil Crosby espouses education and training as a key ingredient in quality improvement.

These seven points summarize the supervisor's role in quality improvement. In addition, supervisors can use the following strategies for improving quality:

- Emphasize quality among workers and make sure they know it is your top priority.
- Make the quality commitment visual by displaying examples of good and poor quality, displaying quality awards or records, and displaying quality reminders such as "Quality First" posters. If posters are used, however, make sure they are changed and updated regularly so that they don't begin to blend into the wall.
- Make sure that all employees understand the specifications for all jobs. Then encourage them to exceed the standards.
- Personally conduct periodic inspections beyond those normally done by QC personnel or individual employees and conduct them at critical points in the process.
- When you identify problems or defects, take the necessary action to correct them immediately.
- If statistical process control (SPC) has been implemented at your company, monitor the control charts continually and act immediately when a process falls outside the limits. If SPC has not been initiated at your company, recommend it to higher management.
- Always inspect the **first article** or first part in a batch before initiating the process to produce the remaining parts.
- Make sure all employees are aware of their role in quality improvement and work with them to build a commitment to this role.

EMPLOYEE'S ROLE IN QUALITY IMPROVEMENT

Regardless of the approach a company takes to quality, the key players in the quality improvement equation are the individual employees who do the work. They are in the best position to recommend improvements and no improvement can be made without their support. The employee's role in quality is to: (1) perform his or her work in a manner that meets or exceeds quality standards; and (2) make suggestions for improving work methods on a continual basis.

Whether or not employees are able to play their roles effectively depends on management in general and their supervisor specifically. Higher management and supervisors can help employees be positive agents for improving quality by doing the following:

- *Personalize quality* so that employees know why it is important to them as individuals. This will lead to quality-conscious employees who are personally committed to quality.

- *Provide the necessary training* so that employees have the skills to produce quality products or deliver quality services.
- *Communicate continually* so that employees understand what is expected of them, the specifications they must meet or exceed, and what to do when confronted by a problem. Make the communication two-way and *listen, listen, listen.*
- *Provide up-to-date tools and equipment* so that employees are able to produce quality products or deliver quality services.
- *Condition employees to view quality through the customer's eyes.* Encourage employees to work as if they will have to personally deliver the product to the customer.
- *Empower employees to make quality improvement.* Continually solicit their input and act on it appropriately. This can be done verbally one-on-one or in groups, or through the use of a formal **suggestion system**.

Providing a mechanism through which employees can suggest improvements and get action on their suggestions is critical. For example, according to the National Association of Suggestion Systems (NASS), in 1989 American Airlines saved over $40 million by implementing 2,600 employee suggestions.[9]

According to Bob Schwarz, author of *The Suggestion System: A Total Quality Process*, there are seven key elements that together will encourage employee participation in suggestion programs:[10]

- All suggestions receive a formal response.
- All suggestions are responded to immediately.
- Performance of each department in generating and responding to suggestions is monitored by management.
- System costs and savings are reported.
- Recognition and awards are handled promptly.
- Good ideas are implemented.
- Personality conflicts are minimized.

ATTRIBUTES OF A GOOD QUALITY PROGRAM

Supervisors carry the weight of responsibility for quality in their units, but in reality quality can be only as good as the company quality program. Therefore, supervisors have an interest in their company's quality program being a good one. How can supervisors analyze their company's quality program and know where to recommend improvements?

Tom Peters answers these questions in his book *Thriving on Chaos.* According to Peters, a good quality program has the following attributes:[11]

- *Program is based on management's obsession with quality.* Top management is committed to quality as the first priority in both word and deed.
- *Program is based on a guiding system or ideology.* The system is what is used to implement the quality commitment on a day-to-day basis.

- *Quality is measured.* **Measurement** is critical to quality. Before something can be improved there must be a baseline or starting point.
- *Quality is rewarded.* A good quality program has built-in incentives that allow those who improve quality to share in the rewards.
- *Everyone is trained in quality assessment methods and technologies.* All employees from the executive level to the operational level should be trained in cause-and-effect analysis, problem solving, group problem solving, and basic statistical process control.
- ***Multiple-function teams** are used.* Quality problems typically involve more than one department. Consequently, the best results will come from quality improvement teams that include members from all departments that might be touched by a quality problem. According to Peters, "most quality improvement opportunities lie outside the natural work group."[12]
- *Even small improvements are considered significant.* A lot of small improvements add up to major improvements. Even isolated small improvements have value. Progress can be made in small increments.
- *There is* **constant stimulation.** Newness and variety are stimulating to workers. It is particularly important to make changes and introduce stimulation with the announced intention of improving quality.
- *There is a* **parallel structure** *devoted to quality improvement.* Cross-department quality committees and companywide quality teams devoted solely to quality improvement can be effective at keeping workers in all departments focused on quality. Because they consist of workers from different departments, workers who themselves must produce a quality product, such teams can be more effective than the QC department, which is sometimes viewed as a bureaucratic adversary.
- *Everyone is involved, including suppliers, distributors, and customers.* Distributors and suppliers must be held to quality standards and customer feedback must be solicited continually and acted on.
- *Quality improvements decrease costs.* Quality products are generally seen as costing more. This is true, for example, when a higher cost material is used in place of a lower cost material (i.e., metal gears in a sewing machine instead of plastic or hand-sewn leather seats in an automobile instead of fabric). However, it should not be true when comparing the quality of two products of the same material or, in other words, when comparing apples to apples and oranges to oranges. In these cases costs should go down because improved quality is typically the result of simplifying the design of the product or the process used to produce it. Simplification typically enhances quality and simultaneously reduces cost.
- *Quality improvement efforts never end.* The improvement of quality is not a one-time phenomenon. What is the best on the market today may not be tomorrow. Quality records are like athletic records. As soon as they are set, someone else is working to break them. This means efforts to improve quality must be ongoing. Never let good enough be good enough.

These twelve characteristics give supervisors a yardstick for measuring the structure of their company's quality program. This is important because the quality program itself should be scrutinized continually and updated as necessary. There are hundreds of different quality improvement strategies that can be used as part of a company's quality program and these can be changed continually. What does not change are the characteristics of a good quality program. Regardless of the individual strategies used, a good quality program should have the twelve characteristics set forth in this section.

CONTEMPORARY QUALITY-RELATED CONCEPTS

In the modern industrial setting, there are a number of contemporary quality-related concepts that combine new philosophies and new technologies. Supervisors should be familiar with these contemporary concepts, which include just-in-time production or JIT, manufacturing resource planning or MRPII, statistical process control or SPC, and total quality control or TQC (Figure 18–5).

Just-in-Time Production

JIT is not a specific approach to quality; rather, it is a unique approach to production that is *dependent* on quality. With JIT, suppliers deliver their goods to the production site just-in-time to be incorporated into the production process. For this type of production system to work, suppliers must deliver their goods on time every time and the goods must meet all specifications. Such an approach requires a quality control system that identifies defects at the source and takes immediate corrective action.

For example, say an automobile manufacturer has implemented a JIT product system with its supplier of air conditioning units. The units are delivered as needed and installed as part of the automobile manufacturer's assembly process. If an assembly worker spots a defective unit, he or she immediately informs the supervisor, who retrieves the unit and immediately notifies the purchasing department of the problem. Purchasing lets the vendor know there is a problem so that it can be corrected immediately. In the meantime, the worker who spotted the problem has installed a quality unit and the assembly process has moved forward.

A JIT production system depends on quality from suppliers and close working relationships among workers, supervisors, quality control, personnel, purchasing personnel, and vendors. It is not uncommon for companies that implement JIT production systems to involve representatives from all of these groups, as well as others, in joint quality training programs.

Figure 18–5
Contemporary approaches to quality.

Contemporary Approaches to Quality

- Just-in-time production (JIT)
- Manufacturing resources planning (MRPII)
- Statistical process planning (SPC)
- Total quality control (TQC)

Manufacturing Resource Planning (MRPII)

Manufacturing resource planning (MRPII) was discussed in the previous chapter on productivity. It is included in this section only to make the point that its success is dependent on quality. MRPII assumes that the benefits of JIT and a total quality control (TQC) program are in place. The resource predictions produced by MRPII can be rendered inaccurate and unreliable by quality problems.

For example, if through MRPII it has been predicted that 300 subassemblies will be needed from a supplier to complete a given job, that supplier must deliver 300 subassemblies, all of which meet specifications. If 300 are delivered and 10 are defective, the MRPII prediction is rendered inaccurate. This is an important fact for supervisors to understand. If your company considers the adoption of MRPII without first having JIT and TQC in place, the resources and effort will be wasted.

Statistical Process Control (SPC)

In the 1980s this American concept came back home from Japan and emerged as the most widely used of the various statistical quality control techniques. SPC involves the use of statistically based control charts to let the operator of a process monitor the process. As readings are taken they are plotted on a control chart that shows the control limits for normal operation.

For example, if a part is supposed to have a diameter of 2.25 inches plus or minus 0.03, the control limits for normal operation are 2.28 (upper limit) and 2.22 (lower limit). The operator of the machine that is producing the part makes periodic checks of the diameter of parts and plots the results on the control chart. As long as the plots fall within the control limits, the process is operating normally. If plots fall outside the limits, the process is adjusted immediately.

SPC offers several advantages: (1) it gives control of quality to the process operator (the person who knows the process best); (2) it reduces the work of QC personnel; (3) it reduces the cost of scrap and rework; and (4) it reduces the potential for sending defective parts to the customer. Figure 18–6 is an example of a control chart with various plots. Notice that one of the plots fell outside the control limits, adjustments were made, and the process returned to an in-control status.

Some supervisors resist SPC, at least at first, because the word statistical frightens them. While it is true that a basic understanding of algebra and elementary statistics is helpful, supervisors do not have to be accomplished statisticians in order to use SPC. The most important knowledge a supervisor can have about SPC is the knowledge of where and how it can be an asset in an organization.

Next a supervisor needs to be familiar with the steps involved in implementing SPC. According to Loye Dockendorf, an IBM statistician, these steps are: (1) analyzing the process to determine capabilities and what represents normal operation; (2) correcting external causes as opposed to natural causes; and (3) setting up control charts.[13] Finally, once an SPC system is established, supervisors need to know how to interpret the data it yields.

The key role supervisors can play with regard to SPC is to make sure that informa-

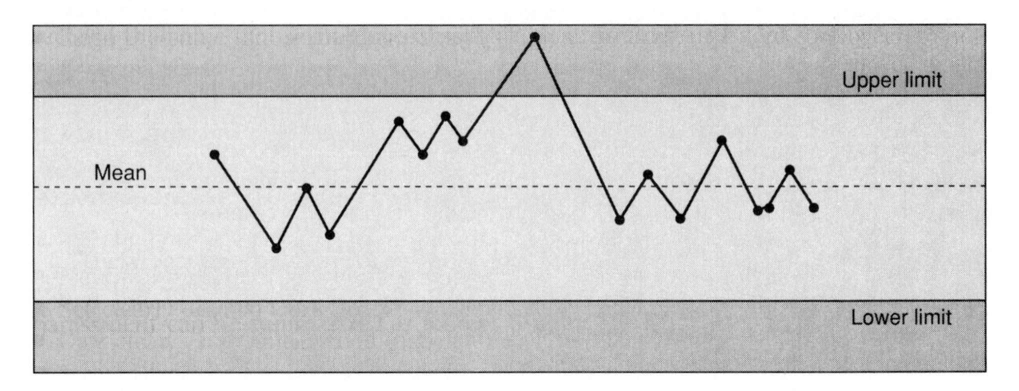

Figure 18–6
Sample control chart for SPC.

tion yielded by the system is acted on and acted on promptly. Another key role is to ensure that employees get the education and training they need, on a continual basis, to be effective users of SPC.

Total Quality Control (TQC)

TQC is a comprehensive approach to quality in which all the quality efforts of the various units in an organization are integrated to accomplish the goal of complete customer satisfaction in the most economical manner possible. TQC is not just a different approach to quality, it is a different approach to management. This is why the concept is sometimes called total quality management (TQM).

Implementing TQC requires a total commitment of all employees from executive management to operational workers. According to Leland C. Hunter, senior vice president for Florida Power & Light Company, TQC is simple to sell to employees.[14] Hunter says managers should simply explain to employees that they cause 15 percent of all defects and management causes 85 percent, then ask them to help identify and solve the mistakes both they and management are making.

Supervisors can play a key role in TQC by serving as the communications link between employees and management. They are responsible for making sure that effective feedback mechanisms are in place and that employees use them. Supervisors are also well positioned to let management know when policies and practices are inhibiting the realization of the benefits of TQC. Hunter calls TQC "a totally new style of management" that requires top management to make a commitment and stick to it.

BUILDING A QUALITY WORKFORCE

Quality is the result of people's efforts; it cannot be just ordered and mandated. If quality is produced by the workforce, and it is, then the key to producing quality products or providing quality services is to build a quality workforce.

American quality consultant Phil Crosby developed a fourteen-point program for building a quality workforce:[15]

- *Obtain the total commitment of management.* Top management must decide that quality is going to be the top priority and then must communicate that message to all employees.
- *Establish a quality improvement team.* Name representatives from all departments, let the team select a chairperson, define the mission, and define individual roles.
- *Establish* **quality measurement standards.** Determine what the level of quality is now and make records. This will establish a basis upon which to measure improvement.
- *Establish cost estimates of quality.* What percentage of the bottom line will be spent on quality efforts? Work with the cost accounting department to establish estimates.
- *Help employees develop quality awareness.* When the base levels for quality have been established, share the information with all employees and help them understand it.
- *Undertake corrective and improvement measures.* Begin making improvements immediately and measure them.
- *Implement a zero defects program under the leadership of an ad hoc committee established especially for this purpose.* Build companywide awareness of and support for a zero defects program (this is covered in the next section of this chapter).
- *Provide training, particularly for supervisors.* All employees will need training in order to produce quality. This is especially true of supervisors. Supervisors play a critical role in a company's quality program.
- *Implement a companywide Zero Defects Day.* This is a special day to spread the word and reinforce it. Such a day will help build a positive quality attitude.
- *Set goals.* Supervisors should work with their units to set goals that are both realistic and challenging. Then the quality of work should be measured against the goals.
- *Remove causes of errors and defects.* Empower all employees to identify causes of errors and defects and work with them to remove the causes.
- *Recognize both quality and improvement.* Establish a program for highly visible recognition of effective contributions to quality. Visibility and recognition is more important than financial reward.
- *Establish a* **quality council.** The council should be broad-based consisting of quality professionals and chairpersons of the individual quality teams. It is responsible for evaluating and continually upgrading the overall quality program.
- *Repeat all the previous steps.* Establishing a quality program does not happen overnight, nor it is a one-time activity. Establishing a quality program is an ongoing process.

These steps can be used by any company to implement a quality program. The supervisor plays a key role both in knowing the process and in participating in it. Supervisors can work with higher management to follow this process.

INSPECTION SYSTEMS AND QUALITY[16]

There is a saying among quality professionals that "you cannot inspect quality into a product." This is true. The purpose of inspection is to identify defects so that the causes can be identified and corrected. Japanese quality expert Shigeo Shingo places inspection systems in three categories: Judgment inspections, informative inspections, and source inspections.

Judgment Inspections

Judgment inspections are after-the-fact inspections that separate products into two categories: defective and acceptable. This is the type of inspection system that has traditionally been used in many industrial firms. Acceptable products are passed along to the customer and defective products are scrapped. The customer pays for both the acceptable and the defective products which, of course, drives up the cost of products delivered. This, in turn, makes the producer of the product less competitive.

An even worse problem with judgment inspections is that it is not uncommon for defective products to be passed along to the customer. This happens because it is typically too expensive to inspect 100 percent of the products produced. As a result, defective products may slip through the system and go to the customer. When this happens, the producer of the product loses credibility and, possibly, the customer.

The most glaring weakness of judgment inspections is that they do nothing to reduce the defect rate. Therefore, supervisors in companies that rely on judgment inspections should work with management to encourage the adoption of informative or, ideally, source inspections.

Informative Inspections

Informative inspection systems are superior to judgment inspection systems because they reduce the defect rate. When a defect occurs, information feeds back to the machine or process and corrective measures are undertaken immediately. This does not prevent the defect that triggered the feedback, but it does prevent additional defects.

Shingo divides informative inspection systems into three categories: (1) statistical quality control systems; (2) successive check systems; and (3) self-check systems. Such systems have the following characteristics: use of control charts to detect defects, use of statistics to establish control limits that distinguish between in-control and out-of control operation, and use of statistics to determine the number of samples to be taken to determine out-of-control operation.

There are two critical weaknesses with statistical quality control techniques: (1) since they rely on sampling, even though the samples are statistically derived, they are still not 100 percent foolproof; and (2) the time that elapses between when a defect is detected and when corrective action is fully accomplished can be too long.

Successive check systems are designed to accomplish 100 percent inspection. The work is inspected by a worker other than the one doing it. Immediate feedback is given that results in immediate corrective action. Figure 18–7 illustrates how successive check inspection works.

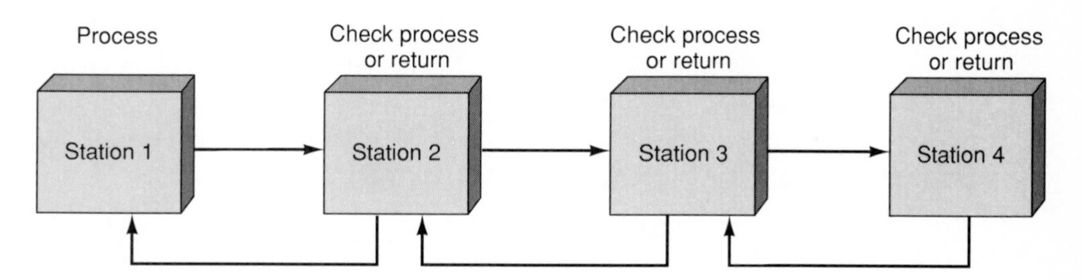

Figure 18–7
How successive check inspection systems work.

The figure shows four workstations through which the product passes. The worker at Station 1 completes his portion of the process and passes the product to the worker at Station 2. This worker performs an immediate inspection. If the product passes the inspection, she performs her part of the process. If not, the product is returned to the previous station for immediate corrective action. This is repeated at every station until the product passes through all four stations.

Self-check systems are an improvement on successive check systems in that the individual worker checks his or her own work, thereby making detection, feedback, and corrective action truly immediate. The obvious weaknesses inherent in self-check systems are that people who inspect their own work may not be critical enough and, when in a hurry, may forget to check at all.

These weaknesses limit the use of self-check systems to situations in which inspections are physical (i.e., visual checks). Where devices can be used to physically check a part, self-check systems can be used effectively. Supervisors should examine their work processes to determine if informative inspections can be used and, if so, what type is the most appropriate.

Source Inspection Systems

Source inspections eliminate defects by identifying errors, feeding back, and taking corrective action before an error causes a defect. Notice the distinction drawn between errors and defects. An error is not a defect; it is a mistake that will lead to a defect if not corrected in time.

Source inspections require shorter work/check cycles and, of course, the checks must be made at the source of the work by the person doing the work. Like self-check systems, source inspections are best suited to those situations that involve physical as opposed to sensory checks.

QUALITY IN A NONPRODUCTION SETTING

Quality is not limited to production settings; it is just as important in the other departments of a company. Errors that cause defects must be corrected and the work redone. Errors that become defects increase the cost of the department's overhead.

Consequently, quality is just as important in the supportive departments as it is on the shop floor. All the various quality improvement strategies set forth in this chapter apply in both production and nonproduction settings.

MALCOLM BALDRIGE NATIONAL QUALITY AWARD

During the mid-1980s, in response to intense competition from abroad, the U.S. Congress became interested in promoting quality in American industry. Its vehicle for promoting quality, the Malcolm Baldrige National Quality Award, was established in August 1987 by Public Law 100-107. According to the Department of Commerce, "The award, named for the late Commerce Secretary Malcolm Baldrige, . . . promotes national awareness about the importance of improving total quality management and recognizes quality achievements of U.S. companies."[17]

Companies applying for the Baldrige Award are evaluated in seven categories: leadership, information and analysis, strategic quality planning, human resources, quality assurance of products and services, quality results, and customer satisfaction (Figure 18–8). The award program is administered by the National Institute of Standards and Technology (NIST) with assistance from the private sector. Only two awards per year are given in each of three categories: large manufacturers, large service companies, and small businesses.[18]

Winners in the large manufacturers category have included Motorola, the Nuclear Fuel Division of Westinghouse Electric Corporation, Xerox Business Products and Systems, IBM–Rochester, and Cadillac Motor Car Division. Federal Express was the first company to win in the service category. A small business winner, Wallace Co. was named in 1990. Wallace, a Houston-based distributor of pipe, valves, fittings, actuated valves, and plastic-lined pipe, employs a total of 280 people.

The types of criteria used to evaluate applicants for the Baldrige Award are summarized in Figure 18–9. By examining the criteria closely one can identify several areas where supervisors can play a key role. Supervisors can help establish quality values under the leadership category. Under the information and analysis category, supervisors can play a key role in analyzing quality data and acting on what the analysis tells them.

Figure 18–8
Major criteria for the Baldrige Award.

Areas of Evaluation of Applicants for the Malcolm Baldrige National Quality Award

- Leadership
- Information and analysis
- Strategic quality planning
- Human resources
- Quality assurance of products and services
- Quality results
- Customer satisfaction

Figure 18–9
How an applicant is evaluated for the Baldrige Award.

Examination Criteria
Malcolm Baldrige National Quality Award

Category	Maximum Points
1. Leadership	100
■ Senior executive leadership (30)	
■ Quality values (20)	
■ Management for quality (30)	
■ Public responsibility (20)	
2. Information and Analysis	60
■ Scope and management of quality data and information (35)	
■ Analysis of quality data and information (25)	
3. Strategic Quality Planning	90
■ Strategic quality planning process (40)	
■ Quality leadership indicators in planning (25)	
■ Quality priorities (25)	
4. Human Resource Utilization	150
■ Human resource management (30)	
■ Employee involvement (40)	
■ Quality education and training (40)	
■ Employee recognition and performance measurement (20)	
■ Employee well-being and morale (20)	
5. Quality Assurance of Products and Services	150
■ Design and introduction of quality products and services (30)	
■ Process and quality control (25)	
■ Continuous improvement (25)	
■ Quality assessment (15)	
■ Documentation (10)	
■ Quality assurance of support services and processes (25)	
■ Quality assurance of suppliers (20)	
6. Quality Results	150
■ Quality of products and services (50)	
■ Comparison of quality results (35)	
■ Business process, operational, and support service quality improvement (35)	
■ Supplier quality improvement (30)	

Figure 18–9
(Continued)

Examination Criteria
Malcolm Baldrige National Quality Award

Category	Maximum Points
7. Customer Satisfaction	300
■ Knowledge of customer requirements and expectations (50)	
■ Customer relationship management (30)	
■ Customer service standards (20)	
■ Commitment to customers (20)	
■ Complaint resolution (30)	
■ Customer satisfaction determination (50)	
■ Customer satisfaction results (50)	
■ Customer satisfaction comparison (50)	

Under the category of human resource utilization supervisors can involve employees in improvement efforts, provide quality training, measure employee performance, and help ensure appropriate recognition for employees. Under the category of quality assurance of products and services, supervisors can play key roles in quality control, improvement, assessment, and documentation. In the category of quality results, supervisors can be particularly helpful in comparing quality results. Finally, regarding customer satisfaction, supervisors can make sure that their employees are aware of customer requirements and expectations and customer satisfaction results. Whether or not a company actually applies for the Baldrige Award, supervisors can still use the criteria in Figure 18–9 to help improve quality continually in their organizations.

SUMMARY

1. Quality can be defined as a measure of the extent to which a product or service conforms to expectations.
2. Key quality-related terms and concepts include quality control, quality assurance, acceptable quality level, inspection, total quality management, zero quality control systems, statistical quality control, and Poka-Yoke systems.
3. The traditional philosophy of quality is that it is expensive. A more contemporary philosophy is that poor quality costs. In a competitive marketplace, poor quality will lead to a loss of marketshare. The cost of quality can be divided into two categories: (1) building in quality, and (2) correcting defects.
4. The success of the Japanese in the areas of quality and productivity can be attributed to both work motivation and work methods: no inventory, small production lot sizes, zero quality control systems, preventive maintenance, people-oriented productivity improvement, and work flow process improvement.

5. W. Edwards Demming listed seven points for the supervisor for improving quality: do not accept commonly applied levels of defects; continually improve the system; focus on helping employees to do a better job; provide the necessary tools and techniques; communicate; eliminate barriers among departments; and institute a comprehensive program of education and training.
6. Management and supervisors can help employees be positive agents for improving quality by: personalizing quality; providing the necessary training; communicating continually; providing up-to-date tools and equipment; conditioning employees to view quality through the eyes of the customer; and empowering employees to make quality improvements.
7. Bob Schwarz sets forth seven key elements that will encourage employee participation in suggestion programs: all suggestions should receive a formal response; respond to all suggestions immediately; monitor the response record of all departments; report system costs and savings; handle recognition and awards promptly; implement good ideas; and minimize personality conflicts.
8. Contemporary quality-related concepts include just-in-time (JIT) production, manufacturing resources planning (MRPII), statistical process control (SPC), and total quality control (TQC).
9. The three basic types of inspection systems are judgment, informative, and source inspections. Judgment inspections are after-the-fact inspections that separate products into two categories: defective and acceptable. They do not reduce defects. Informative inspection systems identify defects and feedback quickly so that corrective action can be taken. They reduce defects. Source inspections eliminate defects by identifying errors immediately at the source and triggering immediate corrective action.
10. All of the quality measures used in a production setting can also be used in other settings such as accounting, engineering, marketing/sales, data processing, and maintenance.

KEY TERMS AND PHRASES

Acceptable quality level
Constant stimulation
Cost of quality
Defects
Dumbing down
Empowering employees
First article
Flexible manufacturing cell (FMC)
Informative inspections
Inventory
Japanese miracle
Judgment inspections
Just-in-time (JIT)
Manufacturing resources planning (MRPII)
Multiple-function team
Parallel structure
Personalizing quality
Poka-Yoke systems
Production lot sizes
Productivity
Quality council
Quality improvement
Quality measurement standards
Quality
Quality assurance

Quality control
Source inspections
Statistical process control (SPC)
Statistical quality control
Suggestion system
Things gone wrong (TGWs)
Total quality control (TQC)
Total quality management (TQM)
Work methods
Work flow process improvement
Work motivation
Zero quality control systems

CASE STUDY: The Family Approach to Quality[19]

The following real-world case study contains an example of one company's approach to quality. Read the case study carefully and answer the accompanying questions.

Patagonia is a widely recognized innovator among garment manufacturers. What began as a small family business has grown into an industry leader employing over 500. In spite of its size, Patagonia still takes the family approach to quality. Patagonia's employees ("family members") are its primary source of improvement ideas.

The underlying philosophy of Patagonia's quality program is to have happy, fully empowered employees. To accomplish this, the company's benefits include on-site day care and kindergarten, employee discounts, subsidized wilderness vacations, and liberal health benefits. The company's "Opportunity for Improvement" (OFI) program is designed to empower employees to make improvements.

The backbone of the OFI program is a companywide suggestion system. Employees can pick up a suggestion form at a variety of locations throughout the company. The form asks these questions:

- What task or process needs improvement?
- Why?
- How do you recommend implementing improvements?

Employees keep a copy of their suggestions, send one to their supervisor, and another to a recording office where it is entered into a database. The supervisor must acknowledge receipt of the suggestion within a day. Supervisors solve the problem if they can or work with a quality improvement team (QIT) if help is needed. Rewards for suggestions that result in improvements are token gifts such as movie tickets or dinner for two.

The primary goal of the OFI process and the QITs is to involve employees in quality improvement and empower them so they know they can make a difference. Patagonia attributes most of its increased profits to its unique family approach to quality improvement.

1. How does Patagonia's family approach to quality improvement compare with the Japanese approach?

2. What do you see as the supervisor's role in the Patagonia system?
3. What additions or deletions (if any) would you make in Patagonia's system? Why?

REVIEW QUESTIONS

1. Tom Peters, author of *Thriving on Chaos,* summarizes his conclusions relative to quality in several brief statements. Reduce Peters's conclusions to a brief quality philosophy.
2. Define the term *quality* and list two key points relating to quality.
3. Explain the following quality-related terms: quality control; acceptable quality level; inspection; statistical quality control.
4. Justify or refute the following statement: "Quality costs too much."
5. Compare and contrast the quality philosophies of Japanese and U.S. companies in terms of work motivation in the following areas: salary systems; employment; labor unions.
6. Compare and contrast the quality and philosophies of Japanese and U.S. companies in terms of work methods in the following areas: inventory; quality control; productivity.
7. W. Edwards Demming gave seven points for supervisors to improve quality. List and explain the three strategies you feel are the most important.
8. Explain how to make quality visible.
9. There are several strategies supervisors can use to help employees be positive agents for improving quality. Explain the three you feel are the most important and why.
10. Bob Schwarz listed seven key elements that will encourage employee participation in suggestion programs. Explain the three you feel are the most important and why.
11. Explain the criteria you would use to evaluate a company's quality program.
12. How does the concept of JIT relate to quality improvement in the modern industrial setting?
13. Explain how the concept of TQC relates to MRPII.
14. Give a supervisor's view of how SPC works.
15. What role can supervisors play in ensuring the success of TQC?
16. Explain from a supervisor's perspective how to build a quality workforce.
17. Compare and contrast judgment, informative, and source inspections.

SIMULATION ACTIVITIES

The following activities simulate the types of quality-related problems faced by supervisors. Apply what you have learned in this and previous chapters to solve these problems.

1. Your company is in the process of developing a new quality philosophy. During a meeting a fellow supervisor says, "We can't afford all the money it's going to take to have that level of quality." Explain how you would refute this argument.
2. The workers in your unit keep reading about how Japan is leading the world in

manufacturing. One of them comes to you and asks, "What are the Japanese doing that's so different? How about explaining it to us?" How would you explain the differences to your unit?

3. Higher management has formed an ad hoc task force of supervisors to develop supervisory strategies for enhancing quality. You have been elected chair of the committee. What strategies will your committee recommend? Put your strategies in the form of a checklist.
4. You have been asked by your manager to get your employees involved in quality. Develop a plan for doing so.
5. Develop a plan for implementing a companywide suggestion program and explain your strategies.

ENDNOTES

1. Peters, T. *Thriving on Chaos* (New York: Harper & Row, 1987), p. 83.
2. Ibid., p. 78.
3. Ibid., pp. 82–83.
4. Shingo, S. *Zero Quality Control, Source Inspection and the Poka-Yoke System* (Stamford, CT: Productivity Press, 1986), pp. 265, 266.
5. Ibid., pp. 266–269.
6. Office of Technology Assessment. *Making Things Better: Competing in Manufacturing* (Washington, D.C.: U.S. Congress, 1990), pp. 5–7.
7. Peters, T. *Thriving on Chaos,* p. 84.
8. Bittel, L. and Newstrom, J. *What Every Supervisor Should Know.* (New York: McGraw-Hill, 1990), p. 411.
9. McDermott, B. *"Employees are Best Source of Ideas for Constant Improvement," Total Quality Newsletter,* July–August 1990, Volume 1, Number 4, p. 5.
10. Ibid., pp. 5, 6.
11. Peters, T. *Thriving on Chaos,* pp. 85–98.
12. Ibid., p. 93.
13. McDermott, B. "Nontechnical Managers Can Lead Statistical Control Process," *Total Quality Newsletter,* July–August 1990, Volume 1, Number 4, p. 4.
14. Hunter, L. C. "Accepting the Concepts of Total Quality Can Be Humbling," *Total Quality Newsletter,* July–August 1990, Volume 1, Number 4, p. 7.
15. Peters T. *Thriving on Chaos,* pp. 85–98.
16. Shingo, S. *Zero Quality Control,* pp. 57–60, 82–83.
17. U.S. Department of Commerce News. "Four Companies Named Winners of Baldrige Quality Award," *Commerce News,* October 10, 1990, p. c-1.
18. Ibid., p. c-2.
19. McDermott, B. (Ed.) "Favorable Work Environment Lays Foundation for Quality," *Total Quality Newsletter,* October 1990, Volume 1, Number 6, p. 384.

PART FIVE

The Supervisor as a Legal Advisor

CHAPTER NINETEEN

Equal Opportunity and the Supervisor

CHAPTER OBJECTIVES

After studying this chapter, you will be able to explain or describe the following topics:

- The Changing Demographics of the Workplace
- Federal Legislation and Executive Orders Relating to Equal Opportunity
- How Various Categories of Employees Are Protected
- Sexual Harassment as a Special Concern
- The Agencies that Enforce Equal Opportunity Legislation
- The Concept of Affirmative Action
- How to Promote Diversity in the Workplace

CHAPTER OUTLINE

- Changing Demographics of the Workplace
- Federal Legislation
- Special Categories and Concerns
- Family Concerns and the Supervisor
- Sexual Harassment and the Supervisor
- Enforcement Agencies
- Affirmative Action and the Supervisor
- Promoting Diversity in the Workplace
- Summary
- Key Terms and Phrases
- Case Study Application Problem
- Review Questions
- Simulation Activities
- Endnotes

The one word that best characterizes the modern workplace is diversity. This chapter will help prospective and practicing supervisors learn how to turn diversity into a strength.

CHANGING DEMOGRAPHICS OF THE WORKPLACE

Modern supervisors face the challenge of **diversity** to a greater extent than did their predecessors. Over the past twenty years the size and composition of the workforce in the United States has changed radically. According to management consultant Seymour L. Wolfbein, changes in the size and composition of the workforce are the result of the following two major factors:[1]

- Population **demographics** changes with regard to age, sex, race, color, and ethnicity.
- Changes in lifestyle, spending patterns, attitudes, and value systems regarding work, child care, and health care.

Workforce of Today and Tomorrow[2]

Perhaps the most significant trend in the last decade of this century will be a continual increase in the number of women in the workforce. Between now and the year 2000, the number of women entering the workforce will double that of the number of men. Over 70 percent of the labor force increases during this period will be accounted for by women. **Minorities** will also account for a much larger percentage of the growth in the labor force.

The number of young workers (16–24 years old) will also increase, reversing a decade-long downturn. The number of older workers (over 60) reentering the workforce will show a slight increase during this period.

Figure 19–1
Workforce demographics in the United States, present and future.

Civilian Labor Force of the U.S. (In thousands)

	1990	2000
Men	68,309	74,324
Women	56,896	66,810
White	107,449	118,981
Black	13,771	6,465
Asian & Other	3,986	5,688
Hispanic	9,789	14,321

Wolfbein summarizes his projections of who will enter the workforce by the turn of the century as follows:

- One-third minorities (blacks, Hispanics, Asians/others)
- One-third women (white non-Hispanic)
- One-third men (white non-Hispanic)

Figure 19–1[3] summarizes the composition of the projected civilian labor force in the United States between 1990 and 2000.

From this table it should be clear that modern supervisors must be able to deal with diversity. This means working effectively with women, minorities, older workers, disabled workers, pregnant workers, workers with different cultural heritages and religious beliefs, as well as those workers who have historically made up the workforce.

FEDERAL LEGISLATION

In the mid-1960s, in an attempt to move the country toward accomplishing equal opportunity and equal access in the workplace, the federal government became actively involved. Several key pieces of federal legislation have been established that have a direct impact on employment practices (see Figure 19–2). Supervisors should be familiar with these laws and executive orders and how they relate to their interaction with employees. They should also be familiar with the guidelines set forth by the Equal Employment Opportunity Commission (EEOC) and the Office of Federal Contract Compliance Programs (OFCCP).

The various laws in Figure 19–2 pertain to **equal employment opportunity**, fair compensation, employee health and safety, and labor/management relations. Those applying specifically to equal employment opportunity and compensation are explained in this section. Those dealing with health/safety and labor/management relations are covered in Chapters 20 and 21, respectively.

Equal Opportunity Laws and the Supervisor

The most fundamental equal opportunity law with which supervisors must deal is the **Civil Rights Act** of 1964 and the amendments to the act of 1972. The original act prohibits discrimination in employment based on race, sex, age, national origin, or religion. Title VII of the act extends this provision by specifically prohibiting such practices when making employment decisions (i.e., hiring, promotions, compensation, layoffs, etc.).[4] The 1972 amendments to the Civil Rights Act gave the Equal Employment Opportunity Commission (EEOC) more latitude in charging public and private employers with fifteen or more employees with noncompliance.

The lawsuits that resulted led the courts to develop two concepts for challenging employment practices: disparate impact and disparate treatment,[5] which we discussed in Chapter 9. Supervisors should be familiar with both of these concepts. Both are used to show that employment practices have an *adverse impact.*

Fair Employment Practices Legislation
Chronological Summary

1926	Railway Labor Act (RLA)
1931	Davis-Bacon Act: prevailing wage laws
1935	OASDHI and amendments: mandated fringe benefits
1935	National Labor Relations Act
1935	Unemployment Compensation
1936	Walsh-Healy Act: prevailing wage laws
1938	Fair Labor Standards Act (FLSA)
1947	Taft-Hartley: labor/management relations
1959	Landrum-Griffin Act: labor/management reporting and disclosure
1963	Equal Pay Act: amendment to FLSA
1964	Civil Rights Act (Title VII)*
1967	Age Discrimination in Employment Act (revised in 1978)
1970	Occupational Safety and Health Act (OSHA)
1972	Equal Employment Opportunity Act (EEOA)
1973	Vocational Rehabilitation Act (amended in 1980)
1974	Employment Retirement Income Security Act (ERISA)
1974	Vietnam Era Veterans Readjustment Act
1974	Freedom of Information Act
1974	Privacy Act
1977	Minimum Wage Law (revised in 1990)
1978	Civil Service Act (Title VII)
1978	Pregnancy Discrimination Act (amendment to Title VII of the Civil Rights Act)
1990	Americans with Disabilities Act

* The Civil Rights Act of 1990 was passed by Congress but vetoed by President Bush on October 22, 1990. An attempt to override the president's veto failed. As of June 1991, Congress was debating a new Civil Rights Act.

Figure 19–2
Historical development of legislation promoting fair employment practices.

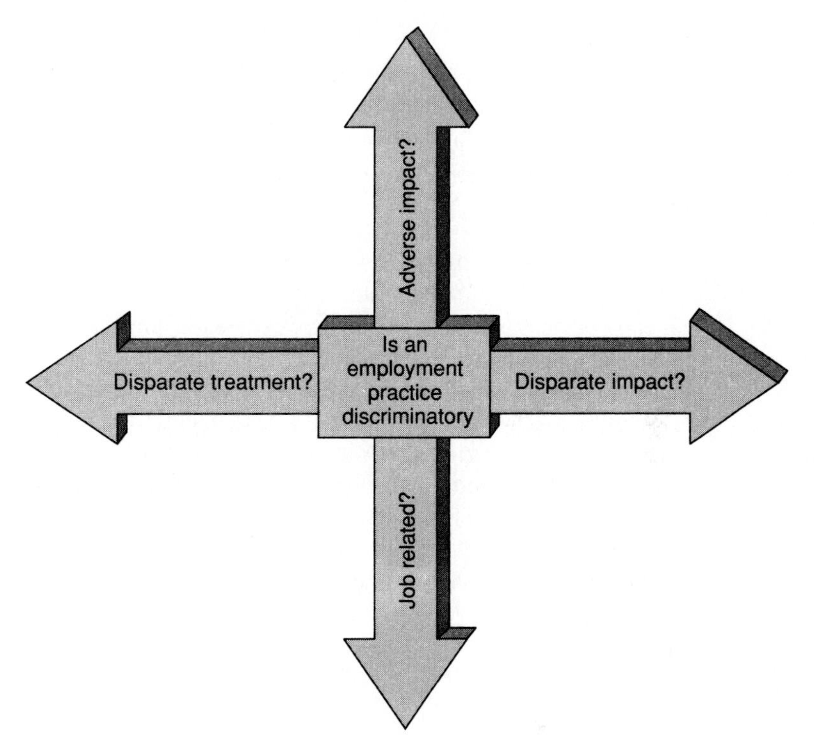

Figure 19–3
Key concepts in equal employment opportunity.

Disparate impact means that a particular staffing practice has an adverse effect on a group of people (i.e., women, older workers, blacks, etc.). The burden of proof rests with the employee to show disparate impact. However, if this can be done the burden shifts to the employer to justify the practice. An employer's justifications are evaluated according to the theory of **job relatedness** or the extent to which the practice relates directly to the job (Figure 19–3).

Recall the case study in Chapter 15, in which the automobile battery manufacturer did not allow women to perform certain jobs. The practice was based on the company's fear that the unborn babies of pregnant women might be harmed. This practice was challenged as having a disparate impact on women. In March 1990, the U.S. Supreme Court held that the practice was discriminatory.

Disparate treatment is used to challenge an employment practice that is perceived as being unfair to an individual. The key concept in disparate treatment cases is **employer intent.** Did the employer apply the practice in question with the intention of discriminating?

The U.S. Supreme Court has developed a four-step test that is now used for determining disparate treatment. For example, say a person applies for a job and is turned down. As a result he or she files a complaint claiming disparate treatment. The test would be applied as follows:

1. Is the complainant a minority? (If yes, go to the next step.)
2. Did the complainant apply for the job? (If yes, go to the next step.)
3. Was the complainant's application turned down? (If yes, go to the next step.)
4. Did the employer continue to seek other applicants? (If yes, the complainant has shown adverse impact.)

At this point the burden shifts to the employer to show why the employment decision concerning the complainant was not discriminatory. The concept of job relatedness becomes critical here. Supervisors should be mindful of this concept in their daily decisions about and interactions with employees.

The trend appears to be toward strengthening the Civil Rights Act. The Civil Rights Act of 1990 sponsored by Senator Edward Kennedy of Massachusetts and Representative Augustus Hawkins of California passed both the House of Representatives and the Senate with strong majorities (273–154 in the House and 62–34 in the Senate).

However, siding with the U.S. Chamber of Commerce and other business interests, President George Bush vetoed the act on July 22, 1990. Opponents of the bill claimed it would encourage lawsuits, make it more difficult for employers to defend against discrimination charges, and force businesses to adopt quota systems. In vetoing the bill, President Bush proposed his own version that would limit the total award of damages any one complainant could win in a lawsuit to $150,000. With strong sentiment for the new civil rights bill in both the House and Senate, continued strenghtening of the act is likely.

In addition to the Civil Rights Act several other laws are aimed at preventing discrimination and promoting diversity in the workplace. Prominent among these are the Age Discrimination Act of 1967 (amended in 1978), the Vocational Rehabilitation Act of 1973, and the Vietnam Era Veterans Readjustment Act of 1974. The Age Discrimination Act gives special protection to persons between the ages of 40 and 70 and the Vocational Rehabilitation and Vietnam Era Readjustment Acts give special protection to persons with physical and mental handicaps.

All the legislation outlined in this section, taken together, represents an active attempt on the part of the federal government to eliminate discrimination and promote diversity in the workplace. What it means to supervisors is that it is imperative to base all decisions in the workplace on job relatedness.

Compensation Laws and the Supervisor[6]

Today's compensation laws have their foundation in the Fair Labor Standards Act (FLSA) of 1938, which is enforced by the U.S. Department of Labor. This act established child labor standards as well as wage and hour guidelines. Two key sections in the law define exempt and nonexempt employees (Figure 19–4).

Nonexempt employees are those entitled to overtime pay and at least the minimum wage. Such employees have come to be called *hourly workers* or *wage earners*. **Exempt employees** are professional- and management-level personnel. Such employees are called *salaried* and are not normally covered by the overtime and minimum wage

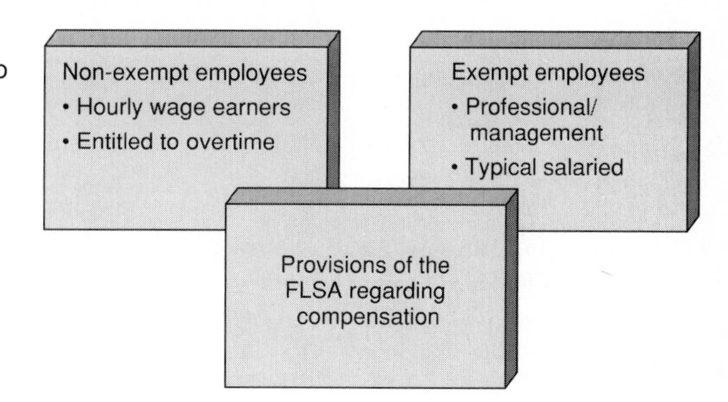

Figure 19–4
Key concepts relating to compensation.

provisions of the FLSA. As the first level of management, supervisors typically fall into this category.

The Equal Pay Act of 1963, enforced by the EEOC, requires equal pay for equal work done by men and women. Equal work is defined as "jobs that require equal preparation, knowledge, skills, effort, and responsibility that are performed under similar working conditions." There are four allowable exceptions to the provisions of the Equal Pay Act. Equal pay is not required when differences are based on one of the following (Figure 19–5):

- Seniority system
- Merit system

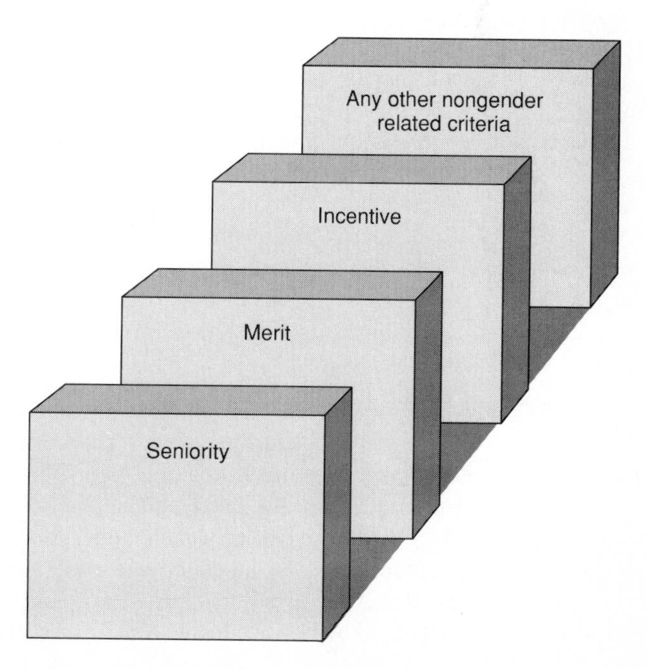

Figure 19–5
Equal pay is not required when differences are based on these factors.

- Incentive system
- Any nongender-based criteria

The effect of this act has been to prevent employers from paying men more simply because they are men.

The Pregnancy Discrimination Act of 1978 defines pregnancy as a disability entitling the pregnant employee to disability benefits. The Employee Retirement Income and Security Act (ERISA) of 1974 protects the private pensions of employees. All these laws mean that the modern supervisor should base all compensation-related recommendations solely on job performance.

SPECIAL CATEGORIES AND CONCERNS

The modern supervisor must be able to function effectively in a diverse work environment. Fairness toward all employees is important. However, certain groups require special attention. These groups include minorities, women, pregnant workers, older employees, the disabled, and the disadvantaged. Religion, nationality, and child care are also issues of concern to supervisors. This section explains the role of supervisors in relation to these special categories and areas of concern.

Minorities and the Supervisor[7]

In companies that have a definite and clearly stated commitment to equal employment opportunity, supervisors can play a pivotal role. According to Edwin B. Flippo, the supervisor's introduction and orientation of new minority employees is of utmost importance. They must be oriented to the workplace and introduced to fellow employees who may or may not make them feel welcome.

Consequently, supervisors should begin by establishing open and frank communication with minority employees. They need to know that the supervisor will not tolerate treatment by other employees that makes minorities uncomfortable or that adversely effects job performance. Supervisors who learn of such treatment should take immediate corrective action in accordance with applicable company policies.

In dealing with new minority employees, supervisors should expect acceptable performance, assist in making improvements, criticize in a constructive manner, and discipline using the same approach that would be applied with any other employee.

Female Employees and the Supervisor

Women now outnumber men in the workplace and their number continues to grow.[8] As this trend continues, not only will there be more female workers, there will be more women performing jobs traditionally held by men. This presents a special challenge to supervisors, particularly male supervisors.

The first step in effectively supervising female employees is to understand that biases can be deeply ingrained and subtly exhibited. This was illustrated in a study conducted by Benson Rosen and Thomas Jerdee, the results of which were published in the *Harvard Business Review.*[9] In this study, identical situations were presented to subjects. However, changing from a male to female name in the scenario produced significantly different responses. The attitudes of study subjects toward hiring, training, and transferring employees were markedly different depending on whether the employee in question was male or female.

In dealing with female employees, male supervisors should apply the same rule of thumb that should be used in dealing with any employee: focus on the employee's job performance only. This rule applies to supervisors of either gender working with employees of either gender.

In addition to focusing on job performance, supervisors can play a positive role in promoting equal employment opportunity for women (Figure 19–6) by:

- Encouraging management to eliminate **gender-based job titles.** For example, the title "draftsman" can be changed to "drafter" or "drafting technician."
- Making training opportunities available to women and men on an equal basis.
- Arranging in-house internships that allow women to gain the experience and skills necessary to perform jobs traditionally held by men.

Figure 19–6
Strategies for improving interaction with women on the job.

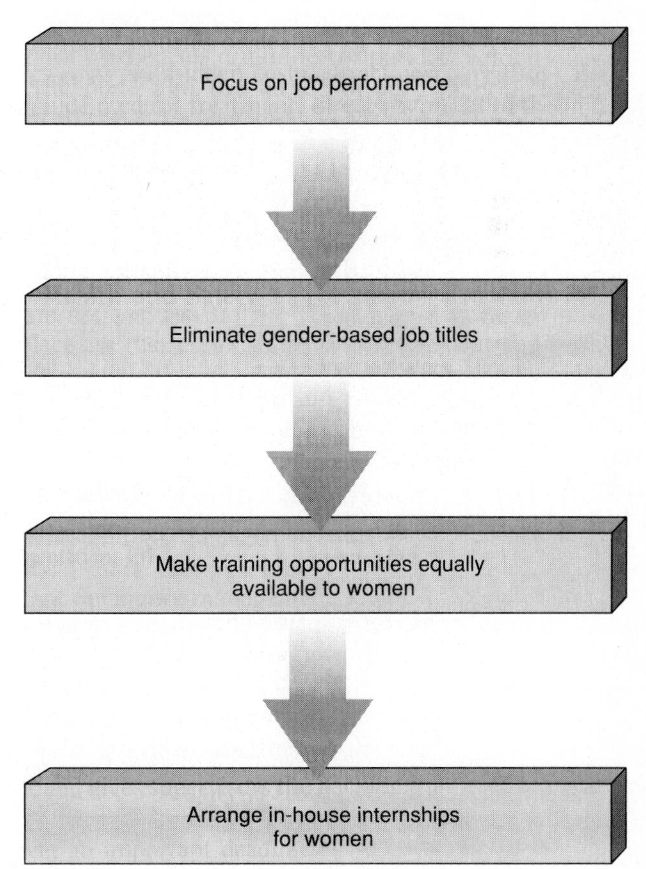

Pregnant Workers and the Supervisor

Pregnant women have employment-related rights that go beyond those accorded female employees in general. The Pregnancy Discrimination Act of 1978 (an amendment to the Civil Rights Act) prohibits discrimination in hiring, promotion, supervision, termination, or any other term of employment as a result of pregnancy, childbirth, or related medical conditions.[10]

The supervisor's role in dealing with pregnant workers can be summarized as follows:

- Communicate openly and continually to ensure that: (1) the pregnant worker knows her rights; and (2) the supervisor is aware of the worker's needs.
- Make provisions to accommodate the pregnant worker's temporary disability such as relief from heavy lifting. These provisions should be the same as those that would be made for any temporarily disabled employee.

Older Employees and the Supervisor

The number of older workers reentering the workforce will increase between 1990 and 2000.[11] This means supervisors must be able to deal effectively with older workers. People in this category (forty to seventy years old) are protected by the Age Discrimination in Employment Act of 1967 (as amended in 1978).

Most litigation relating to this act has centered around the questions of termination and forced retirement, although more recently the issue of medical benefits has begun to be a topic for litigation. The outcomes of court cases relating to forced retirement clearly favor the employee.

Using the Age Discrimination in Employment Act as the basis, numerous older workers have filed suits against employers and won. Companies that have been charged with age discrimination include Consolidated Edison, National Broadcasting, and Standard Oil.[12] In a 1974 age discrimination case, Standard Oil of California was required to pay $2 million to 264 employees. According to George P. Sape, vice president of a New York–based employee relations firm, when age discrimination cases go to court the employee is likely to win.[13]

More recently, retired employees have begun to sue past employers over changes to their retirement benefits, particularly medical benefits. Such cases are likely to increase as the cost of health care continues to rise. In 1987, a retired maintenance foreman filed suit against General Motors on behalf of 84,000 retirees when the company asked him to pay a greater share of his medical insurance premiums.[14] According to the complainant, General Motors promised him lifelong medical benefits and was now reneging on the promise. This case is typical of those growing out of the expectations of retired workers and the need of employees to cut costs.

The supervisor's role with regard to older workers has become clear as a result of age discrimination litigation. It can be summarized as follows:

- Never make promises to older workers in an attempt to entice them to retire.

- Treat older workers in the same manner as other workers and apply the same job performance standards.
- Make all work-related decisions concerning older workers based on job performance only.

Disabled Workers and the Supervisor

Disabled or handicapped workers have a physical or mental impairment that limits some major life function. Such workers are protected from discrimination by the Vocational Rehabilitation Act of 1973 which requires employers to: (1) take affirmative action to employ the disabled; and (2) make reasonable accommodations for handicapped workers.

The Americans with Disabilities Act of 1990 extends even further the accommodations aspects of equal opportunity and equal access for both disabled workers and disabled customer/clients. The main provisions of the bill are summarized in Figure 19–7.

The supervisor's role with regard to equal employment opportunity for disabled employees can be summarized as follows:

- Disregard disabilities when hiring, promoting, suspending, laying off, or terminating employees.
- Work closely with disabled employees to understand their special needs and to plan for the appropriate accommodations.
- Communicate the needs of disabled employees to higher management and serve as a liaison in making sure their needs are met.

Disadvantaged People and the Supervisor

Disadvantaged people are those who, because of educational or cultural limitations, are unable to participate in the economic mainstream. They tend to be among the long-

Main Provisions of the Americans with Disabilities Act

- All businesses must make their facilities accessible to the disabled unless costs are excessive.
- Public transportation must accommodate passengers in wheelchairs.
- Businesses with 15 or more employees must disregard disabilities in hiring and make appropriate accommodations unless the costs are excessive.
- Telephone companies must accommodate speech- and hearing-impaired people.

Figure 19–7
Provisions of the Americans with Disabilities Act.

term or **hard-core unemployed.**[16] As was shown in Chapter 16, this is a rapidly growing segment of the population.

Of the various laws discussed in this chapter relating to fair employment practices, none of them offer protection or special status for disadvantaged people. Consequently, there are no pressures from the federal government to hire the disadvantaged. Instead, the government attempts to encourage opportunities for people in this group through a variety of incentive programs.

The Job Training Partnership Act (**JTPA**) is the federal government's most widely used vehicle for attempting to make hard-core unemployed people employable. The JTPA established and funds Private Industry Councils (**PICs**) in local communities in every state. A PIC is responsible for identifying disadvantaged members of the community and working with local employers or educational institutions to get them positively and permanently placed in the workforce. PIC funding is used to pay training costs; to provide counseling, advising, and personal support; and to subsidize (on a limited basis) transportation and child-care costs during training.

Training may be provided by an educational institution or in an on-the-job training capacity. Businesses that hire and train PIC clients are reimbursed as much as half of the employee's wages for a specified period of time. This is the incentive the government provides to stimulate the process.

A number of major industrial firms have implemented programs designed to recruit, employ, and train disadvantaged people. These companies include Lockheed Aircraft, Boeing, Eastman Kodak, and Westinghouse. James Koch summarizes the lessons these and other companies have learned about employing disadvantaged workers.[17] According to Koch, success in employing the disadvantaged requires the following (Figure 19–8):

- A supportive work environment
- Job-specific training
- Commitment by top management
- Supervisors with strong skills in interpersonal relations

Elchavan Cohn and Morgan Lewis found that companies with the highest retention rates of disadvantaged workers were those that, at least temporarily, relaxed the rules on attendance and tardiness,[18] which tend to be more of a problem for disadvantaged workers than productivity and quality. Their output and quality can equal that of other workers, but poor attendance and tardiness are common among the disadvantaged, particularly in the early stages of employment.

This is due in part to the number of single-parent families in the disadvantaged population. Typically single mothers, these workers experience child-care problems, problems with sick children, and a multitude of other personal difficulties and have little or no help in dealing with them. This leads to tardiness and absenteeism. Companies that plan to employ and retain such workers must be prepared to be flexible with regard to tardiness and absenteeism.

Since the disadvantaged population is growing, many companies will be forced to

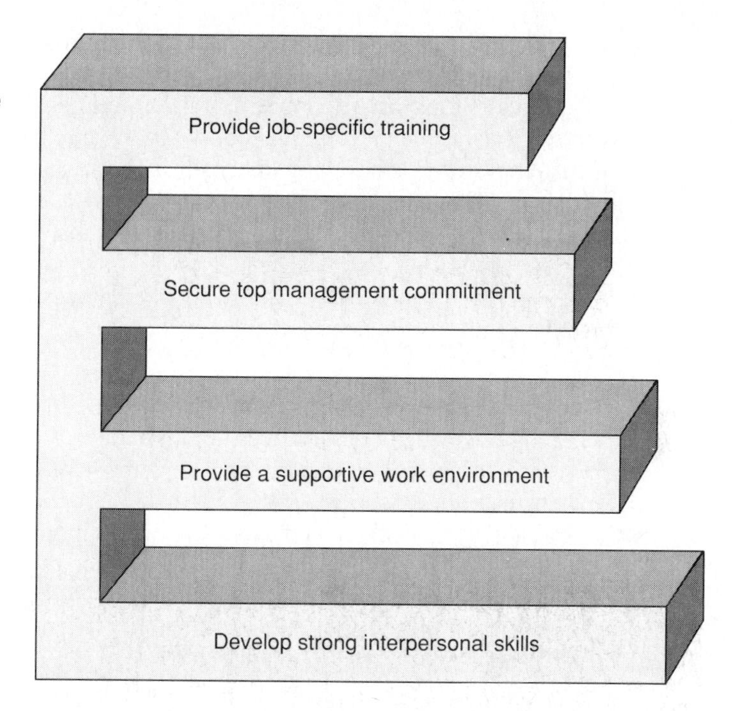

Figure 19–8
Strategies to improve the chances of success in the employment of disadvantaged people.

hire disadvantaged workers in order to meet their staffing needs. These facts present supervisors with a difficult challenge: how to build a world-class workforce composed, at least partially, of disadvantaged workers.

The supervisor's role in this regard and with regard to equal employment opportunity for the disadvantaged can be summarized as follows:

- Work with all members of the team to be open to and supportive of disadvantaged members.
- Provide and/or arrange job-specific training. This includes working with representatives of the local Private Industry Council, technical schools, community colleges, or universities to arrange the needed training.
- Develop the interpersonal skills needed to work effectively with disadvantaged employees and to build a supportive work environment for them.

Religious Concerns and the Supervisor[19]

There have been few civil rights cases filed on the basis of religious discrimination. However, in a diverse society the potential for such cases does exist. For example, an Orthodox Jewish employee charged a city government with violating the religious accommodation provision of the Civil Rights Act when it scheduled his civil service examination on a Saturday and refused to move it to another day. The Civil Rights Act, as amended in 1972, requires employees to accommodate the religious beliefs, observances, and practices of employees when such accommodations do not cause undue cost or serious inconvenience.

The Equal Employment Opportunity Commission developed guidelines for what constitutes **reasonable accommodations.** These guidelines suggest that employees assist the employer in identifying alternatives that accommodate religious observances and practices. They also recommend the adoption of such concepts as floating holidays, flexible work schedules, flexible work days, transfers, and switching of job assignments.

The supervisor's role with regard to equal employment opportunity from a religious perspective can be summarized as follows:

- Get to know employees well enough to be able to perceive conflicts between work and their religious beliefs, observances, and practices.
- Work with employees to identify and implement reasonable accommodations when conflicts arise.
- Serve as a liaison between employees and higher management in communicating the needs of the employee and the needs of the company.

Nationality Concerns and the Supervisor

The EEOC defines discrimination on the basis of nationality as denial of equal employment opportunity based on: (1) a person's country of origin; (2) the country of origin of a person's ancestors; (3) a person's cultural heritage; or (4) language characteristics that may be associated with a given nationality (Figure 19–9).

Of particular importance to supervisors are the provisions in the EEOC's guidelines that protect employees from **ethnic jokes,** racial slurs, graffiti, or any other physical or verbal behavior that would serve to create a hostile work environment. Employers are held responsible not just for their actions in this regard but also for those of employees and nonemployees. This gives supervisors a heavy burden of responsibility. Not only must they not behave in an offensive manner, they must also ensure that the fellow workers of an ethnic minority and nonemployees who visit the company behave appropriately.

For example, a supervisor should respond immediately if anyone tells an ethnic joke, blurts out a racial slur, or writes offensive graffiti on a wall. The appropriate response is straightforward, open, and objective. The message should be clear that such behavior is unacceptable and will not be tolerated.

The supervisor's role with regard to equal employment opportunity for ethnic minorities can be summarized as follows:

- Work to build a **supportive work environment** that values diversity within the work unit.
- Respond immediately in a calm, objective, but firm manner to behavior—verbal, physical, or otherwise—that might be offensive to a team member.

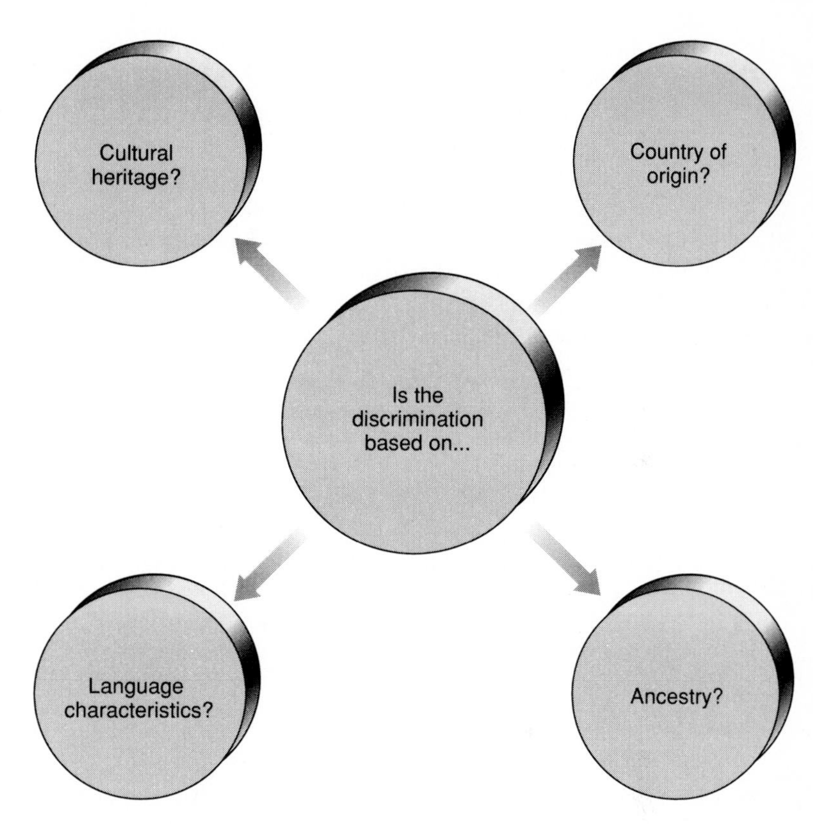

Figure 19–9
EEOC's definition of discrimination on the basis of nationality.

FAMILY CONCERNS AND THE SUPERVISOR

Hugh L. McColl, Jr., chairman of NCNB Corporation, said, "You don't clock out of parenting when you clock in at the office." Managers who have not acknowledged that fact are living in a fantasy land. The number of mothers who are entering the workforce is increasing rapidly. In 1970, 30 percent of the American mothers with children under six years of age were in the workforce. By 1987, this number had increased to 57 percent. Mothers of younger children are following the same trend. In 1970, 24 percent of mothers with children under age one were in the workforce. By 1987, the number had increased to 51 percent.

Over 6 million families in the United States are supported by single mothers. In over 25 million two-parent families, both parents work. Almost 8 million families in the U.S. are composed of children under six years of age and a working mother.

Women are an important and permanent part of the workforce. Consequently, businesses will have to confront the special needs these workers, particularly those with children, bring to the job. In its publication *The Family and the Workplace in Florida,* the Florida Chamber of Commerce says, "Nor can business afford to perpetuate policies that shortchange the children of their employees. Today's children are tomorrow's workers."[20] Owen B. Butler, former chairman of Procter & Gamble, said, "Business is, in the final analysis, totally dependent on how well our children are born and how well they are educated, how well they are raised."[21]

Family-supportive personnel policies are fast becoming a necessity in the modern workplace. Such policies, where they are adopted, are having a positive impact on recruiting qualified workers as well as on absenteeism, tardiness, morale, quality, and productivity. Such policies can include flexible work schedules, **child-care** alternatives, family leave, prenatal health programs, parenting courses, and a number of other options (Figure 19–10).

Investing in such **family-supportive programs** can pay substantial dividends for companies; failing to make such an investment can cost. Consider the following results from a study conducted by Robert Half International, a job recruitment company:[22]

- Employees of American Bankers Insurance Group have a 17 percent companywide absentee rate. Employees who use ABIG's on-site child-care center have only a 7 percent absentee rate.
- Before Aetna Life & Casualty adopted a family-leave program, 23 percent of the female employees who gave birth left the company for good. After the family-leave program was instituted, this number dropped to 12 percent.

Figure 19–10
Examples of family-supportive policies.

Family supportive policies = Flexible work schedules; Childcare alternatives; Family leave; Prenatal health programs; Parenting courses

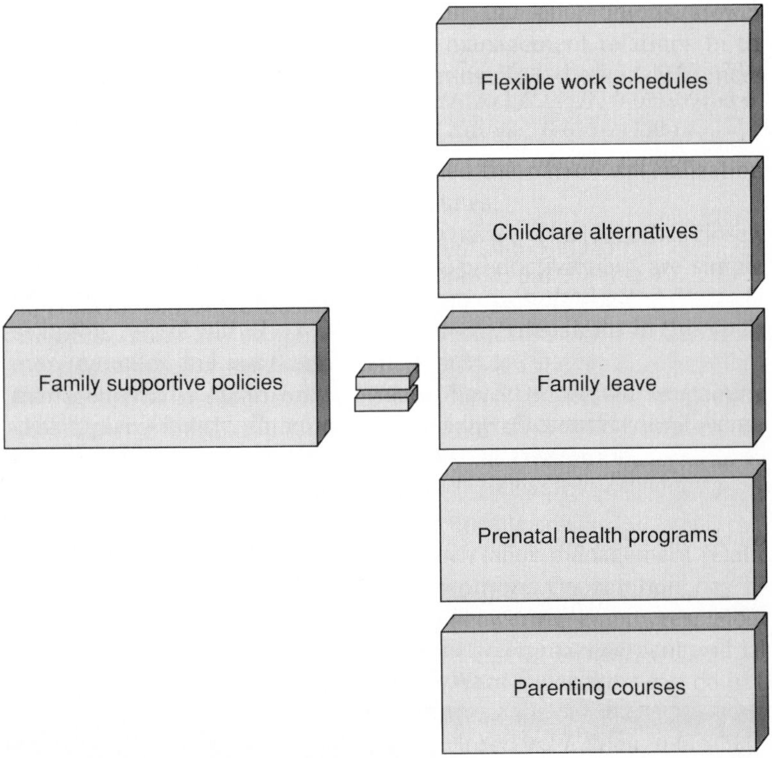

- Eight out of ten workers said they would sacrifice rapid career advancement to have more time with their families.
- Over half of the male study subjects indicated they would give up part of their salary to have more time with their families.

In 1987, Aetna Life & Casualty determined that 21 percent of the female employees who left technical positions, among the most difficult positions to fill, did so for reasons related to family obligations. *Pediatrician Magazine* published a report in which it was estimated that businesses lose from 6 to 29 days per year per parent because children under 6 are ill.[23]

Clearly, the modern supervisor should be aware of the mutual need of employers and employees for family-supportive personnel policies and practices. Supervisors should also be familiar with the more widely used of such practices, summarized here:

- **Employer-sponsored child care.** The company provides a child-care facility at or near the workplace for children of employees. In New York, several employers formed a partnership through which they place child-care workers in the homes of employees with unusual child-care needs. These companies include Colgate-Palmolive, Consolidated Edison, and Home Box Office.
- **Child-care allowances.** Employers pay employees a fixed hourly rate per child to help offset the cost of child care.
- **Flextime.** Employees are allowed to build a work schedule that accommodates the needs of their families as long as the expected total of weekly hours are worked. This practice was first tried in the United States by Hewlett-Packard and is now used by 12 percent of all U.S. workers.
- **Voluntary parttime.** Full time employees are allowed to convert to part-time status to accommodate family needs. Typically, the part-time status is temporary and the hours are worked during peak load times.
- **Job sharing.** Two mutually supportive employees (e.g., a husband and wife) who do the same job are allowed to split the hours for the job. For example, each employee might work 20 hours per week or any combination of hours that equals the total work week. Benefits and pay are assigned in proportion to hours worked. Roscella Company of Pella, Iowa, uses job sharing for approximately 5 percent of its assembly-line and clerical workers.
- **Work-at-home.** Employees are allowed to stay at home and do work historically done on the job. There are limitations on the types of jobs that can be done at home. However, computer and telecommunications technology have made this a viable practice for an increasing number of jobs.
- **Family leave.** A parent (mother or father) is allowed to take extended leave to accommodate family needs.

In addition to these practices, which tend to accommodate the needs of families with young children, there are also family-supportive practices aimed at accommodating other family needs, for example, elder-care, substance abuse, and dependent-care programs. McDonnell-Douglas has a Workers Involvement Network Team to identify needs for family-supportive programs. The team consists of members from the company's various departments who solicit ideas from employees, research them, and make recommendations to management. NCNB Corporation has a work/family issues task force to identify employee needs and recommend ways to meet them. The key role supervisors can play with regard to the family aspects of equal employment opportunity is summarized as follows:

- Communicate with employees well enough to know their family-related needs.
- Know the various types of family-supportive programs that might meet the needs of employees.
- Communicate both the needs of employees and potential ways to meet those needs to higher management.

SEXUAL HARASSMENT AND THE SUPERVISOR[24]

With the rapid growth in the number of women entering the workforce, there is increasing potential for sexual harassment, particularly in those instances in which women are pursuing historically male-dominated career fields. Consequently, modern supervisors should be aware of the problem, how to deal with it, and how to prevent it.

What is Sexual Harassment?

There are two types of **sexual harassment** as defined by the courts. The first is **quid pro quo harassment** in which sexual favors are demanded in return for job benefits (e.g., promotion, pay increase, better working conditions, or other special considerations). The second type is **condition of work harassment** in which an offensive work environment is created. Condition of work harassment is the result of sex-related behavior that is unwelcome and/or demeaning to the extent that it creates an intimidating, offensive, or hostile work environment (Figure 19–11).

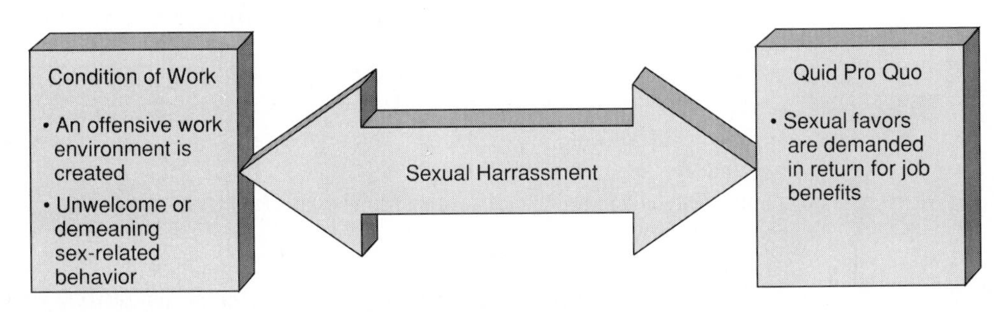

Figure 19–11
Two types of sexual harassment as defined by the courts.

EEOC Guidelines on Sexual Harassment

In November 1980 the Equal Employment Opportunity Commission introduced guidelines that deal specifically with sexual harassment. These guidelines are abbreviated and summarized as follows:

- Harassment on the basis of sex is a violation of Section 703 of Title VII of the Civil Rights Act. Unwelcome sexual advances constitute harassment when: (1) submission to the advances can be interpreted as a condition of employment; (2) acceptance or rejection of advances is used as the basis for employment decisions; (3) this conduct affects job performance or creates a hostile work environment.
- In determining sexual harassment the EEOC examines all circumstances and facts of the case including the nature and context of the advances.
- An employer is responsible for the acts of its employees and agents, even when the behavior in question is forbidden by company policy.
- An employer may be liable for the actions of employees and nonemployees if it knows or should know of incidents of sexual harassment and does not take immediate and appropriate corrective action.
- An employer should take appropriate steps to discourage and prevent sexual harassment.
- An employer may be held liable for unlawful sex discrimination against other qualified employees who are denied an employment opportunity when another employee submits to sexual advances.

What Is the Extent of the Problem?

According to researcher Barbara A. Gutek, between 21 and 53 percent of the women surveyed in a 1981 study had been the target of sexual harassment at least once in their working lives; the figures for male respondents ranged from 9 to 37 percent.[25] In 1982, the city of Seattle, Washington, surveyed 4,608 city employees concerning sexual harassment in the workplace. Of the 1,683 respondents, 64 percent were male and 34 percent were female. The findings of this study were as follows:[26]

- Over 50 percent of the female and 26 percent of the male respondents had experienced sexual harassment on the job.
- The most common form of harassment was unwelcome sexual teasing, jokes, or comments. The next most common form was suggestive looks or gestures.
- The third most common form of sexual harassment took the form of unwanted deliberate touching.

What is the Supervisor's Role?

The EEOC guidelines on sexual harassment plainly encourage prevention. The guidelines recommend that employers take the following action to prevent sexual harassment:

- Raise the subject of sexual harassment in an affirmative manner.
- Express strong disapproval of sexual harassment.
- Develop appropriate sanctions against sexual harassment.
- Inform employees of their rights relative to sexual harassment and how to raise the issue in accordance with Title VII of the Civil Rights Act.
- Develop methods to sensitize employees with regard to sexual concerns.

These recommendations apply to higher management. However, supervisors can play a key role in carrying them out. The supervisor's role with regard to sexual harassment can be summarized as follows (Figure 19–12):

- Set a positive example. Never make sexual advances toward an employee.
- Make sure all employees are familiar with the company's policies regarding sexual harassment.
- Make sure all employees are familiar with how to raise the issue of sexual harassment in accordance with Title VII of the Civil Rights Act.
- Help develop and participate in programs to sensitize employees to sexual harassment concerns.
- Communicate openly with employees so that they feel comfortable discussing sexual harassment concerns, questions, and issues.

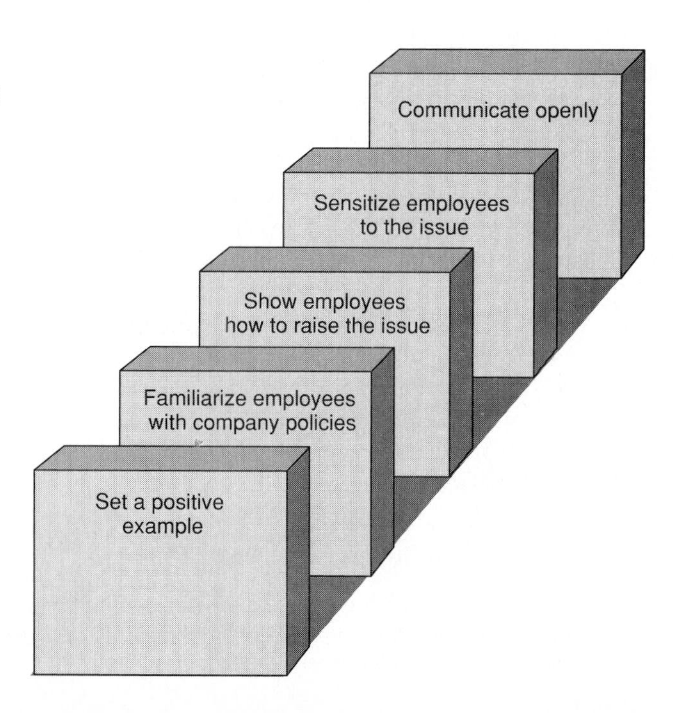

Figure 19–12 Supervisor's role in preventing sexual harassment in the workplace.

ENFORCEMENT AGENCIES

The federal government has three agencies that enforce equal employment opportunity legislation. As might be expected, there is some overlap and occasionally these agencies issue conflicting directives. The Equal Employment Opportunity Commission enforces the Civil Rights Act and its various amendments including the Age Discrimination in Employment Act, Equal Pay Act, and Pregnancy Discrimination Act.

The Office of Federal Contract Compliance Programs (OFCCP) and the U.S. Civil Service Commission enforce the Vocational Rehabilitation Act as well as equal opportunity–related executive orders. In doing so, these agencies and the EEOC are able to apply a variety of enforcement penalities including court-ordered affirmative action, cancellation of contracts, award of back pay, assessment of fines, and imprisonment.

AFFIRMATIVE ACTION AND THE SUPERVISOR

Companies that fall within the jurisdiction of the Office of Federal Contract Compliance Programs (OFCCP) are required to have **affirmative action** programs. Such programs are required to have the following elements:

- An analysis of the company's workforce with regard to the hiring of minorities and women.
- A comparison of the ethnic and gender composition of the workforce with that of a specifically defined available labor pool.
- Goals for correcting employment disparities (i.e., too few minorities or too few women).
- Recruiting, training, and promotion programs for accomplishing the goals.
- Continual monitoring of progress toward accomplishing the goals.

The analysis element of a company's affirmative action plan results in a factual breakdown of the composition of the company's labor force. This element answers such questions as: How many men are employed and what percentage of the total labor force do they represent? How many women are employed and what percentage of the total labor force do they represent? The same types of questions relating to blacks, Hispanics, Caucasians, Asians, and other groups (male and female) should be answered in this section of the plan.

The comparison section of the plan compares the data from the first section with the same information for the specified recruiting region. For example, the percentage of blacks in the company's workforce is compared with the percentage of blacks in the recruiting region. If the percentage of blacks or any other group in the workforce is drastically different from its counterpart in the recruiting region, the company will typically set a goal for correcting the disparity. These goals constitute the third section of the plan.

The fourth section contains written plans and strategies for accomplishing the goals set in the previous section. Strategies should include innovative approaches to recruiting; training programs for enhancing the skills, employability, and advancement potential of underrepresented classes of people; and promotion programs for moving people into employment categories where they are underrepresented.

The final section contains written evidence of the company's continual monitoring efforts. If a company sets a goal of having, within five years, a workforce in which 42 percent of the employees are female, what is the percentage after one year? Two? Three? This section should clearly show that all goals set in section three are monitored and the results are recorded objectively.

What Is the Supervisor's Role in Affirmative Action?

Supervisors do not develop affirmative action plans for their companies, although they might serve on committees that help develop them. However, supervisors do have a role to play with regard to affirmative action. The supervisor's first responsibility is to understand the concept and what it is designed to accomplish: It is an active attempt on the part of the federal government to move companies toward the provision of equal employment opportunities for all categories of people. It is important for supervisors to understand affirmative action and set an example of supporting it because there is much potential for misunderstanding and resentment among employees.

Employees may view affirmative action as **favored treatment** for minorities and women or as setting racial and gender **quotas**. Such viewpoints, left unchecked, can damage morale. They can also lead team members to suspect that new employees were hired based solely on race or gender considerations. If this happens it may make it difficult for new employees to be accepted into the unit.

Supervisors can play a key role in the success of affirmative action programs by doing the following:

- Becoming familiar with the company's affirmative action plan and familiarizing employees with it.
- Communicating with employees openly, objectively, and frankly concerning the purpose, goals, and strategies of the company's affirmative action plan.
- Assisting in the development and implementation of affirmative action training programs.
- Keeping the company's affirmative action goals in mind when participating in the staffing process.
- Stressing the value of diversity in the workplace and the role affirmative action can play in bringing it about.

PROMOTING DIVERSITY IN THE WORKPLACE

The next frontier beyond affirmative action might be called promoting diversity in the workplace (Figure 19–13). Its potential for success is probably greater than that of affirmative action because while affirmative action is a government-mandated concept, companies that are promoting diversity in the workplace are doing so primarily out of

Figure 19–13
The next frontier beyond affirmative action is promoting diversity.

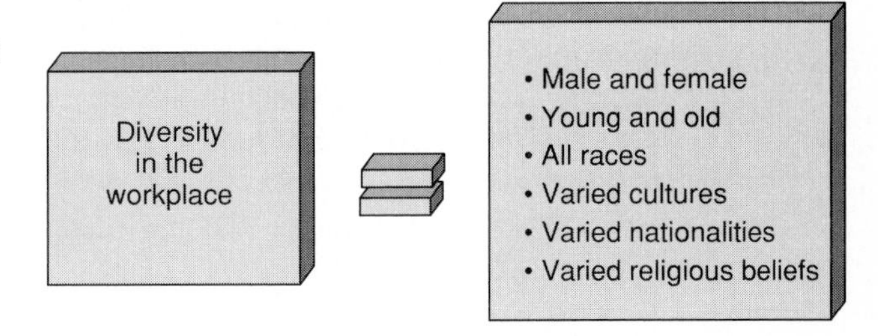

economic necessity. The economic necessity is created by the fact that almost 85 percent of the new entrants in the workforce between now and 2000 will be women, minorities, and immigrants.

Promoting diversity requires that "organizations make whatever changes are necessary in their systems, structures, and management practices to eliminate any subtle barriers that might keep people from reaching their full potential."[27]

Promoting diversity goes beyond the elimination of outright sexism and racism to deal with the more subtle forms of behavior that inhibit equal employment opportunity. It involves asking such questions as: Does this company do everything possible to ensure that all employees have equal opportunities to achieve their full potential? Does this company unintentionally hold employees back? Is the company's culture so heavily defined by traditional white male values that others are held back in a de facto manner?[28]

Some companies make their diversity promotion efforts an extension of existing affirmation action efforts. Xerox Corporation is an example of such a company.[29] Xerox uses two different approaches to promote diversity: (1) training to help managers understand how unconscious behavior can inhibit equal employment opportunity; and (2) identifying key jobs that would be expected to be found on the resumes of successful managers and then giving minorities and women opportunities to hold these jobs.

Other companies consciously separate their diversity promotion efforts from existing affirmative action efforts. Such a company is the Semiconductor Product Sector of Motorola.[30] Motorola established a diversity task force that interviewed 400 employees in a series of focus groups to identify inhibiting factors. The task force then examined the policies, practices, and programs of 50 other companies in an attempt to identify ways of overcoming the inhibitors.

The result of the task force's study was a four-component diversity program that is now used at Motorola. The components are: (1) training designed to help employees understand and appreciate diversity; (2) a task force for dealing with family-supportive issues; (3) a task force for recommending flexible scheduling strategies; and (4) a committee responsible for identifying ways the company can work with educational institutions to help educate future workers.

Regardless of the approach, the key to success in promoting diversity is to let employees identify the factors they feel hold them back. This means the supervisor can play a pivotal and ongoing role in the promotion of diversity in the workplace. This role involves taking advantage of the supervisor's daily interaction with employees to solicit their input concerning inhibiting factors. Supervisors have opportunities every day to identify such factors by observing, asking, and listening. Information learned should be passed along to diversity task forces, if they exist, or to higher management if they do not.

SUMMARY

1. Changing demographics in the workplace are the result of two factors: (1) changing population demographics; and (2) changing lifestyles, spending patterns, attitudes, and value systems.
2. The most significant demographic trend concerning the workforce is the growing number of women and minorities entering the workforce.
3. Important federal legislation concerning equal employment opportunity includes the Civil Rights Act of 1964 (amended in 1972), the Age Discrimination Act of 1967 (amended in 1978), the Vocational Rehabilitation Act of 1973, and the Vietnam Era Veterans Readjustment Act of 1974.
4. Two key concepts for challenging employment practices are disparate impact and disparate treatment. Disparate impact applies to groups of people; disparate treatment applies to individuals. The key in determining disparate treatment is employer intent.
5. The Fair Labor Standards Act established child labor standards and wage and hour guidelines. Key concepts in the law are exempt and nonexempt. Nonexempt employees are entitled to at least minimum wage and overtime pay.
6. In dealing with minorities, it is important to provide a positive introduction and a thorough orientation. Open and frank communication and equal treatment are critical.
7. In dealing with women it is important to understand that biases can be deeply ingrained and subtle. Supervisors should encourage the elimination of gender-based job titles, make training opportunities available equally, and arrange in-house internships that will allow women to gain the experience they need to succeed in jobs traditionally held by men.
8. The keys to dealing with pregnant employees are communication and accommodation. When dealing with older employees it is important to apply the same performance standards applied to other workers and to make decisions based on job performance.
9. In dealing with disabled workers it is important to understand their special needs and to work cooperatively in planning accommodations for them.
10. In dealing with disadvantaged workers it is important to build a supportive work environment, arrange job-specific training, and communicate continually.
11. In dealing with religious issues, supervisors should get to know the religious observances of employees that might conflict with work practices and work with employees to plan reasonable accommodations.

12. Studies have shown that employees value family-supportive policies and practices. Such practices will become increasingly important in the modern workplace. They include employer-sponsored child care, child-care allowances, flexible scheduling, job sharing, work-at-home, voluntary part time, and family leave.
13. Two categories of sexual harassment are quid pro quo and condition of work harassment. Sexual harassment is a serious problem in the modern workplace. Supervisors can reduce sexual harassment by setting a positive example, helping develop sensitivity training programs, and communicating openly and frankly on the subject.
14. Federal agencies that enforce equal employment opportunity are the Equal Employment Opportunity Commission (EEOC), Office of Federal Contract Compliance Program (OFCCP), and the U.S. Civil Service Commission.
15. Affirmative action requires an employer to analyze the composition of its workforce; compare the composition of its workforce with the composition of the community from which the labor pool is drawn; set goals for correcting disparities; provide recruiting, training, and promotion programs; and monitor results continually.
16. Promoting diversity requires that organizations make any and all changes necessary to eliminate barriers to equal employment opportunity. The key to success in promoting diversity is to let employees identify the barriers that hold them back.

KEY TERMS AND PHRASES

Affirmative action
Child-care allowances
Child care
Civil Rights Act
Condition of work harassment
Demographics
Disabled workers
Disadvantaged people
Disparate treatment
Disparate impact
Diversity
EEOC
Employer intent
Employer-sponsored child care
Equal employment opportunity
ERISA
Ethnic jokes
Exempt employees
Family leave
Family-supportive programs
Favored treatment
Flextime
FLSA
Gender-based job titles
Hard-core unemployed
Job sharing
Job relatedness
JTPA
Minorities
Nonexempt employees
OFCCP
PIC
Quid pro quo harassment
Quotas
Reasonable accommodations
Sexual harassment
Supportive work environment
Voluntary part time
Work-at-home

CASE STUDY: Valuing Diversity in the Workplace[31]

The following real-world case study contains an example of an equal opportunity problem of the type confronted by supervisors on the job. Read the case study carefully and answer the accompanying questions.

For years the Snohomish County, Washington, Public Utility Division (SCPUD), a public electric utility, had gotten by unaffected by the changing demographics of the workplace. Most workers were white males with similar backgrounds. However, in the mid-1980s modern demographics caught up with SCPUD and problems began to occur. In response, management took care to develop a comprehensive affirmative action plan and to regularly share information with all employees.

Unfortunately, early efforts at helping employees value diversity did not work. In spite of the best efforts of management, employees were turned off by the statistics, compliance overtones, and unaudited affirmative action goals. At this point management decided to switch its emphasis from affirmative action to workforce diversity.

The new approach had three major areas of concentration:

- Education and training programs to increase employee awareness of cultural issues and differences.
- A new emphasis on the business' need to adapt to a more diverse workforce rather than emphasizing affirmative action mandates.
- Opportunities for employees to develop the skills needed to deal effectively with fellow employees and customers in an increasingly diverse environment.

This new approach, which is ongoing, has been well received by employees and appears to be producing the desired results.

1. Does the new approach at SCPUD appear to meet applicable affirmative action guidelines?
2. What do you see as the supervisor's role in the new SCPUD program?
3. What additions, deletions, or changes would you recommend to improve the new SCPUD program? Why?

REVIEW QUESTIONS

1. What are the two factors behind the changing demographics of the workplace?
2. Explain the term *diverse workplace.*
3. Construct a chart that summarizes key legislation concerning equal employment opportunity and the federal agencies that enforce it.
4. Compare and contrast the concepts of disparate impact and disparate treatment.
5. How does the concept of job relatedness affect rulings on disparate impact?

6. How does the concept of employer intent affect rulings on disparate treatment?
7. Explain the current disposition of the Civil Rights Act.
8. Besides the Civil Rights Act, what other laws focus on preventing discrimination in the workplace?
9. Contrast and compare the concepts of exempt and nonexempt employees as established by the FLSA.
10. What are the four bases for not requiring equal pay?
11. Explain briefly the supervisor's role in dealing with the following special categories of employees: minorities, women, disabled, disadvantaged.
12. How can the PIC help supervisors with regard to disadvantaged employees?
13. What are the four factors that can enhance the success of employers in dealing with disadvantaged people?
14. Explain the concept of family-supportive policies/practices.
15. Explain the following family-supportive practices: child-care allocations, flextime, job sharing, family leave.
16. Compare and contrast quid pro quo and condition of work sexual harassment.
17. How does the EEOC define sexual harassment? What is the extent of the problem?
18. Explain the supervisor's role in preventing and dealing with sexual harassment in the workplace.
19. What are the five components that must be included in a company's affirmative action plan?
20. How does the promotion of diversity differ from affirmative action?

SIMULATION ACTIVITIES

The following activities are provided to give you opportunities to develop supervisory skills by applying the material presented in this chapter and previous chapters to solve simulated supervisory problems.

1. One of your employees approaches you wanting to know if a friend has grounds for a disparate treatment case. Describe the tests you would recommend the friend apply.
2. When you became a supervisor five years ago, all your employees were men. Now one-third are women. Problems are beginning to arise. Explain what you can do to turn the situation around and make it positive.
3. Your company has asked you to chair a committee to develop a plan for bringing disadvantaged people into the workforce. Develop an outline for such a plan. Then work with fellow students to fill out the plan.
4. Your company has gone to six-day weeks (Monday–Saturday) to fulfill a long list of back orders. A key employee complains to you that working on Saturday conflicts with her religious beliefs. How do you plan to handle this situation?
5. You have received three complaints of sexual harassment in one week in your unit. Explain how you plan to handle the situation.

ENDNOTES

1. Wolfbein, S. L. *To the Year 2000/A New Look*. Monograph, 1990, Volume 3, Number 1. Philadelphia, PA: Vocational Research Institute, p. 1.
2. Ibid., pp. 5–8.
3. U.S. Department of Labor, Bureau of Labor Statistics. *Monthly Labor Review,* November 1989.
4. Anderson, H. J. and Levin-Epstein, M. *Primer of Equal Employment Opportunity,* 2nd ed. (Washington, D.C.: The Bureau of National Affairs, Inc., 1982), p. 32.
5. Baysinger, R. A. "Disparate Treatment and Disparate Impact, There Is a Difference: The Continuing Evaluation of the 1964 Civil Rights Act," in Schuler, R. S., Youngblood, A. and Huber, V. L. *Reading in Personnel and Human Resources Management* (St. Paul, MN: West Publishing Company, 1987), p. 73.
6. *Wage and Hour Manual,* BNA Policy and Practice Series (Washington, D.C.: The Bureau of National Affairs, Inc., 1986), p. 16.
7. Flippo, E. B. *Personnel Management* (New York: McGraw-Hill, 1984), p. 72.
8. Wolfbein, S. L. *To the Year 2000/A New Look,* p. 8.
9. Rosen, B. and Jerdee, T. H. "Sex Stereotyping in the Executive Suite," *Harvard Business Review,* March–April 1974, Volume 52, Number 2, pp. 45–58.
10. Flippo, E. B. *Personnel Management,* p. 74.
11. Wolfbein, S. L. *To the Year 2000/A New Look,* pp. 5–8.
12. "Wounded Executives Fight Back on Age Bias," *Business Week,* July 21, 1980, p. 109.
13. Ibid.
14. Garland, S. B. "The Retiring Kind are Getting Militant About Benefits," *Nation's Business,* May 28, 1990, p. 29.
15. McKee, B. A. "Planning for the Disabled." *Nation's Business,* November 1990, pp. 24–26.
16. Hodgson, J. D. and Brenner, M. H. "Successful Experience: Training Hard-Core Unemployed," *Harvard Business Review,* September–October 1968, Volume 46, Number 5, p. 150.
17. Koch, J. L. "Employing the Disadvantaged: Lessons from the Past Decade," *California Management Review,* Fall 1974, Volume 17, Number 1, pp. 68–77.
18. Cohn, E. and Lewis, M. V. "Employers' Experience in Retaining Hard-Core Hires," *Industrial Relations,* February 1975, Volume 14, Number 1, p. 58.
19. "Religious Accommodations and the Courts: An Overview," *Personnel,* May–June 1981, Volume 58, Number 3, p. 48.
20. *The Family and the Workplace in Florida: A Business Agenda* (Tallahassee, FL: The Florida Chamber of Commerce, 1990), pp. 1–20.
21. Ibid., p. 12.
22. Ibid., p. 3.
23. Ibid., p. 16.
24. Webb, S. L. *Sexual Harassment . . . Shades of Grey* (Seattle, WA: Pacific Resource Development Group, 1988), p. 36.

25. Ibid., p. 6.
26. Ibid.
27. Geber, B. "Managing Diversity," *Training,* July 1990, p. 24.
28. Ibid.
29. Ibid., p. 25.
30. Ibid.
31. Elshult, S. and Little, J. "The Case for Valuing Diversity," *HR Magazine,* June 1990, pp. 50, 51.

CHAPTER TWENTY

Employee Health and Safety and the Supervisor

CHAPTER OBJECTIVES

After studying this chapter you will be able to explain or define the following topics:

- The Historical Development of Health and Safety Programs
- The Legal Foundation of Health and Safety Programs
- The Policy Aspects of Health and Safety
- Safety Training Methods
- Accident Prevention Techniques
- How to Conduct Accident Investigations
- Worker's Compensation
- The Impact of AIDS on Health and Safety

CHAPTER OUTLINE

- Historical Development of Health and Safety Programs
- Legal Foundation of Health and Safety Programs
- Policy Aspects of Health and Safety
- Safety Training
- Accident Prevention Techniques
- Accident Investigation and Reporting
- Worker's Compensation
- AIDS and Employee Health and Safety
- AIDS Education and the Supervisor
- Summary
- Key Terms and Phrases
- Case Study Application Problem
- Review Questions
- Simulation Activities
- Endnotes

Safe, healthy employees are more likely to be productive employees. Supervisors play a key role in promoting a safe and healthy workplace. This chapter gives supervisors the information they need to have a positive impact on health and safety in the workplace.

HISTORICAL DEVELOPMENT OF HEALTH AND SAFETY PROGRAMS

Today's concern for the health and safety of industrial workers contrasts sharply with that of companies during the Industrial Revolution. Since the early years of American industry, the employee health and safety pendulum has swung from a position characterized by little concern and no regulation to great concern and substantial regulation. During the Industrial Revolution the factory floor and equipment on it were designed to promote productivity with little or no concern for worker health and safety. Processes and circumstances that could force the closure of a plant today were common practice in earlier years. Accidents resulting in serious injuries and even death were considered a normal part of the workday.

Prior to the Civil War, laws and government regulations supporting health and safety in the workplace were almost nonexistent. Employers were not liable for accidents.

Much of the credit for the beginning of health and safety consciousness in American industry goes to organized labor. Labor unions had to fight hard for better working conditions and compensation for the families of accident victims. The most difficult battles between management and labor over working conditions are associated with the mining, manufacturing, and food processing industries. Figure 20–1 summarizes some

Early Health and Safety Legislation/Organizations

1869: Pennsylvania legislature passes a mining safety law requiring two exits from all mines.

1877: Massachusetts legislature passes a safeguarding of harzarous machines law.

1877: Employer's Liability Law passes making employers potentially liable for injuries to workers.

1907: U.S. Department of the Interior creates the Bureau of Mines to monitor mining safety.

1911: Wisconsin legislature passes the first worker's compensation law.

1913: National Council of Industrial Safety is created. In 1915 the name is changed to the National Safety Council

Figure 20–1
Historical development of health and safety legislation and organizations.

of the earliest legislation and organizations that promoted health and safety in the workplace.

After World War I, safety consciousness in U.S. industry began to grow. Then, during World War II, facing severe labor shortages, industrial firms learned how costly it could be to lose the services of skilled employees because of accidents. This led to the realization that a safe working environment could actually save more money in reduced compensation, medical bills, and absenteeism than it cost to provide.

During the 1960s several pieces of legislation were passed to promote health and safety in the workplace for specific types of workers. Prominent among these laws were the Federal Metal and Nonmetallic Mine Safety Act, the Federal Coal Mine Health and Safety Act, and the Contract Workers and Safety Standards Act. These laws were positive steps toward a safer work environment. However, their impact was limited to mining workers and employees of companies with federal government contracts.

Even with the growing safety consciousness among employers, the late 1960s saw over 14,000 workers killed on the job in the United States. Impaired by weak, overly specific legislation and budget restraints, the states were not making progress toward reversing an upward trend in on-the-job accidents. As a result, the federal government passed the Occupational Safety and Health Act of 1970. The act is still the most broadly applicable legislation regulating health and safety practices in the workplace.

LEGAL FOUNDATION OF HEALTH AND SAFETY PROGRAMS[1]

The Occupational Safety and Health Act represents an attempt on the part of the federal government to ensure a safe and healthy working environment for the broadest possible base of workers. It applies to most employers in this country and its various possessions including the federal government. State and local governments and industries covered by other federal acts (e.g., Federal Mine Safety and Health Act of 1977) are exempt. The act clearly establishes on-the-job health and safety to be a responsibility of management.

Administration and Enforcement

Administration and enforcement of the act are the responsibility of the U.S. Secretary of Labor and the Occupational Safety and Health Review Commission. Administrative and enforcement duties are divided as follows:

- The U.S. Secretary of Labor conducts investigations of suspected wrongdoing and prosecutes alleged violators as necessary.
- The Occupational Safety and Health Review Commission **arbitrates** cases of wrongdoing that are challenged and rules in those cases as necessary.

In addition to investigations, prosecutions, and arbitration, there is a need for an agency to carry out the day-to-day tasks required in the act. This agency is the Occupational Safety and Health Administration, or **OSHA**. OSHA is responsible for the following duties:

Setting health and safety standards
Revising health and safety standards
Revoking health and safety standards
Inspecting companies
Issuing citations
Assessing penalties
Petitioning for court action against flagrant violators
Providing training for employers and employees
Awarding grants to promote health and safety

OSHA has ten regional offices for carrying out its responsibilities. Within each region are smaller area offices that bring OSHA services to the local level. Figure 20–2 contains the addresses of the regional OSHA offices.

When OSHA takes action against an employer, the employer has the right to contest the action. Such challenges are heard and acted on by the Occupational Safety and Health Review Commission. Penalties assessed by OSHA are held in abeyance until the commission issues a final ruling, which can be appealed within sixty days to the U.S. Court of Appeals.

Employer and Employee Rights Under OSHA

Employers are responsible for complying with OSHA's regulations. One requirement is that employers display the poster shown in Figure 20–3, or a facsimile of it, prominently in the workplace. Employers can stay abreast of the most recent regulations by securing a new copy of the *Code of Federal Regulations* (**CFR**) each year from the nearest OSHA office. The CFR is divided into 50 parts or titles. Title 29 contains the regulations set forth by the U.S. Department of Labor and is the title with which industrial supervisors should be familiar. For example, the OSHA poster shown in Figure 20–3 is contained in Title 29, Part 1903.2(a)(1) of the CFR.

Both employers and employees have clearly defined rights. Those of employers are summarized as follows:[2]

1. Seek off-site consultation from the nearest OSHA office.
2. Receive free on-site consultation service from OSHA to help identify hazardous conditions and take corrective measures.
3. Receive proper identification from the OSHA compliance officer before an inspection takes place.
4. Receive advice from the OSHA compliance officer concerning the reason for the inspection.
5. Receive opening and closing sessions with the OSHA compliance officer if an inspection takes place.
6. Receive protection of proprietary trade secrets observed by an OSHA compliance officer.

Regional OSHA Offices

All street addresses should be preceded by:
U.S. Department of Labor
Occupational Safety and Health Administration

Region I
JFK Building, Room 1804
Boston, MA 02203
617/223-6712

Region II
1515 Broadway
(1 Astor Plaza), Room 3445
New York, NY 10036
212/971-5941

Region III
15220 Gateway Center
3535 Market Street
Philadelphia, PA 19104
215/596-1201

Region IV
1375 Peachtree St. NE, Suite 587
Atlanta, GA 30309
404/526-3573

Region V
230 S. Dearborn, 32nd Floor
Chicago, IL 60604
312/353-4716

Region VI
555 Griffin Square Bldg, Rm. 602
Dallas, TX 75202
214/749-2477

Region VII
Federal Building, Rm. 3000
911 Walnut Street
Kansas City, MO 64106
816/374-5861

Region VIII
Federal Building, Room 15010
1961 Stout Street
Denver, CO 80202
303/837-3883

Region IX
9470 Federal Building
450 Golden Gate Avenue
Post Office Box 36017
San Francisco, CA 94102
415/556-0584

Region X
6048 Federal Office Building
909 First Avenue
Seattle, WA 98174
206/442-5930

Figure 20–2
Supervisors should know how to contact their nearest OSHA office.

7. Request a conference with the area director when a **citation** is issued. A notice of intent to challenge can be filed within fifteen working days from receipt of a citation.
8. If unable to comply with a standard within the required time, a request for an extension can be filed with OSHA. This is called a petition for modification of abatement (PMA).

JOB SAFETY & HEALTH PROTECTION

The Occupational Safety and Health Act of 1970 provides job safety and health protection for workers by promoting safe and healthful working conditions throughout the Nation. Requirements of the Act include the following:

Employers

All employers must furnish to employees employment and a place of employment free from recognized hazards that are causing or are likely to cause death or serious harm to employees. Employers must comply with occupational safety and health standards issued under the Act.

Employees

Employees must comply with all occupational safety and health standards, rules, regulations and orders issued under the Act that apply to their own actions and conduct on the job.

The Occupational Safety and Health Administration (OSHA) of the U.S. Department of Labor has the primary responsibility for administering the Act. OSHA issues occupational safety and health standards, and its Compliance Safety and Health Officers conduct jobsite inspections to help ensure compliance with the Act.

Inspection

The Act requires that a representative of the employer and a representative authorized by the employees be given an opportunity to accompany the OSHA inspector for the purpose of aiding the inspection.

Where there is no authorized employee representative, the OSHA Compliance Officer must consult with a reasonable number of employees concerning safety and health conditions in the workplace.

Complaint

Employees or their representatives have the right to file a complaint with the nearest OSHA office requesting an inspection if they believe unsafe or unhealthful conditions exist in their workplace. OSHA will withhold, on request, names of employees complaining.

The Act provides that employees may not be discharged or discriminated against in any way for filing safety and health complaints or for otherwise exercising their rights under the Act.

Employees who believe they have been discriminated against may file a complaint with their nearest OSHA office within 30 days of the alleged discrimination.

Citation

If upon inspection OSHA believes an employer has violated the Act, a citation alleging such violations will be issued to the employer. Each citation will specify a time period within which the alleged violation must be corrected.

The OSHA citation must be prominently displayed at or near the place of alleged violation for three days, or until it is corrected, whichever is later, to warn employees of dangers that may exist there.

Proposed Penalty

The Act provides for mandatory penalties against employers of up to $1,000 for each serious violation and for optional penalties of up to $1,000 for each nonserious violation. Penalties of up to $1,000 per day may be proposed for failure to correct violations within the proposed time period. Also, any employer who willfully or repeatedly violates the Act may be assessed penalties of up to $10,000 for each such violation.

Criminal penalties are also provided for in the Act. Any willful violation resulting in death of an employee, upon conviction, is punishable by a fine of not more than $10,000, or by imprisonment for not more than six months, or by both. Conviction of an employer after a first conviction doubles these maximum penalties.

Voluntary Activity

While providing penalties for violations, the Act also encourages efforts by labor and management, before an OSHA inspection, to reduce workplace hazards voluntarily and to develop and improve safety and health programs in all workplaces and industries. OSHA's Voluntary Protection Programs recognize outstanding efforts of this nature.

Such voluntary action should initially focus on the identification and elimination of hazards that could cause death, injury, or illness to employees and supervisors. There are many public and private organizations that can provide information and assistance in this effort, if requested. Also, your local OSHA office can provide considerable help and advice on solving safety and health problems or can refer you to other sources for help such as training.

Consultation

Free consultative assistance, without citation or penalty, is available to employers, on request, through OSHA supported programs in most State departments of labor or health.

More Information

Additional information and copies of the Act, specific OSHA safety and health standards, and other applicable regulations may be obtained from your employer or from the nearest OSHA Regional Office in the following locations:

Atlanta, Georgia
Boston, Massachusetts
Chicago, Illinois
Dallas, Texas
Denver, Colorado
Kansas City, Missouri
New York, New York
Philadelphia, Pennsylvania
San Francisco, California
Seattle, Washington

Telephone numbers for these offices, and additional area office locations, are listed in the telephone directory under the United States Department of Labor in the United States Government listing.

Washington, D.C.
1985
OSHA 2203

William E. Brock, Secretary of Labor

U.S. Department of Labor
Occupational Safety and Health Administration

Under provisions of Title 29, Code of Federal Regulations, Part 1903.2(a)(1) employers must post this notice (or a facsimile) in a conspicuous place where notices to employees are customarily posted.

Figure 20–3
OSHA poster for display in the workplace.
Source: U.S. Department of Labor (OSHA).

9. Assist in developing safety and health standards through participation with OSHA Standards Advisory Committees, through standards writing organizations, and through public comment and public hearings.
10. Use Small Business Administration loans to bring the company into compliance, if applicable.

The rights of employees with regard to health and safety are also clearly set forth by OSHA. Examples of employees' rights are summarized as follows:[3]

1. An employer cannot be punished, harassed, or reassigned for job safety or health activities, such as complaining to the union or OSHA or participating in union or OSHA inspections or conferences.
2. Employees can privately confer with and answer questions from an OSHA compliance office in connection with a workplace inspection.
3. During an OSHA inspection, an authorized workers' representative may be given an opportunity to accompany the compliance officer to aid the inspection. An authorized worker has the right to participate in the opening and closing inspection conferences and be paid his or her normal wage for the time.
4. Employees can contact OSHA to make an inspection when employers fail to correct a hazard. OSHA will not tell the employer who requested an inspection without the employee's permission.
5. An employee can notify OSHA or a compliance officer in writing of a potential violation before or during a workplace inspection.
6. An employee can give OSHA information that could influence proposed penalties by OSHA against the employer.
7. If OSHA denies the inspection request of an employee, the employee must be informed of the reasons in writing by OSHA. The employee may request an OSHA hearing should he or she object to the OSHA decision.
8. If an OSHA compliance officer fails to cite an employer concerning an alleged violation submitted in writing by an employee, OSHA must furnish the employee with a written statement of the reason for the disposition.
9. Employees have the right to review an OSHA citation against their employer. The employer must post a copy of the citation at the location where the violation took place.
10. Employees can appear to view or be a witness in a contested enforcement matter before the Occupational Safety and Health Review Commission.
11. If OSHA fails to take action to rectify a dangerous hazard and an employee is injured, that employee has the right to bring action against OSHA to seek appropriate relief.
12. If an employee disagrees with the amount of time OSHA gives the employer to correct a hazard, he or she can ask for review by the Occupational Safety and Health Review Commission within fifteen days of when the citation was issued.
13. Employees may ask OSHA to adopt a new standard or to modify or revoke a current standard.

14. Employees may take action for or against any proposed federal standards and may appeal any final OSHA decision.
15. An employer must inform employees when applying for a variance of an OSHA standard.
16. Employees must be given the opportunity to view or take part in a variance hearing and have a right to appeal OSHA's final decision.
17. Employees have the right to access to all information available in the workplace pertaining to employee protections and obligations under OSHA and standards and regulations.
18. Employees involved in hazardous operations have a right to information from the employer regarding toxicity, conditions of exposure, and precautions for safe use of all **hazardous materials** in the establishment.
19. The employer must inform an employee who might be overexposed to any harmful materials, and the employee must be told of corrective action undertaken.
20. If an OSHA compliance officer determines that an imminent danger exists, he or she must tell affected employees of the danger and further inform them that prompt action will be taken if the employer fails to eliminate the danger.
21. If an employee requests access to records covering his or her exposure to toxic materials or harmful physical agents that require monitoring, the employer must comply with the request.
22. Employees must be given the opportunity to observe the monitoring or measuring of hazardous materials or harmful physical agents, if OSHA standards require monitoring.
23. An employee can make a request in writing to NIOSH (National Institute for Occupational Safety and Health) for a determination of whether or not a substance used in the workplace is harmful.
24. If an employee asks to review the Log and Summary of Occupational Injuries (OSHA No. 200), the employer must comply with the request.
25. An employee is entitled to a copy of the Notice of Contest and to participate in hearings on contested hearings.

OSHA Inspections

Companies that fall under the authority of OSHA may be inspected without prior notice any time work is taking place. An OSHA inspection consists of an opening conference, a walk-through inspection, and a closing conference. A designated company representative and an employee can participate in the conferences and inspection.

OSHA conducts investigations in five different categories (Figure 20–4):

- *Imminent danger investigation.* An imminent danger is a condition that if not corrected immediately might result in a serious injury or death to a worker. Because of the nature of an imminent danger condition, complaints alleging such a condition are typically investigated within twenty-four hours.
- *Catastrophic accident investigation.* An accident that results in a fatality calls for a

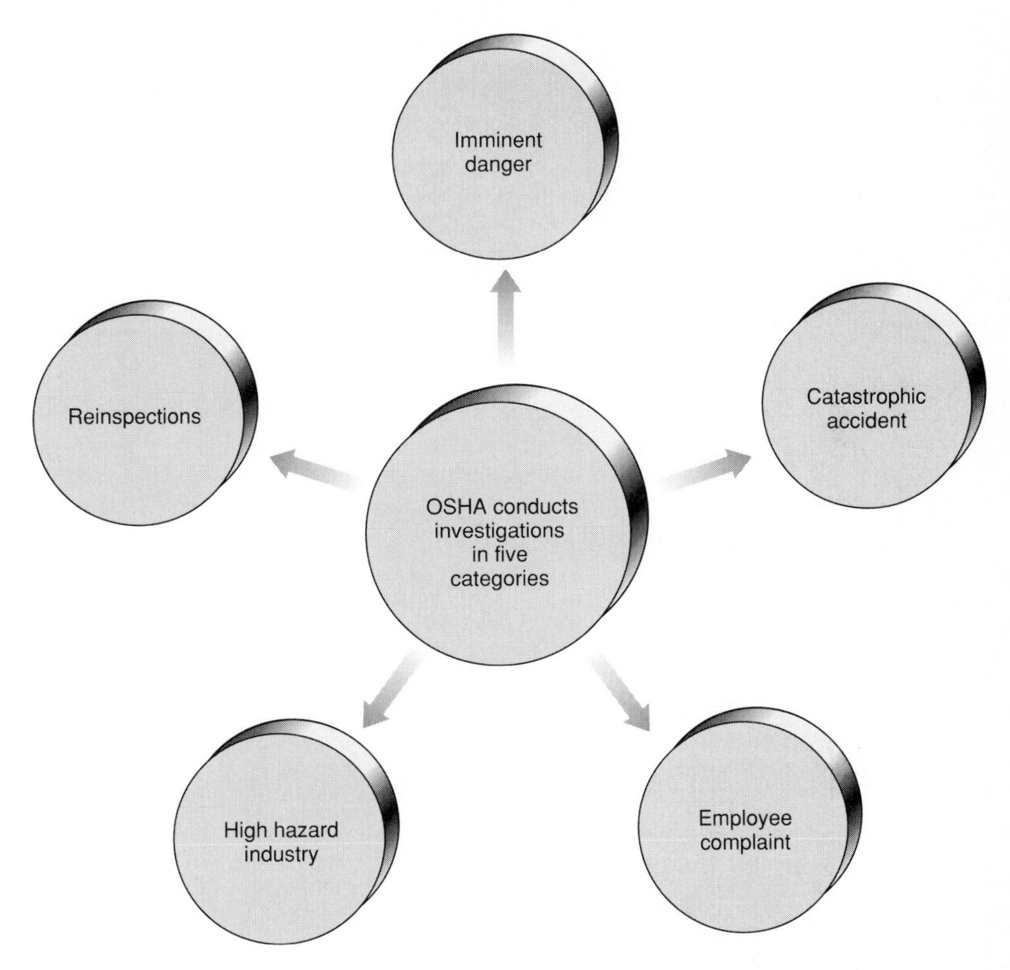

Figure 20–4
Types of OSHA inspections.

special investigation with special reporting requirements. Such an investigation is known as a **catastrophic accident investigation.**

- *Employee complaint investigation.* Employees who feel they are working in an unsafe or unhealthy environment are free to report the situation to an OSHA office. When this happens, the resultant investigation is known as an **employee complaint investigation.** Such investigations typically focus narrowly on the specific complaint.
- *High hazard industry inspection.* High priority is given to inspections of conditions in industries that have traditionally had high accident, injury, or death rates.
- *Reinspections.* Employers who are cited for violations are typically given a specified amount of time to correct the situation. At the end of this period another **inspection** or **reinspection** takes place.

OSHA Violations

There are four types of violations for which an employer can be cited during an OSHA inspection. These categories of violations are summarized as follows (Figure 20–5):

- *Imminent danger violation.* An **imminent danger violation** is issued when an OSHA inspection reveals a condition that is so unsafe or so unhealthy that serious injury or death might result and the employer does not correct it immediately. Imminent danger violations can result in court action to shut a company down.
- *Serious violation.* A **serious violation** citation is issued when a condition exists that might result in death or serious injury and the employer knew or should have known of the condition.
- *Nonserious Violation.* A **nonserious violation** citation is issued when a condition exists that might result in injury, but not serious injury or death or when a hazard exists that the employer legitimately did not know about.
- *De minimis violation.* A ***de minimis* violation** citation is issued when an OSHA standard is violated but there is no danger of injury, death, or illness.

POLICY ASPECTS OF HEALTH AND SAFETY

The supervisor's role with regard to health and safety in the workplace should be defined by management policy. Modern industrial companies should have written health and safety policies and those policies should clearly define the supervisor's responsibilities and authority.

A safety policy, regardless of the type and size of company, should have the following characteristics:

- Should be developed with input from employees at all levels an in all departments.
- Should have the full support of top management and reflect the attitude of management toward health and safety.

Figure 20–5
Employers may be cited for these four types of OSHA violations.

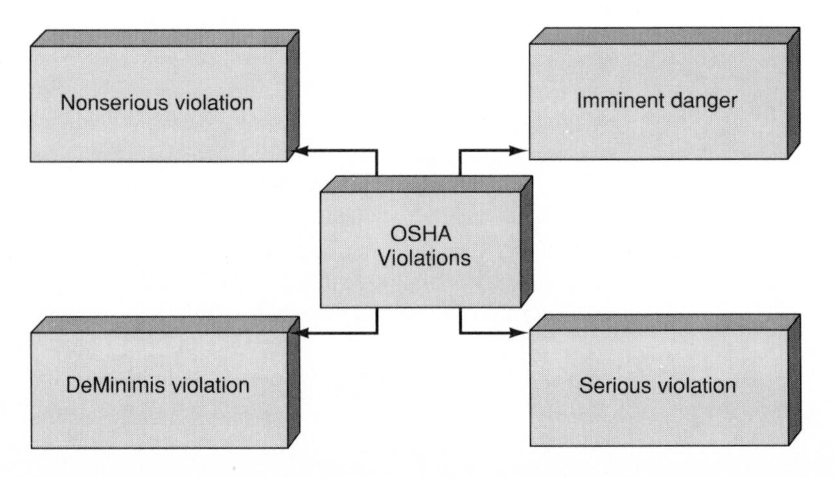

- Should be put in writing, shared with all employees, and widely publicized throughout the company.
- Should be simply stated and easy to understand.
- Should clearly delineate responsibility for health and safety.
- Should provide the foundation for the development of a plan for creating and maintaining a safe working environment.

There are two fundamental reasons for having a comprehensive health and safety policy. The first has to do with legislative mandates and the second with economics. These reasons are summarized as follows:

- *Legislative mandates and policy.* The OSHA act sets forth very specific health and safety standards. In addition to the specific standards, the act contains an all-encompassing *general-duty clause* that covers conditions not specifically covered by the minimum standards. Complying with the act begins with the development of a written health and safety policy.
- *Economics and policy.* It has been well established that good health and safety are good business. A safe and healthy working environment is the most economical environment. Accidents resulting in injuries or death cost the employer. The costs fall into two categories: direct costs and hidden costs. **Direct costs** are those associated with insurance. They include medical treatment, direct payments to the injured worker, worker's compensation costs, and the cost of lost time of the injured worker. **Hidden costs** include the cost of temporarily replacing the injured worker, the cost of paperwork associated with the accident, the cost of repairing damaged equipment, and the cost of reworking damaged products (Figure 20–6).

Assigning Responsibility for Health and Safety

Health and safety in the workplace are the responsibility of management and workers. Management is responsibility for providing a safe working environment and employees are responsible for working in a safe manner.

Supervisors should be assigned responsibility for the work environment in their units and for the safety of their employees. This means supervisors should also have the authority necessary to make decisions about safety, discipline employees, and correct unsafe conditions in their units. In addition, management must back decisions made by the supervisor concerning safety.

This is why it is so important for higher management to put its **safety philosophy** and commitment in writing. For example, higher management at E. I. DuPont de Nemours and Company states its safety policy as follows: "We will not make, handle, use, sell, transport, or dispose of a product unless we can do so safely and in an environmentally sound manner."[4]

Such a statement put in writing and widely distributed among employees clearly shows top management support and gives supervisors the backing they need to insist on safe work practices. It also protects supervisors when a higher manager might press for unsafe practices in order to meet an important deadline.

Figure 20–6
Direct costs and hidden costs of accidents.

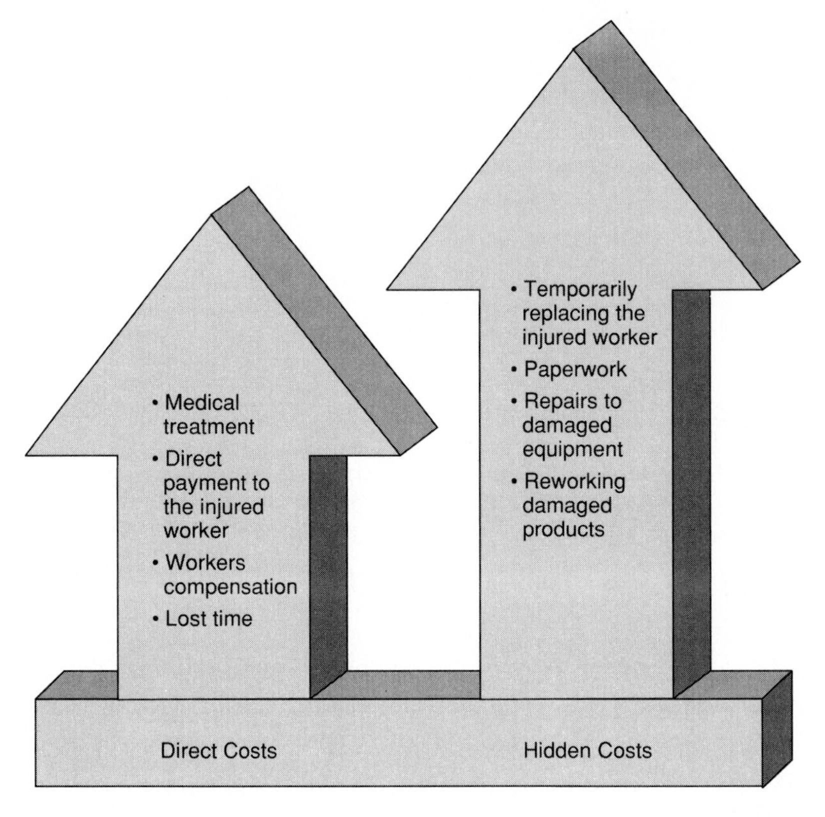

In addition to the philosophy statement, it is important to develop **safety principles** that operationalize the philosophy. Those too should be written down and circulated among all employees. E. I. DuPont de Nemours and Company operationalizes its safety philosophy with the following principles:[5]

- "All injuries and occupational illnesses can be prevented."
- "Management is directly responsible for preventing injuries and illnesses."
- "Safety is a condition of employment."
- "Training is an essential element for safe workplaces."
- "Safety audits must be conducted."
- "All deficiencies must be corrected promptly."
- "It is essential to investigate all unsafe practices and incidents."
- "Safety off the job is just as important as safety on the job."
- "It's good business to prevent illnesses and accidents."
- "People are the most critical element in the success of a safety and health program."

Such principles give supervisors the guidance they need to make decisions when deadlines, productivity, budget, or quality concerns appear to conflict with health and safety practices. Supervisors who work in companies that have not put their safety philosophy on paper should encourage higher management to do so. In developing a health and safety philosophy and corresponding principles, it is important to involve a broad cross-section of employees from all levels.

SAFETY TRAINING

A written health and safety philosophy accompanied by principles that operationalize it set the tone and lay the foundation. Once these steps have been taken, the next step is to provide health and safety training. Training builds on the foundation and gives employees the knowledge they need to be safe workers.

It is important to train new employees so that they learn to work safely from the outset. When beginning a safety program in an established organization that has not trained previously, treat all employees as new employees. It is also important to provide training that is appropriate to the level of employee and the type of work performed.

OSHA specifies minimum requirements for **safety training** in Parts 1910 through 1926. OSHA's guidelines for training recommend a model consisting of the following steps:[6]

Determining if training is needed

Assessing training needs

Setting training goals and objectives

Developing learning materials

Conducting the training

Evaluating the training

Using evaluation data to improve training

Supervisors are interested primarily in the type of training that is appropriate for their employees. Generally speaking, such training should cover the following subjects:

- Orientation to the company's safety policy and principles
- General housekeeping procedures
- Emergency procedures
- Proper use of applicable equipment
- Orientation to hazardous materials present in the company and proper handling of these materials
- Accident reporting procedures
- **Accident follow-up** procedures

The method of instruction used can range from simple one-on-one conversations between supervisor and employee to discussion groups to formal instruction. As explained in Chapter 16, it is important to combine both visual and verbal material when presenting instruction. According to the Society of Manufacturing Engineers, student retention of instruction occurs as follows:[7]

10 percent of what is read
20 percent of what is heard
30 percent of what is seen
50 percent of what is seen and heard
70 percent of what is seen and spoken
90 percent of what is said while doing

There is an important message in this for supervisors regarding how safety instruction should be presented. Just reading safety materials or hearing safety lectures or seeing safety videos is not sufficient. Employees need to participate in active learning that involves seeing, hearing, speaking, and doing. Role-playing and simulation can be effective ways of integrating these various methods.

ACCIDENT PREVENTION TECHNIQUES

Accident prevention requires an ongoing program consisting of a variety of techniques. Here are some techniques supervisors can use to prevent accidents in their units:

- *Involve all employees in an ongoing hazard identification program.* Employees should be empowered to identify hazards associated with their work and to make recommendations for eliminating them.
- *Involve employees in developing safe job procedures.* Once hazards are identified, supervisors and employees should work together to find productive but safe ways to perform the job in question.
- *Teach employees how to properly use personal protective equipment and devices and monitor to make sure they do.* It is important to ensure that employees learn how to use appropriate personal protective equipment before beginning a job. It is equally important for supervisors to make sure that employees follow through and apply what has been learned. When pressed to meet a deadline, the natural human tendency is to take shortcuts. Shortcuts taken with personal protective equipment can cause accidents. Therefore, supervisors must monitor as well as train.
- *Teach employees good housekeeping practices and require their use.* One of the most effective ways to prevent accidents is to maintain a clean, well-organized, orderly workplace. These things result from good housekeeping. Supervisors should teach good housekeeping and monitor to ensure it is practiced.
- *Teach employees the fundamentals of safe work practices (i.e., safe lifting, proper dress, safety glasses, etc.).* General safe work practices are perhaps the most important practices to remember. Some of the most frequently occurring accidents result from such simple mistakes as improper bending and lifting. Supervisors should monitor general work practices closely (Figure 20–7).

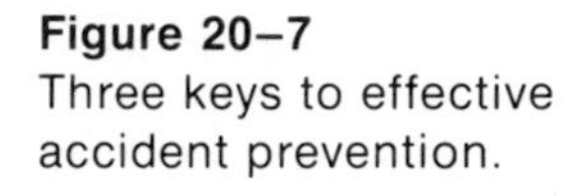

Figure 20–7
Three keys to effective accident prevention.

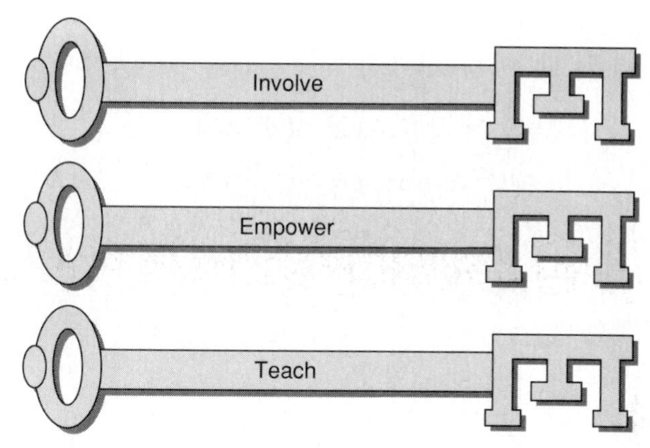

ACCIDENT INVESTIGATION AND REPORTING

In spite of your prevention efforts, an accident may occur. It is important to determine the cause. Why did the accident happen? Answering this question is the purpose of accident investigation. By investigating the cause of an accident, supervisors may be able to prevent a reoccurrence. For this reason, it is important to investigate every accident, no matter how small. It is also important to conduct the investigation immediately. Time can obscure the facts. The accident scene can change, witnesses can forget what they saw, or unrelated factors can creep in and obscure what really happened.

Figure 20–8 contains a checklist supervisors can use as a guide in conducting on-the-spot accident investigations. This checklist will help supervisors determine if the accident was caused by factors that could have been prevented and what those factors are. Occasionally the causal factor will be failure by employees to observe mandatory

Figure 20–8
Checklist for conducting on-the-spot accident investigations.

Accident Investigation Checklist

- Who was involved in the accident? Who are the witnesses?
- Where did the accident take place? At what time?
- What work task was being performed at the time of the accident? What equipment was involved?
- Was the injured employee qualified to perform the task or operate the equipment involved?
- Had the injured employee completed work-related and safety training?
- Were all required safety precautions and procedures being followed?
- Was the required personal protective equipment being used?
- Has a similar accident occurred before?

safety precautions. When this is the case it should be noted, but the purpose is to prevent future accidents, not assign blame.

Writing the Accident Report

Once the investigation has been completed an **accident report** should be written. The report format can follow the investigation checklist unless the company uses a standard accident report form. Regardless of the format, supervisors should remember several rules when writing accident reports. Prominent among these are the following:

- Be brief and stick to the facts.
- Be objective and impartial.
- Be comprehensive; leave out no facts.
- State clearly what employees and what equipment were involved.
- List any procedures, processes, or precautions that were not being observed at the time of the accident.
- List causal factors and any contributing factors.
- Make brief, clear, concise recommendations for corrective measures.

WORKER'S COMPENSATION

Worker's compensation is a concept with which supervisors should be familiar. Unlike OSHA, worker's compensation is not a federal law. Rather, it is based in state laws that vary from state to state. The purpose of worker's compensation is to give injured workers recourse without the need for lengthy, expensive court action. Regardless of the form they take, all worker's compensation laws are designed to finance the rehabilitation of injured employees and to minimize the employee's loss of ability to work (Figure 20–9).

- *Income replacement.* Income lost due to injury or disease must be replaced. This applies to both present and future income. Replacement should be prompt and approximately equivalent to the amount of income lost. This is typically the first objective of a worker's compensation law.
- *Rehabilitation of the worker.* **Rehabilitation** refers to restoring the injured worker's ability to participate competitively in the workplace. The compensation should cover the cost of both medical and vocational rehabilitation. This is typically an objective of worker's compensation laws.
- *Accident prevention.* Like any health and safety program, one of the objectives of worker's compensation laws is to prevent accidents. As with insurance, worker's compensation rates assessed employers are affected by accident rates. Therefore, worker's compensation laws provide a strong incentive for employers to maintain high safety standards. This is an objective of most worker's compensation plans.
- *Cost distribution.* **Cost distribution** serves to motivate employers to make continual safety improvements because it allocates cost based on safety performance. In other

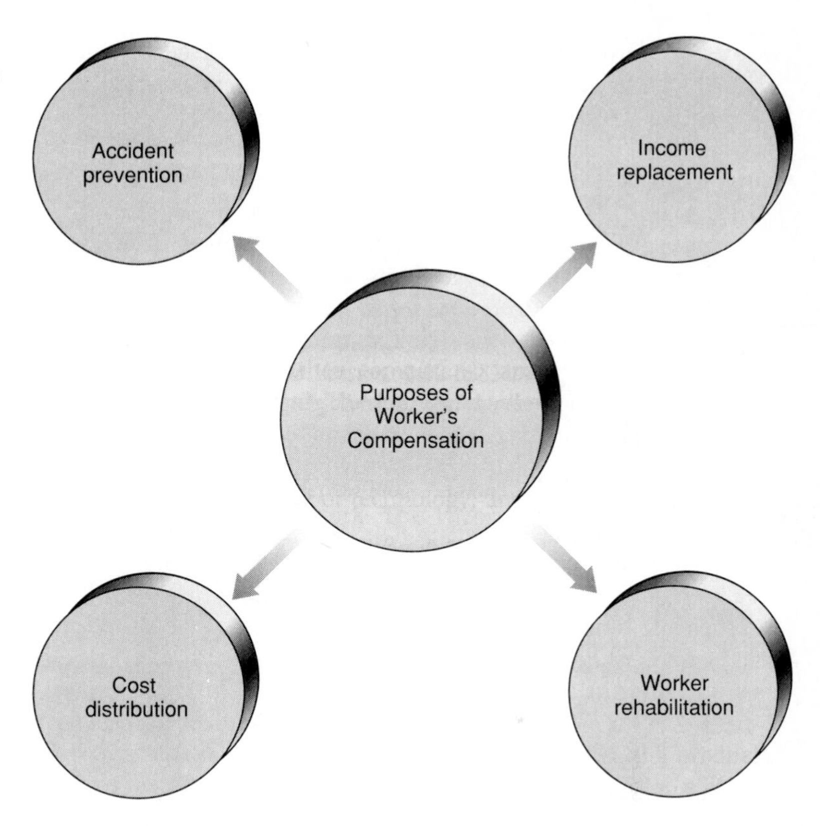

Figure 20–9
Regardless of the form they take, all worker's compensation laws are designed to serve these basic purposes.

words, the better an industry's safety record, the lower its worker's compensation rates, at least in theory.

Worker's Compensation Benefits[8]

Employers are required to carry worker's compensation insurance. It is typically obtained through one of three options: (1) a private insurance company; (2) self-insurance; or (3) state insurance funds. Benefits typically include the following:

- Payment for medical treatment.
- Income for the injured worker during the period of disability.
- Funeral and burial costs in the event of death.
- Income for dependents of the injured worker as appropriate.
- Income to cover the costs of in-home medical care (available in some states).
- Payment of the cost of a prosthesis (available in some states).

Employer Liability Beyond Worker's Compensation

One of the main objectives of the enactment of worker's compensation laws has been to make appropriate recourse to injured workers without the need for court action. In most cases this works. However, there are instances in which employers can be liable beyond the coverage of worker's compensation. Supervisors should be familiar with these instances. The potential areas of **employer liability** are summarized as follows:

- *Criminal liability.* Employers can be held **criminally liable** if they fail to train employees in how to properly handle on-the-job hazards, fail to provide personal protective equipment, or fail to respond appropriately to the complaints of workers about hazardous conditions.
- *Aggravation of injuries.* Employers can be held liable if they aggravate injuries suffered by a worker. This is particularly important to supervisors because an injury that is aggravated by personnel attempting to help an injured employee can result in the employer being held liable.
- *Product liability.* An employer can be held liable if a product it produces causes an injury to an employee who uses it.
- *Intentional assault.* Employers can be held liable if a supervisor or higher level manager attacks and injures an employee.
- *Losses to the immediate family.* Employers can be held liable for losses felt by members of the immediate family as a result of the worker's injuries. Losses to immediate family members may include loss of consortium, companionship, or peace of mind.

AIDS AND EMPLOYEE HEALTH AND SAFETY

AIDS has become one of the most significant health and safety issues modern supervisors are likely to face. It is critical that supervisors know how to deal properly and appropriately with this controversial disease. The major concerns of supervisors with regard to AIDS are: (1) legal considerations; and (2) employee education.

Legal Considerations

The legal considerations relating to AIDS in the workplace grow out of several pieces of federal legislation, including the following: the Rehabilitation Act of 1973; the Occupational Safety and Health Act of 1970; and the Employee Retirement Income Security Act of 1974.

The *Rehabilitation Act of 1973* was enacted to protect handicapped people, including handicapped workers. Section 504 of the act makes discrimination on the basis of a handicap unlawful. Any agency, organization, or company that receives federal funding falls within the purview of the act. Such entities may not discriminate against handicapped individuals who are *otherwise qualified.*

Through various court actions, this concept has been well defined. A handicapped person is otherwise qualified when he or she can perform what the courts have described as the **essential functions** of the job.

When the worker's handicap is a contagious disease such as AIDS, it must be shown that there is no significant risk of the disease being transmitted in the workplace. If there is a significant risk, the infected worker is not considered otherwise qualified. Employers and the courts must make these determinations on a case-by-case basis.

There is one final concept associated with the Rehabilitation Act with which supervisors should be familiar: the concept of **reasonable accommodation**. In determining if a handicapped worker can perform the essential functions of a job, employers are required to make reasonable accommodations to help the worker. This concept applies to workers with any type of handicapping condition including a communicable disease such as AIDS. What constitutes reasonable accommodation, like what constitutes "otherwise qualified," must be determined on a case-by-case basis.

The concepts growing out of the Rehabilitation Act of 1973 give the supervisor added importance when dealing with AIDS-infected employees. The supervisor's knowledge of the various jobs in his or her unit will be essential in helping company officials make an "otherwise qualified" decision. The supervisor's knowledge that AIDS is transmitted only in an exchange of body fluids coupled with his or her knowledge of the job tasks in question will be helpful in determining the likelihood of AIDS being transmitted to other employees. Finally, the supervisor's knowledge of the job tasks in question will be essential in determining what constitutes reasonable accommodation and what the actual accommodations should be.

In arriving at what constitutes reasonable accommodation, supervisors should be aware that employers are not required to make fundamental changes that alter the nature of the job or result in undue costs or administrative burdens. Clearly, good judgment and a thorough knowledge of the job is required when attempting to make reasonable accommodations for an AIDS-infected employee. Modern supervisors should expect to be actively involved in such efforts (Figure 20–10).

The Occupational Safety and Health Act of 1970 requires that employers provide a safe workplace free of hazards. The act also prohibits employers from retaliating against an employee who refuses to work in an environment he or she believes may be unhealthy (Section 654). This poses a special problem for supervisors of AIDS-infected employees. Other employees may attempt to use Section 654 of the act as the basis for refusing to work with such employees. For this reason, it is important that companies educate their employees about AIDS and how it is transmitted. If employees know how

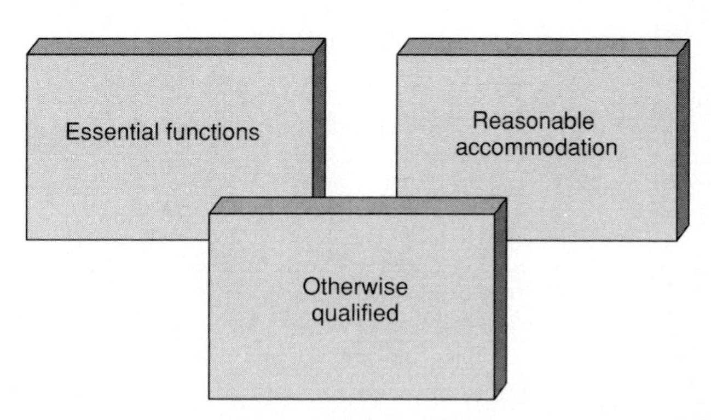

Figure 20–10
Key concepts of the Rehabilitation Act of 1973 relating to AIDS.

AIDS is transmitted they will be less likely to exhibit an irrational fear of working with an infected colleague.

Even when a comprehensive AIDS education program is provided, employers should not automatically assume that an employee's fear of working with the infected individual is irrational. Employers have an obligation to treat each such case individually. Does the complaining employee have a physical condition that puts him or her at greater risk of contracting AIDS than other employees? If so, the fears may not be irrational. However, a fear of working with an AIDS-infected co-worker is usually irrational, making it unlikely that Section 654 could be used successfully as the basis of a refusal to work.

The Employee Retirement Income Security Act (ERISA) of 1974 protects the benefits of employees by prohibiting actions taken against them based on their eligibility for benefits. This means that employers covered by ERISA cannot terminate an employee with AIDS or one who is suspected of having AIDS as a way of avoiding expensive medical costs. Under ERISA, it is irrelevant whether the employee's condition is considered a handicap since the act applies to all employees regardless of condition.

AIDS EDUCATION AND THE SUPERVISOR

The public is becoming more knowledgeable about AIDS and how the disease is spread. However, this is a slow process and AIDS is a complex, controversial disease. Unfortunately, many people still respond to the disease out of ignorance and with inaccurate information. For this reason it is imperative that a company's education and training effort include an **AIDS education** program.

A well-planned AIDS education program can serve several purposes: (1) it can give management the facts needed to develop policy and make informed decisions with regard to AIDS; (2) it can result in changes in behavior that will make employees less likely to contract or spread the disease; (3) it can prepare management and employees to respond appropriately when a worker falls victim to the disease; and (4) it can decrease the likelihood of legal problems resulting from an inappropriate response to an AIDS-related issue. Consequently, modern supervisors should be prepared to participate in the development of AIDS education programs.

Planning an AIDS Education Program

The first step in planning an AIDS education program is to decide what purpose it is to serve. This should be a broad conceptual statement that encapsulizes management's reason for providing the education program. A statement of purpose for an AIDS education program might read as follows:

> The purpose of this AIDS education program is to deal with the disease in a positive, active manner that is in the best interests of the company and its employees.

The next step in the planning process involves developing goals that translate the statement of purpose into more specific terms. The goals should tell specifically what the AIDS education program will do. Sample goals are as follows:

- The program will change employee behaviors that might otherwise promote the spread of AIDS.
- The program will help company management develop a rational, appropriate AIDS policy.
- The program will help managers make responsible decisions concerning AIDS issues.
- The program will help employees protect themselves from the transmission of AIDS.
- The program will alleviate the fears of employees concerning working with an AIDS-infected co-worker.
- The program will help managers respond appropriately and humanely to the needs of AIDS-infected workers.

Once goals have been set, a program is developed to meet the goals. The various components of the program might include confidential, one-on-one counseling, referral, posters, a newsletter, classroom instruction, self-paced multimedia instruction, group discussion sessions, printed materials, or a number of other approaches.

The final step is to decide how the effectiveness of the AIDS education program will be measured. As with any education program, this is determined by the goals of the program. The evaluation techniques described in Chapter 16 also apply in the case of AIDS education programs.

SUMMARY

1. The most fundamental, broadly applied, and far-reaching piece of legislation relating to health and safety in the workplace is the Occupational Safety and Health Act of 1970.
2. Administration and enforcement of the Occupational Safety and Health Act are the responsibility of the U.S. Secretary of Labor and the Occupational Safety and Health Review Commission.
3. The Occupational Safety and Health Administration is responsible for setting health and safety standards, revising health and safety standards, revoking health and safety standards, inspecting companies, issuing citations, assessing penalties, petitioning the court for action for flagrant violators, providing training for employers and employees, and awarding grants to promote health and safety.
4. Companies that fall under the authority of OSHA may be inspected any time work is taking place and without prior notice. The types of inspection conducted are imminent danger investigations, catastrophic accident investigations, employee complaint investigations, high hazard industry inspections, and reinspections. There are four types of violations for which an employer can be cited during an OSHA inspection: imminent danger violations, serious violations, nonserious violations, and *de minimis* violations.
5. There are two fundamental reasons for having a comprehensive health and safety policy. The first has to do with legislative mandates in the Occupational Safety and Health Act. The second is because having a comprehensive health and safety policy makes good business sense.

6. Supervisors should assume responsibility for the work environment in their unit and the safety of their employees. This means supervisors should also have the authority to make decisions about safety, to discipline employees, and to correct unsafe conditions in their units.
7. Written principles that operationalize a company's health and safety policy are critical. Such principles give supervisors the guidance they need to make decisions when deadlines, productivity, budget, or quality concerns appear to conflict with health and safety practices.
8. The Occupational Safety and Health Administration publishes guidelines for safety training. These guidelines suggest the following steps: (1) determining if training is needed, (2) assessing training needs, (3) setting training goals and objectives, (4) developing learning materials, (5) conducting the training, (6) evaluating the training, and (7) using evaluation data to improve the training.
9. Health and safety training in the workplace should cover at least the following topics: (1) orientation to the company's safety policies and principles, (2) general housekeeping procedures, (3) emergency procedures, (4) proper use of applicable equipment, (5) proper lifting procedures, (6) orientation to hazardous materials in the company and proper handling of these materials, (7) accident reporting procedures, and (8) accident follow-up procedures.
10. Accident prevention requires an ongoing program consisting of a variety of techniques. These techniques include: (1) involving all employees in an ongoing hazard program, (2) involving employees in developing safe job procedures, (3) teaching employees how to properly use personal protective equipment and devices, (4) monitoring to make sure they follow through and use them, (5) teaching employees good housekeeping practices and requiring their use, and (6) teaching employees fundamentals of general safe work practices.
11. When writing accident reports, supervisors should apply the following guidelines: (1) be brief and stick to the facts, (2) be objective and impartial, (3) be comprehensive, (4) state clearly what employees and what equipment was involved, (5) list any procedures, processes, or precautions that were not being observed at the time of the accident, (6) list causal factors and any contributing factors, and (7) make brief, clear, concise recommendations for corrective measures.
12. Worker's compensation laws typically cover both occupational injuries and diseases. The specific objectives of worker's compensation are: (1) income replacement, (2) rehabilitation of the worker, (3) accident prevention, and (4) cost distribution.
13. Worker's compensation benefits typically include: (1) payment for medical treatment, (2) income for the injured worker during the period of disability, (3) funeral and burial costs in the event of a death, (4) income for dependents of the injured worker as appropriate, (5) income to cover the costs of in-home medical care, and (6) payment of the cost of a prosthesis.
14. The employer can, on occasion, be held liable beyond the scope of the worker's compensation law in the following areas: (1) criminal liability, (2) aggravation of injuries, (3) product liability, (4) intentional assault, and (5) losses to the immediate family.

15. The legal considerations relating to AIDS in the workplace grow out of several pieces of federal legislation, including: (1) the Rehabilitation Act of 1973, (2) the Occupational Health and Safety Act of 1970, and (3) the Employee Retirement Income Security Act of 1974.
16. In planning an AIDS education program, the first step is to develop a statement of purpose. The next steps are to develop goals that translate the statement of purpose into more specific terms and to develop a program that will allow the goals to be accomplished. The final step involves deciding how to measure the effectiveness of the AIDS education program.

KEY TERMS AND PHRASES

Accident report
Accident follow-up
Aggravation of injuries
AIDS education
Arbitration
Catastrophic accident investigation
Citation
Cost distribution
Criminal liability
CFR
De minimis violation
Direct cost
Employee complaint investigation
Employer liability
Essential functions
Hazardous materials
Hidden cost
High hazard industry inspection
Imminent danger violation
Imminent danger investigation
Income replacement
Inspection
Intentional assault
Nonserious violation
OSHA
Product liability
Reasonable accommodation
Rehabilitation
Reinspection
Safety training
Safety philosophy
Safety principles
Serious violation
Worker's compensation

CASE STUDY: The Self-Directed Safety Program[9]

The following real-world case study contains an actual health and safety situation of the type supervisors might confront on the job. Read the case study carefully and answer the accompanying questions.

General Electric (GE) uses the concept of the *self-directed workforce* to promote safety. Employees have the responsibility for developing safety standards, conducting inspections, and working with OSHA personnel. The self-directed workforce concept is based on the philosophy that well-trained employees are able to make the best decisions concerning how employees as a group do their work. Consequently, GE trains employees to make decisions, solve problems, and take the initiative.

The self-directed concept was originally initiated at GE to improve productivity. It worked so well the decision was made to see if the concept could also be used to reduce accident rates and lower accident-related costs. GE wanted its employees to understand the need for safety and feel ownership in the company's safety policies and practices.

A key part of the self-directed safety program at GE is the *participative safety management team.* Each team chooses its own leader. Team members are trained to conduct walk-through inspections using the same criteria and methods used by OSHA inspectors.

Workers observed in the act of an unsafe practice are handed a wallet-size safety awareness card that carries the following message:

Safety Awareness
The best accident prevention

I just noticed you doing something that could have caused an accident. Think about what you've done in the last few minutes and you will probably recall what I saw.

I am giving you this card as a part of our campaign to make all of us safety conscious. Keep it until you see someone doing something in an unsafe way and then pass it on.

P.S. I hope I don't get it back!

Safety problems discovered with machines, equipment, or facilities are recorded and the list of corrective actions needed is given to higher management for action. The list is monitored until all corrective actions have been accomplished. The result has been that workers know that management places a high priority on safety and that their voice is heard.

A safety incentive program originally developed by one participatory safety management team reduced injuries from eight or nine per week to an average of 2.4, with a number of weeks in which only one injury occurred.

However, even these excellent results did not completely solve all safety problems. Ninety-five percent of GE's employees went along with the self-directed safety program, but a hardcore 5 percent still worked unsafely. This group always comprised the same employees.

One participatory safety management team took the initiative and developed a safety performance policy that established standards for allowable numbers and types of injuries. Three recordable injuries are allowed per year. On the occasion of the fourth injury the employee is required to meet with his or her supervisor to discuss safety. The supervisor can recommend additional safety training that must be done on the employee's time. The training is optional, but if the employee waives it he or she can be removed from the workplace and placed on layoff status.

1. What do you see as the role of the supervisor in the self-directed safety program?
2. What changes to the program, if any, would you recommend and why?
3. What ways other than the one currently used by GE could you recommend for handling hardcore safety violators?

REVIEW QUESTIONS

1. Explain briefly what prompted the U.S. government to pass the Occupational Safety and Health Act of 1970.
2. Explain how the Occupational Safety and Health Act is administered and enforced.
3. The Occupational Safety and Health Administration has several specific responsibilities. List at least four of these.
4. The *Code of Federal Regulations* is updated annually and contains a list of employer rights under the Occupational Safety and Health Act. List four of these rights.
5. Explain what rights a worker has if he or she disagrees with the amount of time OSHA gives the employer to correct a hazard.
6. Explain the following types of OSHA inspections: imminent danger investigation; catastrophic accident investigation; high hazard industry inspection.
7. Explain the following types of OSHA violations: imminent danger violation; nonserious violation; *De minimis* violation.
8. There are two fundamental reasons for having a comprehensive health and safety policy. Explain these two reasons from the perspective of an industrial supervisor.
9. Differentiate between the concepts of direct costs and indirect costs.
10. What is the supervisor's responsibility with regard to health and safety in the workplace?
11. Why is it important to have a written health and safety policy that is operationalized by health and safety principles?
12. List the seven steps recommended by OSHA for health and safety training.
13. What are the general subject areas that should be covered in any health and safety education program?
14. List and explain three accident prevention techniques that can be used by modern industrial supervisors.
15. List four of the rules to apply when writing accident reports.
16. Explain the specific objectives of worker's compensation legislation.
17. List the types of benefits typically provided by worker's compensation insurance.
18. Explain three potential areas of employer liability beyond worker's compensation.
19. How do the following concepts contained in the Rehabilitation Act of 1973 apply to AIDS-infected workers: otherwise qualified; essential functions; reasonable accommodation?
20. What purposes will a well-planned AIDS education program serve?
21. Explain the various steps in planning and developing an AIDS education program.

SIMULATION ACTIVITIES

The following activities are provided to give you opportunities to develop supervisory skills by applying the material presented in this chapter and previous chapters in solving simulated supervisory problems. The activities may be completed by individuals or in groups.

1. You are the most senior supervisor at the Gulf Coast Electrical Power Company. Each year for the past five years your company's accident rate has increased markedly. The company has no written health and safety philosophy statement, no principles, no education and training, and only haphazard record-keeping. You have been asked to chair a committee of workers, supervisors, and higher level managers to make recommendations as to what the company should do to stem the flow of increasing accidents. Formulate the recommendations you will make to your committee.
2. Your company has a health and safety program. However, it does not seem to be working well and it is not popular with the employees. Develop a plan for evaluating the program and making recommendations for improvement.
3. You have been selected to chair a committee of employees that will be responsible for developing health and safety training goals and objectives for the company. Select any type of industrial company and develop a comprehensive set of training goals and objectives for that company in the area of health and safety.
4. You have recently been appointed as the new supervisor of the packaging and shipping unit of Microtech Computer Company. The accident rate in your unit is the highest in the company. Develop a plan for preventing future accidents.
5. A serious accident has occurred in your unit and you did not witness it. However, as the supervisor, you are required to write an accident report. Develop an accident report that meets all the criteria of a well-written report. You may create the circumstances of the accident.
6. You are the shift supervisor in a plastics recycling plant. One of your employees approaches you at the shift change and asks to speak with you in private. The employee tells you confidentially that he has just tested positive for AIDS. How should you handle this situation with regard to the employee, the other employees in your unit, and your company?

ENDNOTES

1. *All About OSHA,* OSHA Publication 2056, U.S. Department of Labor, Occupational Safety and Health Administration.
2. *Tool and Manufacturing Engineers Handbook,* Volume 5, *Manufacturing Management* (Dearborn, MI: Society of Manufacturing Engineers; 1988), p. 12–4.
3. Ibid., p. 12–5.
4. Ibid., p. 12–12.
5. Ibid., p. 12–12.

6. *Voluntary Training Guidelines,* 49FR30290 (Des Plaines, IL: OSHA, Office of Training and Education, 1984).
7. *Tool and Manufacturing Engineers Handbook,* Volume 5, *Manufacturing Management,* p. 12–4.
8. Ibid., p. 12–24.
9. Jenkins, J. A. "Self-Directed Work Force Promotes Safety," *HR Magazine*, February 1990, pp. 54–56.

CHAPTER TWENTY-ONE

Labor Relations and the Supervisor

CHAPTER OBJECTIVES

After studying this chapter, you will be able to define or explain the following topics:

- Historical Perspective of Labor Unions in the United States
- Legal Foundation for Labor Unions
- Current Status of Labor Unions in the United States
- Labor Unions in the United States vs. Those in Japan and Germany
- How Unions are Organized
- Why Employees Join Unions
- The Supervisor's Role with Regard to Union Contracts
- The Supervisor's Role in the Decertification Process
- The Supervisor's Role in Avoiding Union Organization
- The Nature of Future Labor/Management Relations

CHAPTER OUTLINE

- Labor Unions in the United States: A Historical Perspective
- Legal Foundation for Labor Unions
- Labor Unions in the United States: Current Status
- Labor Unions in Japan and Germany
- How Unions Are Organized
- Why Employees Join Unions
- Union Contracts and the Supervisor
- Decertification and the Supervisor
- Supervisor's Role in the Company's Staying Union-Free
- Future of Labor/Management Relations
- Summary
- Key Terms and Phrases
- Case Study Application Problem
- Review Questions
- Simulation Activities
- Endnotes

Traditionally, labor and management relations have been adversarial in nature. However, socioeconomic conditions in this country and the world are beginning to change this. The evolution of today's global economy has turned competition among industrial firms into an international matter. U.S. companies find themselves competing against companies from other industrialized countries such as Japan, Germany, Korea, Taiwan, Denmark, to name just a few. This has an effect on labor/management relations.

Intense **competition** has seriously eroded the position of American industry in the international **marketplace**. Once the world's leader in such key manufacturing areas as automobiles, consumer electronics, and computers, the United States is now struggling to regain substantial losses in market share in these areas to foreign competition. In addition, foreign competition has forced many small- and medium-sized firms out of business and forced many large firms to merge in order to survive. In some cases, these circumstances have cast unions in a bad light. Unions have been viewed by some as organizations that focus too much attention on improving salaries and benefits and too little on improving productivity and quality.

The net result has been that the past twenty years have been difficult for both **labor** and **management**. These mutual difficulties are causing both sides to question their traditional adversarial relationship and experiment with **cooperation**. Cooperation is leading to more involvement of workers in all aspects of planning, doing, and evaluating work as well as in making work-related decisions and in the improvement of productivity and quality. General Electric and Goodyear are American firms that have had success with the cooperative approach.[1]

However, not all attempts at cooperation are meeting with success. Deeply ingrained mistrust between labor and management can be difficult to overcome. Some labor leaders are suspicious of management attempts at promoting cooperation. They are also concerned that higher levels of productivity might backfire and lead to the elimination of jobs. Managers, particularly supervisors, are sometimes concerned that greater levels of employee involvement will diminish their status and importance.

Changing socioeconomic circumstances have brought **labor/management relations** to the edge of a new era. This new era may be characterized by union/management cooperation, but this will not happen without a concerted effort by both sides. Modern supervisors must be prepared to be a positive force in promoting cooperation between labor and management.

LABOR UNIONS IN THE UNITED STATES: A HISTORICAL PERSPECTIVE

Unions in this country were born in the late 1800s. The driving force that moved unions forward and caused them to become highly structured, highly formalized organizations was the Industrial Revolution. The evolution of work processes from manual to mechanized operations threatened the security of workers on the one hand and depersonalized their jobs on the other. Workers also began to express a greater interest in sharing more equitably in the fruits of their labor.

As a result workers began to see unions as a way to gain strength through numbers. This led to a rapid growth in union membership that continued through the end of

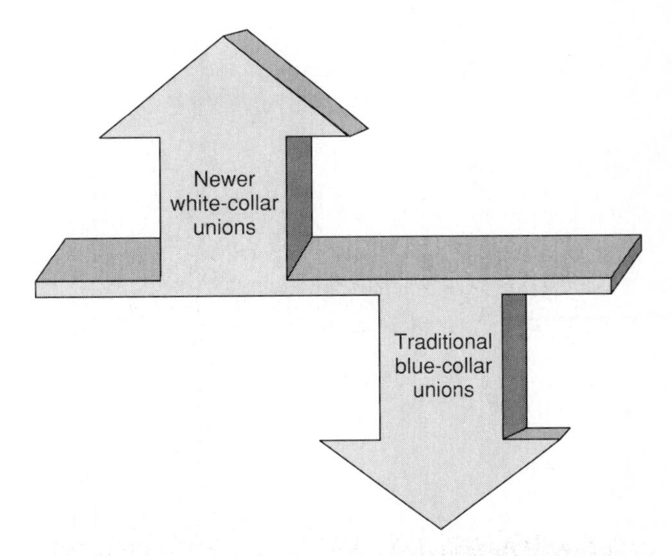

Figure 21–1
Current trends in unions in the United States.

World War II (1945). At this point union membership began to level off. It stayed level until the 1980s, at which time membership in the more traditional **blue-collar unions** began to decline. However, professional or **white-collar unions** are growing. The net result is that as of 1990 approximately 22 million workers in the United States belonged to unions (Figure 21–1).

LEGAL FOUNDATION FOR LABOR UNIONS

Labor unions are well founded in the laws of this country. The legislation relating to organized labor dates back to the late 1800s. Since that time several key pieces of legislation have defined the rights of both labor and management relative to organized labor.

Public attitudes toward unions have swung back and forth over the years. The best way to understand the ebb and flow of public opinion regarding unions is to examine union-related legislation. The legislation with which supervisors should be familiar is summarized below (Figure 21–2):

- *Sherman Antitrust Act (1890).* The **Sherman Antitrust Act** was intended to prevent businesses from engaging in acts that might have a restraining effect on free trade. It also affected unions in that it put constraints on their growth.
- *Clayton Act (1914).* On the one hand, the **Clayton Act** removed unions from the Sherman Antitrust Act. But on the other, it held that strikes amounted to *restraint of trade.* In addition, it allowed companies to undertake contracts in which employees had to agree not to join unions. Such contracts came to be known as *yellow dog* contracts.
- *Norris-La Guardia Act (1932).* Generally perceived as being prolabor because it prohibited yellow dog contracts, the **Norris-La Guardia Act** did not require manage-

Figure 21–2
Steps in the development of union-related legislation.

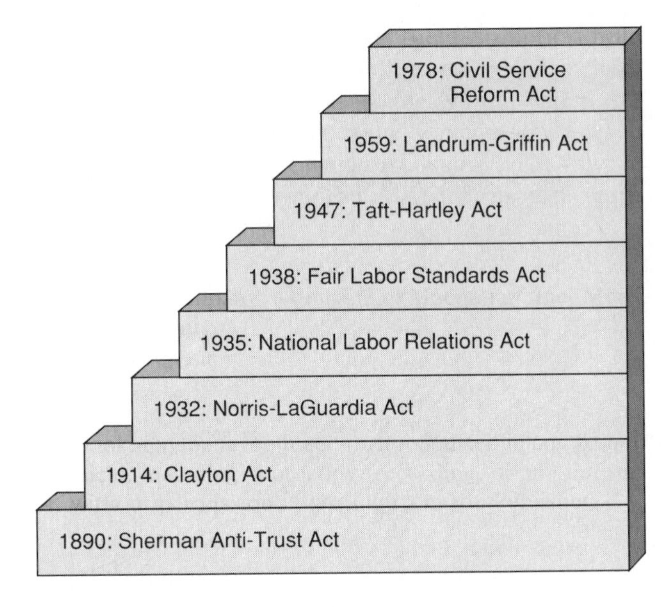

ment to bargain with unions. Union officials viewed it as good legislation that did not go far enough.

- *National Labor Relations Act (1935).* Know as the Wagner Act, the **National Labor Relations Act** was decidedly prolabor. It required management to refrain from interfering with workers who wished to join unions and to recognize the worker's right to unionize. It also required management to bargain in good faith with legally constituted (elected) unions. The Wagner Act created the **National Labor Relations Board (NLRB)** and gave it responsibility for overseeing union elections and for ruling on charges of unfair labor practices. This is still the most important piece of federal legislation to date concerning labor/management relations.
- *Fair Labor Standards Act (1938).* Decidedly prolabor, the **Fair Labor Standards Act** established the concepts of *exempt* and *nonexempt* employees. Nonexempt employees are those to whom the provisions of the FLSA apply. These provisions require payment of a minimum wage to nonexempt employees and time-and-a-half pay for overtime (hours worked in excess of 40 hours per week). Exempt employees are salaried and include professional and executive-level personnel.
- *Taft-Hartley Act (1947).* With passage of the **Taft-Hartley Act** the public policy pendulum swung back in favor of management. Taft-Hartley eliminated the practice of *closed-shop unions.* A closed shop is a company that cannot hire nonunion workers. It also required unions to give management at least sixty days notice before striking and gave the federal government the power to issue an injunction if it was determined that a strike might be detrimental to the national interest.
- *Landrum-Griffin Act (1959).* The **Landrum-Griffin Act** was aimed at curbing perceived union corruption at higher levels. It required full disclosure of union bylaws

and financial records and protected individual members from intimidation by union officials.

- *Civil Service Reform Act (1978).* One part of the **Civil Service Reform Act** gave federal civil service workers the right to join unions. This provision accounts in part for the rapid growth of white-collar unions since 1978. However, strikes by unionized federal workers are prohibited under this act.

At this point, the public policy pendulum regarding unions is in the middle. The National Labor Relations Act represents the prolabor end of the labor/management relations spectrum; the Taft-Hartley Act established the opposite end. Currently, factors greater than legislation are at work that will shape the labor/management relations of the future. An important factor affecting these relations is the mutual need of labor and management to survive. For those companies that must compete either directly or indirectly in the international marketplace, the best chance of surviving and prospering may be in cooperation.

LABOR UNIONS IN THE UNITED STATES: CURRENT STATUS

Labor unions have had more than their share of image-damaging problems in recent years. Newspaper headlines showing union officials indicted on charges of corruption and racketeering have been all too common since the 1950s. Stories associating labor leaders with criminals or with questionable political deals have also received extensive media coverage. In more recent years, strikes by providers of critical services (i.e., teachers, air traffic controllers, public transportation personnel, and airline pilots) have, deservedly or not, cast labor unions in a bad light in the eyes of the public.

However, more than just bad publicity has affected the current status of labor unions. Labor unions have historically found their members from blue-collar workers. The declining manufacturing sector has left unions with fewer blue-collar workers from which to draw. The declining blue-collar figures have been partially offset by marked increases in white-collar unionism, but only partially. Another factor contributing to the declining status of labor unions has been the passage of legislation that protects the rights and job security of workers. These laws (see Chapter 19) have provided the types of protections workers could get only from unions in the past. As a result there is less of a perceived need for union membership. And membership in the traditional blue-collar unions is declining. For example, when union memberships were at their highest levels in this country (mid-1950s) they accounted for approximately 35 percent of the workforce. By the late 1980s this number had fallen to less than 20 percent. Further, it has been projected that by the year 2000 union memberships will account for only 13 percent of the workforce.[2]

Commenting on declining union membership, Peter Drucker said:

> Today's young worker, feeling keenly that he is a loser and a reject, resists, understandably, the union leader's authority even more than he resists the rest of the bosses. As a result, union leaders are increasingly losing control over their own members, are repudiated by

> them, resisted by them, disavowed by them. This, in turn, makes the union increasingly weak. For a union is impotent if it cannot deliver the union member's vote and behavior, cannot guarantee observance of a contract agreement, and cannot count on the members' support for the leader's position and actions.[3]

With this trend one might think the modern supervisor need not be concerned about unions, but this is not the case. K. Ropp stresses that in spite of the downward trend labor unions, which represent over 20 million workers in this country, are still a force to be reckoned with.[4] As a result, modern supervisors should know how to work effectively in both unionized and nonunionized settings.

LABOR UNIONS IN JAPAN AND GERMANY

Two of the most successful competitors in the international marketplace over the past decade have been Japan and Germany. Japan has been particularly successful in the areas of automobile, computer, and consumer electronics manufacturing; Germany has been particularly successful in the areas of automobile and aircraft manufacturing. Because of their success in the marketplace, it is sometimes assumed that Japanese and German companies do not have to deal with labor unions. This is not the case.

In fact, both countries have long traditions of craft and labor unions. However, there are key differences between the nature of labor/management relations in these countries and those in the United States. T. Peters summarizes the key differences as follows:[5]

- Craft unions in Japan and Germany have not pursued the narrow specialization of work that has characterized unions in the United States.
- Craft unions and management in Japan and Germany have worked together closely to develop new technologies that enhance the workers' productivity but are simple to operate and are worker friendly. Unions and management in the United States have cooperated less in this regard with the result that automated systems in this country are typically more complex and less worker friendly.
- Unions and management in Japan and Germany have cooperated in gaining a competitive advantage by jointly pursuing a philosophy of constant improvement. Historically, unions and management in this country have had an adversarial relationship.

The key word in summarizing the difference between labor/management relations in the United States and in other countries is *cooperation.* Cooperation has been achieved in Japan and Germany by involving and empowering employees. Modern supervisors can help increase the level of cooperation between management and labor by involving and empowering employees in their units. What supervisors can do in this regard is covered in more detail later in this chapter.

HOW UNIONS ARE ORGANIZED

Modern supervisors should be well versed in how unions are organized. Supervisors are in a pivotal position regarding unions. As the level of management closest to workers, supervisors are ideally situated to prevent some of the concerns and insecurities that lead to unionization. Correspondingly, they are also positioned to make mistakes that might trigger the process. Knowing how unions are organized and the do's and don'ts of both labor and management regarding the process is important to modern supervisors.

Organizing a union is a four-step process (Figure 21–3). These steps are explained in the following paragraphs.

Initial Contact: Step 1

Initial contact might happen when a dissatisfied employee contacts an existing union or it might involve a union contacting employees of a company. Both approaches are acceptable under the guidelines of the National Labor Relations Board. It is not uncom-

Figure 21–3
The organization process.

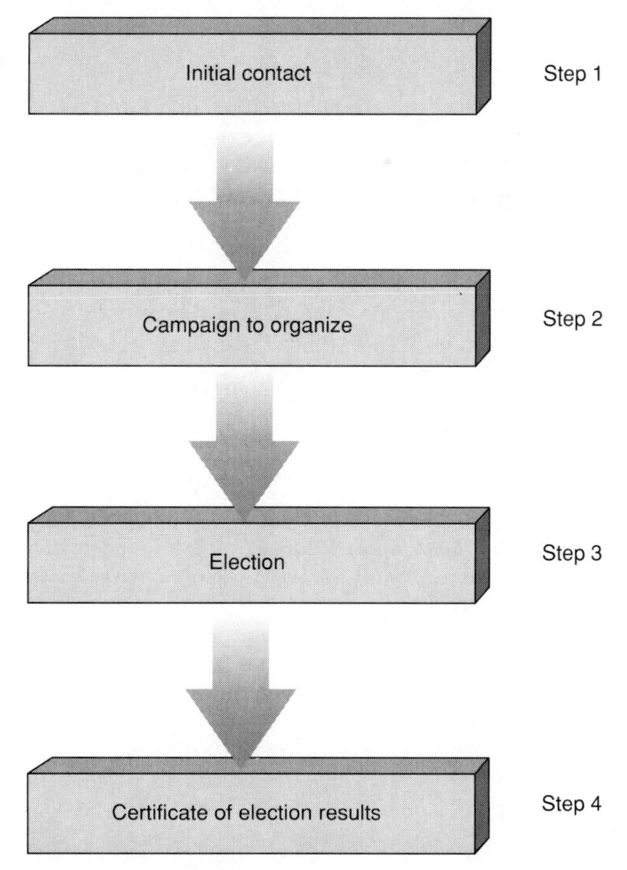

mon for unions to target specific nonunion companies, particularly if a company is the only one in a given area that has remained nonunion. However, dissatisfied employees frequently make the initial contact. This is one reason why the supervisor's role in keeping employees satisfied is so critical (Figure 21–4).

Campaign to Organize: Step 2

The next official step is for the union to send in representatives to conduct an **organizational campaign.** However, the presence of a union representative can tip off management that a campaign is being organized, giving them an opportunity to organize one of their own. Consequently, union officials typically respond to the initial contact by quietly feeling out other employees to determine the level of interest in a union. Before undertaking an organization campaign, a union needs to know if there is a real chance of winning. To organize a campaign and lose the election can make future attempts even more difficult.

If unofficial feelers reveal a sufficient level of interest, union organizers go in and begin the campaign. Depending on the company's solicitation rules, union organizers may or may not be allowed on-site. If they are allowed to solicit on-site, the organizers go into carefully specified areas of the company facility during nonworking hours. If not, they work through prounion employees.

The goal of organizers is to convince employees to sign authorization cards. If 30 percent of the employees in the bargaining unit sign cards, the union can petition the NLRB for an election. The **bargaining unit** consists of a group of employees who have such things in common as type of work, working conditions, or other similar interests.

Figure 21–4
How initial contact is made.

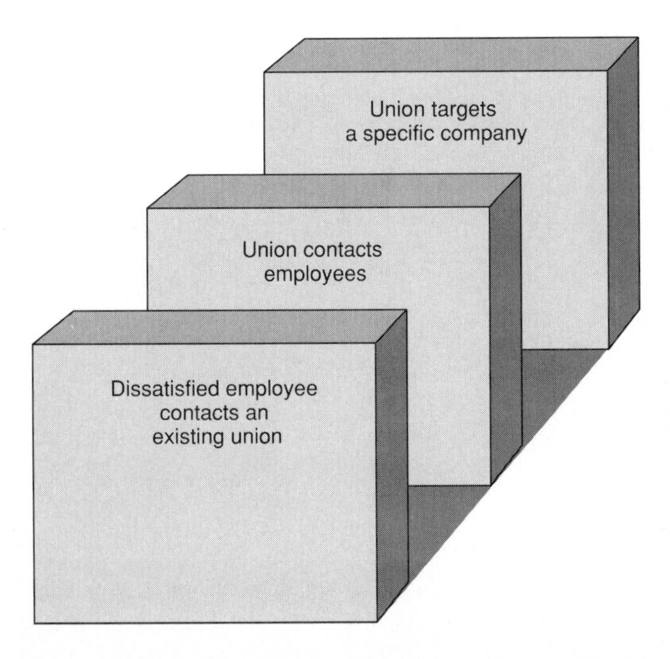

Figure 21–5
Key percentages in the organization process.

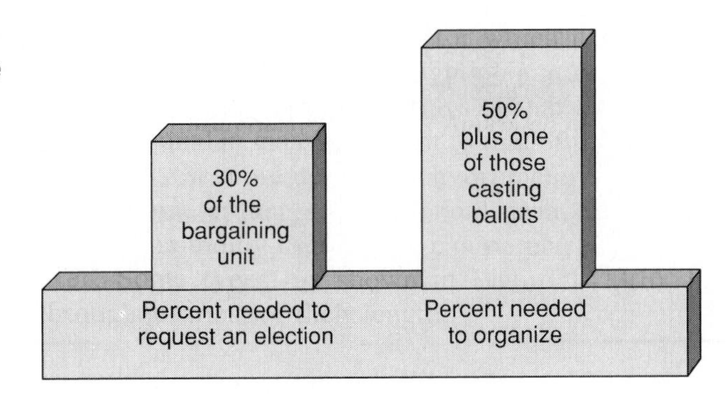

Managers, supervisors, independent contractors, and confidential employees (employees who are privy to confidential information about the company, e.g., the CEO's executive secretary) are excluded from the bargaining unit by the National Labor Relations Act.

Authorization cards state that employees authorize the union to represent them in collective bargaining. However, authorization cards are not binding. Employees who sign cards can still vote against the union.

Election: Step 3

If union organizers obtain the necessary 30 percent or more they may request an **election.** When this happens, the NLRB undertakes a verification process to ensure that the 30 percent has been obtained properly. It will also clearly define the bargaining unit that is to be represented. A company's workforce will not necessarily all belong to the same bargaining unit. For example, machinists may not fit into the same bargaining unit as delivery drivers. If the NLRB determines that the organizing campaign was properly conducted and the 30 percent requirement has been met or exceeded, it will set a date and conduct an election.

Certification of Election Results: Step 4

During the election, all members of the bargaining unit are given the opportunity to vote. If 50 percent plus one of those casting ballots vote for the union, the union wins the right to represent the bargaining unit. Fewer than 50 percent plus one means the company wins and the workforce is not organized. The NLRB **certifies** the results of the election either way. If the NLRB certifies that the union has lost the election, at least twelve months must pass before another election can take place (Figure 21–5).

WHY EMPLOYEES JOIN UNIONS

Experts on labor unions point to several points in the organization process that are critical if a company is to remain union-free. However, in reality, the battle is probably won or lost even before the organization process begins. A primary factor behind a decision to join a union is **employee dissatisfaction.** As a rule, satisfied employees do not join unions. Consequently, supervisors need to know what they can do to promote and enhance employee satisfaction.

V. G. Scarpello and J. Ledvinka conducted a study in 1987 to determine how supervisors can increase employees' job satisfaction. They found that employee satisfaction increases when supervisors do the following:[6]

- Exhibit a high level of technical skill
- Set clearly defined work goals
- Give clear, easily understood instructions
- Clearly define employees' duties and responsibilities
- Back up their employees with other managers
- Appraise performance fairly and objectively
- Give employees adequate time to do the job
- Give employees sufficient time to learn the job
- Inform employees about work-related changes before they occur
- Behave consistently toward employees
- Assist employees as necessary to get the job done
- Recognize employees appropriately for their ideas
- Listen to employees
- Be understanding of the problems of employees
- Follow through when solving problems
- Be fair when employees make mistakes
- Be concerned about the career development of employees
- Acknowledge the good work of employees

Job dissatisfaction by itself will not necessarily force the issue. Unhappy employees who feel they have a voice in correcting the factors about which they are dissatisfied are less likely to join a union than those who are unhappy and feel powerless to do anything about it. According to J. M. Brett, the most likely candidate to join a union is the employee who feels both dissatisfied *and* powerless (Figure 21–6).[7]

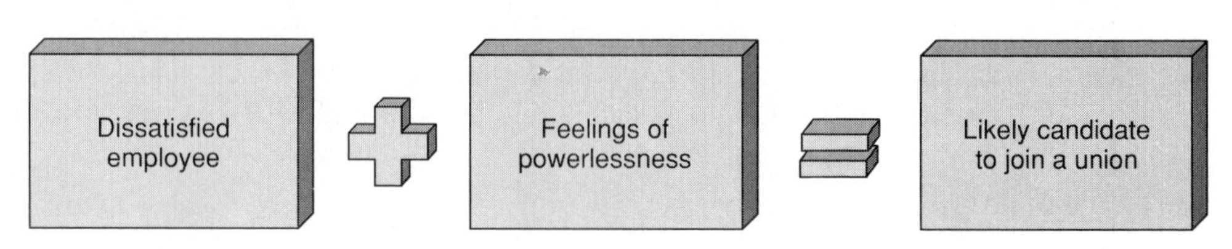

Figure 21–6
Why employees join unions.

UNION CONTRACTS AND THE SUPERVISOR

In unionized companies the labor/management relationship is set forth in a contract. Rights and responsibilities are clearly spelled out and both parties must abide by the provisions of the contract until they are renegotiated. Since the **union contract** is a legally binding document, a breach of contract by either labor or management can have serious implications. For this reason, it is important that supervisors understand union contracts and what they mean.

Union bargaining is limited to **negotiating** such factors as vacation time, holidays, overtime provisions, work hours, seniority, benefits, health and safety matters, strike guidelines, and others relating to wages, benefits, and working conditions. Unions are excluded from and may not negotiate matters relating to strategic planning, marketing, product decisions, or corporate goals.

Unions have three options when negotiating the membership guidelines of a contract: open shop, agency shop, and union shop (Figure 21–7). If the **open-shop** option is agreed to, individual workers are free to decide whether or not they wish to join the union. Union leaders complain that open shops allow employees to reap the benefits of the union without joining. Such complaints led to the establishment of the **agency-shop** option. In an agency shop employees are free to choose or reject union membership. However, those who choose not to join must pay a service fee to the union. The final option is the **union shop.** In a union shop all workers must join the union within a certain time frame as set forth in the contract.

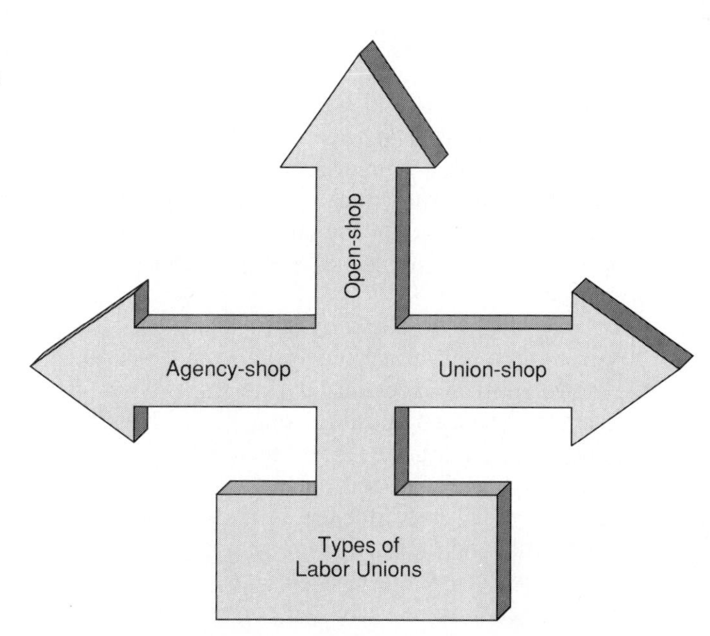

Figure 21–7
Three options unions have for establishing membership criteria.

The open-shop option is becoming the most widely used as economic conditions and state legislation make the union- and agency-shop options less viable. According to R. L. Sauer and K. Voelker, the following states have passed legislation that makes them **right-to-work states** in which the union- and agency-shop options are illegal:[8]

Alabama	Nevada
Arizona	North Carolina
Arkansas	North Dakota
Florida	South Dakota
Georgia	Tennessee
Iowa	Texas
Kansas	Utah
Louisiana	Virginia
Mississippi	Wyoming
Nebraska	

Once negotiations have been completed and a contract is signed, primary responsibility for carrying out the contract falls to the supervisor and the shop steward. The supervisor is management's representative at the operational level. The **shop steward** is the union's representative at the operative level.

Operationalizing the Union Contract

There are several rules of thumb supervisors may find helpful when operationalizing a union contract. They are explained in order of importance in the following paragraphs:

- *Study the contract and learn all its provisions.* When carrying out a union contract ignorance can lead to serious problems. Supervisors should assume that the shop steward and every employee will know every provision of the contract. In planning and scheduling work, supervisors must be aware of the leave, holiday, vacation, and overtime provisions of the contract. In assigning work it is necessary to know the health and safety provisions. Study the contract until every provision is committed to memory.
- *Develop a positive working relationship with the shop steward.* The most difficult task to accomplish in labor/management relations is the building of mutually supportive, mutually beneficial partnerships. What often happens is that labor and management become opposing camps in an ongoing battle. When the supervisor and shop steward trust and respect each other, this is less likely to happen. Time spent by supervisors building and nurturing a positive, trusting relationship with the shop steward will be time invested well. Such a relationship is the best way to keep minor complaints and problems from escalating into full-scale grievances.
- *Be fair and consistent in dealing with employees.* It is important for supervisors to treat all employees fairly and consistently regardless of whether the workforce is unionized or not. However, in a unionized setting this fundamental rule of supervi-

sion takes on even greater significance. Fairness is a matter of observing all provisions in both the letter and the spirit of the contract and of applying them equitably. **Consistency** is a matter of applying all provisions to all employees in the same manner all the time. Consistency is another fundamental rule of supervision that takes on even greater significance in a union setting.

Handling Grievances

A **grievance** is a formal complaint by a union member that some provision of the contract has been violated. When an employee has a complaint, the first step is a meeting in which the supervisor, shop steward, and complaining employee discuss the problem. When a positive relationship exists between the shop steward and a supervisor who is viewed by employees as being fair and consistent, the complaint can often be resolved at this point.

If a resolution is not reached, higher management and union officials will become involved. If a resolution is still not reached, an outside arbitrator is brought in. The arbitrator is a neutral party who, after listening to both sides, arrives at a decision. The arbitrator's decision is final and binding.

J. M. Ivancevich and W. F. Glueck set forth the following guidelines for handling grievances:[9]

- Treat every grievance, regardless of how small, seriously and as a matter of importance.
- Get the shop steward involved immediately and work closely with him or her.
- Collect all the information that can be found relating to the grievance.
- Give the employee your decision concerning his or her grievance promptly.
- Regardless of how the grievance is resolved, once it is resolved, move on. Do not harbor bad feelings or grudges.

DECERTIFICATION AND THE SUPERVISOR

Unions are voted in and they can be voted out. Voting to remove a union as the bargaining agent for employees is known as **decertification**. In recent years, decertification of unions has become increasingly common.[10] The decertification process gives employees the ability to remove a union they do not feel is representing their best interests. It works in much the same way as the organization process. The steps in the decertification process are shown in Figure 21–8.

Dissatisfied union members must convince at least 30 percent of the workers in the bargaining unit to sign authorization cards. If this happens, NLRB can be petitioned for a decertification election. The NLRB will verify whether or not the 30 percent requirement has been properly met. If it has, an election will be scheduled. The same 50 percent plus one needed in the union election is also needed to decertify. As with the union election this is 50 percent plus one of those who actually vote. The NRLB certifies the results of the election and declares the union either decertified or still the bargaining agent.

Figure 21–8
The decertification process.

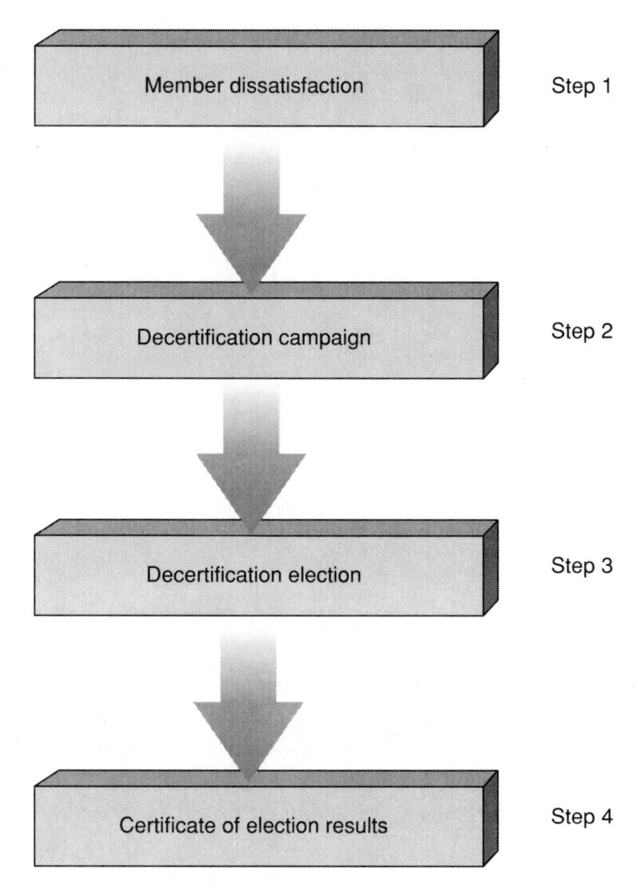

There are strict guidelines as to what supervisors can and cannot do with regard to the decertification of a union. The following guidelines apply to all levels of management with regard to decertification of unions:[11]

- Managers *may not* assist employees in any way in obtaining, distributing, or completing authorization cards.
- Managers *may not* initiate conversations with employees concerning decertification or the decertification process.
- Managers *may not* advocate for or promote decertification.
- Managers *may not* make promises or threats that might serve to encourage the decertification process.
- Managers *may not* interrogate or question employees concerning their union activities, interests, or preferences.
- Managers *may not* undertake or otherwise make use of surveillance to gather information concerning the union activities of employees.

Supervisors and all other managers should comply with these guidelines strictly and completely. A violation of even one guideline can lead to the decertification process being ruled invalid, causing the election petition to be denied or the election itself to be set aside. There are, however, actions managers can take concerning decertification. These actions are summarized below:[12]

- If asked about the decertification process, managers may refer employees to the NLRB.
- Managers may respond to specific, unsolicited questions from employees.
- Managers may provide employees who request it with a list of employee names and addresses.

SUPERVISOR'S ROLE IN THE COMPANY'S STAYING UNION-FREE

There are sound reasons why companies might wish to remain union-free. Perhaps the most important reason is flexibility. Managers of a nonunionized workforce have more flexibility than those who must manage in accordance with a labor contract. Flexibility is essential in the rapidly changing, intensely competitive modern marketplace.

When asked to comment on why he prefers a nonunion workforce, L. W. Looney, CEO of Looney & Son Manufacturing Company in Crestview, Florida, responded, "If you have to contend with a union in today's marketplace, while you're bargaining, your competition is shipping."[13] Supervisors can play an important role in keeping their companies union-free by promoting two key concepts: employee satisfaction and empowerment.

The strategies supervisors can use to promote employee satisfaction were explained earlier in this chapter. In addition to following these guidelines, supervisors must also help employees feel truly **empowered.** An empowered employee is one who has a voice in all matters relating to his or her job and whose input is both actively sought and seriously considered. Strategies for empowering employees are summarized as follows:

- Seek the input of employees and listen to it.
- Demonstrate real interest in the suggestions and opinions of employees.
- Involve all employees affected by a decision in making it.
- Keep employees fully informed of all information relating to their jobs.

FUTURE OF LABOR/MANAGEMENT RELATIONS

One cannot predict the future with any degree of authority. However, those who predict the imminent demise of labor unions in the United States may be premature in their conclusions. After all, even the countries that have become noted for productivity and competitiveness (Japan and Germany, for example) still have unions. A more likely scenario for the future is a greater degree of cooperation between management and labor.

The forces driving the trend toward labor/management cooperation are more practical than philosophical. The most prominent of these forces are: (1) intense competition in the marketplace; and (2) equal employment opportunity legislation. In the United States these seemingly unrelated factors are converging in such a way as to promote labor/management cooperation.

Impact of Competition

As was discussed in Chapters 17 and 18 the United States used to lead the world in production and sales volume in such key areas as automobiles, computers, and consumer electronics. This is no longer the case. Japan, Germany, Korea, and Taiwan have emerged not just as competitors, but as leaders in the international marketplace in these and other areas.

The greater the market share of these countries worldwide, the smaller that of the United States. As a country's market share declines, so does its workforce. The threat of declining market share is therefore of immediate interest to both management and labor.

To understand just how intense the competition facing U.S. firms has become, consider the state of the automobile industry. As foreign competitors, particularly the Japanese, moved ahead in the marketplace, U.S. manufacturers began to get serious about productivity and quality. However, improvements by U.S. firms simply matched corresponding improvements made by foreign competitors. As a result, they maintain their lead and the market share of the U.S. firms continues to dwindle.

An example of how the United States fares in international competition can be seen in the race to develop a better automobile engine. The pressing need for greater fuel economy led to the emergence of the four-cylinder engine as the most widely used option. This had the desired impact on fuel consumption, but at a cost of decreased horsepower. Clearly the consumer, at least in this country, wants an automobile engine with the gas mileage of a four-cylinder engine and the horsepower of the old eight-cylinder engine. For years this combination was thought to be impossible. U.S. manufacturers responded to the need with a six-cylinder engine. This compromise provides more power than four-cylinder engines and better gas mileage than eight-cylinder engines. The Japanese, however, are more creative. Their designers continued to work on a four-cylinder design with the power of an eight-cylinder. In January 1991, Japanese automobile manufacturers were close to completing their prototype of a four-cylinder engine that matches the horsepower of traditional eight-cylinder engines but gets four times the gas mileage. It has more horsepower than the six-cylinder option favored by U.S. automobile manufacturers, but uses one-third less gas. Such competitive innovations as this are providing strong incentives for cooperation between management and labor in U.S. industry. For both it could be the key to survival.

Impact of Legislation

Workers have traditionally looked to unions for job security; better wages, benefits, and working conditions; and an avenue for redress of grievances. The federal legislation discussed in Chapter 19, as well as important state laws, have made the need for union protection in these areas less necessary. Federal and state laws as well as enforcement agencies now govern equal employment opportunity, compensation/benefits, health and safety, and labor/management relations. This has led unions to begin exploring other ways to protect workers. On way is to be a positive force in keeping U.S. employers productive and, in turn, competitive.

Will Cooperation Work?

In order for cooperation to work, efforts on both sides of the table must be sincere. However, evidence suggests that when labor and management can sincerely agree on a common working philosophy, true cooperation can result. Peters cites Dayton Power and Light, an electric utility company, as an example. Facing increasingly stiff competition, DP&L began early negotiations with its union for the purpose of making all parties aware of how intense the competition had become. After eighteen months of negotiating, DP&L and its union agreed to a shared philosophy statement that, in turn, reduced the size of the new labor contract from 200 pages to just fourteen. As a result, workers and managers stopped bickering over trivial details in the contract and began cooperating to improve productivity.[14]

T. Nicholson and R. Manning cite General Motors' Saturn plant in Spring Hill, Tennessee as an example of the trend toward labor/management cooperation.[15] GM designed Saturn to compete with the various Japanese subcompacts that have had such success in the marketplace. The contract agreed to by GM and the United Auto Workers (UAW) established a model for cooperation between management and labor. Among the provisions of the contract were the following:

- All employees, management, and labor received a salary as opposed to an hourly wage.
- Executive dining rooms and reserved parking places were eliminated.
- Workers were divided into self-managed work teams. Each team was given responsibility for staffing, assigning work, scheduling, budgeting, controlling absenteeism, health and safety, ordering supplies, quality, equipment maintenance, and vacation scheduling.
- At least 80 percent of the workers were to be UAW members who had been laid off at other GM plants.

Under this contract, management and labor cooperate as partners. They work together, eat together, and share a common goal: to survive and prosper in a tough marketplace. Writing for *Fortune* magazine, A. Taylor described the innovative cooperative partnership at GM's Saturn plant as a case of "Back to the Future."[16] The model likely will not be limited to the automobile industry.

SUMMARY

1. Unions in the United States were born in the late 1800s as work processes evolved from manual to mechanization. Union membership grew through World War II and then leveled off. In recent years union membership has declined.

2. The following legislation provides the legal foundation for unions in the United States: Sherman Antitrust Act of 1890, Clayton Act of 1914, Norris-La Guardia Act of 1932, National Labor Relations Act of 1935, Fair Labor Standards Act of 1938, Taft-Hartley Act of 1947, Landrum-Griffin Act of 1959, and the Civil Service Reform Act of 1978.
3. Membership in traditional blue-collar unions is declining while membership in white-collar unions is growing.
4. Unions in Japan and Germany are craft unions. As such, they have not pursued the narrow specialization of work that has characterized unions in the United States. They work closely with management to improve productivity and competitiveness.
5. Organizing for union representation is a four-step process: initial contact, organization campaign, election, and certification of election results.
6. Workers join unions when they are dissatisfied and insecure and feel powerless to do anything about it.
7. There are three types of union contracts: open shop, union shop, and agency shop. In carrying out a union contract, supervisors should know all provisions of the contract, develop a positive working relationship with the shop steward, and treat all workers fairly and consistently.
8. In handling grievances, supervisors should treat all complaints seriously, involve the shop steward from the start, collect all available information, give the employee the decision promptly, and move on to other matters.
9. The union decertification process is similar to the organization process. It too has four steps: member dissatisfaction, decertification campaign, decertification election, and certification of election results.
10. To help keep an organization union-free, supervisors should focus their attention on enhancing employee satisfaction and empowerment.
11. The future of labor/management relations are likely to be characterized by increased levels of cooperation as market pressures and worker-supportive legislation make the adversarial approach less viable.

KEY TERMS AND PHRASES

Agency shop
Bargaining unit
Blue-collar union
Civil Service Reform Act
Clayton Act
Competition
Consistency
Cooperation
Decertification
Election
Election certification
Employee empowerment
Employee dissatisfaction
Fair Labor Standards Act
Grievance
Initial contact
Labor
Labor/management relations
Landrum-Griffin Act
Management
Marketplace
National Labor Relations Act

Negotiations
National Labor Relations Board (NLRB)
Norris-La Guardia Act
Open shop
Organizational campaign
Right-to-work state
Sherman Antitrust Act
Shop steward
Taft-Hartley Act
Union contract
Union shop
White-collar union

CASE STUDY: Violence on the Picket Line[17]

The following real-world case study contains an actual labor relations problem of the type supervisors may face on the job. Read the case study carefully and answer the accompanying questions.

In August 1985 a labor/management conflict at the Hormel meat-packing plant in Austin, Minnesota eventually erupted in violence. The conflict began when Hormel responded to economic hardships by cutting wages. Although other meat-packing firms had applied similar cost-cutting techniques, the union at Hormel's Austin plant decided to fight the wage-cutting measure.

The union argued that Hormel had posted record earnings during the year and did not need to reduce wages. Hormel's management countered that low demand for its products made it difficult if not impossible to maintain or increase wages. Hormel threatened to shift work to other plants in its system where wages were lower, which could have meant layoffs for employees of the Austin plant.

With neither side willing to give on the key issues, the union decided to call a strike, sending its 1,500 members to the picket lines. Hormel's management responded to the strike by hiring nonunion workers to replace the strikers. As the replacements reported for work, violence broke out on the picket lines that eventually became so intense the National Guard had to be called in.

1. How would you assess Hormel's handling of this situation?
2. What could Hormel's management have done differently that might have prevented the strike?
3. What role could a supervisor play in helping prevent conflicts such as this one?

REVIEW QUESTIONS

1. Explain briefly the impact of the following pieces of federal legislation on labor/management relations: National Labor Relations Act; Fair Labor Standards Act; Taft-Hartley Act.
2. At this point in time, where is the public policy pendulum regarding labor/management relations? Explain.

3. Support or refute the following statement: "Labor unions in the United States are dying."
4. How do labor unions in Japan differ from those in the United States?
5. Explain each step in the union organization process.
6. List five strategies supervisors can use to enhance employee job satisfaction.
7. Support or refute the following statement: "Happy employees don't join unions."
8. Give four examples of factors that might be negotiated as part of the bargaining process.
9. Compare and contrast the three types of union contracts.
10. Is your current state of residence a right-to-work state? What is the impact of a right-to-work law?
11. List and explain three rules of thumb supervisors should follow in operationalizing a union contract.
12. List the rules supervisors should follow in handling grievances.
13. What can a supervisor legally do with regard to the union decertification process?
14. Name four things a supervisor should *not* do regarding decertification.
15. Explain the supervisor's role in keeping his or her organization union-free.
16. Why is the trend toward greater levels of cooperation between labor and management?

SIMULATION ACTIVITIES

The following activities are provided to give you opportunities to develop supervisory skills by applying the material presented in this chapter and previous chapters in solving simulated supervisory problems.

1. You have just been promoted to supervisor and a new union contract has just been signed. Develop a step-by-step plan for operationalizing the contract.
2. An employee in your unit comes to you with the following complaint: "I have more seniority but you've been giving more overtime to John Smith. I want to file a grievance." How will you handle this situation?
3. Two employees in your unit approach you after work with the following request: "We want to petition for a decertification election. Will you help us?" How will you handle this situation?
4. You have just been promoted to supervisor. During the interview a higher manager made the following comment: "Our last union election was too close for comfort. There will probably be another in about ten months. We are expecting your unit to vote no." How should you proceed?

ENDNOTES

1. Bernstein, A. and Rothman, M. "Steelmakers Want to Make Teamwork an Institution," *Business Week,* May 11, 1987, p. 84.
2. Hoerr, J., Glaberson, W. G., Moskowitz, D. B., Cahan, V., Pollock, M. A., and Tasini,

J. "Beyond Unions: A Revolution in Employee Rights Is in the Making," *Business Week,* July 8, 1986, pp. 72–77.
3. Drucker, P. *Management* (New York: Harper & Row, 1985), pp. 172–176.
4. Ropp, K. "State of Unions," *Personnel Administrator,* 1987, Volume 32, Number 7, pp. 36–40.
5. Peters, T. *Thriving on Chaos* (New York: Harper & Row, 1987), p. 345.
6. Scarpello, V. G. and Ledvinka, J. *Personnel/Human Resource Management* (Boston: PWS-Kent Publishing Company, 1988), p. 592.
7. Brett, J. M. "Why Employees Join Unions," *Organizational Dynamics,* 1980, Number 8, pp. 47–59.
8. Sauer, R. L. and Voelker, K. *Labor Relations: Structure and Process* (Columbus, OH: Merrill, 1987), p. 305.
9. Ivancevich, J. M. and Glueck, W. F. *Foundations of Personnel* (Plano, TX: Business Publications, 1983), p. 567.
10. Anderson, J. C., O'Reilly, C. A., and Bushman, G. "Union Decertifications in the U.S.: 1947–1977," *Industrial Relations,* 1980, Number 19, pp. 100–107.
11. Coleman, F. J. "Once a Union Not Always a Union," *Personnel Journal,* 1985, Number 64, pp. 42–45.
12. Ibid.
13. Interview with author on December 19, 1990 in Fort Walton Beach, Florida.
14. Peters, T. *Thriving on Chaos,* p. 456.
15. Nicholson, T. and Manning, R. "Saturn Gets a Launching Pad," *Newsweek,* August 5, 1987, p. 42.
16. Taylor, A. "Back to the Future of Saturn," *Fortune,* August 1, 1988, pp. 63–72.
17. Pitzer, M. J. and Bernstein, A. "The Union vs. the Union at Hormel," *Business Week,* February 17, 1986, p. 36.

INDEX